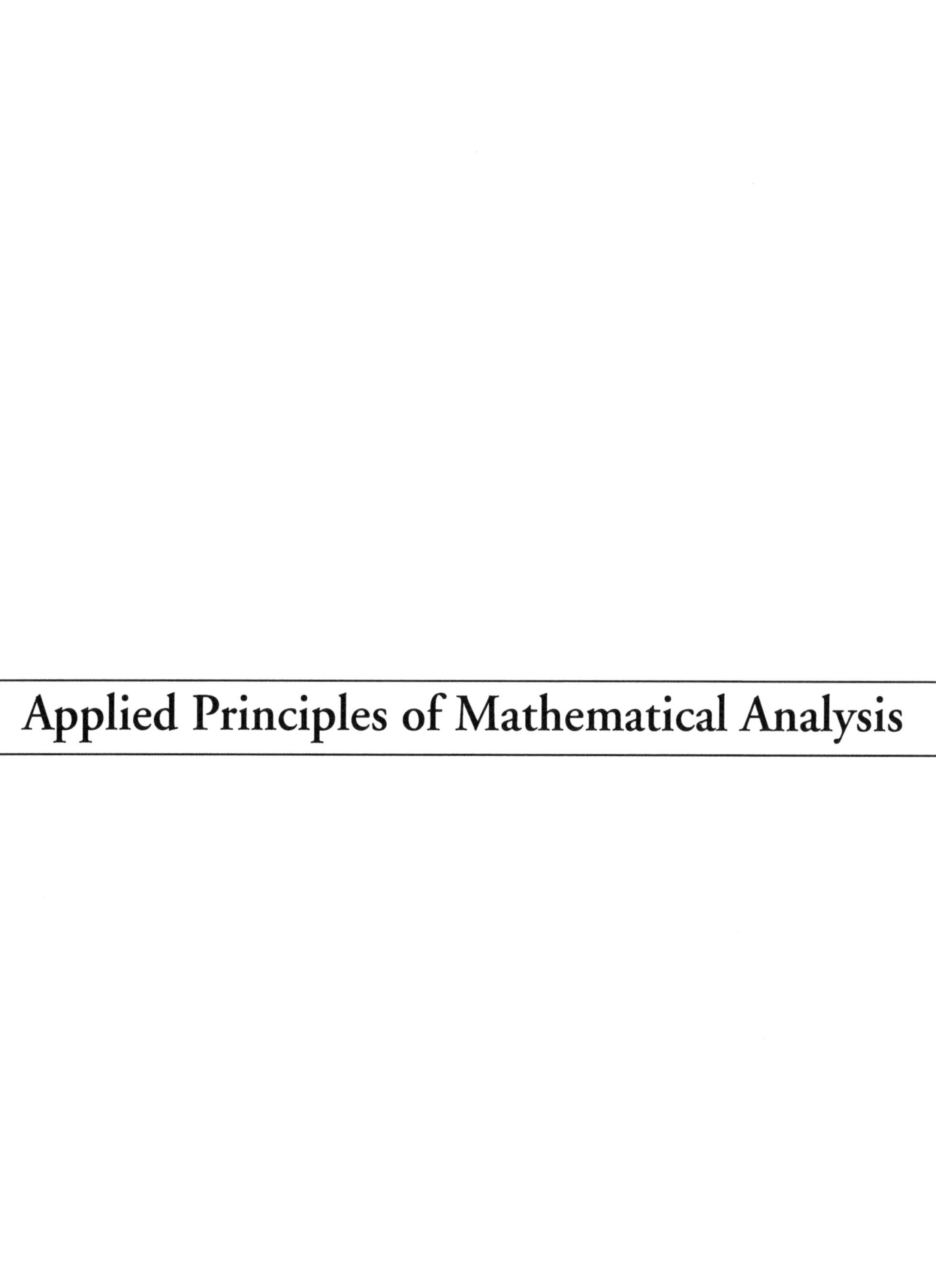

Applied Principles of Mathematical Analysis

Applied Principles of Mathematical Analysis

Editor: Casper Geller

New York

Published by NY Research Press
118-35 Queens Blvd., Suite 400,
Forest Hills, NY 11375, USA
www.nyresearchpress.com

Applied Principles of Mathematical Analysis
Edited by Casper Geller

International Standard Book Number: 978-1-63238-730-1 (Hardback)

Cataloging-in-Publication Data

Applied principles of mathematical analysis / edited by Casper Geller.
p. cm.
Includes bibliographical references and index.
ISBN 978-1-63238-730-1
1. Mathematical analysis. 2. Algebra. 3. Mathematics. I. Geller, Casper.
QA300 .A67 2020
515--dc23

Contents

Preface VII

Chapter 1 **Compact Operators on the Bergman Spaces with Variable Exponents on the Unit Disc of $\mathbb{C}$** 1
Dieudonne Agbor

Chapter 2 **Toward a Theory of Normalizing Function of Interestingness Measure of Binary Association Rules** 12
Armand Armand, André Totohasina and Daniel Rajaonasy Feno

Chapter 3 **Ordered Structures of Constructing Operators for Generalized Riesz Systems** 20
Hiroshi Inoue

Chapter 4 **Convolutions of Harmonic Functions with Certain Dilatations** 28
Om P. Ahuja and Jay M. Jahangiri

Chapter 5 **On the Differentiability of Weak Solutions of an Abstract Evolution Equation with a Scalar Type Spectral Operator on the Real Axis** 32
Marat V. Markin

Chapter 6 **Extension of Kirk-Saliga Fixed Point Theorem in a Metric Space with a Reflexive Digraph** 46
Karim Chaira, Abderrahim Eladraoui, Mustapha Kabil and Samih Lazaiz

Chapter 7 **Greedy Expansions with Prescribed Coefficients in Hilbert Spaces** 52
Artur R. Valiullin, Albert R. Valiullin and Vladimir V. Galatenko

Chapter 8 **Coincidence Point Theorems for (α, β, γ)-Contraction Mappings in Generalized Metric Spaces** 58
Chaiporn Thangthong and Anchalee Khemphet

Chapter 9 **Application of ADM using Laplace Transform to Approximate Solutions of Nonlinear Deformation for Cantilever Beam** 65
Ratchata Theinchai, Siriwan Chankan and Weera Yukunthorn

Chapter 10 **On Surface Completion and Image Inpainting by Biharmonic Functions: Numerical Aspects** 70
S. B. Damelin and N. S. Hoang

Chapter 11 **Hopf Bifurcation Analysis of a New SEIRS Epidemic Model with Nonlinear Incidence Rate and Nonpermanent Immunity** 78
M. P. Markakis and P. S. Douris

Chapter 12 **A New Approach to Approximate Solutions for Nonlinear Differential Equation** 91
Safia Meftah

Chapter 13 **Univalent and Starlike Properties for Generalized Struve Function** 99
Aaisha Farzana Habibullah, Adolf Stephen Bhaskaran and Jeyaraman Muthusamy Palani

Chapter 14 **On Solving of Constrained Convex Minimize Problem using Gradient Projection Method** 106
Taksaporn Sirirut and Pattanapong Tianchai

Chapter 15 **Oscillatory and Asymptotic Behavior of a First-Order Neutral Equation of Discrete Type with Variable Several Delay under Δ Sign** 116
Radhanath Rath and Chittaranjan Behera

Chapter 16 **Optimal Form for Compliance of Membrane Boundary Shift in Nonlinear Case** 124
Y. Yousfi, I. Hadi and A. Benbrik

Chapter 17 **An Extension Theorem for a Sequence of Krein Space Contractions** 133
Gerald Wanjala

Chapter 18 **Best Proximity Point Theorems for a Berinde MT-Cyclic Contraction on a Semisharp Proximal Pair** 140
Chalongchai Klanarong and Tanadon Chaobankoh

Chapter 19 **New Extension of Beta Function and its Applications** 147
Mehar Chand, Hanaa Hachimi and Rekha Rani

Chapter 20 **(p, q)-Growth of an Entire GASHE Function and the Coefficient $\beta n = |a_n| / \Gamma(n + \mu + 1)$** 172
Mohammed Harfaoui, Abdellah Mourassil and Loubna Lakhmaili

Chapter 21 **Solving Volterra Integrodifferential Equations via Diagonally Implicit Multistep Block Method** 180
Nur Auni Baharum, Zanariah Abdul Majid and Norazak Senu

Chapter 22 **New Modified Adomian Decomposition Recursion Schemes for Solving Certain Types of Nonlinear Fractional Two-Point Boundary Value Problems** 190
Sekson Sirisubtawee and Supaporn Kaewta

Chapter 23 **A 4-Point Block Method for Solving Higher Order Ordinary Differential Equations Directly** 210
Nazreen Waeleh and Zanariah Abdul Majid

Permissions

List of Contributors

Index

Preface

Over the recent decade, advancements and applications have progressed exponentially. This has led to the increased interest in this field and projects are being conducted to enhance knowledge. The main objective of this book is to present some of the critical challenges and provide insights into possible solutions. This book will answer the varied questions that arise in the field and also provide an increased scope for furthering studies.

Mathematical analysis is a domain of mathematics that deals with limits and other related theories such as, measure, infinite series, differentiation, integration, and analytical functions. All of these theories are often studied in the context of real and complex numbers along with their functions. The main branches of mathematical analysis include real analysis, complex analysis and functional analysis. The fundamental concepts of this field are metric spaces and sequences and limits. Mathematical analysis has evolved from calculus that includes elementary techniques and concepts of analysis. It can be applied to any space of those mathematical objects that have a topological space or a metric space. This book contains some path-breaking studies in the field of mathematical analysis. It studies and analyzes various principles of mathematical analysis. It is appropriate for students seeking detailed information in this area as well as for experts.

I hope that this book, with its visionary approach, will be a valuable addition and will promote interest among readers. Each of the authors has provided their extraordinary competence in their specific fields by providing different perspectives as they come from diverse nations and regions. I thank them for their contributions.

Editor

1

Compact Operators on the Bergman Spaces with Variable Exponents on the Unit Disc of $\mathbb{C}$

Dieudonne Agbor

Department of Mathematics, Faculty of Science, University of Buea, P.O. Box 63, Buea, Cameroon

Correspondence should be addressed to Dieudonne Agbor; dieu_agb@yahoo.co.uk

Academic Editor: Nageswari Shanmugalingam

We study the compactness of some classes of bounded operators on the Bergman space with variable exponent. We show that via extrapolation, some results on boundedness of the Toeplitz operators with general L^1 symbols and compactness of bounded operators on the Bergman spaces with constant exponents can readily be extended to the variable exponent setting. In particular, if S is a finite sum of finite products of Toeplitz operators with symbols from class BT, then S is compact if and only if the Berezin transform of S vanishes on the boundary of the unit disc.

1. Introduction and Statement of Results

Variable Lebesgue spaces are a generalization of the Lebesgue spaces that allow the exponents to be a measurable function and thus the exponent may vary. These spaces have many properties similar to the normal Lebesgue spaces, but they also differ in surprising and subtle ways. For this reason, the variable Lebesgue spaces have an intrinsic interest, but they are also very important in applications to partial differential equations and variational integrals with nonstandard growth conditions. See [1] for more details on the variable Lebesgue spaces.

Let Δ denote the unit disc in $\mathbb{C}$ and dA the normalized Lebesgue measure on Δ. For $1 \leq p < \infty$, the Bergman space $A^p = A^p(\Delta, dA)$ is the space of all analytic functions, f, on Δ such that

$$\|f\|_p^p = \int_\Delta |f(z)|^p \, dA(z) < \infty. \qquad (1)$$

Let P be the Bergman projection from L^2 onto A^2. Then P is an integral operator given by

$$P(f)(z) = \int_\Delta \frac{f(w)}{(1 - z\overline{w})^2} dA(w), \qquad (2)$$

for each $z \in \Delta$ and $f \in L^2$. Here, the function $K(z, w) = K_w(z) = 1/(1 - z\overline{w})^2$ is the reproducing kernel for A^2. For $f \in L^\infty$, the Toeplitz operator with symbol f is defined on A^p by

$$T_f g = P(fg), \quad g \in A^p. \qquad (3)$$

Toeplitz operators are amongst the most widely studied classes of concrete operators and have attracted a lot of interest in recent years. The behaviour of these operators on the Hardy spaces, Bergman spaces, and Fock spaces has been studied widely and a lot of results are available in the literature. The characterization of compactness has been studied in [2–8] just to cite a few.

Given $\Omega \subset \mathbb{R}^n$, a measurable function $p : \Omega \to [1, \infty)$ will be called a variable exponent. If p is a variable exponent then we denote

$$p_+ = p_\Omega^+ := \operatorname*{ess\,sup}_{x \in \Omega} p(x),$$
$$p_- = p_\Omega^- := \operatorname*{ess\,inf}_{x \in \Omega} p(x). \qquad (4)$$

Let $\mathscr{P}(\Omega)$ denote the set of all variable exponents for which $p_+ < \infty$.

For a complex-valued measurable function $\phi : \Omega \to \mathbb{C}$, we define the modular $\rho_{p(\cdot)}$ by

$$\rho_{p(\cdot)}(\phi) := \int_\Omega |\phi(x)|^{p(x)} dx \qquad (5)$$

and the norm

$$\|\phi\|_{p(\cdot)} := \inf\left\{\lambda > 0 : \rho_{p(\cdot)}\left(\frac{\phi}{\lambda}\right) \leq 1\right\}. \quad (6)$$

Let $p(\cdot) \in \mathscr{P}(\Omega)$. Then the Lebesgue variable exponent space $L^{p(\cdot)}$ is the set of all complex-valued measurable functions $\phi : \Omega \to \mathbb{C}$ for which $\rho_{p(\cdot)}(\phi) < \infty$. If we equip $L^{p(\cdot)}$ with the norm given in (6), then $L^{p(\cdot)}$ becomes a Banach space. We note here that the condition $\rho_{p(\cdot)}(\phi) < \infty$ is not enough in general to define the variable exponent Lebesgue space (e.g., see chapter 2 of [1]).

It is known (e.g., see chapter 2 of [1]) that the dual of $L^{p(\cdot)}$ is $L^{p'(\cdot)}$, where $1/p(\cdot) + 1/p'(\cdot) = 1$. A straightforward computation shows that

$$\begin{aligned}(p'(\cdot))_+ &= (p_-)',\\ (p'(\cdot))_- &= (p_+)'.\end{aligned} \quad (7)$$

For simplicity, we will omit one set of parenthesis and write the left-hand side of each equality as $p'(\cdot)_+$ and $p'(\cdot)_-$. Throughout this work, we shall use $p'(\cdot)$ as the conjugate exponent of $p(\cdot)$ and if p is a constant in $(1,\infty)$ we shall use p' as the conjugate exponent of p. In other words, to study these spaces, some regularity conditions are imposed on the exponents. A function $p : \Omega \to \mathbb{C}$ is said to be log-Hölder continuous on Ω if there exists a positive constant $C_{\log}$ such that

$$|p(x) - p(y)| \leq \frac{C_{\log}}{\log(1/|x-y|)}, \quad (8)$$

for all $x, y \in \Omega$ with $|x - y| < 1/2$. It follows that

$$|p(x) - p(y)| \leq \frac{2lC_{\log}}{\log(2l/|x-y|)}, \quad (9)$$

for all $x, y \in \Omega$ with $|x - y| < l$. We denote by $\mathscr{P}^{\log}(\Omega)$ the exponents in $\mathscr{P}(\Omega)$ that are log-Hölder continuous on Ω. For $p(\cdot) \in \mathscr{P}^{\log}(\Omega)$ and a given measurable function, f, define

$$\|f\|^*_{p(\cdot)} = \sup\left\{\left|\int_\Omega f(x)\, g(x)\, dx\right| : \|g\|_{p(\cdot)} \leq 1\right\}. \quad (10)$$

Theorem 2.34 of [1] shows that there exist constants C_1 and C_2, depending on $p(\cdot)$, such that

$$C_1 \|f\|_{p(\cdot)} \leq \|f\|^*_{p(\cdot)} \leq C_2 \|f\|_{p(\cdot)}. \quad (11)$$

The next result which establishes a relationship between the Lebesgue spaces with exponents p_-, p_+, and $p(\cdot)$ will be very useful in the rest of the work. It is Corollary 2.50 of [1].

Lemma 1. *Suppose $p(\cdot) \in \mathscr{P}^{\log}(\Omega)$ and $|\Omega| < \infty$. Then there exist constants c_1 and c_2 such that*

$$c_1 \|f\|_{p_-} \leq \|f\|_{p(\cdot)} \leq c_2 \|f\|_{p_+}. \quad (12)$$

The study of variable exponent Bergman space, $A^{p(\cdot)}$, which is the space of analytic functions in $L^{p(\cdot)}$, has been introduced in [9]. There it was shown, amongst other things, that the Bergman projector P is bounded from $L^{p(\cdot)}$ onto $A^{p(\cdot)}$. Also in [10], the authors studied Carleson measures in such spaces.

In this paper, we will extend the results in [3, 7] on boundedness and compactness of operators for the Bergman spaces with constant exponents to the Bergman spaces with variable exponents.

For $z \in \Delta$, let φ_z be the analytic map of Δ onto Δ given by $\varphi_z(w) = (z - w)/(1 - \bar{z}w)$. We define the operator U_z on A^2 by

$$U_z f = (f \circ \varphi_z)\varphi_z', \quad z \in \Delta. \quad (13)$$

Then U_z is a unitary operator on A^2. We shall show later that U_z is bounded on $A^{p(\cdot)}$. For S, a bounded operator on $A^{p(\cdot)}$, we define S_z by $S_z = U_z S U_z$.

If S is a bounded operator on $A^{p(\cdot)}$, then the Berezin transform of S is the function $\widetilde{S}$ on Δ defined by

$$\widetilde{S}(z) = \langle Sk_z, k_z\rangle, \quad (14)$$

where $k_z(w) = (1-|z|^2)K_z$ is the normalized Bergman kernel which also belongs to $A^{p(\cdot)}$ and $\langle\, ,\, \rangle$ is the inner product of A^2. We let

$$\begin{aligned}\widetilde{f}(z) &:= \widetilde{T_f}(z) = \langle fk_z, k_z\rangle\\ &= \int_\Delta f(w)\,|k_z(w)|^2\, dA(w)\end{aligned} \quad (15)$$

and set

$$BT := \left\{f \in L^1 : \|f\|_{BT} = \sup_{z\in\Delta}\widetilde{|f|}(z) < \infty\right\}. \quad (16)$$

Our first result gives some conditions for the boundedness of Toeplitz operators with L^1 symbols on the variable Bergman spaces.

Theorem 2. *Suppose $p(\cdot) \in \mathscr{P}^{\log}(\Delta)$, $1 < p_0 \leq p_- \leq p_+ < \infty$, and $p_1 = \min(p_0, p_0')$. Suppose $f \in L^1$ is such that*

$$\begin{aligned}C_1 &= \sup_{z\in\Delta}\|P(f \circ \varphi_z)\|_q < \infty,\\ C_2 &= \sup_{z\in\Delta}\|P(\overline{f} \circ \varphi_z)\|_q < \infty\end{aligned} \quad (17)$$

for all $q > (p_1 + 1)/(p_1 - 1)$. Then T_f is bounded on $A^{p(\cdot)}$.

We note here that this result was proved in [3] in the Bergman spaces A^p, where p is a constant. We also have the following result on compactness.

Theorem 3. *Suppose $p(\cdot) \in \mathscr{P}^{log}(\Delta)$, $1 < p_0 \leq p_- \leq p_+ < \infty$, and $p_1 = \min(p_0, p_0')$. If S is a bounded operator on $A^{p(\cdot)}$ such that*

$$C_1 = \sup_{z\in\Delta} \|S_z 1\|_q < \infty,$$
$$C_2 = \sup_{z\in\Delta} \|S_z^* 1\|_q < \infty \quad (18)$$

for some $q > (p_1+1)/(p_1-1)$, then the following are equivalent:

(1) *S is compact on $A^{p(\cdot)}$,*

(2) *$\widetilde{S}(z) \to 0$ as $z \to \partial\Delta$,*

(3) *for every $s \in [1, q)$, $\|S_z 1\|_s \to 0$ as $z \to \partial\Delta$,*

(4) *$\|S_z 1\|_1 \to 0$ as $z \to \partial\Delta$.*

This theorem is well known in the Bergman spaces with constant exponents; for example, see [3, 7]. However, the techniques here are different from those used in either of the papers for both the proof of boundedness and compactness. This is because their proofs depend on the use of Schur's test which does not hold in the variable Lebesgue space. However, using the Muckenhoupt weights we were able to develop some Schur-like tests from where we obtain the theory that builds upon the Rubio de Francia theory of extrapolation from the theory of weighted norm inequalities. The advantage of this approach is that it quickly yields to sufficient conditions for these operators to be bounded on variable Lebesgue spaces. Through such techniques, we are also able to obtain some norm estimates for bounded operators on the space $A^{p(\cdot)}$.

Similar to the work of Miao and Zheng [7], we consider the case of the algebra of Toeplitz operators generated by symbols in the class BT. To be precise, we have the following.

Theorem 4. *Suppose $p(\cdot) \in \mathscr{P}^{log}(\Delta)$ and S is a finite sum of finite products of Toeplitz operators with symbols in the class BT. Then S is compact on $A^{p(\cdot)}$ if and only if $\widetilde{S}(z) \to 0$ as $z \to \partial\Delta$.*

This paper is organized as follows: in Section 2, we will study some basic concepts on the Muckenhoupt weights. Section 3 deals with the variable Bergman spaces and the proof of Theorem 2. In Section 4, we study some norm estimates on these spaces and in Section 5 we give the proof of the compactness results.

2. Muckenhoupt Weights A_1

Definition 5. Let Ω be a set. Then the function $d : \Omega \times \Omega \to \mathbb{R}^+$ is said to be a pseudodistance on Ω if it satisfies the following:

(1) $d(x, y) = 0$ if and only if $x = y$;

(2) $d(x, y) = d(y, x)$;

(3) there exists a positive constant $K \geq 1$, such that, for all $x, y, z \in \Omega$,

$$d(x, y) \leq K(d(x, z) + d(z, y)). \quad (19)$$

For $x \in \Omega$ and $r > 0$, the set $B(x, r) = \{y \in \Omega : d(x, y) < r\}$ is called a pseudoball with centre x and radius r. If μ is a measure on Ω, then the triple (Ω, d, μ) is called a homogeneous space if Ω is endowed with the topology generated by the collection $\{B(x, r) : x \in \Omega,\ r > 0\}$ (that is, the topology generated by the pseudoballs) and μ satisfies the doubling property; there exists a constant δ such that, for all $x \in \Omega$ and $r > 0$, we have

$$0 < \mu(B(x, 2r)) \leq \delta\mu(B(x, r)) < \infty. \quad (20)$$

We now turn our attention to the case when $\Omega = \Delta$. By lemma 2.2 of [11], it is shown that the distance function d given on Δ by

$$d(z, w) = \begin{cases} ||z| - |w|| + \left|1 - \dfrac{z\overline{w}}{|z|\,|w|}\right| & \text{if } z, w \in \Delta^*, \\ |z| + |w| & \text{otherwise} \end{cases} \quad (21)$$

is a pseudodistance on Δ, where $\Delta^* = \Delta \setminus \{0\}$. It is known (see [12]) that, at the boundary of Δ, d becomes the Koranyi distance. Also by Lemma 2 of [12], we have that for any pseudoball $B(w, r)$, $w \neq 0$, and $r \in (0, 2)$ we have that

$$|B(w, r)| \approx r^2, \quad (22)$$

where $|B|$ denotes the Lebesgue area measure of set B. Also observe that the pseudoball $B(0, 1) = \Delta$. It is known that (see [12]) (Δ, d, dA) is a homogeneous space.

Let f be a locally integrable function in Δ. Then the Hardy-Littlewood maximal function relative to the pseudodistance d is given by

$$\mathrm{M}f(z) = \sup_B \frac{1}{|B|} \int_B |f(w)|\, dA(w), \quad (23)$$

where the supremum is taken over all pseudoballs containing z.

Suppose $0 < \omega(z) < \infty$ almost everywhere on Δ. Then we say that ω is in the Muckenhoupt weight A_1 if

$$[\omega]_{A_1} = \operatorname*{ess\,sup}_{z\in\Delta} \frac{M\omega(z)}{\omega(z)} < \infty. \quad (24)$$

There are two equivalent definitions which are useful in practice. First, $\omega \in A_1$, if for almost every $z \in \Delta$,

$$M\omega(z) \leq [\omega]_{A_1}\, \omega(z). \quad (25)$$

It follows that if $\omega \in A_1$ then

$$[\omega]_{A_1}\, \omega(z) \geq M\omega(z) \geq \omega(\Delta), \quad (26)$$

and thus

$$1 \leq \frac{[\omega]_{A_1}\, \omega(z)}{\omega(\Delta)}. \quad (27)$$

Alternatively, $w \in A_1$ if for every pseudoball B we have that

$$\frac{1}{|B|} \int_B \omega(z)\, dA(z) \leq [\omega]_{A_1} \operatorname*{ess\,inf}_{u\in B} \omega(u). \quad (28)$$

For more details on the Muckenhoupt weights, see chapter 9 of [13] or chapter 4 of [1].

We will need some results on extrapolation. The following proposition is Theorem 5.24 of [1].

Proposition 6. *Let $\Omega \in \mathbb{R}^n$ and suppose there is some $p_0 \geq 1$ and the family $\mathcal{F}$ such that for all $\omega \in A_1$,*

$$\int_{\Omega} F(x)^{p_0} \omega(x)\, dx \leq C_0 \int_{\Omega} G(x)^{p_0} \omega(x)\, dx, \quad (F,G) \in \mathcal{F}. \tag{29}$$

Given $p(\cdot) \in \mathcal{P}^{log}(\Omega)$, if $1 \leq p_0 \leq p_- \leq p_+ < \infty$ and the maximal operator is bounded on $L^{(p(\cdot)/p_0)'}(\Omega)$, then

$$\|F\|_{p(\cdot)} \leq C_{p(\cdot)} \|G\|_{p(\cdot)}, \quad (F,G) \in \mathcal{F}, \tag{30}$$

where $C_{p(\cdot)} = CC_0$ and C is some positive constant depending on the dimension of Ω.

The following is Theorem 3.16 of [1].

Proposition 7. *Let $p \in \mathcal{P}^{log}(\Omega)$. Then the Hardy-Littlewood maximal operator function is bounded in $L^{p(\cdot)}(\Omega)$ and we have*

$$\|Mf\|_{p(\cdot)} \leq C \|f\|_{p(\cdot)}. \tag{31}$$

3. Variable Exponent Bergman Spaces

Given $p(\cdot) \in \mathcal{P}^{\log}(\Delta)$, we define the variable exponent Bergman space $A^{p(\cdot)}$ as the space of all analytic functions on Δ that belong to the variable exponent Lebesgue space $L^{p(\cdot)}$ with respect to the area measure dA on the unit disc. With this definition $A^{p(\cdot)}$ is a closed subspace of $L^{p(\cdot)}$. By Theorem 4.4 of [9], the Bergman projection, P, given by (2) is bounded from $L^{p(\cdot)}$ onto $A^{p(\cdot)}$ for any $p(\cdot) \in \mathcal{P}^{\log}(\Delta)$. It is, thus, necessary to study the behaviour of Toeplitz operators on such spaces. Similar to the definition of Toeplitz operators on the Bergman spaces with constant exponent, we define the Toeplitz operator with symbol $f \in L^{\infty}$ on $A^{p(\cdot)}$ by

$$T_f g = P(fg), \quad g \in A^{p(\cdot)}. \tag{32}$$

Lemma 8. *The operator U_z is bounded on $A^{p(\cdot)}$ for $p(\cdot) > 1$.*

Proof. Let $1 < p_0 \leq p_- \leq p_+ < \infty$ and $\omega \in A_1$. Then

$$\int_{\Delta} |U_z f(\xi)|^{p_0} \omega(\xi)\, dA(\xi) \leq \int_{\Delta} |U_z f(\xi)|^{p_0} M\omega(\xi)\, dA(\xi) \tag{33}$$

since $\omega(\xi) \leq M\omega(\xi)$ for almost every $\xi \in \Delta$. Now, for $0 < \epsilon < \omega(\Delta)$ there is a pseudoball B containing ξ such that

$$M\omega(\xi) \leq \frac{\omega(B)}{|B|} + \epsilon \leq \frac{\omega(\Delta)}{|B|} + \omega(\Delta). \tag{34}$$

It follows that

$$\int_{\Delta} |U_z f(\xi)|^{p_0} M\omega(\xi)\, dA(\xi) \leq \left(\frac{\omega(\Delta)}{|B|} + \omega(\Delta)\right) \int_{\Delta} |U_z f(\xi)|^{p_0}\, dA(\xi). \tag{35}$$

Also

$$\begin{aligned} \int_{\Delta} |U_z f(\xi)|^{p_0} dA(\xi) &= \int_{\Delta} \left|f \circ \varphi_z(\xi) \varphi_z'(\xi)\right|^{p_0} dA(\xi) \\ &= \int_{\Delta} \left|f(\xi) \varphi_z'(\varphi_z(\xi))\right|^{p_0} \left|\varphi_z'(\xi)\right|^2 dA(\xi) \\ &= \int_{\Delta} |f(\xi)|^{p_0} \left|\varphi_z'(\varphi_z(\xi))\right|^{p_0 - 2} dA(\xi) \\ &\leq \frac{2^{p_0}}{\left(1 - |z|^2\right)^{p_0 - 2}} \int_{\Delta} |f(\xi)|^{p_0} dA(\xi) \\ &\leq \frac{[\omega]_{A_1} 2^{p_0}}{\omega(\Delta)\left(1 - |z|^2\right)^{p_0 - 2}} \int_{\Delta} |f(\xi)|^{p_0} \omega(\xi)\, dA(\xi), \end{aligned} \tag{36}$$

where the last inequality comes from (27). This shows that

$$\int_{\Delta} |U_z f(\xi)|^{p_0} \omega(\xi)\, dA(\xi) \leq \frac{(|B| + 1)[\omega]_{A_1} 2^{p_0}}{|B|\left(1 - |z|^2\right)^{p_0 - 2}} \int_{\Delta} |f(\xi)|^{p_0} \omega(\xi)\, dA(\xi). \tag{37}$$

It follows that the family $\{(|U_z f|, |f|) : f \in L^{p_0}\}$ satisfies inequality (29). Also, by Proposition 7 the maximal function Mf belongs to $L^{(p(\cdot)/p_0)'}$. Thus by Proposition 6 U_z is bounded on $A^{p(\cdot)}$.

Remark 9. We just want to give an alternative argument to obtain the estimate (35), and this argument has different effects and may be useful in applications.

We recall that if ω is locally integrable in Δ, then

$$\lim_{r \to 0} \frac{1}{|B(z,r)|} \int_{B(z,r)} |\omega(z) - \omega(\xi)|\, dA(\xi) = 0. \tag{38}$$

The proof of this statement can easily be adapted from that of Theorem 1.3 of [14]. We use this statement as follows:

Let $\omega \in A_1$ and $p_0 > 1$. Then for any $\epsilon > 0$, we can find $R > 0$ such that

$$\frac{1}{|B(z,r)|} \int_{B(z,r)} |\omega(z) - \omega(\xi)|\, dA(\xi) < \epsilon\omega(\Delta), \tag{39}$$

for all $r \in (0, R)$. Now, if we fix such R then for $0 < r < R$ we have

$$\int_{\Delta} |U_z f(\xi)|^{p_0} \omega(\xi)\, dA(\xi) = \int_{\Delta} |U_z f(\xi)|^{p_0} \Bigg(\omega(\xi) - \frac{1}{|B(\xi,r)|} \int_{B(\xi,r)} \omega(\zeta)\, dA(\zeta) + \frac{1}{B(\xi,r)} \int_{B(\xi,r)} \omega(\zeta)\, dA(\zeta)\Bigg) dA(\xi)$$

$$\begin{aligned}&\leq \int_{\Delta} |U_z f(\xi)|^{p_0} \frac{1}{|B(\xi,r)|} \\ &\cdot \int_{B(\xi,r)} |\omega(\xi) - \omega(\zeta)| \, dA(\zeta) \, dA(\xi) \\ &+ \int_{\Delta} |U_z f(\xi)|^{p_0} \frac{1}{|B(\xi,r)|} \int_{B(\xi,r)} \omega(\zeta) \, dA(\zeta) \, dA(\xi) \\ &\leq \omega(\Delta)(\epsilon + C_r) \int_{\Delta} |U_z f(\xi)|^{p_0} \, dA(\xi),\end{aligned} \tag{40}$$

where the constant C_r does not depend on $\omega \in A_1$, and thus (35) holds. We may also use this same argument in some parts of the proofs of Theorem 2 and Proposition 17 by replacing similar statements that give rise to the estimates (34) in the proof of Lemma 8.

Let

$$Tf(z) = \int_{\Delta} f(w) K(z,w) \, dA(w), \quad z \in \Delta, \tag{41}$$

where $K : \Delta \times \Delta \to \mathbb{C}$ is a kernel function. We give a Schur-type lemma that will be useful in our work.

Lemma 10. *Let $\omega \in A_1$, $p > 1$, and $1/p + 1/p' = 1$. If there exist positive constants C_1 and C_2 that depends on $[\omega]_{A_1}$ and not ω, and a nonnegative measurable function, h, such that*

$$\int_{\Delta} |K(\xi,z)| h(z)^{p'} \, dA(z) \leq C_1 h(\xi)^{p'} \tag{42}$$

for almost every $\xi \in \Delta$ and

$$\omega(\Delta)^{-1} \int_{\Delta} |K(\xi,z)| h(\xi)^p \omega(\xi) \, dA(z) \leq C_2 h(z)^p \tag{43}$$

for almost every $z \in \Delta$, then

$$\begin{aligned}&\int_{\Delta} |Tf(z)|^p \omega(z) \, dA(z) \\ &\quad \leq C_1 C_2 \int_{\Delta} |f(z)|^p \omega(z) \, dA(z).\end{aligned} \tag{44}$$

Proof. Using Hölder's inequality, we have

$$\begin{aligned}|Tf(z)| &\leq \int_{\Delta} |K(\xi,z)| h(\xi) h(\xi)^{-1} |f(\xi)| \, dA(\xi) \\ &\leq C_1' h(z) \left\{ \int_{\Delta} |K(\xi,z)| h(\xi)^{-p} |f(\xi)|^p \, dA(\xi) \right\}^{1/p} \\ &\leq C_1' h(z) \left\{ \int_{\Delta} |K(\xi,z)| h(\xi)^{-p} |f(\xi)|^p \right. \\ &\quad \left. \cdot \frac{[\omega]_{A_1} \omega(\xi)}{\omega(\Delta)} dA(\xi) \right\}^{1/p},\end{aligned} \tag{45}$$

where the second inequality comes from (42) and the third inequality is from (27). Also, Fubini's theorem gives

$$\begin{aligned}&\int_{\Delta} |Tf(z)|^p \omega(z) \, dA(z) \\ &\quad \leq C_1 \int_{\Delta} \frac{[\omega]_{A_1} h(z)^p}{\omega(\Delta)} \left\{ \int_{\Delta} |K(\xi,z)| h(\xi)^{-p} |f(\xi)|^p \right. \\ &\quad \left. \cdot \omega(\xi) \, dA(\xi) \right\} \omega(z) \, dA(z) \\ &\quad = C_1 [\omega]_{A_1} \int_{\Delta} |f(\xi)|^p h(\xi)^{-p} \int_{\Delta} |K(\xi,z)| h(z)^p \\ &\quad \cdot \frac{\omega(z)}{\omega(\Delta)} dA(z) \, \omega(\xi) \, dA(w) \leq C_1 C_2 [\omega]_{A_1} \\ &\quad \cdot \int_{\Delta} |f(z)|^p \omega(z) \, dA(z),\end{aligned} \tag{46}$$

where we have used (43) to get the last inequality.

Proposition 11. *Let $p(\cdot) \in \mathcal{P}^{log}(\Delta)$, $p \in (1,\infty)$ be such that $1 < p \leq p_- \leq p_+ < \infty$. Suppose the function $K : \Delta \times \Delta \to \mathbb{C}$ satisfies the hypothesis of Lemma 10. Then there is a constant C such that*

$$\|Tf\|_{p(\cdot)} \leq C C_1 C_2 \|f\|_{p(\cdot)}, \tag{47}$$

for all $f \in L^{p(\cdot)}$.

Proof. Let $\omega \in A_1$. Then by Lemma 10, we obtain (29) for the family $\{(|Tf|, |f|) : f \in L^{p_0}(\Delta)\}$, where $1 < p_0 \leq p_- \leq p_+ < \infty$. Also by Proposition 7, the maximal function Mf belongs to $L^{(p(\cdot)/p_0)'}(\Delta)$ and thus by Proposition 6 we get the required estimates.

In the application of Lemma 10, we may assume that $\operatorname{ess\,inf}_{z\in\Delta}\omega(z) \geq 1$ for $\omega \in A_1$, as the following lemma shows.

Lemma 12. *Let $\omega \in A_1$ be such that $\operatorname{ess\,inf}_{z\in\Delta}\omega(z) < 1$ and $\mu = \delta\omega$ for any $\delta > 1$ such that $\operatorname{ess\,inf}_{z\in\Delta}\mu(z) \geq 1$. If the hypothesis of Lemma 10 holds for the weight μ, then the conclusion of Lemma 10 holds for the weight ω.*

Proof. By (27), we have that $1 \leq ([\mu]_{A_1}/\mu(\Delta))\mu(z)$ for almost every $z \in \Delta$. Now by Proposition 9.1.5 of [13], we have that $[\mu]_{A_1} = [\delta\omega]_{A_1} = [\omega]_{A_1}$. It follows that $1 \leq ([\omega]_{A_1}/\mu(\Delta))\mu(z)$ for almost every $z \in \Delta$. Thus we have that the constants $C_1 = [\omega]_{A_1}$ and C_2 are independent of μ and, hence, independent of ω. Now since the hypothesis of Lemma 10 holds for the weight μ, we have that

$$\int_{\Delta} |Tf(z)|^p \mu(z) \, dA(z) \leq C_1 C_2 \left\| f\mu^{1/p} \right\|_p^p. \tag{48}$$

Thus,

$$\delta \int_{\Delta} |Tf(z)|^p \omega(z) \, dA(z) \leq \delta C_1 C_2 \left\| f\omega^{1/p} \right\|_p^p. \tag{49}$$

It follows that

$$\int_{\Delta} |Tf(z)|^p \omega(z)\, dA(z) \le C_1 C_2 \left\| f\omega^{1/p} \right\|_p^p, \quad (50)$$

which gives the result.

The next lemma will be used frequently and is well known; see, for example, Lemma 3.10 of [15] for the proof.

Lemma 13. *Suppose $a < 1$ and $a + b < 2$. Then*

$$\sup_{z\in\Delta} \int_{\Delta} \frac{dA(w)}{\left(1-|w|^2\right)^a |1-\overline{z}w|^b} < \infty. \quad (51)$$

Proof of Theorem 2. Let $g \in A^{p(\cdot)}$. Then

$$T_f g(z) = \langle T_f g, K_z \rangle = \langle g, T_{\overline{f}} K_z \rangle = \int_{\Delta} g(\xi)\, \overline{T_{\overline{f}} K_z}(\xi)\, dA(\xi). \quad (52)$$

Now, let $h(z) = (1-|z|^2)^{-\epsilon}$ for

$$\max\left(\frac{2}{qp_0}, \frac{2}{qp_0'}\right) < \epsilon < \min\left(\frac{1}{q'p_0'}, \frac{1}{q'p_0}\right). \quad (53)$$

Then using the identity

$$T_{\overline{f}} K_z(\xi) = P\left(\overline{f}\circ\varphi_z\right)\left(\varphi_z(\xi)\right) K_z(w), \quad z, \xi \in \Delta \quad (54)$$

we have

$$\begin{aligned} &\int_{\Delta} \left|T_{\overline{f}} K_z(\xi)\right| h(\xi)^{p_0'} dA(\xi) \\ &= \int_{\Delta} \left|P\left(\overline{f}\circ\varphi_z\right)\left(\varphi_z(\xi)\right) K_z(\xi)\right| \left(1-|z|^2\right)^{-\epsilon p_0'} dA(\xi) \\ &= \frac{1}{\left(1-|z|^2\right)^{\epsilon p_0'}} \int_{\Delta} \frac{\left|P\left(\overline{f}\circ\varphi_z\right)(\xi)\right| dA(\xi)}{\left(1-|\xi|^2\right)^{\epsilon p_0'} |1-\overline{z}\xi|^{(2-2\epsilon p_0')}} \\ &\le \frac{\left\|P\left(\overline{f}\circ\varphi_z\right)\right\|_q}{\left(1-|z|^2\right)^{\epsilon p_0'}} \left(\int_{\Delta} \frac{dA(\xi)}{\left(1-|\xi|^2\right)^{\epsilon p_0' q'} |1-\overline{z}\xi|^{(2-2\epsilon p_0')q'}}\right)^{1/q'}. \end{aligned} \quad (55)$$

Also,

$$\sup_{z\in\Delta} \int_{\Delta} \frac{dA(\xi)}{\left(1-|\xi|^2\right)^{\epsilon p_0' q'} |1-\overline{z}\xi|^{(2-2\epsilon)p_0' q'}} < \infty \quad (56)$$

provided that $\epsilon < 1/p_0'q'$ and $\epsilon p_0'q' + 2q' - 2\epsilon p_0'q' < 2$. That is, $2/p_0'q < \epsilon < 1/p_0'q'$ which holds from the choice of ϵ.

Now, observe that

$$T_{\overline{f}} K_z(\xi) = \langle T_{\overline{f}} K_z, K_\xi \rangle = \langle K_z, T_f K_\xi \rangle = \overline{T_f K_\xi(z)}. \quad (57)$$

Thus, for each $\omega \in A_1$, we have

$$\begin{aligned} &\int_{\Delta} \left|T_{\overline{f}} K_z(\xi)\right| h(z)^{p_0} \frac{\omega(z)}{\omega(\Delta)} dA(z) \\ &= \int_{\Delta} \left|T_f K_\xi(z)\right| h(z)^{p_0} \frac{\omega(z)}{\omega(\Delta)} dA(z) \\ &\le \frac{1}{\omega(\Delta)} \int_{\Delta} \left|T_f K_\xi(z)\right| h(z)^{p_0} M\omega(z)\, dA(z) \end{aligned} \quad (58)$$

since $\omega(z) \le M\omega(z)$ for almost every $z \in \Delta$. For any $0 < \epsilon_1 < \omega(\Delta)$, there is a pseudoball B containing z such that

$$M\omega(z) \le \frac{\omega(B)}{|B|} + \epsilon_1 \le \frac{\omega(\Delta)}{|B|} + \epsilon_1 \le \frac{\omega(\Delta)}{|B|} + \omega(\Delta). \quad (59)$$

Substitute this in (58) to obtain

$$\begin{aligned} &\int_{\Delta} \left|T_f K_\xi(z)\right| h(z)^{p_0} \frac{\omega(z)}{\omega(\Delta)} dA(z) \\ &\le \left(|B|^{-1} + 1\right) \int_{\Delta} \left|T_f K_\xi(z)\right| h(z)^{p_0} dA(z). \end{aligned} \quad (60)$$

Using the identity (54), we have

$$\begin{aligned} &\int_{\Delta} \left|T_{\overline{f}} K_z(\xi)\right| h(\xi)^{p_0} \frac{\omega(\xi)}{\omega(\Delta)} dA(\xi) \le C' \int_{\Delta} \left|T_f K_\xi(z)\right| \\ &\cdot \frac{1}{\left(1-|z|^2\right)^{\epsilon p_0}} dA(z) \\ &= C' \int_{\Delta} \left|P\left(\overline{f}\circ\varphi_\xi\right)\left(\varphi_\xi(z)\right) K_\xi(z)\right| \\ &\cdot \left(1-|z|^2\right)^{-\epsilon p_0} dA(\xi) = \frac{C'}{\left(1-|\xi|^2\right)^{-\epsilon p_0}} \\ &\cdot \int_{\Delta} \left|P\left(\overline{f}\circ\varphi_\xi\right)(z) K_\xi(z)\right| \left(1-|z|^2\right)^{-\epsilon p_0} dA(\xi) \\ &= \frac{C'}{\left(1-|\xi|^2\right)^{-\epsilon p_0}} \int_{\Delta} \frac{\left|P\left(f\circ\varphi_\xi\right)(z)\right| dA(z)}{\left(1-|z|^2\right)^{\epsilon p_0} |1-\overline{z}\xi|^{(2-2\epsilon p_0)}}, \end{aligned} \quad (61)$$

where the last equality is from the change of variable $z = \varphi_\xi$. By Hölder's inequality, we have that

$$\begin{aligned} &\int_{\Delta} \frac{\left|P\left(\overline{f}\circ\varphi_\xi\right)(z)\right| dA(z)}{\left(1-|z|^2\right)^{\epsilon p_0} |1-\overline{z}\xi|^{(2-2\epsilon p_0)}} \le \left\|P\left(\overline{f}\circ\varphi_z\right)\right\|_q \\ &\cdot \left(\int_{\Delta} \frac{dA(\xi)}{\left(1-|w|^2\right)^{\epsilon p_0 q'} |1-\overline{z}\xi|^{(2-2\epsilon p_0)q'}}\right)^{1/q'}, \end{aligned} \quad (62)$$

$$\sup_{\xi\in\Delta} \int_{\Delta} \frac{dA(\xi)}{\left(1-|\xi|^2\right)^{\epsilon p_0 q'} |1-\overline{z}\xi|^{(2-2\epsilon p_0)q'}} < \infty$$

if $\epsilon < 1/p_0q'$ and $\epsilon p_0q' + 2q' - 2\epsilon p_0q' < 2$; that is, $2/qp_0 < \epsilon < 1/p_0q'$. Now if $\epsilon > 0$ is chosen to satisfy (53) we see that the hypothesis of Lemma 10 is satisfied and thus for every $\omega \in A_1$ we have

$$\begin{aligned} &\int_{\Delta} \left|T_f g(z)\right|^p \omega(z)\, dA(z) \le C \sup_{z\in\Delta} \left\|P\left(f\circ\varphi_z\right)\right\|_q \\ &\cdot \sup_{z\in\Delta} \left\|P\left(\overline{f}\circ\varphi_z\right)\right\|_q \int_{\Delta} |g(z)|^p \omega(z)\, dA(z) \\ &= CC_1C_2 \int_{\Delta} |g(z)|^p \omega(z)\, dA(z), \end{aligned} \quad (63)$$

where the constant C does not depend on ω but on $[\omega]_{A_1}$. Finally we apply Proposition 11 to $\{(|T_f g|, |g|) : g \in L^{p_0}\}$ to obtain the result.

For $f \in BT$ it is easy to show that

$$\sup_{z\in\Delta} \|P(f \circ \varphi_z)\|_q < \infty, \qquad \sup_{z\in\Delta} \|P(\overline{f} \circ \varphi_z)\|_q < \infty \tag{64}$$

for any $q > 0$. Thus, from Theorem 2, we immediately have the following.

Corollary 14. *Suppose $f \in BT$. Then*

(1) *T_f is bounded on $A^{p(\cdot)}$,*

(2) *$T_{f\circ\varphi_z}$ is bounded on $A^{p(\cdot)}$, and*

$$\|T_{f\circ\varphi_z}\|_{p(\cdot)} \le C\|f\|_{BT}. \tag{65}$$

Proof. (1) is an immediate consequence of Theorem 2.

(2) follows from the fact that $\|f \circ \varphi_z\|_{BT} = \|f\|_{BT}$ and $\|T_{f\circ\varphi_z}\|_{p(\cdot)} \le C\|f\circ\varphi_z\|_{BT}$, which is given by assertion (1).

We also have the following estimate for operators in the Toeplitz algebra. To be precise, we have the following.

Lemma 15. *Let $S = T_{f_1}\cdots T_{f_n}$, $f_i \in BT$, $i = 1,\dots,n$. Then*

$$\sup_{z\in\Delta} \|S_z 1\|_{p(\cdot)} < \infty, \qquad \sup_{z\in\Delta} \|S_z^* 1\|_{p(\cdot)} < \infty \tag{66}$$

for any $P \in \mathcal{P}^{log}(\Delta)$.

Proof. By assertion (2) of Corollary 14, we have that

$$\|S_z 1\|_{p(\cdot)} = \|T_{f_1\circ\varphi_z}\cdots T_{f_n\circ\varphi_z}\|_{p(\cdot)} \le C\|f_1\|_{BT}\cdots\|f_n\|_{BT}. \tag{67}$$

Also, since each $\overline{f_j} \in BT$ and $\|\overline{f_j}\|_{BT} = \|f_j\|_{BT}$, we have that

$$\|S_z 1\|_{p(\cdot)} = \|T_{\overline{f_1}\circ\varphi_z}\cdots T_{\overline{f_n}\circ\varphi_z}\|_{p(\cdot)} \le C\|f_1\|_{BT}\cdots\|f_n\|_{BT}. \tag{68}$$

This completes the proof of the lemma.

4. Norm Estimates

Lemma 16. *Let $1 < p_0 < \infty$ and $p_1 = \min(p_0, p_0')$ and suppose that S is a bounded operator on $A^{p(\cdot)}$ and $q > (p_1 + 1)/(p_1 - 1)$. If $\epsilon > 0$ satisfies (53), then for all $\omega \in A_1$ we have the following:*

$$\int_\Delta \frac{|(SK_z)(\xi)|}{(1-|\xi|^2)^{\epsilon p_0'}} dA(\xi) \le \frac{C_1\|S_z 1\|_q}{(1-|z|^2)^{\epsilon p_0'}} \tag{69}$$

for all $z \in \Delta$ and

$$\omega(\Delta)^{-1}\int_\Delta \frac{|(SK_z)(\xi)|}{(1-|z|^2)^{\epsilon p_0}}\omega(z)\,dA(z) \le \frac{C_2\|S_\xi^* 1\|_q}{(1-|\xi|^2)^{\epsilon p_0}} \tag{70}$$

for all $\xi \in \Delta$, where the constant C_2 does not depend on ω.

Proof. Fix $z \in \Delta$. Then

$$SK_z = \frac{SU_z 1}{|z|^2 - 1} = \frac{U_z S_z 1}{|z|^2 - 1} = \frac{((S_z 1)\circ\varphi_z)\varphi_z'}{|z|^2 - 1}, \tag{71}$$

where the second equality comes from the definition of S_z and the third equality from the definition of U_z. Thus,

$$\begin{aligned}\int_\Delta \frac{|(SK_z)(\xi)|}{(1-|\xi|^2)^{\epsilon p_0'}} dA(\xi) &= \frac{1}{1-|z|^2}\int_\Delta \frac{|(S_z 1)\circ\varphi_z(\xi)|\,|\varphi_z'(\xi)|}{(1-|\xi|^2)^{\epsilon p_0'}} dA(\xi)\\ &= \frac{1}{(1-|z|^2)^{\epsilon p_0'}}\int_\Delta \frac{|(S_z 1)(\xi)|}{(1-|\xi|^2)^{\epsilon p_0'}|1-\overline{z}\xi|^{2-2\epsilon p_0'}} dA(\xi)\\ &\le C\frac{\|S_z 1\|_q\|f_z\|_{q'}}{(1-|z|^2)^{\epsilon p_0'}},\end{aligned} \tag{72}$$

where $f_z(\xi) = 1/(1-|\xi|^2)^{\epsilon p_0'}|1-\overline{z}\xi|^{2-2\epsilon p_0'}$. By the choice of ϵ, we have that $\sup_{z\in\Delta}\|f_z\|_{q'} < \infty$ and (69) holds.

To prove (70), replace S by S^* in (69), interchange ξ and ξ in (69), and then use the equation

$$S^* K_\xi(z) = \langle S^* K_\xi, K_z\rangle = \langle K_\xi, SK_z\rangle = \overline{SK_z(\xi)}. \tag{73}$$

Finally, we use the same argument as in the proof of Theorem 2 to obtain that there is a pseudoball B containing z such that $\omega(z) \le M\omega(z) \le \omega(\Delta)(|B|^{-1} + 1)$ and thus

$$\begin{aligned}&\omega(\Delta)^{-1}\int_\Delta \frac{|(SK_z)(\xi)|}{(1-|z|^2)^{\epsilon p_0}}\omega(z)\,dA(z)\\ &\quad\le (|B|^{-1}+1)\int_\Delta \frac{|(S^* K_\xi)(z)|}{(1-|z|^2)^{\epsilon p_0}} dA(z).\end{aligned} \tag{74}$$

A similar argument as the one used to obtain the estimate (69) will give us (70).

Proposition 17. *Let $p \in P^{log}(\Delta)$, $1 < p_0 \le p_- \le p_+ < \infty$, and $p_1 = \min(p_0, p_0')$ and suppose that S is a bounded operator on $A^{p(\cdot)}$. If*

$$C_1 = \sup_{z\in\Delta}\|S_z 1\|_q < \infty, \qquad C_2 = \sup_{z\in\Delta}\|S_z^* 1\|_q < \infty \tag{75}$$

for some $q > (p_1 + 1)/(p_1 - 1)$, then there is a constant C such that

$$\|S\|_{p(\cdot)} \leq CC_1C_2. \tag{76}$$

Proof. For $f \in A^{p(\cdot)}$ and $w \in \Delta$, we have

$$\begin{aligned}(Sf)(\xi) &= \langle Sf, K_\xi \rangle = \langle f, S^* K_\xi \rangle \\ &= \int_\Delta f(z) \overline{(S^* K_\xi)(z)} dA(z) \\ &= \int_\Delta f(z) (SK_z)(\xi) dA(z),\end{aligned} \tag{77}$$

where the last equation follows from (69). Given that $\epsilon > 0$ that satisfies (53), we have by (69) that

$$\int_\Delta \frac{|(SK_z)(\xi)|}{(1-|\xi|^2)^{\epsilon p_0'}} dA(w) \leq \frac{C\|S_z 1\|_q}{(1-|z|^2)^{\epsilon p_0'}}. \tag{78}$$

In a similar manner, we use (73) and (70) to get that

$$\begin{aligned}&\omega(\Delta)^{-1} \int_\Delta \frac{|(SK_z)(\xi)|}{(1-|z|^2)^{\epsilon p_0}} \omega(z) dA(z) \\ &= \omega(\Delta)^{-1} \int_\Delta \frac{|(S^* K_\xi)(z)|}{(1-|z|^2)^{\epsilon p_0}} \omega(z) dA(z) \\ &\leq \frac{C\|S_\xi^* 1\|_q}{(1-|\xi|^2)^{\epsilon p_0}}.\end{aligned} \tag{79}$$

$$\int_\Delta |Sf(z)|^p \omega(z) dA(z) \leq C \int_\Delta |f(z)|^p \omega(z) dA(z)$$

for all $\omega \in A_1$ where the constant C depends on $[\omega]_{A_1}$ and not on ω. We now apply Proposition 11 to get the required result.

Lemma 18. *(a) $\|K_z\|_{p(\cdot)}$ is equivalent to $(1-|z|^2)^{-2+2/p(z)}$ for all $z \in \Delta$.*

(b) $K_z/\|K_z\|_{p(\cdot)} \to 0$ weakly in $A^{p(\cdot)}$ as $z \to \partial\Delta$.

Proof. The assertion (a) is just Theorem 3.5 of [10].

(b) If $f \in A^{p'(\cdot)}$, then assertion (a) implies

$$\left\langle f, \frac{K_z}{\|K_z\|_{p(\cdot)}} \right\rangle \cong f(z) (1-|z|^2)^{-2+2/p(z)}. \tag{80}$$

Thus if f is a bounded function in $A^{p'(\cdot)}$, then $\langle f, K_z/\|K_z\|_{p(\cdot)} \rangle \to 0$ as $z \to \partial\Delta$. The assertion follows from the fact that polynomials are dense in $A^{p'(\cdot)}$.

5. Compact Operators on $A^{p(\cdot)}$

Theorem 19. *Let $p \in \mathcal{P}^{log}(\Delta)$ be such that $1 < p_- \leq p_+ < \infty$ and suppose that*

$$\int_\Delta \|K(\cdot, z)\|_{p(\cdot)}^{p'(\cdot)_+} dA(z) < \infty. \tag{81}$$

Then the operator T given by (41) is compact on $L^{p(\cdot)}(\Delta)$.

Proof. Firstly we observe that if (81) holds and $g \in L^{p'(\cdot)}$, then the function $w \mapsto F(w)$, where

$$F(w) = \int_\Delta K(z, w) g(z) dA(z) \tag{82}$$

belongs to $L^{p'(\cdot)}(\Delta)$. Indeed,

$$\begin{aligned}&\int_\Delta |F(w)|^{p'(\cdot)_+} dA(w) \\ &= \int_\Delta \left| \int_\Delta g(z) K(z, w) dA(z) \right|^{p'(\cdot)_+} dA(w) \\ &\leq C_{p(\cdot)} \int_\Delta \|g\|_{p'(\cdot)}^{p'(\cdot)_+} \|K(\cdot, w)\|_{p(\cdot)}^{p'(\cdot)_+} dA(w) < \infty.\end{aligned} \tag{83}$$

It follows from Lemma 1 that $\|F\|_{p'(\cdot)} \leq \|F\|_{p'(\cdot)_+} < \infty$. Thus for $g \in L^{p'(\cdot)}$, we see that $F \in L^{p'(\cdot)}$. Now, suppose (φ_n) is a bounded sequence in $L^{p(\cdot)}$ such that $\varphi_n \to \varphi$ weakly in $L^{p(\cdot)}$. For $\epsilon > 0$ and any $g \in L^{p'(\cdot)}$, we can find that $N \in \mathbb{N}$ such that for $n \geq N$ we have

$$\left| \int_\Delta (\varphi_n(w) - \varphi(w)) \overline{g}(w) dA(w) \right| < \epsilon. \tag{84}$$

We will show that $\|T\varphi_n - T\varphi\|_{p(\cdot)} \to 0$ as $n \to \infty$. Now, given ϵ, we fix N such that (84) holds. It follows for any $n \geq N$ and (11) that

$$\begin{aligned}&\|T\varphi_n - T\varphi\|_{p(\cdot)} \leq C_{p(\cdot)} \|T\varphi_n - T\varphi\|_{p(\cdot)}^* \\ &= \sup \left\{ \left| \int_\Delta (T\varphi_n - T\varphi)(w) \overline{g}(w) dA(w) \right| : \|g\|_{p'(\cdot)} \leq 1 \right\} \\ &= \sup \left\{ \left| \int_\Delta \int_\Delta (\varphi_n(w) - \varphi(w)) K(w, z) dA(w) \overline{g}(z) dA(z) \right| : \right. \\ &\left. \|g\|_{p'(\cdot)} \leq 1 \right\} \\ &= \sup \left\{ \left| \int_\Delta (\varphi_n(w) - \varphi(w)) \overline{\int_\Delta K(z, w) g(z) dA(z)} dA(w) \right| : \right. \\ &\left. \|g\|_{p'(\cdot)} \leq 1 \right\} = \sup \left\{ \left| \int_\Delta (\varphi_n(w) - \varphi(w)) \overline{F(w)} dA(w) \right| : \right. \\ &\left. \|g\|_{p'(\cdot)} \leq 1 \right\} < \epsilon.\end{aligned} \tag{85}$$

Thus $\|T\varphi_n - T\varphi\|_{p(\cdot)} \to 0$ as $n \to \infty$. Finally, it is shown in Corollary 2.81 of [1] that the variable Lebesgue space $L^{p(\cdot)}$ is reflexive if and only if $1 < p_- \leq p_+ < \infty$. We thus conclude that T is compact, since $L^{p(\cdot)}$ is reflexive.

We will need the power series formula for the Berezin transform of the bounded operator S on A^2. From the definition of the reproducing kernel, we get that

$$k_z(w) = (1-|z|^2) \sum_{n=0}^{\infty} (n+1) w^n \overline{z}^n \tag{86}$$

for $z, w \in \Delta$. To compute $\widetilde{S}(z) = \langle Sk_z, k_z \rangle$, we first compute Sk_z by applying S to both sides of (86) and then take the inner product with k_z, again using (86), to obtain

$$\widetilde{S}(z) = \left(1 - |z|^2\right)^2 \sum_{m,j=0}^{\infty} (j+1)(m+1)\left\langle Sw^j, w^m \right\rangle \overline{z}^j z^m. \tag{87}$$

Lemma 20. *Suppose S is a bounded operator on $A^{p(\cdot)}$ such that*

$$\sup_{z\in\Delta} \left\|S_z 1\right\|_q < \infty \tag{88}$$

for some $q > 1$. Then $\widetilde{S}(z) \to 0$ as $z \to \partial\Delta$ if and only if for every $t \in [1, q)$, $\|S_z 1\|_t \to 0$ as $z \in \partial\Delta$.

Proof. Suppose for every $t \in [1, q)$, $\|S_z 1\|_t \to 0$ as $z \to \partial\Delta$. In particular, $\|S_z 1\|_1 \to 0$ as $z \in \partial\Delta$. Thus

$$\left|\widetilde{S}(z)\right| = \left|\langle S_z 1, 1\rangle\right| \le \left\|S_z 1\right\|_1 \longrightarrow 0 \tag{89}$$

as $z \in \partial\Delta$.

Suppose $\widetilde{S}(z) \to 0$ as $z \to \partial\Delta$. Fix $t \in [1, q)$. We will show that $\|S_z 1\|_t \to 0$ as $z \in \partial\Delta$.

For $z \in \Delta$, $j, m = 0, 1, \ldots$, an easy computation shows that

$$\int_\Delta \left(S_z w^j\right) \overline{w^m} dA(w) = \int_\Delta \left(SU_z w^j\right) \overline{U_z w^m} dA(w). \tag{90}$$

Since

$$\begin{aligned} U_z w^j &= \left(\varphi_z(w)\right)^j \varphi_z'(w) \\ &= -\left(\varphi_z(w)\right)^j \left(1 - |z|^2\right) K_z(w), \end{aligned} \tag{91}$$

we have that

$$\begin{aligned} \int_\Delta \left(SU_z w^j\right) \overline{U_z w^m} dA(w) &= \left(1 - |z|^2\right)^2 \\ &\cdot \int_\Delta S\left[\left(\varphi_z\right)^j K_z\right](w) \overline{\left(\varphi_z(w)\right)^m K_z(w)} dA(w). \end{aligned} \tag{92}$$

It follows from (90) and (92) and Hölder's inequality that

$$\begin{aligned} \left|\left\langle S_z w^j, w^m \right\rangle\right| &= \left|\int_\Delta \left(S_z w^j\right) \overline{w^m} dA(w)\right| = \left(1 - |z|^2\right)^2 \\ &\cdot \left|\int_\Delta S\left[\left(\varphi_z\right)^j K_z\right](w) \overline{\left(\varphi_z(w)\right)^m K_z(w)} dA(w)\right| \\ &\le C_{p(\cdot)} \left(1 - |z|^2\right)^2 \left\|S\left[\varphi_z^j K_z\right]\right\|_{p(\cdot)} \left\|\varphi_z^m K_z\right\|_{p'(\cdot)} \\ &\le C_{p(\cdot)} \left(1 - |z|^2\right)^2 \|S\|_{p(\cdot)} \left\|\varphi_z^j K_z\right\|_{p(\cdot)} \left\|\varphi_z^m K_z\right\|_{p'(\cdot)}. \end{aligned} \tag{93}$$

Now, let

$$\begin{aligned} S_1 &= \left\{\lambda > 0 : \int_\Delta \left(\frac{\left|K_z(w)\right|}{\lambda}\right)^{p(w)} dA(w) \le 1\right\}, \\ S_2 &= \left\{\lambda > 0 : \int_\Delta \left(\frac{\left|\varphi_z(w) K_z(w)\right|}{\lambda}\right)^{p(w)} dA(w) \le 1\right\}. \end{aligned} \tag{94}$$

Then $S_1 \subset S_2$ and thus $\inf S_1 \ge \inf S_2$. This shows that

$$\left\|\varphi_z^j K_z\right\|_{p(\cdot)} \le \left\|K_z\right\|_{p(\cdot)} \tag{95}$$

for any $p(\cdot) \in P^{\log}(\Delta)$. This and Lemma 18(a) show that

$$\begin{aligned} &\left|\left\langle S_z w^j, w^m \right\rangle\right| \\ &\le C_{p(\cdot)} \left(1 - |z|^2\right)^2 \|S\|_{p(\cdot)} \left\|K_z\right\|_{p(\cdot)} \left\|K_z\right\|_{p'(\cdot)} \\ &\le C \|S\|_{p(\cdot)}; \end{aligned} \tag{96}$$

that is, $\langle S_z w^j, w^m \rangle$ is uniformly bounded in $z \in \Delta$ and $j, m = 0, 1, \ldots$.

Now, we will show that for every nonnegative integer n

$$\left\langle S_z 1, w^n \right\rangle \longrightarrow 0 \quad \text{as } z \longrightarrow \partial\Delta. \tag{97}$$

If this is not true, then there is a sequence $z_k \in \Delta$ such that

$$\left\langle S_{z_k} 1, w^n \right\rangle \longrightarrow a_{0n} \quad \text{as } z_k \longrightarrow \partial\Delta \tag{98}$$

for some nonzero constant a_{0n} and $n \ge 1$. Since $\langle S_z w^j, w^m \rangle$ is uniformly bounded, we may assume without loss of generality that for each j and m

$$\left\langle S_z w^j, w^m \right\rangle \longrightarrow a_{jm} \quad \text{as } z \longrightarrow \partial\Delta, \tag{99}$$

for some constant a_{jm}. For $z, \xi \in \Delta$, we have

$$\begin{aligned} \widetilde{S}\left(\varphi_z(\xi)\right) &= \widetilde{S_z}(\xi) \\ &= \left(1 - |\xi|^2\right)^2 \sum_{j,m=0}^{\infty} (j+1)(m+1)\left\langle S_z w^j, w^m \right\rangle \overline{\xi}^j \xi^m, \end{aligned} \tag{100}$$

where the second equality comes from (87). Also, note that the power series in (100) converges uniformly for each $\xi \in \Delta$.

For each $\xi \in \Delta$, we know that $\varphi_{z_k}(\xi) \to \partial\Delta$ as $z_k \to \partial\Delta$. Thus $\widetilde{S}(\varphi_z(\xi)) \to 0$ as $z_k \to \partial\Delta$ for each $\xi \in \Delta$. Replacing z by z_k in (100) and taking the limit as $z_k \to \partial\Delta$, we get

$$\left(1 - |\xi|^2\right)^2 \sum_{j,m=0}^{\infty} (j+1)(m+1) a_{jm} \overline{\xi}^j \xi^m = 0 \tag{101}$$

for each $\xi \in \Delta$. If

$$f(\xi) = \sum_{j,m=0}^{\infty} (j+1)(m+1) a_{jm} \overline{\xi}^j \xi^m, \tag{102}$$

then $f(\xi) = 0$ for all $\xi \in \Delta$. This gives

$$\left[\frac{\partial^m}{\partial \xi^m} \frac{\partial^j}{\partial \overline{\xi}^j} f\right](0) = 0 \tag{103}$$

for each j and m. On the other hand, we have

$$\left[\frac{\partial^m}{\partial \xi^m} \frac{\partial^j}{\partial \overline{\xi}^j} f\right](0) = \left((j+1)!\,(m+1)!\right) a_{jm} \tag{104}$$

for each j and m. In particular, $a_{0n} = 0$ which is a contradiction. Hence, we obtain

$$\lim_{z\to\partial\Delta} \langle S_z 1, w^n\rangle = 0. \tag{105}$$

For $\xi \in \Delta$, we have

$$(S_z 1)(\xi) = \sum_{n=0}^{\infty} (n+1) \langle S_z 1, w^n\rangle \xi^n. \tag{106}$$

It is clear that for each fixed $\xi \in \Delta$, the power series above converges uniformly for $z \in \Delta$. This gives

$$\lim_{z\to\partial\Delta} (S_z 1)(\xi) = 0 \tag{107}$$

for each $\xi \in \Delta$. It follows that

$$\lim_{z\to\partial\Delta} |(S_z 1)(\xi)|^t = 0 \tag{108}$$

for each $\xi \in \Delta$. If $s = q/t$ then $s > 1$ and

$$\int_\Delta \left[|(S_z 1)(\xi)|^t\right]^s dA(\xi) = \|S_z 1\|_q^q \le \sup_{z\in\Delta} \|S_z 1\|_q^q < \infty. \tag{109}$$

This shows that $\{|(S_z 1)|^t\}_{z\in\Delta}$ is uniformly integrable. By Vitali's Theorem or Exercise 11 on pages 133-134 of [16], we have that

$$\lim_{z\to\partial\Delta} \|S_z 1\|_{t(\cdot)} = 0. \tag{110}$$

This completes the proof of the lemma.

Proof of Theorem 3. If S is compact on $A^{p(\cdot)}$, then by Lemma 18(b),

$$\left\langle \frac{SK_z}{\|K_z\|_{p(\cdot)}}, \frac{K_z}{\|K_z\|_{p'(\cdot)}} \right\rangle \longrightarrow 0 \tag{111}$$

as $z \to \partial\Delta$. Now by Lemma 18(a), we see that $\widetilde{S}(z)$ is equivalent to $\langle SK_z/\|K_z\|_{p(\cdot)}, K_z/\|K_z\|_{p'(\cdot)}\rangle$ for $z \in \Delta$. Thus $\widetilde{S}(z) \to 0$ as $z \to \partial\Delta$.

Suppose $\widetilde{S}(z) \to 0$ as $z \to \partial\Delta$. By Lemma 20, we have that $\|S_z 1\|_t \to 0$ as $z \to \partial\Delta$ for every $t \in [1, q)$. We will show that S is compact on $A^{p(\cdot)}$. Fix $t \in (1, q)$ in the rest of the proof.

For $f \in A^{p(\cdot)}$, we have that

$$(Sf)(w) = \int_\Delta f(z)(SK_z)(w)\, dA(z). \tag{112}$$

For $0 < r < 1$, we define an operator $S_{[r]}$ on $A^{p(\cdot)}$ by

$$(S_{[r]} f)(w) = \int_{r\Delta} f(z)(SK_z)(w)\, dA(z). \tag{113}$$

Then $S_{[r]}$ is an integral operator with kernel $(SK_z)(w)\chi_{r\Delta}(z)$. We will first show that the operator $S_{[r]}$ is compact on $A^{p(\cdot)}$. By Theorem 19, we only need to show that

$$\begin{aligned}\int_\Delta \|(SK_z)\chi_{r\Delta}(z)\|_{p(\cdot)}^{p'(\cdot)_+} dA(z) \\ = \int_{r\Delta} \|(SK_z)\|_{p(\cdot)}^{p'(\cdot)_+} dA(z) < \infty.\end{aligned} \tag{114}$$

But

$$\begin{aligned}&\int_{r\Delta} \|(SK_z)\|_{p(\cdot)}^{p'(\cdot)_+} dA(z) \\ &\le \int_{r\Delta} \|S\|_{p(\cdot)}^{p'(\cdot)_+} \|K_z\|_{p(\cdot)}^{p'(\cdot)_+} dA(z) \\ &\le C\|S\|_{p(\cdot)}^{p'(\cdot)_+} \int_{r\Delta} \frac{dA(z)}{\left(1-|z|^2\right)^{(2-2/p(z))p'(\cdot)_+}} \\ &\le \frac{Cr^2 \|S\|_{p(\cdot)}^{p'(\cdot)_+}}{\left(1-|r|^2\right)^{2p'(\cdot)_+}}.\end{aligned} \tag{115}$$

This shows that

$$\int_{r\Delta} \|(SK_z)\|_{p(\cdot)}^{p'(\cdot)_+} dA(z) < \infty, \tag{116}$$

and thus $S_{[r]}$ is compact on $A^{p(\cdot)}$. Hence, to prove that S is compact, we only need to show that $\|S - S_{[r]}\|_{p(\cdot)} \to 0$ as $r \to 1^-$.

If $r \in (0,1)$ then $S - S_{[r]}$ is the integral operator with kernel

$$(SK_z)(w)\chi_{\Delta\setminus r\Delta}(z) \tag{117}$$

as can be seen from (77) and (113). The proof of Proposition 17 indicates that

$$\|S - S_{[r]}\|_{p(\cdot)} \le CC_1C_2, \tag{118}$$

where

$$\begin{aligned}C_1 &= \sup\left\{\|S_z 1\|_t : r \le |z| < 1\right\}, \\ C_2 &= \sup\left\{\|S_z^* 1\|_t : z \in \Delta\right\}.\end{aligned} \tag{119}$$

We have shown that $C_1 \to 0$ as $r \to 1^-$ and the hypothesis of the theorem shows that $C_2 < \infty$. Thus, $\|S - S_{[r]}\|_{p(\cdot)} \to 0$ as $r \to 1^-$, which completes the proof.

Proof of Theorem 4. Suppose S is a finite sum of operators of the form $T_{f_1} \cdots T_{f_n}$, where each $f_j \in BT$. By Corollary 14 and Lemma 15, we have that S is bounded on $A^{p(\cdot)}$ and

$$\begin{aligned}&\sup_{z\in\Delta} \|S_z 1\|_q < \infty, \\ &\sup_{z\in\Delta} \|S_z^* 1\|_q < \infty\end{aligned} \tag{120}$$

for all $q > 0$. The conclusion then follows from Theorem 3.

Conflicts of Interest

The author declares that there are no conflicts of interest regarding the publication of this paper.

References

[1] D. V. Cruz-Uribe and A. Fiorenza, *Variable Lebesgue Spaces: Foundations and Harmonic Analysis*, Applied and Numerical Harmonic Analysis, Birkhäuser, Basel, Switzerland, 2013.

[2] A. Dieudonne, "M-Berezin Transform and Approximation of Operators on the Bergman Space Over Bounded Symmetric Domains," *Complex Analysis and Operator Theory*, vol. 11, no. 3, pp. 651–674, 2017.

[3] A. Dieudonne and E. Tchoundja, "Toeplitz operators with L^1 symbols on Bergman spaces in the unit ball of Cn," *Advances in Pure and Applied Mathematics*, vol. 2, no. 1, pp. 65–88, 2011.

[4] S. Axler and D. Zheng, "Compact operators via the Berezin transform," *Indiana University Mathematics Journal*, vol. 47, no. 2, pp. 387–400, 1998.

[5] W. Bauer and J. Isralowitz, "Compactness characterization of operators in the Toeplitz algebra of the Fock space F^p_α," *Journal of Functional Analysis*, vol. 263, no. 5, pp. 1323–1355, 2012.

[6] M. Englis, "Compact Toeplitz operators via the Berezin transform on bounded symmetric domains," *Integral Equations and Operator Theory*, vol. 33, no. 4, pp. 426–455, 1999.

[7] J. Miao and D. Zheng, "Compact operators on Bergman spaces," *Integral Equations and Operator Theory*, vol. 48, no. 1, pp. 61–79, 2004.

[8] M. Mitkovski, D. Suárez, and B. D. Wick, "The Essential Norm of Operators on $A^p{}_\alpha$ (B_n)," *Integral Equations and Operator Theory*, vol. 75, no. 2, pp. 197–233, 2013.

[9] G. R. Chacón and H. Rafeiro, "Variable exponent Bergman spaces," *Nonlinear Analysis. Theory, Methods & Applications. An International Multidisciplinary Journal*, vol. 105, pp. 41–49, 2014.

[10] G. R. Chacón, H. Rafeiro, and J. . Vallejo, "Carleson measures for variable exponent Bergman spaces," *Complex Analysis and Operator Theory*, vol. 11, no. 7, pp. 1623–1638, 2017.

[11] E. Tchoundja, "Carleson measures for the generalized Bergman spaces via a $T(1)$-type theorem," *Arkiv för Matematik*, vol. 46, no. 2, pp. 377–406, 2008.

[12] D. Békollé, "Inégalités á poids pour les projecteur de Bergman dans la boule unité de Cn," *Studia Math*, vol. 71, no. 3, pp. 305–323, 1982.

[13] L. Grafakos, *Modern Fourier Analysis*, Springer, New York, NY, USA, 2nd edition, 2009.

[14] E. M. Stein and R. Shakarchi, *Real Analysis: Measure Theory, Integration and Hilbert Spaces*, Princetown University Press, 2005.

[15] K. Zhu, *Operator Theory in Function Spaces*, American Mathematical Society, 2007.

[16] W. Rudin, *Real and Complex Analysis*, McGraw-Hill Book, 3rd edition, 1987.

2

Toward a Theory of Normalizing Function of Interestingness Measure of Binary Association Rules

Armand Armand,[1] André Totohasina,[2] and Daniel Rajaonasy Feno[3]

[1]*Lycee Mixte Antsiranana, Madagascar*
[2]*Department of Mathematics and Informatics, University of Antsiranana, Madagascar*
[3]*Department of Mathematics, Informatics and Applications, University of Toamasina, Madagascar*

Correspondence should be addressed to André Totohasina; andre.totohasina@gmail.com

Academic Editor: Theodore E. Simos

Regarding the existence of more than sixty interestingness measures proposed in the literature since 1993 till today in the topics of association rules mining and facing the importance these last one, the research on normalization probabilistic quality measures of association rules has already led to many tangible results to consolidate the various existing measures in the literature. This article recommends a simple way to perform this normalization. In the interest of a unified presentation, the article offers also a new concept of normalization function as an effective tool for resolution of the problem of normalization measures that have already their own normalization functions.

1. Introduction

1.1. Definitions and Notations. We always put ourselves in the framework of a context of binary data mining (see, for example, [1–3], which illustrate the importance of association rules mining based on choosing some quality measure) $D = (T, I, \mathcal{R})$, where I is a nonempty finite set of attributes or variables, T a finite set of entities or objects, $\mathcal{R}$ a binary relation from T to I, and P discrete uniform probability in the probabilistic space $(T, \mathcal{P}(T))$ [4, 5].

In the next sections, we use the following notation for two itemsets X, Y: $X' = \{e \in T / \forall x \in X; e\mathcal{R}x\}$, i.e., the set of all transactions containing the X pattern that is the dual of a pattern X of I $(X \subseteq I)$ [4, 6, 7]. $n = |T|$ represents the size of the total sample size; $n_X = |X'|$ represents the number of transactions satisfying pattern X; $n_{XY} = |X' \cap Y'|$ represents the number of transactions satisfying both X and Y; $n_{\overline{X}} = |T| - |T - T_{X'}|$, where $\overline{X}$ is the logical negation of X; $n_X/n = |T_{X'}|/|T|$ represents the support of the pattern X.

Hereafter, our work is divided into three sections. Section 2 gives the definition of normalization function. Section 3 recommends the raw results of normalizing function of some probabilistic quality measures. Finally Section 4, [5, 8–10] sets out the conclusion and perspectives.

2. Normalizing Function F_n

2.1. Motivations. The theory and practice of normalization probabilistic quality measures (see, for example, [4, 5, 11], (Totohasina et al. [12]), [6, 7]) have been resolving included in the list of tools for problems concerning the data mining. This is done in the view of regrouping [3, 4, 6, 12–17] different existing measures available from the literature. Let us notice that [4] proves existence of infinity quality measures through the concept of the so-called normalized quality measure under five conditions, but recently [16] still proposes a novel interestingness measure. By opening the door to the possibility of creating definitions of new concepts in the context of data mining, perhaps, this will bring to following a new reflection among researchers in this field. This is the normalizing function. What is meant by a normalizing function? The following section will attempt to answer such question. Remember that this paper is the logical continuation of the paper [18].

2.2. Proposal Approach

Definition 1. We consider a probabilistic measure of interest μ. We call normalizing function of μ the piecewise continuous function that can normalize directly the measure μ and is defined as

$$F_n(x) = \begin{cases} F_n^f(x), & \textit{if } X \textit{ favors } Y \textit{ and } x \in \left]x_{ind}; x_{imp}\right[; \\ F_n^d(x), & \textit{if } X \textit{ disfavors } Y \textit{ and } x \in \left]x_{inc}; x_{ind}\right[. \end{cases} \quad (1)$$

It takes the following particular values:

$$F_{an}(x) = \begin{cases} x_f x + y_f, & \textit{if } X \textit{ favors } Y; \\ x_d x + y_d, & \textit{if } X \textit{ disfavors } Y. \end{cases} \quad (2)$$

$$F_{shnd}(x) = \begin{cases} \dfrac{x_f}{x+m} + y_f, & \textit{if } X \textit{ favors } Y; \\ x_d x + y_d, & \textit{if } X \textit{ disfavors } Y. \end{cases} \quad (3)$$

$$F_{shng}(x) = \begin{cases} x_f x + y_f, & \textit{if } X \textit{ favors } Y; \\ \dfrac{x_d}{x+m} + y_d, & \textit{if } X \textit{ disfavors } Y. \end{cases} \quad (4)$$

$$F_{hn}(x) = \begin{cases} \dfrac{x_f}{x+m} + y_f, & \textit{if } X \textit{ favors } Y; \\ \dfrac{x_d}{x+m} + y_d, & \textit{if } X \textit{ disfavors } Y, \end{cases} \quad (5)$$

where x is value of μ; x_{imp} is value of μ at logical implication; x_{ind} is value of μ at independence; x_{inc} is value of μ at incompatibility; F_n^f is normalizing function, if X favors Y, that is to say, in case where $P(Y'/X') > P(Y')$; F_n^d is normalizing function, if X disfavors Y, that is to say, in case where $P(Y'/X') < P(Y')$; F_{an} is normalizing function of measure μ, that is, affine normalizable; F_{hn} is normalizing function of measure μ, that is, homographic normalizable; F_{shnd} is normalizing function of measure μ, that is, semihomographic-normalizable to right; F_{shng} is μ normalizing function of measure μ, that is, semihomographic-normalizable to left; x_f, y_f, x_d, and y_d are μ of normalization coefficient x; m is real and $\overline{\mathbb{R}} = [-\infty; +\infty]$.

The definition of normalizing function is thus determined according to the quality measures. Note that this is a function as any other, with respect to the variable x; it is a numerical function of a real variable; only, this variable x is as follows: $x = f_1(n, P(X'), P(Y'), P(Y' \cap X'))$. The four coefficients of normalization of probabilistic measures quality have to meet the following conditions: $x_f = f_2(n, P(X'), P(Y'))$, $y_f = f_3(n, P(X'), P(Y'))$, $x_d = f_4(n, P(X'), P(Y'))$, and $y_d = f_5(n, P(X'), P(Y'))$, where f_1, f_2, f_3, f_4, and f_5 are the five real functions. We note in passing that there is also a group of measures with the same normalization coefficients. The normalizing function is also one of the means that allow to give an interpretation of a normalized measurement after providing the respective values of normalization coefficients x_f, y_f, x_d, and y_d. It also expresses an opportunity of value for all normalized measures. According to what we have just written, any normalization of function reflecting the objective in normalization must necessarily have the following properties.

Property 2 (necessary conditions). P_1: F_n^f is a continuous function, positive and strictly increasing on the interval $]x_{ind}; x_{imp}[$ and realizes a bijection on the interval $]x_{ind}; x_{imp}[\longrightarrow]0; 1[$; that is to say, F_n^f must have the following: a limit 1 to the point x_{imp}; that is to say, $\lim_{x \longrightarrow x_{imp}} F_n^f(x) = 1$; a limit 0 to the point x_{ind}; that is to say, $\lim_{x \longrightarrow x_{ind}} F_n^f(x) = 0$.

P_2: F_n^d is a continuous function, negative, and strictly increasing on the interval $]x_{inc}; x_{ind}[$ and realizes a bijection on the interval $]x_{inc}; x_{ind}[\longrightarrow]-1; 0[$; that is to say, F_n^d must have the following: a limit 0 to the point x_{ind}; that is to say, $\lim_{x \longrightarrow x_{ind}} F_n^d(x) = 0$; a limit -1 to the point x_{inc}; that is to say, $\lim_{x \longrightarrow x_{inc}} F_n^d(x) = -1$. As a recap, we have the following.

P_3: F_n is a continuous function at the point x_{ind} and increasing on the interval $]x_{inc}; x_{imp}[$ and realizes a bijection on the interval $]x_{inc}; x_{imp}[\longrightarrow]-1; 1[$; that is to say, F_n must have the following: a limit 1 to the point x_{imp}; that is to say, $\lim_{x \longrightarrow x_{imp}} F_n(x) = 1$; a limit 0 to the point x_{ind}; that is to say, $\lim_{x \longrightarrow x_{ind}} F_n(x) = 0$ and $F_n(x_{ind}) = 0$; limit -1 to the point x_{ind}; that is to say, $\lim_{x \longrightarrow x_{inc}} F_n(x) = -1$.

3. Applications

We recall in Table 1 the respective definitions of the various measures that lead to the results below:

(1) **Cost multiplying**: $x = P(X' \cap Y')P(\overline{Y'})/P(X' \cap \overline{Y'})P(Y') = (1 - P(Y'))P(Y'/X')/(1 - P(Y'/X'))P(Y')$ which is such that $x_{imp} = +\infty$, $x_{ind} = 1$, $x_{inc} = 0$. Using research theories of normalization coefficients x_f, y_f, x_d, and y_d in [18] we have $x_f = -1$, $y_f = 1$, $x_d = 1$, and $y_d = -1$. As a result, by replacing x, x_f, y_f, x_d, and y_d by their values in the expression (3) its normalizing function is such that

$$F_{shnd}(x) = \begin{cases} -\dfrac{1}{x} + 1, & \text{if } X \text{ favors } Y \text{ and } x \in [1; +\infty[\\ x - 1, & \text{if } X \text{ disfavors } Y \text{ and } x \in [0; 1]. \end{cases} \quad (6)$$

It is easy to see that this function is continuous piecewise, in particular point $x_{ind} = 1$.

Finally, it represents a function that exposed the necessary and sufficient conditions for the normalized and continuous measures.

Its tables of variation are respectively presented in Figures 1 and 2.

The graphic illustration and the variation table of F_n of the measure Cost Multiplying reveal that F_n^f is strictly increasing, is positive, and realizes a bijection on

TABLE 1: Probabilistic quality measures of expression.

N^o	Measures μ	Expressions of $x = \mu(X \longrightarrow Y)$
1	Cost multiplying	$x = \dfrac{P(X' \cap Y')\,P(\overline{Y}')}{P(X' \cap \overline{Y}')\,P(Y')}$
2	Example counter-example	$x = \dfrac{P(X' \cap Y') - P(X' \cap \overline{Y}')}{P(X' \cap Y')}$
3	Informal gain	$x = \log \dfrac{P(X' \cap Y')}{P(X')\,P(Y')}$
4	Odd-Ratio	$x = \dfrac{P(X' \cap Y')\,P(\overline{X}' \cap \overline{Y}')}{P(\overline{X}' \cap Y')\,P(X' \cap \overline{Y}')}$
5	Conviction	$x = \dfrac{P(X')\,P(\overline{Y}')}{P(X' \cap \overline{Y}')}$
6	Sebag	$x = \dfrac{P(X' \cap Y')}{P(X')\,P(\overline{Y}')}$
7	M_{GK}	$x = \begin{cases} \dfrac{P(Y'/X') - P(Y')}{1 - P(Y')} \\ \dfrac{P(Y'/X') - P(Y')}{P(Y')} \end{cases}$
8	Rucel and Rao Index	$x = P(X' \cap Y')$
9	Confidence or precision	$x = P\left(\dfrac{Y'}{X'}\right)$
10	Recall	$x = P\left(\dfrac{X'}{Y'}\right)$
11	Interest or Lift	$x = \dfrac{P(Y'/X')}{P(Y')}$
12	Laverage	$x = P\left(\dfrac{Y'}{X'}\right)$
13	Centered confidence	$x = P\left(\dfrac{Y'}{X'}\right) - P(Y')$
14	Confirmed confidence	$x = 1 - 2P\left(\dfrac{\overline{Y}'}{X'}\right)$
15	Certainty factor	$x = \dfrac{P(Y'/X') - P(Y')}{1 - P(Y')}$
16	Gras implication	$x = \sqrt{n}\dfrac{P(X' \cap Y') - P(X')\,P(Y')}{\sqrt{P(X')\,P(Y')}}$
17	Piatesky-Shapiro	$x = n(P(X' \cap Y')$
18	Cosinus	$x = \dfrac{P(X' \cap Y')}{\sqrt{P(X')\,P(Y')}}$
19	Loevinger	$x = 1 - \dfrac{P(X' \cap \overline{Y}')}{P(X')\,P(\overline{Y}')}$
20	Cohen ou Kappa	$x = 2\dfrac{P(X' \cap Y') - P(X')\,P(Y')}{P(X') + P(Y') - 2P(X')\,P(Y')}$
21	Addiction	$x = \left\lvert P(\overline{Y}') - P\left(\dfrac{\overline{Y}'}{X'}\right) \right\rvert$
22	Novelty	$x = P(X'Y') - P(X')\,P(Y')$
23	Czekanowski-Dice	$x = \dfrac{2P(X' \cap Y')}{P(X' \cap Y') + 1 - P(\overline{X}' \cap \overline{Y}')}$
24	Relative risk	$x = \dfrac{P(X'/Y')}{P(X'/\overline{Y}')}$
25	Negative reliability	$x = P\left(\dfrac{\overline{X}'}{\overline{Y}'}\right)$

x	0	1	$+\infty$
$F'_{shnd}(x)$	+		+
$F_{shnd}(x)$	Repulsion zone -1 ↗	0 ↗	1 Attraction zone

FIGURE 1: Change in measure normalizing function "Cost Multiplying".

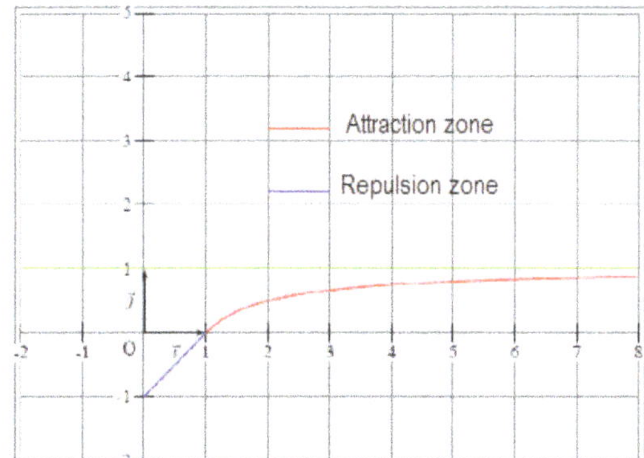

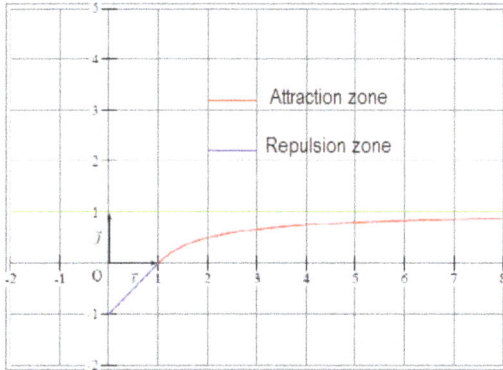

FIGURE 2: Geometric interpretation of the normalization function of the "Cost Multiplying" measure.

$[1;+\infty[\longrightarrow [0;1]$ and even f_d is strictly increasing and negative and realizes a bijection on $[0;1] \longrightarrow [-1;0]$. These results show that F_n is strictly increasing and it realizes a bijection on $[0;+\infty[\longrightarrow [-1;1]$.

Note that, in the following, for the search of the four normalization coefficients, the same principle is used with the measure "dimension multiplier" for the other following measures because they are already expressed in Table 1.

(2) **Example counter-example: its normalizing function is such that** using the expression (4) for $m = 0$ with $x_f = P(Y')/(1-P(Y'))$, $y_f = (1-2P(Y'))/(1-P(Y'))$, $x_d = (2P(Y')-1)/P(Y')$ and $y_d = -1$ so

$$F_{shng}(x) = \begin{cases} \dfrac{P(Y')}{1-P(Y')}x + \dfrac{1-2P(Y')}{1-P(Y')}, & \text{if } X \text{ favors } Y \\ \dfrac{2P(Y')-1}{P(Y')}\dfrac{1}{x} - 1, & \text{if } X \text{ disfavors } Y. \end{cases} \quad (7)$$

(3) **Informal gain: its normalizing function is such that** using expression (4) for $m = 1$ with $x_f = -1/\log P(Y')$, $y_f = 0$, $x_d = 1$, and $y_d = -1$ so

$$F_{shng}(x) = \begin{cases} \dfrac{-1}{\log P(Y')}x, & \text{if } X \text{ favors } Y \\ \dfrac{1}{x+1} - 1, & \text{if } X \text{ disfavors } Y. \end{cases} \quad (8)$$

(4) **Odd-ratio: its normalizing function is such that** using expression (3) for $m = 0$ with $x_f = -1$, $y_f = 1$, $x_d = 1$, and $y_d = -1$ so

$$F_{shnd}(x) = \begin{cases} \dfrac{1}{x} + 1, & \text{if } X \text{ favors } Y \text{ and } x \in [1;+\infty[\\ x - 1, & \text{if } X \text{ disfavors } Y \text{ and } x \in [0;1]. \end{cases} \quad (9)$$

(5) **Conviction: its normalizing function is such that** using expression (5) for $m = 0$ with $x_f = -1$, $y_f = 1$, $x_d = -(1-P(Y'))/P(Y')$, and $y_d = (1-P(Y'))/P(Y')$ so

$$F_{hn}(x) = \begin{cases} \dfrac{-1}{x} + 1, & \text{if } X \text{ favors } Y \\ -\dfrac{1-P(Y')}{P(Y')}\dfrac{1}{x} + \dfrac{1-P(Y')}{P(Y')}, & \text{if } X \text{ disfavors } Y. \end{cases} \quad (10)$$

(6) **Sebag: its normalizing function is such that** using expression (3) for $m = 0$ with $x_f = -P(Y')/(1-P(Y'))$, $y_f = 1$, $x_d = (1-P(Y'))/P(Y')$, and $y_d = -1$ so

$$F_{shnd}(x) = \begin{cases} -\dfrac{P(Y')}{1-P(Y')}\dfrac{1}{x} + 1, & \text{if } X \text{ favors } Y \\ \dfrac{1-P(Y')}{P(Y')}x - 1, & \text{if } X \text{ disfavors } Y. \end{cases} \quad (11)$$

(7) M_{GK}: **its normalizing function is such that** using expression (2) with $x_f = 1$, $y_f = 0$, $x_d = 1$, and $y_d = 0$ so

$$F_{an}(x) = x \quad \text{avec } x = \begin{cases} \dfrac{P(Y'/X') - P(Y')}{1-P(Y')}, & \text{if } X \text{ favors } Y \text{ and } x \in [0;1] \\ \dfrac{P(Y'/X') - P(Y')}{P(Y')}, & \text{if } X \text{ disfavors } Y \text{ and } x \in [-1;0]. \end{cases} \quad (12)$$

(8) **Support; its normalizing function is such that** using expression (2) with $x_f = 1/P(X')(1-P(Y'))$, $y_f = -P(Y')/(1-P(Y'))$, $x_d = 1/P(X')P(Y')$, and $y_d = -1$ so

$$F_{an}(x) = \begin{cases} \dfrac{1}{P(X')(1-P(Y'))}x - \dfrac{P(Y')}{1-P(Y')}, & \text{if } X \text{ favors } Y \\ \dfrac{1}{P(X')P(Y')}x - 1, & \text{if } X \text{ disfavors } Y. \end{cases} \quad (13)$$

(9) **Confidence: its normalizing function is such that** using expression (2) with $x_f = 1/(1-P(Y'))$, $y_f = -P(Y')/(1-P(Y'))$, $x_d = 1/P(Y')$, and $y_d = -1$ so

$$F_{an}(x) = \begin{cases} \dfrac{1}{1-P(Y')}x - \dfrac{P(Y')}{1-P(Y')}, & \text{if } X \text{ favors } Y \\ \dfrac{1}{P(Y')}x - 1, & \text{if } X \text{ disfavors } Y. \end{cases} \quad (14)$$

(10) **Recall: its normalizing function is such that** using expression (2) with $x_f = P(Y')/P(X')(1-P(Y'))$, $y_f = -P(Y')/(1-P(Y'))$, $x_d = 1/P(X')$, and $y_d = -1$ so

$$F_{an}(x) = \begin{cases} \dfrac{P(Y')}{P(X')(1-P(Y'))}x - \dfrac{P(Y')}{1-P(Y')}, & \text{if } X \text{ favors } Y. \\ \dfrac{1}{P(X')}x - 1, & \text{if } X \text{ disfavors } Y \end{cases} \quad (15)$$

(11) **Lift: its normalizing function is such that** using expression (2) with $x_f = P(Y')/(1-P(Y'))$, $y_f = -P(Y')/(1-P(Y'))$, $x_d = 1$, and $y_d = -1$ so

$$F_{an}(x) = \begin{cases} \dfrac{P(Y')}{1-P(Y')}x - \dfrac{P(Y')}{1-P(Y')}, & \text{if } X \text{ favors } Y \\ x - 1, & \text{if } X \text{ disfavors } Y. \end{cases} \quad (16)$$

(12) **Laverage: its normalizing function is such that** using expression (2) with $x_f = P(Y')/(1-P(Y'))$, $y_f = -P(Y')(P(X')-1)/(1-P(Y'))$, $x_d = 1/P(Y')$, and $y_d = -P(Y')(P(X')-1)/P(Y')$ so

$$F_{an}(x) = \begin{cases} \dfrac{P(Y')}{1-P(Y')}x - \dfrac{P(Y')(P(X')-1)}{1-P(Y')}, & \text{if } X \text{ favors } Y \\ \dfrac{1}{P(Y')}x - \dfrac{P(Y')(P(X')-1)}{P(Y')}, & \text{if } X \text{ disfavors } Y. \end{cases} \quad (17)$$

(13) **Centered confidence: its normalizing function is such that** using expression (2) with $x_f = 1/(1-P(Y'))$, $y_f = 0$, $x_d = 1/P(Y')$, and $y_d = 0$ so

$$F_{an}(x) = \begin{cases} \dfrac{1}{1-P(Y')}x, & \text{if } X \text{ favors } Y \\ \dfrac{1}{P(Y')}x, & \text{if } X \text{ disfavors } Y. \end{cases} \quad (18)$$

(14) **Featured confirmed confidence: its normalizing function is such that** using expression (2) with $x_f = 1/2(1-P(Y'))$, $y_f = (1-2P(Y'))/2P(Y')$, $x_d = 1/2P(Y')$, and $y_d = (1-2P(Y'))/2P(Y')$ so

$$F_{an}(x) = \begin{cases} \dfrac{1}{2(1-P(Y'))}x + \dfrac{1-2P(Y')}{2(1-P(Y'))}, & \text{if } X \text{ favors } Y \\ \dfrac{1}{2P(Y')}x + \dfrac{1-2P(Y')}{2P(Y')}, & \text{if } X \text{ disfavors } Y. \end{cases} \quad (19)$$

(15) **Certainty factor: its normalizing function is such that** using expression (2) with $x_f = 1$, $y_f = 0$, $x_d = (1-P(X'))/P(Y')$, and $y_d = 0$ so

$$F_{an}(x) = \begin{cases} x, & \text{if } X \text{ favors } Y \\ \dfrac{1-P(X')}{P(Y')}x, & \text{if } X \text{ disfavors } Y. \end{cases} \quad (20)$$

(16) **Gras implication index: its normalizing function is such that** using expression (2) with $x_f = \sqrt{P(X')P(Y')}/\sqrt{n}P(X')(1-P(Y'))$, $y_f = 0$, $x_d = (\sqrt{P(X')P(Y')}/\sqrt{n}P(X')P(Y'))x$, and $y_d = 0$ so

$$F_{an}(x) = \begin{cases} \dfrac{\sqrt{P(X')P(Y')}}{\sqrt{n}P(X')(1-P(Y'))}x, & \text{if } X \text{ favors } Y \\ \dfrac{\sqrt{P(X')P(Y')}}{\sqrt{n}P(X')P(Y')}x, & \text{if } X \text{ disfavors } Y. \end{cases} \quad (21)$$

(17) **Piatetsky-Shapiro: its normalizing function is such that** using expression (2) with $x_f = 1/nP(X')(1-P(Y'))$, $y_f = 0$, $x_d = 1/nP(X')P(Y')$, and $y_d = 0$ so

$$F_{an}(x) = \begin{cases} \dfrac{1}{nP(X')(1-P(Y'))}x, & \text{if } X \text{ favors } Y \\ \dfrac{1}{nP(X')P(Y')}x, & \text{if } X \text{ disfavors } Y. \end{cases} \quad (22)$$

(18) **Cosinus: its normalizing function is such that** using expression (2) with $x_f = \sqrt{P(X')P(Y')}/P(X')(1-$

$P(Y'))$, $y_f = -P(X')/(1 - P(Y'))$, $x_d = \sqrt{P(X')P(Y')}/P(X')P(Y')$, and $y_d = -1$ so

$$F_{an}(x) = \begin{cases} \frac{\sqrt{P(X')P(Y')}}{P(X')(1-P(Y'))}x - \frac{P(X')}{1-P(Y')}, & \text{if } X \text{ favors } Y \\ \frac{\sqrt{P(X')P(Y')}}{P(X')P(Y')}x - 1, & \text{if } X \text{ disfavors } Y. \end{cases} \quad (23)$$

(19) **Loevinger: its normalizing function is such that** using expression (2) with $x_f = 1$, $y_f = 0$, $x_d = (1 - P(Y'))/P(Y')$, and $y_d = 0$ so

$$F_{an}(x) = \begin{cases} x, & \text{if favors } Y \\ \frac{1-P(Y')}{P(Y')}x, & \text{if } X \text{ disfavors } Y. \end{cases} \quad (24)$$

(20) **Cohen ou Kappa: its normalizing function is such that** using expression (2) with $x_f = (P(X') + P(Y') - 2P(X')P(Y'))/2P(X')(1 - P(Y'))$, $y_f = 0$, $x_d = (P(X')+P(Y')-2P(X')P(Y'))/2P(X')P(Y')$, and $y_d = 0$ so

$$F_{an}(x) = \begin{cases} \frac{P(X')+P(Y')-2P(X')P(Y')}{2P(X')(1-P(Y'))}x, & \text{if } X \text{ favors } Y \\ \frac{P(X')+P(Y')-2P(X')P(Y')}{2P(X')P(Y')}x, & \text{if } X \text{ disfavors } Y. \end{cases} \quad (25)$$

(21) **Addiction: its normalizing function is such that** using expression (2) with $x_f = 1/(1 - P(Y'))$, $y_f = 0$, $x_d = 1/P(Y')$, and $y_d = -1$ so

$$F_{an}(x) = \begin{cases} \frac{1}{1-P(Y')}x, \\ \frac{1}{P(Y')}x - 1. \end{cases} \quad (26)$$

(22) **Novelty: its normalizing function is such that** using expression (2) with $x_f = 1/P(X')(1 - P(Y'))$, $y_f = 0$, $x_d = 1/P(X')P(Y')$, and $y_d = 0$ so

$$F_{an}(x) = \begin{cases} \frac{1}{P(X')(1-P(Y'))}x \\ \frac{1}{P(X')P(Y')}x. \end{cases} \quad (27)$$

(23) **Czekanowski-Dice ou F-measure: its normalizing function is such that** using expression (2) with $x_f = (P(X') + P(Y'))/2P(X')(1 - P(Y'))$, $y_f = -2P(X')P(Y')/2P(X')(1 - (P(Y'))$, $x_d = (P(X') + P(Y'))/2P(X')P(Y')$, and $y_d = -1$ so

$$F_{an}(x) = \begin{cases} \frac{P(X')+P(Y')}{2P(X')(1-P(Y'))}x - \frac{2P(X')P(Y')}{2P(X')(1-(P(Y'))} \\ \frac{P(X')+P(Y')}{2P(X')P(Y')}x - 1. \end{cases} \quad (28)$$

(24) **Relative risk: its normalizing function is such that** using expression (2) with $x_f = (P(Y') - P(X'))/(1 - P(Y'))$, $y_f = -(P(Y') - P(X'))/(1 - P(Y'))$, $x_d = 1$, and $y_d = -1$ so

$$F_{an}(x) = \begin{cases} \frac{P(Y')-P(X')}{1-P(Y')}x - \frac{P(Y')-P(X')}{1-P(Y')} \\ x - 1,. \end{cases} \quad (29)$$

(25) **Negative reliability: its normalizing function is such that** using expression (2) with $x_f = 1/P(X')$, $y_f = -(1 - P(X') - P(Y') + P(Y')P(X'))/P(X')(1 - P(Y'))$, $x_d = (1 - P(Y'))/P(X')(Y')$, and $y_d = -(1 - P(X') - P(Y') + P(Y')P(X'))/P(X')P(Y')$ so

$$F_{an}(x) = \begin{cases} \frac{1}{P(X')}x - \frac{1-P(X')-P(Y')+P(Y')P(X')}{P(X')(1-P(Y'))} \\ \frac{1-P(Y')}{P(X')(Y')}x - \frac{1-P(X')-P(Y')+P(Y')P(X')}{P(X')P(Y')}. \end{cases} \quad (30)$$

Researches out on the expressions of measurement normalizing functions prove that certain measures have identical normalizing functions, leading to what state the theorem below.

Theorem 3. *(i) All measures having the same baseline* $(x_{imp}, x_{ind}, x_{inc})$ *have the same normalizing function.*

(ii) All measures affine normalizable are homographic normalizable, but the converse is false.

Proof. (i) Suppose the four possible forms of F_n.

(a) **μ is affine normalizable:**

$$F_{an}(x) = \begin{cases} x_f x + y_f, & \text{if } X \text{ favors } Y \\ x_d x + y_d, & \text{if } X \text{ disfavors } Y \end{cases} \quad (31)$$

with

$$x_f = \frac{1}{x_{imp} - x_{ind}},$$
$$y_f = -\frac{1}{x_{imp} - x_{ind}} x_{ind}, \quad (32)$$
$$x_d = \frac{1}{x_{ind} - x_{inc}}$$

and

$$y_d = -\frac{1}{x_{ind} - x_{inc}} x_{ind} \quad (33)$$

so we have

$$F_{an}(x) = \begin{cases} \frac{1}{x_{imp} - x_{ind}} x - \frac{1}{x_{imp} - x_{ind}} x_{ind}, & \text{if } X \text{ favors } Y \\ \frac{1}{x_{ind} - x_{inc}} x - \frac{1}{x_{ind} - x_{inc}} x_{ind}, & \text{if } X \text{ disfavors } Y. \end{cases} \quad (34)$$

(b) μ **is to right semihomographic normalizable**: in the c ase where $x_{imp} = +\infty$, that is to say, $\lim_{x \longrightarrow +\infty}(x_f/(x+m)) = 0$, the expression is used:

$$F_{shnd}(x) = \begin{cases} \frac{x_f}{x+m} + y_f, & \text{if } X \text{ favoris } Y \\ x_d x + y_d, & \text{if } X \text{ disfavors } Y \end{cases} \quad (35)$$

with

$$x_f = -(x+m), \quad y_f = 1, \quad x_d = \frac{1}{x_{ind} - x_{inc}} \quad (36)$$

and

$$y_d = -\frac{1}{x_{ind} - x_{inc}} x_{ind} \quad (37)$$

so

$$F_{shnd}(x) = \begin{cases} -\frac{x_{ind} + m}{x+m} + 1, & \text{if } X \text{ favors } Y \\ \frac{1}{x_{ind} - x_{inc}} x + \frac{1}{x_{ind} - x_{inc}} x_{inc}, & \text{si } X \text{ disfavors } Y. \end{cases} \quad (38)$$

(c) μ **is to left semihomographic normalizable**: in the case where $x_{inc} = -\infty$; that is to say, $\lim_{x \longrightarrow -\infty}(x_f/(x_{inc} + m)) = 0$. This time we use the following expression:

$$F_{shng}(x) = \begin{cases} x_f x + y_f, & \text{if } X \text{ favors } Y \\ \frac{x_d}{x+m} + y_d, & \text{if } X \text{ disfavors } Y \end{cases} \quad (39)$$

such that

$$x_f = \frac{1}{x_{imp} - x_{ind}}, \quad y_f = -\frac{1}{x_{imp} - x_{ind}} x_{ind}, \quad x_d = x_{ind} + m \quad (40)$$

and

$$y_d = -1 \quad (41)$$

so

$$F_{shnd}(x) = \begin{cases} \frac{1}{x_{imp} - x_{ind}} x - \frac{1}{x_{imp} - x_{ind}} x_{ind}, & \text{if } X \text{ favors } Y \\ \frac{x_{ind} + m}{x+m} - 1, & \text{if } X \text{ disfavors } Y. \end{cases} \quad (42)$$

(d) μ **is homographic normalizable**: in the case where $x_{inc} = -\infty$, $x_{imp} = +\infty$ and $x_{ind} \in \mathbb{R}^*$; that is to say,

$$\lim_{x \longrightarrow -\infty} \frac{x_f}{x_{inc} + m} = 0 \quad (43)$$

and

$$\lim_{x \longrightarrow -\infty} \frac{x_f}{x_{inc} + m} = 0. \quad (44)$$

Here we use the following expression:

$$F_{hn}(x) = \begin{cases} \frac{x_f}{x+m} + y_f, & \text{if } X \text{ favors } Y \\ \frac{x_d}{x+m} + y_d, & \text{si } X \text{ disfavors } Y \end{cases} \quad (45)$$

where

$$x_f = -(x_{ind} + m), \quad y_f = 1, \quad x_d = (x_{ind} + m) \quad (46)$$

and

$$y_d = -1, \quad (47)$$

so

$$F_{hn}(x) = \begin{cases} -\frac{x_{ind} + m}{x+m} + 1, & \text{if } X \text{ favors } Y \\ \frac{x_{ind} + m}{x+m} - 1, & \text{if } X \text{ disfavors } Y. \end{cases} \quad (48)$$

(ii) (a) It is seen that if $x \in \mathbb{R}$ and $(x+m) \in \mathbb{R}^*$, then the four terms of the function of normalizations are well defined; therefore, there is no problem for calculating the normalization coefficients.

(b) We see that if $x = \infty$, then we can always get a projective application $\overline{\mathbb{R}}$ in $\mathbb{R}$ for the functions F_{shn}, F_{shnd}, and F_{shng} and therefore the four coefficients of normalization are always calculable; by constraint, if $x = \infty$, then you can never get a projective application over the interval $\overline{\mathbb{R}}$ in $\mathbb{R}$ for the function F_{an}; therefore, we can not calculate these four coefficients. The theorem is stated.

Table 1 recalls the respective definitions of the various measures.

4. Conclusion and Perspectives

This study showed that normalization of probabilistic quality measures with a homographic homeomorphism is more powerful than the normalization homeomorphism refines initiated by André Totohasina. Indeed, we showed that any measure affine-normalizable is homographic normalizable, while the converse is false. Besides, this work has explained the process of normalization by homographic function and combination with an affine function by trying to sweep the present main measures in the literature with the aim of a presentation easier to understand. The database has several branches; the purpose of this research is the normalization of quality measures. We always say, in the context of the database, the study on association rules knows an important development, added to the measures called interest; yet probabilistic quality measures have an important place in the context of data mining. Thereafter, the probabilistic measure of quality and its normalization must be complement. As shown by research on normalization probabilistic quality measures realizing the normalization operation requires passing through a relatively complex theory. We can consider several possible ways to carry out its standardization process. In our opinion, the use of normalizing function seems the simplest way.

In future work, we understand the positive impact of consideration of these normalizing functions in the development of the bases of the rules in search of binary data.

Conflicts of Interest

The authors declare that there are no conflicts of interest regarding the publication of this paper.

References

[1] R. Agrawal, T. Imieliński, and A. Swami, "Mining association rules between sets of items in large databases," in *Proceedings of the ACM SIGMOD International Conference on Management of Data (SIGMOD '93)*, pp. 207–216, 1993.

[2] M. Kamber and R. Shinghal, "Proposed interestingness measure for characteristic rules," *Review*, vol. 20, no. 1, p. 21, 1996.

[3] D. Grissa, *Etude comportementale des mesures dinterêt dextraction de connaissance [Ph.D. thesis]*, Universite Clermont-Ferrand II, 2013.

[4] A. Totohasina, "Towards a theory unifying implicative interestinge mesures and critial values consideration in MGK," in *Educ. Matem. Pesq*, vol. 16 no 3, pp. 881–900, Sao Paulo, 2014.

[5] G. Yongmei and B. Fuguang, "The Research on Measure Method of Association Rules Mining," *Journal of Database Theory and Application*, vol. 8, no. 2, pp. 245–258, 2015.

[6] H. F. Rakotomalala, B. B. Ralahady, and A. Totohasina, "A Novel cohesitive implicative classification based on MGK and application on diagonostic on informatics literacy of studens of higher education in Madagascar," in *Third International Congress on Information and Communication Technology*, vol. 797 of *Advances in Intelligent Systems and Computing*, pp. 161–174, Springer Singapore, Singapore, 2019.

[7] B. Shekar and R. Natarajan, "A Transaction-Based Neighbourhood-Driven Approach to Quantifying Interestingness of Association Rules," in *Proceedings of the Fourth IEEE International Conference on Data Mining (ICDM'04)*, pp. 194–201, Brighton, UK.

[8] X. Wu, C. Zhang, and S. Zhang, "Efficient mining of both positive and negative association rules," *ACM Transactions on Information and System Security*, vol. 22, no. 3, pp. 381–405, 2004.

[9] L. Nguyen, B. Vo, and T.-P. Hong, "CARIM: An efficient algorithm for mining class-association rules with interestingness measures," *International Arab Journal of Information Technolog*, vol. 12, no. 6A, pp. 627–634, 2015.

[10] P. Tan, V. Kumar, and J. Srivastava, "Selecting the right interestingness measure for association patterns," in *Proceedings of the the eighth ACM SIGKDD international conference*, p. 32, Edmonton, Alberta, Canada, July 2002.

[11] Dr. Niket Bhargava and Manoj. Shukla, "Survey of Interestingness Measures for Association Rules Mining: Data Mining, Data Science for Business Perspective," *IRACST - International Journal of Computer Science and Information Technology & Security (IJCSITS)*, vol. 6, no. 2, pp. 2249–2249-9555, 2016.

[12] J. Diatta, H. Ralambondrainy, and A. Totohasina, "Towards a unifying probabilistic implicative normalized quality measure for association rules," *Studies in Computational Intelligence*, vol. 43, pp. 237–250, 2007.

[13] A. Totohasina, *Contribution a l'étude des mesures de qualité des regles d'associations: normalisation sous cinq contraintes et cas de MGK: propriétés, bases composites des regles et extension en vue d'applications en statistique et en sciences physiques [Ph.D. thesis]*, Université, dAntsiranana, Madagascar: Mathématiques Informatique, Madagascar, 2008.

[14] P. Bemarisika and A. Totohasina, "Optimized mining of potential positive and negative association rules," *Lecture Notes in Computer Science (including subseries Lecture Notes in Artificial Intelligence and Lecture Notes in Bioinformatics): Preface*, vol. 10440, pp. 424–432, 2017.

[15] A. Totohasina and H. Ralambondrainy, "ION: A pertinent new measure for mining information from many types of data," in *Proceedings of the 1st IEEE International Conference on Signal-Image Technology and Internet-Based Systems, SITIS 2005*, pp. 202–207, Cameroon, December 2005.

[16] M. Hahsler and K. Hornik Vienna, *New Probabilistic Interest Measures for Association Rules*, University of Economics and Business Administration, Augasse 26, A-1090 Vienna, Austria, 2018.

[17] L. Lin, M.-L. Shyu, and S.-C. Chen, "Association rule mining with a correlation-based interestingness measure for video semantic concept detection," *International Journal of Information and Decision Sciences*, vol. 4, no. 2-3, pp. 199–216, 2012.

[18] A. Armand, *Totohasina and Feno Daniel Rajaonasy. An extension of Totohasina's normalization theory of quality measures of association rules*, 2018, In press.

3

Ordered Structures of Constructing Operators for Generalized Riesz Systems

Hiroshi Inoue

Center for Advancing Pharmaceutical Education, Daiichi University of Pharmacy, 22-1 Tamagawa-cho, Minami-ku, Fukuoka 815-8511, Japan

Correspondence should be addressed to Hiroshi Inoue; h-inoue@daiichi-cps.ac.jp

Academic Editor: Seppo Hassi

A sequence $\{\varphi_n\}$ in a Hilbert space $\mathcal{H}$ with inner product $< \cdot, \cdot >$ is called a generalized Riesz system if there exist an ONB $\boldsymbol{e} = \{e_n\}$ in $\mathcal{H}$ and a densely defined closed operator T in $\mathcal{H}$ with densely defined inverse such that $\{e_n\} \subset D(T) \cap D((T^{-1})^*)$ and $Te_n = \varphi_n$, $n = 0, 1, \cdots$, and $(\boldsymbol{e}, T)$ is called a constructing pair for $\{\varphi_n\}$ and T is called a constructing operator for $\{\varphi_n\}$. The main purpose of this paper is to investigate under what conditions the ordered set C_φ of all constructing operators for a generalized Riesz system $\{\varphi_n\}$ has maximal elements, minimal elements, the largest element, and the smallest element in order to find constructing operators fitting to each of the physical applications.

1. Introduction

Generalized Riesz systems can be used to construct some physical operators (non-self-adjoint Hamiltonian, generalized lowering operator, generalized raising operator, number operator, etc.) [1–3]. Then these operators provide a link to *quasi-Hermitian quantum mechanics*, and its relatives. Many researchers have investigated such operators both from the mathematical point of view and for their physical applications [4–9]. Let $\{\varphi_n\}$ be a generalized Riesz system with a constructing pair $(\boldsymbol{e}, T)$. Then, putting $\psi_n^T = (T^{-1})^* e_n$, $n = 0, 1, \cdots$, $\{\varphi_n\}$ and $\{\psi_n^T\}$ are biorthogonal sequences, that is, $< \varphi_n, \psi_m^T >= \delta_{nm}$, $n, m = 0, 1, \cdots$. For any $\{\alpha_n\} \subset \mathscr{C}$ we can define the operators: $H_\varphi^{\boldsymbol{\alpha}} := TH_{\boldsymbol{e}}^{\boldsymbol{\alpha}}T^{-1}$, $A_\varphi^{\boldsymbol{\alpha}} := TA_{\boldsymbol{e}}^{\boldsymbol{\alpha}}T^{-1}$, and $B_\varphi^{\boldsymbol{\alpha}} := TB_{\boldsymbol{e}}^{\boldsymbol{\alpha}}T^{-1}$, where $H_{\boldsymbol{e}}^{\boldsymbol{\alpha}} := \sum_{n=0}^{\infty} \alpha_n e_n \otimes \overline{e}_n$, $A_{\boldsymbol{e}}^{\boldsymbol{\alpha}} := \sum_{n=0}^{\infty} \alpha_{n+1} e_{n+1} \otimes \overline{e}_n$, and $B_{\boldsymbol{e}}^{\boldsymbol{\alpha}} := \sum_{n=0}^{\infty} \alpha_{n+1} e_{n+1} \otimes \overline{e}_n$ are standard self-adjoint Hamiltonian, lowering operator, and raising operator for $\{e_n\}$, respectively, where for $x, y \in \mathcal{H}$, $(x \otimes y)\xi :=< \xi, y > x$, $\xi \in \mathcal{H}$. Since

$$H_\varphi^{\boldsymbol{\alpha}} \varphi_n = \alpha_n \varphi_n,$$

$$A_\varphi^{\boldsymbol{\alpha}} \varphi_n = \begin{cases} 0, & n = 0 \\ \alpha_n \varphi_{n-1}, & n = 1, 2, \cdots \end{cases} \tag{1}$$

$$B_\varphi^{\boldsymbol{\alpha}} \varphi_n = \alpha_{n+1} \varphi_{n+1}, \quad n = 0, 1, \cdots,$$

$H_\varphi^{\boldsymbol{\alpha}}$, $A_\varphi^{\boldsymbol{\alpha}}$, and $B_\varphi^{\boldsymbol{\alpha}}$ are called the non-self-adjoint Hamiltonian, the generalized lowering operator, and the generalized raising operator for $\{\varphi_n\}$, respectively. The physical operators of the extended quantum harmonic oscillator and the Swanson model are of this form (see Examples 9–11 in Section 3).

From this fact, it seems to be important to consider under what conditions biorthogonal sequences are generalized Riesz systems and in [1–3] we have investigated this problem. In this paper, we shall focus on the following facts: physical operators defined by a generalized Riesz system $\{\varphi_n\}$ depend on constructing pairs; for example, their operators may not be densely defined for some constructing pairs. On the other hand, if there exists a dense subspace $\mathscr{D}$ in $\mathcal{H}$ for a constructing pair $(\boldsymbol{e}, T)$ which is a core for T such that $H_{\boldsymbol{e}}^{\boldsymbol{\alpha}} \mathscr{D} \subset \mathscr{D}$, $A_{\boldsymbol{e}}^{\boldsymbol{\alpha}} \mathscr{D} \subset \mathscr{D}$, and $B_{\boldsymbol{e}}^{\boldsymbol{\alpha}} \mathscr{D} \subset \mathscr{D}$, then they have an algebraic structure; in detail, the O-algebra on $\mathscr{D}$ is defined by the restrictions of the operators $A_\varphi^{\boldsymbol{\alpha}}$ and $B_\varphi^{\boldsymbol{\alpha}}$ to $\mathscr{D}$ [10]. Thus it seems to be important to find a constructing pair fitting to each of the physical applications. From this reason, in this paper we shall investigate the properties of constructing pairs for a generalized Riesz system.

In Section 2, we shall investigate the basic properties of constructing operators. Let $\{\varphi_n\}$ be a generalized Riesz system with a constructing pair $(\boldsymbol{e}, T)$. The constructing operators for $\{\varphi_n\}$ are unique for the fixed ONB $\boldsymbol{e}$ in $\mathcal{H}$ if $\{\varphi_n\}$ is a Riesz basis; that is, T and T^{-1} are bounded, but they are not unique in general. So, we investigate the set $C_{e,\varphi}$ of all constructing operators for $\boldsymbol{e}$. In Proposition 1, we shall show that it is possible to fix an ONB $\boldsymbol{e} = \{e_n\}$ in $\mathscr{H}$ without loss of generality for our study in this paper. Hence, we fix an ONB $\boldsymbol{e}$ in $\mathscr{H}$ and denote $C_{e,\varphi}$ by C_φ for simplicity. We consider the following problem: *Is any sequence $\{\psi_n\}$ which is biorthogonal to $\{\varphi_n\}$ a generalized Riesz system?*

Here we put

$$C_\varphi^N := \left\{T \in C_\varphi; \left(T^{-1}\right)^* e_n = \psi_n,\ n = 0, 1, \cdots\right\}. \quad (2)$$

Then we shall show in Proposition 5 that if $C_\varphi^N \neq \emptyset$, then $\{\psi_n\}$ is a generalized Riesz system and $(\boldsymbol{e}, (T^{-1})^*)$ is a constructing pair for $\{\psi_n\}$ for every $T \in C_\varphi^N$, and the mapping

$$T \in C_\varphi^N \longmapsto \left(T^{-1}\right)^* \in C_\psi^N \quad (3)$$

is a bijection, where C_ψ is the set of all constructing operators for $\{\psi_n\}$ and

$$C_\psi^N := \left\{K \in C_\psi;\ \left(K^{-1}\right)^* e_n = \varphi_n,\ n = 0, 1, \cdots\right\}. \quad (4)$$

Furthermore, we shall show in Proposition 6 that if there exists a bounded operator T_0 in C_φ, then $C_\varphi = \{T_0\}$ and $C_\psi^N = \{(T_0^{-1})^*\}$.

In Section 3, we shall consider the ordered set C_φ with order $\subset$ and investigate under what conditions the ordered set C_φ has a maximal element, a minimal element, the smallest element, and the largest element. First we have shown that if $D_\varphi :=$ linear span $\{\varphi_n\}$ is dense in $\mathscr{H}$, then $C_\varphi = C_\varphi^N$ and there exists the smallest element of C_φ, and furthermore if D_φ and $D(\varphi) := \{x \in \mathscr{H};\ \sum_{k=0}^{\infty} | < x, \varphi_k > |^2 < \infty\}$ is dense in $\mathscr{H}$, there exist the smallest element of C_φ and the largest element of C_ψ^N, and in particular, if $\{\varphi_n\}$ and $\{\psi_n\}$ are regular biorthogonal sequences in $\mathscr{H}$, that is, both D_φ and D_ψ are dense in $\mathscr{H}$, then $C_\varphi = C_\varphi^N, C_\psi = C_\psi^N$, and C_φ has the smallest element and the largest element. Next we shall consider the case when D_φ is not necessarily dense in $\mathscr{H}$. In Theorem 14, we shall show that for a subset $\mathscr{F}$ of C_φ if there exists a closed operator A in $\mathscr{H}$ such that $T \subset A$ for all $T \in \mathscr{F}$, then $\mathscr{F}$ has a maximal element, and furthermore, if there exists a closed operator B in $\mathscr{H}$ such that $(T^{-1})^* \subset B$ for all $T \in \mathscr{F}$, then $\mathscr{F}$ have a maximal element and a minimal element.

For the existence of the smallest element of C_φ and of the largest element of C_φ, we shall show in Theorem 16 that if there exist closed operators A and B in $\mathscr{H}$ such that $T \subset A$ and $(T^{-1})^* \subset B$ for all $T \in C_\varphi$, then C_φ has the smallest element and the largest element. Furthermore, for a biorthogonal pair $(\{\varphi_n\}, \{\psi_n\})$ of generalized Riesz systems satisfying $C_\varphi = C_\varphi^N$ and $C_\psi = C_\psi^N$, we shall show in Theorem 18 that C_φ and C_ψ have the smallest element and the largest element, respectively, if and only if there exist closed operators A and B in $\mathscr{H}$ such that $T \subset A$ and $K \subset B$ for all $T \in C_\varphi$ and $K \in C_\psi$. These results seem to be useful to find fitting constructing operators for each physical model because every closed operator T in $\mathscr{H}$ satisfying $T_S \subset T \subset T_L$ belongs to C_φ, where T_S is the smallest element of C_φ and T_L is the largest element of C_φ.

2. The Basic Properties of Constructing Operators

In this section, we shall investigate the basic properties of constructing operators. Let $\{\varphi_n\}$ be a generalized Riesz system with a constructing pair $(\boldsymbol{e}, T)$. It is easily shown that if $\{\varphi_n\}$ is a Riesz basis, then the constructing operator T for $\{\varphi_n\}$ is unique for $\boldsymbol{e}$ (see Proposition 1 in detail). But, in general, the constructing operators for $\{\varphi_n\}$ are not unique, and so we put

$$C_{e,\varphi} := \{T; (\boldsymbol{e}, T) \text{ is a constructing pair for } \{\varphi_n\}\}. \quad (5)$$

First, we investigate the relationship between $C_{e,\varphi}$ and $C_{f,\varphi}$ for the other ONB $\boldsymbol{f} = \{f_n\}$ in $\mathscr{H}$.

Proposition 1. *Let $T \in C_{e,\varphi}$ and $\boldsymbol{f} = \{f_n\}$ be any ONB in $\mathscr{H}$. Then the following statements hold.*

(1) $(\boldsymbol{f}, TU^*)$ is a constructing pair for $\{\varphi_n\}$, where U is a unitary operator on $\mathscr{H}$ defined by $Ue_n = f_n$, $n = 0, 1, \cdots$, and

$$C_{f,\varphi} = \left\{TU^*;\ T \in C_{e,\varphi}\right\}. \quad (6)$$

(2) For the non-self-adjoint Hamiltonian, the generalized lowering operator, and the generalized raising operator for $\{\varphi_n\}$, we have

$$\begin{aligned} TH_e^\alpha T^{-1} &= TU^* H_f^\alpha U T^{-1}, \\ TA_e^\alpha T^{-1} &= TU^* A_f^\alpha U T^{-1}, \\ TB_e^\alpha T^{-1} &= TU^* B_f^\alpha U T^{-1}. \end{aligned} \quad (7)$$

Proof. (1) This is almost trivial.

(2) This follows from

$$\begin{aligned} D\left(H_f^\alpha\right) &= UD\left(H_e^\alpha\right), \\ H_e^\alpha &= U^* H_f^\alpha U, \\ D\left(A_f^\alpha\right) &= UD\left(A_e^\alpha\right), \\ A_e^\alpha &= U^* A_f^\alpha U, \\ D\left(B_f^\alpha\right) &= UD\left(B_e^\alpha\right), \\ B_e^\alpha &= U^* B_f^\alpha U. \end{aligned} \quad (8)$$

By Proposition 1, we have the following.

Corollary 2. *Let $T \in C_{e,\varphi}$ and $T^* = U|T^*|$ be the polar decomposition of T^*. Then $\boldsymbol{f} := U^*\boldsymbol{e}$ is an ONB in $\mathscr{H}$ and $|T^*| = TU \in C_{f,\varphi}$. Furthermore, we have*

$$TH_e^\alpha T^{-1} = \left|T^*\right| H_f^\alpha \left|T^*\right|^{-1},$$

$$TA_{\boldsymbol{e}}^{\alpha}T^{-1} = |T^*| A_f^{\alpha} |T^*|^{-1},$$
$$TB_{\boldsymbol{e}}^{\alpha}T^{-1} = |T^*| B_f^{\alpha} |T^*|^{-1}. \quad (9)$$

Thus we may fix an ONB $\boldsymbol{e} = \{e_n\}$ in $\mathscr{H}$ without loss of generality for investigating the properties of $C_{\boldsymbol{e},\varphi}$, and so throughout this paper, we fix an ONB $\boldsymbol{e}$ in $\mathscr{H}$ and denote $C_{\boldsymbol{e},\varphi}$ by C_φ for simplicity. Next we consider the following problem: *Suppose that* $(\{\varphi_n\}, \{\psi_n\})$ *is a biorthogonal pair such that* $\{\varphi_n\}$ *is a generalized Riesz system. Then, is* $\{\psi_n\}$ *also a generalized Riesz system?*

Let $T \in C_\varphi$ and $\psi_n^T := (T^{-1})^* e_n$, $n = 0, 1, \cdots$. Then $(\{\varphi_n\}, \{\psi_n^T\})$ is a biorthogonal pair and $\{\psi_n^T\}$ is a generalized Riesz system with a constructing pair $(\boldsymbol{e}, (T^{-1})^*)$. If $\psi_n^T = \psi_n$, $n = 0, 1 \cdots$, then $\{\psi_n\}$ is a generalized Riesz system with a constructing pair $(\boldsymbol{e}, (T^{-1})^*)$. But, the equality $\psi_n^T := (T^{-1})^* e_n = \psi_n$, $n = 0, 1, \cdots$ does not necessarily hold. To consider when this equality holds, we define the operators $T_{\varphi,\boldsymbol{e}}^0$, $T_{\varphi,\boldsymbol{e}}$, and $T_{\boldsymbol{e},\varphi}$ for any sequence $\{\varphi_n\}$ in $\mathscr{H}$ as follows:

$$T_{\varphi,\boldsymbol{e}}^0 := \text{the linear operator defined by } T_{\varphi,\boldsymbol{e}}^0 e_n = \varphi_n, \quad n = 0, 1, \cdots,$$
$$T_{\varphi,\boldsymbol{e}} := \sum_{n=0}^{\infty} \varphi_n \otimes \overline{e}_n, \quad (10)$$
$$T_{\boldsymbol{e},\varphi} := \sum_{n=0}^{\infty} e_n \otimes \overline{\varphi}_n.$$

These operators have played an important role for our studies [3] and also in this paper. By Lemma 2.1, 2.2 in [3] we have the following.

Lemma 3. *(1)* $T_{\varphi,\boldsymbol{e}}^0$ *and* $T_{\varphi,\boldsymbol{e}}$ *are densely defined linear operators in* $\mathscr{H}$ *such that*

$$T_{\varphi,\boldsymbol{e}} \supset T_{\varphi,\boldsymbol{e}}^0,$$
$$T_{\varphi,\boldsymbol{e}}^0 e_n = T_{\varphi,\boldsymbol{e}} e_n = \varphi_n, \quad n = 0, 1, \cdots. \quad (11)$$

(2) $D(T_{\boldsymbol{e},\varphi}) = D(\varphi) := \{x \in \mathscr{H};\ \sum_{n=0}^{\infty} | < x, \varphi_n > |^2 < \infty\}$ *and* $(T_{\varphi,\boldsymbol{e}}^0)^* = T_{\varphi,\boldsymbol{e}}^* = T_{\boldsymbol{e},\varphi}$.

(3) $T_{\varphi,\boldsymbol{e}}^0$ *is closable if and only if* $T_{\varphi,\boldsymbol{e}}$ *is closable if and only if* $D(\varphi)$ *is dense in* $\mathscr{H}$. *If this holds, then*

$$\overline{T}_{\varphi,\boldsymbol{e}}^0 = \overline{T}_{\varphi,\boldsymbol{e}} = \left(T_{\boldsymbol{e},\varphi}\right)^*. \quad (12)$$

From now on, let $(\{\varphi_n\}, \{\psi_n\})$ be a biorthogonal pair.

Lemma 4. *Suppose that* $\{\varphi_n\}$ *is a generalized Riesz system and* $T \in C_\varphi$. *Then the following statements are equivalent.*

(i) $\psi_n^T := (T^{-1})^* e_n = \psi_n$, $n = 0, 1, \cdots$.

(ii) $D_\psi \subset D(T^*)$.

If this holds true, then T *is called natural.*

Proof. (i)$\Longrightarrow$(ii) This is trivial.

(ii)$\Longrightarrow$(i) By definition of $T_{\varphi,\boldsymbol{e}}^0$, we have $T_{\varphi,\boldsymbol{e}}^0 \subset T$. Furthermore, by Lemma 3, (2), we have

$$T^* \subset \left(T_{\varphi,\boldsymbol{e}}^0\right)^* = T_{\boldsymbol{e},\varphi}. \quad (13)$$

Take an arbitrary $n \in \mathbf{N} \cup \{0\}$. Then, since

$$\left\langle T_{\varphi,\boldsymbol{e}}^0 e_k, \psi_n \right\rangle = \langle \varphi_k, \psi_n \rangle = \delta_{kn} = \langle e_k, e_n \rangle \quad (14)$$

for $k = 0, 1, \cdots$, we have $\psi_n \in D((T_{\varphi,\boldsymbol{e}}^0)^*) = D(T_{\boldsymbol{e},\varphi})$ and $T_{\boldsymbol{e},\varphi}\psi_n = e_n$. Hence it follows from (13) that

$$T^* \psi_n = T_{\boldsymbol{e},\varphi} \psi_n = e_n. \quad (15)$$

Thus, we have

$$\psi_n = (T^*)^{-1} e_n = \psi_n^T. \quad (16)$$

This completes the proof.

We denote the set of all natural constructing operators for $\{\varphi_n\}$ by C_φ^N; that is,

$$C_\varphi^N = \left\{T \in C_\varphi; \psi_n^T := \left(T^{-1}\right)^* e_n = \psi_n, n = 0, 1, \cdots\right\}. \quad (17)$$

Then we have the following.

Proposition 5. *Suppose that* $\{\varphi_n\}$ *is a generalized Riesz system. Then the following statements hold.*

(1) If $C_\varphi^N \neq \emptyset$, *then* $\{\psi_n\}$ *is a generalized Riesz system and* $(\boldsymbol{e}, (T^{-1})^*)$ *is a constructing pair for* $\{\psi_n\}$ *for every* $T \in C_\varphi^N$.

(2) Suppose that $\{\psi_n\}$ *is also a generalized Riesz system and put*

$$C_\psi = \{K;\ (\boldsymbol{e}, K) \text{ is a constructing pair for } \{\psi_n\}\},$$
$$C_\psi^N = \left\{K \in C_\psi;\ \varphi_n^K := \left(K^{-1}\right)^* e_n = \varphi_n,\ n = 0, 1, \cdots\right\}. \quad (18)$$

Then the mapping

$$T \in C_\varphi^N \longmapsto \left(T^{-1}\right)^* \in C_\psi^N \quad (19)$$

is a bijection.

(3) Suppose that $T_0 \in C_\varphi^N$ *and* $T \in C_\varphi$ *satisfying* $T \subset T_0$ *or* $T_0 \subset T$. *Then* $T \in C_\varphi^N$. *Similarly, suppose that* $K_0 \in C_\psi^N$ *and* $K \in C_\psi$ *satisfying* $K \subset K_0$ *or* $K_0 \subset K$. *Then* $K \in C_\psi^N$.

Proof. The statements (1) and (2) are easily shown.

(3) Suppose that $T \subset T_0$. Then, since $(T_0^{-1})^* \subset (T^{-1})^*$, it follows that $\psi_n = (T_0^{-1})^* e_n = (T^{-1})^* e_n = \psi_n^T$, $n = 0, 1, \cdots$, which implies that $T \in C_\varphi^N$. Similarly, we can show $T \in C_\varphi^N$ in case that $T_0 \subset T$ and can show $K \in C_\psi^N$ in case that $K \subset K_0$ or $K_0 \subset K$. This completes the proof.

As for the uniqueness of constructing operators for a generalized Riesz system we have the following.

Proposition 6. *Let $\{\varphi_n\}$ be a generalized Riesz system. Then the following statements hold.*

(1) Suppose that $\{\varphi_n\}$ is a Riesz basis, then $\{\psi_n\}$ is also a Riesz basis, $C_\varphi = \{\overline{T}_{\varphi,e}\} = \{T_{e,\psi}^{-1}\}$ and $C_\psi = \{\overline{T}_{\psi,e}\} = \{T_{e,\varphi}^{-1}\}$.

(2) Suppose that D_φ and $D(\varphi)$ are dense in $\mathscr{H}$. Then we have the following.

(i) If there exists an element T_0 of C_φ such that T_0 is bounded, then $C_\varphi = \{T_0\} = \{\overline{T}_{\varphi,e}\}$ and $C_\psi^N = \{(T_0^{-1})^*\} = \{T_{e,\varphi}^{-1}\}$.

(ii) If there exists an element T_0 of C_φ such that T_0^{-1} is bounded, then $C_\psi = \{(T_0^{-1})^*\} = \{\overline{T}_{\psi,e}\}$ and $C_\varphi^N = \{T_0\} = \{T_{e,\psi}^{-1}\}$.

Proof. (1) Since $\{\varphi_n\}$ is a Riesz basis, there exists an element T_0 of C_φ such that T_0 and T_0^{-1} are bounded, which implies that $T = T_0 = \overline{T}_{\varphi,e} = T_{e,\psi}^{-1}$ and $(T^{-1})^* = (T_0^{-1})^* = \overline{T}_{\psi,e} = T_{e,\varphi}^{-1}$ for all $T \in C_\varphi$.

(2) (i) Since $\overline{T}_{\varphi,e} \subset T_0$ and T_0 is bounded, we have $\overline{T}_{\varphi,e} = T_0$. Take an arbitrary $T \in C_\varphi$. Then, since $\overline{T}_{\varphi,e} \subset T$ and $\overline{T}_{\varphi,e}$ is bounded, we have $\overline{T}_{\varphi,e} = T$. Thus, $C_\varphi = \{\overline{T}_{\varphi,e}\}$. We show $C_\psi^N = \{T_{e,\varphi}^{-1}\}$. Take an arbitrary $K \in C_\psi^N$. Since $(K^{-1})^* e_n = \varphi_n = \overline{T}_{\varphi,e} e_n$ and $\overline{T}_{\varphi,e}$ is bounded, it follows that $(K^{-1})^* = \overline{T}_{\varphi,e}$, which implies $K = T_{e,\varphi}^{-1}$. Thus, $C_\psi^N = \{T_{e,\varphi}^{-1}\}$.

(ii) This is similarly shown.

3. Ordered Structures of C_φ

In this section, we shall consider the ordered set C_φ of all constructing operators for a generalized Riesz system $\{\varphi_n\}$ with order $\subset$ and investigate when C_φ has maximal elements, minimal elements, the largest element, and the smallest element. The following result gives a motivation to study the ordered structures of C_φ

Lemma 7. *Suppose that $T, S \in C_\varphi$ and $T \subset S$. Then, for any linear operator A such that $T \subset A \subset S$, the closure $\overline{A}$ of A belongs to C_φ.*

Proof. This is trivial.

For biorthogonal sequences satisfying density-conditions, we have the following.

Proposition 8. *The following statements hold.*

(1) Suppose that D_φ is dense in $\mathscr{H}$. Then, $\{\varphi_n\}$ is a generalized Riesz system and $C_\varphi = C_\varphi^N$, and $\overline{T}_{\varphi,e}$ is the smallest element of C_φ. Furthermore, suppose that $D(\varphi)$ is dense in $\mathscr{H}$. Then, $T_{e,\varphi}^{-1}$ is the largest element of C_ψ^N.

(2) Suppose that D_ψ is dense in $\mathscr{H}$. Then, $\{\psi_n\}$ is a generalized Riesz system and $C_\psi = C_\psi^N$, and $\overline{T}_{\psi,e}$ is the smallest element in C_ψ. Furthermore, suppose that $D(\psi)$ is dense in $\mathscr{H}$. Then, $T_{e,\psi}^{-1}$ is the largest element in C_φ.

(3) Suppose that $(\{\varphi_n\}, \{\psi_n\})$ is regular; that is, both D_φ and D_ψ are dense in $\mathscr{H}$. Then, $\{\varphi_n\}$ and $\{\psi_n\}$ are generalized Riesz systems and $C_\varphi = C_\varphi^N$ and $C_\psi = C_\psi^N$, and $\overline{T}_{\varphi,e}$ is the smallest element in C_φ, $\overline{T}_{\psi,e}$ is the smallest element in C_ψ, $T_{e,\varphi}^{-1}$ is the largest element in C_ψ, and $T_{e,\psi}^{-1}$ is the largest element in C_φ.

Proof. (1) We can show using Lemma 3 that $\{\varphi_n\}$ is a generalized Riesz system with a constructing pair $(\boldsymbol{e}, \overline{T}_{\varphi,e})$ and the constructing operator $\overline{T}_{\varphi,e}$ for $\{\varphi_n\}$ is the smallest element in C_φ. For more detail, refer to [3]. Furthermore, a sequence $\{\psi_n\}$ which is biorthogonal to $\{\varphi_n\}$ is unique. In fact, let $\{\psi_n\}$ and $\{\psi_n'\}$ be any sequences in $\mathscr{H}$ which are biorthogonal to $\{\varphi_n\}$. Then, since $< \psi_n, \varphi_m >= \delta_{nm} =< \psi_n', \varphi_m >$ for $n, m = 0, 1, \cdots$ and D_φ is dense in $\mathscr{H}$, we have $\psi_n = \psi_n'$ for every $n = 0, 1, \cdots$. We show $C_\varphi = C_\varphi^N$. Take an arbitrary $T \in C_\varphi$. Then, $\{\psi_n^T\}$ is biorthogonal to $\{\varphi_n\}$. By the uniqueness of biorthogonal sequences to $\{\varphi_n\}$, we have $\psi_n^T = \psi_n$, $n = 0, 1, \cdots$, which implies that $T \in C_\varphi^N$ and $C_\varphi = C_\varphi^N$. Suppose that D_φ and $D(\varphi)$ are dense in $\mathscr{H}$. We show that $T_{e,\varphi}^{-1}$ is the largest element in C_ψ^N. Since $D(T_{e,\varphi}^{-1}) = T_{e,\varphi} D(T_{e,\varphi}) \supset T_{e,\varphi} D_\psi = D_e$, $T_{e,\varphi}^{-1}$ is a densely defined closed operator in $\mathscr{H}$, and since $D(T_{e,\varphi}) = D(\varphi)$, it has a densely defined inverse $T_{e,\varphi}$. Furthermore, since $T_{e,\varphi}\psi_n = e_n$, $n = 0, 1, \cdots$, $\psi_n = T_{e,\varphi}^{-1} e_n$, $n = 0, 1, \cdots$. Thus we have $T_{e,\varphi}^{-1} \in C_\psi$. Since $((T_{e,\varphi}^{-1})^{-1})^* = T_{e,\varphi}^* = \overline{T}_{\varphi,e} \in C_\varphi$, it follows that $T_{e,\varphi}^{-1} \in C_\psi^N$. Next we show that $T_{e,\varphi}^{-1}$ is the largest element in C_ψ^N. Take an arbitrary $K \in C_\psi^N$. Then $(K^{-1})^* \in C_\varphi$, and so $(K^{-1})^* e_n = \varphi_n$, $n = 0, 1, \cdots$. Hence we have $T_{\varphi,e} \subset (K^{-1})^*$. Thus $K \subset T_{e,\varphi}^{-1}$, and so $T_{e,\varphi}^{-1}$ is the largest element in C_ψ^N.

(2) This is proved at the same way as (1).

(3) Since $D(\varphi) \supset D_\psi$ and $D(\psi) \supset D_\varphi$, it follows that $D(\varphi)$ and $D(\psi)$ are dense in $\mathscr{H}$, which implies by (1) and (2) that the statement (3) holds.

Here we give some physical examples. Let $\{f_n\}$, $n = 0, 1, \cdots$, be an ONB in $L^2(\mathbb{R})$ consisting of the Hermite functions which is contained in the Schwartz space $\mathcal{S}(\mathbb{R})$ of all infinitely differential rapidly decreasing functions on $\mathbb{R}$. We define the moment operator p and the position operator q by

$$D(p) := \text{the set of all differentiable functions } f \text{ on } \mathbb{R} \text{ such that } \frac{df}{dx} \in L^2(\mathbb{R}), \qquad (20)$$

$$(pf)(x) := -i\frac{df}{dx}, \quad f \in D(p)$$

and

$$D(q) := \left\{ f \in L^2(\mathbb{R}); \int_{-\infty}^{\infty} |xf(x)|^2 dx < \infty \right\}, \qquad (21)$$

$$(qf)(x) := xf(x), \quad f \in D(q).$$

Then p and q are self-adjoint operators in $L^2(\mathbb{R})$ and $\mathcal{S}(\mathbb{R})$ is a core for p and q, and furthermore $p\mathcal{S}(\mathbb{R}) \subset \mathcal{S}(\mathbb{R})$ and $q\mathcal{S}(\mathbb{R}) \subset \mathcal{S}(\mathbb{R})$, and $[p,q] := pq - qp = -i\mathbb{1}$ on $\mathcal{S}(\mathbb{R})$. Next we define the standard bosonic operators $a, a^\dagger$ by

$$a = \frac{1}{\sqrt{2}}(q + ip), \qquad a^\dagger = \frac{1}{\sqrt{2}}(q - ip). \tag{22}$$

Then,

$$af_n = \begin{cases} 0, & n = 0 \\ \sqrt{n} f_{n-1} & n = 1, 2, \cdots, \end{cases} \tag{23}$$

$$a^\dagger f_n = \sqrt{n+1} f_{n+1}, \quad n = 0, 1, \cdots$$

and $[a, a^\dagger] = \mathbb{1}$ on $\mathcal{S}(\mathbb{R})$.

Example 9 (the extended quantum harmonic oscillator). The Hamiltonian of this model is the non-self-adjoint operator, introduced in [11, 12],

$$H_\beta := \frac{\beta}{2}(p^2 + q^2) + i\sqrt{2}p = \beta a^\dagger a + (a - a^\dagger) + \frac{\beta}{2}\mathbb{1}, \quad \beta > 0. \tag{24}$$

We put

$$\varphi_0^{(\beta)} := e^{1/\beta^2} f_0\left(x - \frac{\sqrt{2}}{\beta}\right) = \frac{e^{1/\beta^2}}{\pi^{1/4}} e^{-(1/2)(x-\sqrt{2}/\beta)^2}. \tag{25}$$

Then, $\varphi_0^{(\beta)} \in \mathcal{S}(\mathbb{R})$ and $\varphi_0^{(\beta)} = e^{1/\beta^2} U(1/\beta) f_0 = e^{(a+a^\dagger)/\beta} f_0$, where $U(1/\beta)$ is a unitary operator defined by $U(1/\beta) := e^{(1/\beta)(a^\dagger - a)} = e^{-(\sqrt{2}/\beta)ip}$. Hence we can define a sequence $\boldsymbol{\varphi}_\beta := \{\varphi_n^{(\beta)}\}$ in $\mathcal{S}(\mathbb{R})$ by

$$\varphi_n^{(\beta)} := \frac{1}{\sqrt{n!}}\left(a^\dagger + \frac{1}{\beta}\right)^n \varphi_0^{(\beta)}, \quad n = 1, 2, \cdots. \tag{26}$$

Similarly, we define a sequence $\boldsymbol{\psi}_\beta := \{\psi_n^{(\beta)}\}$ in $\mathcal{S}(\mathbb{R})$ as follows:

$$\psi_0^{(\beta)} := e^{1/\beta^2} f_0\left(x + \frac{\sqrt{2}}{\beta}\right), \qquad \psi_n^{(\beta)} := \frac{1}{\sqrt{n!}}\left(a^\dagger - \frac{1}{\beta}\right)^n \psi_0^{(\beta)}, \quad n = 1, 2, \cdots. \tag{27}$$

Then $\{\varphi_n^{(\beta)}\}$ and $\{\psi_n^{(\beta)}\}$ are regular biorthogonal sequences in $L^2(\mathbb{R})$ which are generalized Riesz systems with constructing pairs $(\{f_n\}, e^{(a^\dagger+a)/\beta})$ and $(\{f_n\}, e^{-(a^\dagger+a)/\beta})$, respectively, and

$$\overline{A}_{\boldsymbol{\varphi}_\beta} := \overline{e^{(a^\dagger+a)/\beta} a e^{-(a^\dagger+a)/\beta}} = a - \frac{1}{\beta}, \qquad \overline{B}_{\boldsymbol{\varphi}_\beta} := \overline{e^{(a^\dagger+a)/\beta} a^\dagger e^{-(a^\dagger+a)/\beta}} = a^\dagger + \frac{1}{\beta}. \tag{28}$$

By Proposition 8, $T_{\boldsymbol{\varphi}_\beta, \boldsymbol{f}}$ is the smallest constructing operator and $T_{\boldsymbol{f}, \boldsymbol{\psi}_\beta}^{-1}$ is the largest constructing operator for $\{\varphi_n^{(\beta)}\}$ and $T_{\boldsymbol{\varphi}_\beta, \boldsymbol{f}} \subset e^{(a^\dagger+a)/\beta} \subset T_{\boldsymbol{f}, \boldsymbol{\psi}_\beta}^{-1}$. Similarly, $T_{\boldsymbol{\psi}_\beta, \boldsymbol{f}}$ is the smallest constructing operator and $T_{\boldsymbol{f}, \boldsymbol{\varphi}_\beta}^{-1}$ is the largest constructing operator for $\{\psi_n^{(\beta)}\}$ and $T_{\boldsymbol{\psi}_\beta, \boldsymbol{f}} \subset e^{-(a^\dagger+a)/\beta} \subset T_{\boldsymbol{f}, \boldsymbol{\varphi}_\beta}^{-1}$.

The following example is a modification of the non-self-adjoint Hamiltonian H_β in Example 9 exchanging the momentum operator p with the position operator q.

Example 10. We introduce a non-self-adjoint Hamiltonian

$$H'_\beta := \frac{\beta}{2}(p^2 + q^2) + \sqrt{2}iq = \beta a^\dagger a + i(a + a^\dagger) + \frac{\beta}{2}, \quad \beta > 0. \tag{29}$$

We define sequences $\boldsymbol{\varphi}'_\beta := \{\varphi'_n\}$ and $\boldsymbol{\psi}'_\beta := \{\psi'_n\}$ in $\mathcal{S}(\mathbb{R})$ as follows:

$$\varphi'_0 := e^{-(i/\beta)a^\dagger} f_0, \qquad \varphi'_n := \frac{1}{\sqrt{n!}}\left(a^\dagger + \frac{i}{\beta}\right)^n \varphi'_0, \quad n = 1, 2, \cdots \tag{30}$$

and

$$\psi'_0 := e^{1/\beta^2} e^{(i/\beta)a^\dagger} f_0, \qquad \psi'_n := \frac{1}{\sqrt{n!}}\left(a^\dagger - \frac{i}{\beta}\right)^n \psi'_0, \quad n = 1, 2, \cdots. \tag{31}$$

Then $\boldsymbol{\varphi}'_\beta$ and $\boldsymbol{\psi}'_\beta$ are regular biorthogonal sequences in $L^2(\mathbb{R})$ which are generalized Riesz systems with constructing pairs $(\{f_n\}, T)$ and $(\{f_n\}, T^{-1})$, respectively, where $T := \overline{e^{-(i/\beta)a^\dagger} e^{(i/\beta)a}}$ and $T^{-1} = \overline{e^{-(i/\beta)a} e^{(i/\beta)a^\dagger}}$, and $\overline{TaT^{-1}} = a + i/\beta$, $\overline{Ta^\dagger T^{-1}} = a^\dagger + i/\beta$, $\overline{T^{-1}aT} = a - i/\beta$, and $\overline{T^{-1}a^\dagger T} = a^\dagger - i/\beta$.

Example 11 (the Swanson model). The Swanson Hamiltonian, introduced in [11, 13], is a non-self-adjoint Hamiltonian

$$\begin{aligned} H_\theta &:= \frac{1}{2}(p^2 + q^2) - \frac{i}{2}\tan 2\theta (p^2 - q^2) \\ &= a^\dagger a + \frac{i}{2}\tan 2\theta \left(a^2 + (a^\dagger)^2\right) + \frac{1}{2}\mathbb{1}, \end{aligned} \quad \theta \neq 0 \in \left(-\frac{\pi}{4}, \frac{\pi}{4}\right). \tag{32}$$

We define sequences $\boldsymbol{\varphi}_\theta := \{\varphi_n^{(\theta)}\}$ and $\boldsymbol{\psi}_\theta := \{\psi_n^{(\theta)}\}$ in $L^2(\mathbb{R})$ as follows:

$$\begin{aligned} \varphi_0^{(\theta)} &:= c_0 \sum_{k=0}^{\infty} e^{-i(\tan\theta/2)(a^\dagger)^2} f_0 \\ &= c_0 \sum_{k=0}^{\infty} (-i\tan\theta)^k \sqrt{\frac{(2k-1)!!}{(2k)!!}} f_{2k}, \end{aligned} \tag{33}$$

$$\varphi_n^{(\theta)} := \frac{1}{\sqrt{n!}}\left(\cos\theta a^\dagger + i\sin\theta a\right)^n \varphi_0^{(\theta)}$$

and

$$\psi_0^{(\theta)} := d_0 \sum_{k=0}^{\infty} e^{i(\tan\theta/2)(a^\dagger)^2} f_0 = d_0 \sum_{k=0}^{\infty} (i\tan\theta)^k \sqrt{\frac{(2k-1)!!}{(2k)!!}} f_{2k}, \tag{34}$$

$$\psi_n^{(\theta)} := \frac{1}{\sqrt{n!}} \left(\cos\theta a^\dagger - i\sin\theta a\right)^n \psi_0^{(\theta)},$$

where $(2k)!! = 2k(2k-2)\cdots 4\cdot 2$, $(2k-1)!! = (2k-1)(2k-3)\cdots 3\cdot 1$, and c_0 and d_0 are constants satisfying $< \varphi_0^{(\theta)}, \psi_0^{(\theta)} >= 1$. Then $\boldsymbol{\varphi}_\theta$ and $\boldsymbol{\psi}_\theta$ are regular biorthogonal sequences in $L^2(\mathbb{R})$ contained in $\mathcal{S}(\mathbb{R})$ which are generalized Riesz systems with constructing operators $T_\theta := e^{i(\theta/2)(a^2-(a^\dagger)^2)}$ and T_θ^{-1}, respectively. For the generalized lowering operator $A_\theta := T_\theta a T_\theta^{-1}$ and the raising operator $B_\theta := T_\theta a^\dagger T_\theta^{-1}$, we have

$$A_\theta = (\cos\theta)\, a + i(\sin\theta)\, a^\dagger,$$
$$B_\theta = (\cos\theta)\, a^\dagger + i(\sin\theta)\, a. \tag{35}$$

By Proposition 8, $T_{\boldsymbol{\varphi}_\theta,\boldsymbol{f}}$ (resp., $T_{\boldsymbol{\psi}_\theta,\boldsymbol{f}}$) is the smallest constructing operator and $T_{\boldsymbol{\psi}_\theta,\boldsymbol{f}}^{-1}$ (resp., $T_{\boldsymbol{\varphi}_\theta,\boldsymbol{f}}^{-1}$) is the largest constructing operator for $\boldsymbol{\varphi}_\theta$ (resp., $\boldsymbol{\psi}_\theta$) and every closed operator T (resp., K) in $L^2(\mathbb{R})$ satisfying $T_{\boldsymbol{\varphi}_\theta,\boldsymbol{f}} \subset T \subset T_{\boldsymbol{\psi}_\theta,\boldsymbol{f}}^{-1}$ (resp., $T_{\boldsymbol{\psi}_\theta,\boldsymbol{f}} \subset K \subset T_{\boldsymbol{\varphi}_\theta,\boldsymbol{f}}^{-1}$) is a constructing operator for $\boldsymbol{\varphi}_\theta$ (resp., $\boldsymbol{\psi}_\theta$).

All physical models discussed above are regular cases, but it seems to be mathematically meaningful to study nonregular cases and furthermore the studies may become useful for physical applications in future. Let $\{\varphi_n\}$ be a generalized Riesz system. First we investigate under what conditions C_φ has maximal elements and minimal elements.

Let C be a totally ordered subset of C_φ. Then, it is easily shown that $(T^{-1})^* e_n = (S^{-1})^* e_n$, $n = 0, 1, \cdots$, for any $T, S \in C$. Hence we may put

$$\psi_n^C = \left(T^{-1}\right)^* e_n, \quad T \in C,\ n = 0, 1, \cdots. \tag{36}$$

We have the following statements.

Lemma 12. *Let C be any totally ordered subset of C_φ. The following statements hold.*

(1) Suppose that $\cap_{T\in C} D(T^*)$ is dense in $\mathcal{H}$. Then there exists an upper bounded element G of C.

(2) Suppose that $\cap_{T\in C} R(T)$ is dense in $\mathcal{H}$. Then there exists a lower bounded element S of C.

(3) Suppose that $\cap_{T\in C} D(T^*)$ and $\cap_{T\in C} R(T)$ are dense in $\mathcal{H}$. Then for every linear operator A such that $S \subset A \subset G$, the closure $\overline{A}$ of A belongs to C_φ.

Proof. (1) We put

$$D(G) = \bigcup_{T\in C} D(T),$$
$$Gx = T_0 x, \quad x \in D(G), \tag{37}$$

where T_0 is an operator in C whose domain $D(T_0)$ contains x. Since C is totally ordered, it follows that $D(G)$ is a subspace in $\mathcal{H}$ and $Tx = T_0 x$ for any operators T, T_0 in C whose domains contain x. Hence, G does not depend on the method of choosing $T_0 \in C$ whose domain contains x. Thus G is a well-defined densely defined linear operator in $\mathcal{H}$ such that $T \subset G$ for all $T \in C$. We show that G is closable. Indeed, we may show

$$\bigcap_{T\in C} D(T^*) = D(G^*),$$
$$G^* y = T_0^* y, \tag{38}$$

where $T_0 \in C$ whose domain $D(T_0)$ contains x. Take an arbitrary $y \in \cap_{T\in C} D(T^*)$. Then, we have

$$\langle Gx, y\rangle = \langle T_0 x, y\rangle = \langle x, T_0^* y\rangle \tag{39}$$

for all $x \in D(G)$, where $T_0 \in C$ whose domain contains x. Hence, $y \in D(G^*)$ and $G^* y = T_0^* y$. Since $T \subset G$ for all $T \in C$, $D(G^*) \subset \cap_{T\in C} D(T^*)$ is trivial. Thus, (38) holds. By (38) and the assumption of (1), $D(G^*)$ is dense in $\mathcal{H}$, that is, G is closable. Next we show that G has a densely defined inverse. Suppose that $Gx = 0$, $x \in D(G)$. Then $T_0 x = 0$ for some $T_0 \in C$, and so $x = 0$ since T_0 has an inverse. Thus G has an inverse. Since $R(T) \subset R(G)$ and $R(T)$ is dense in $\mathcal{H}$ for all $T \in C$, it follows that the inverse of G is densely defined, which implies that the closure $\overline{G}$ of G has a densely defined closed operator in $\mathcal{H}$ such that $\overline{G} \supset T$ for all $T \in C$. Finally we show

$$\{e_n\} \subset D(\overline{G}) \cap D\left((G^*)^{-1}\right),$$
$$\overline{G} e_n = \varphi_n, \quad n = 0, 1, \cdots. \tag{40}$$

Clearly, $\{e_n\} \subset D(T) \subset D(\overline{G})$ for all $T \in C$. Next we show that for any n there exists an element y_n of $\cap_{T\in C} D(T^*)$ such that

$$e_n = T^* y_n \tag{41}$$

for all $T \in C$. Indeed, take an arbitrary $T \in C$. Since $e_n \in D((T^*)^{-1}) = R(T^*)$, there exists an element y_n^T of $D(T^*)$ such that $e_n = T^* y_n^T$. Let any $T' \in C$. Since C is totally ordered, either $T' \subset T$ or $T \subset T'$ holds. Suppose that $T' \subset T$. Since $T^* \subset (T')^*$, it follows that $y_n^T, y_n^{T'} \in D((T')^*)$ and

$$\left(T'\right)^* y_n^T = T^* y_n^T = e_n = \left(T'\right)^* y_n^{T'}, \tag{42}$$

which implies that $y_n^T = y_n^{T'}$ since $(T')^*$ has inverse. The equality $y_n^T = y_n^{T'}$ is similarly shown in case that $T \subset T'$. Hence, we have that $y_n := y_n^T \in \cap_{T\in C} D(T^*)$ and $T^* y_n = e_n$ for all $T \in C$. Thus (41) holds. By (38) and (41) we have $e_n \in R(G^*) = D((G^*)^{-1})$. Furthermore, we have $\overline{G} e_n = T e_n = \varphi_n$, $n = 0, 1, \cdots$ for all $T \in C$. Thus we have $\overline{G} \in C_\varphi$ and $\overline{G}$ is an upper bounded element of C.

(2) We put

$$D(S) = \cap_{T\in C} D(T),$$
$$Sx = Tx, \quad x \in D(S), \tag{43}$$

where T is any element of C. Since $\{e_n\} \subset \cap_{T\in C} D(T)$ and $T_1 x = T_2 x$ for all $x \in \cap_{T\in C} D(T)$, S is a well-defined densely defined closed operator in $\mathscr{H}$ such that $S \subset T$ for all $T \in C$. Hence, it is sufficient to show $S \in C_\varphi$. Since $S \subset T$ for all $T \in C$ and T has the inverse, S has the inverse. Furthermore, we may show

$$R(S) = \cap_{T\in C} R(T). \quad (44)$$

In fact, take an arbitrary $y \in \cap_{T\in C} R(T)$. Since C is totally ordered, there exists an element x of $D(S) = \cap_{T\in C} D(T)$ such that $y = Tx$ for all $T \in C$. Hence, $y = Sx \in R(S)$. The inverse inclusion $R(S) \subset \cap_{T\in C} R(T)$ is clear. Hence (44) holds. By the assumption and (44), $R(S) = D(S^{-1})$ is dense in $\mathscr{H}$. Furthermore, since $\{e_n\} \subset D(T)$ for all $T \in C$, we have $\{e_n\} \subset D(S)$ and $Se_n = Te_n = \varphi_n$, $n = 0, 1, \cdots$. Since $S \subset T$, it follows that $\{e_n\} \subset D((T^*)^{-1}) \subset D((S^*)^{-1})$. Thus, $S \in C_\varphi$ and it is a lower bound of C.

(3) This follows from (1) and (2).

For a subset $\mathscr{F}$ of C_φ, we put

$$\mathscr{F}_\psi := \left\{\left(T^{-1}\right)^*;\ T \in \mathscr{F}\right\}. \quad (45)$$

Then we have the following.

Lemma 13. *Let $\mathscr{F}$ be a subset of C_φ. Then, T_0 is a maximal (resp., minimal, the largest and the smallest) element of $\mathscr{F}$ if and only if $(T_0^{-1})^*$ is a minimal (resp., maximal, the smallest and the largest) element of $\mathscr{F}_\psi$.*

Proof. Suppose that T_0 is a maximal element of $\mathscr{F}$. Take an arbitrary $K \in \mathscr{F}_\psi$ satisfying $K \subset (T_0^{-1})^*$. Then we have that $K = (T^{-1})^*$ for some $T \in \mathscr{F}$ and $T_0 \subset T$, which implies by the maximality of T_0 that $T = T_0$ and $K = (T_0^{-1})^*$. Thus, $(T_0^{-1})^*$ is a minimal element of $\mathscr{F}_\psi$. Furthermore, we can similarly show that if $(T_0^{-1})^*$ is a minimal element of $\mathscr{F}_\psi$, then T_0 is a maximal element $\mathscr{F}$. The other statements are similarly shown.

Theorem 14. *Let $\mathscr{F}$ be a subset of C_φ. Then we have the following:*

(1) *The following statements are equivalent:*

(i) *$\mathscr{F}$ has a maximal element.*

(ii) *There exists a closed operator A in $\mathscr{H}$ such that $T \subset A$ for all $T \in \mathscr{F}$.*

(iii) *$\mathscr{F}_\psi$ has a minimal element.*

(2) *The following statements are equivalent:*

(i) *$\mathscr{F}$ has a minimal element.*

(ii) *There exists a closed operator B in $\mathscr{H}$ such that $(T^{-1})^* \subset B$ for all $T \in \mathscr{F}$.*

(iii) *$\mathscr{F}_\psi$ has a maximal element.*

(3) *The following statements are equivalent:*

(i) *$\mathscr{F}$ has a maximal element and a minimal element.*

(ii) *There exist closed operators A and B in $\mathscr{H}$ such that $T \subset A$ and $(T^{-1})^* \subset B$ for all $T \in \mathscr{F}$.*

(iii) *$\mathscr{F}_\psi$ has a maximal element and a minimal element.*

Proof. (1) (i)$\Longrightarrow$(ii) This is trivial.

(ii)$\Longrightarrow$(i) Suppose that there exists a closed operator A in $\mathscr{H}$ such that $T \subset A$ for all $T \in \mathscr{F}$. Then for any totally ordered subset C of $\mathscr{F}$ we have $D(A^*) \subset \bigcap_{T\in C} D(T^*)$. Hence, it follows that $\bigcap_{T\in C} D(T^*)$ is dense in $\mathscr{H}$, which implies by Lemma 12 that $\mathscr{F}$ has an upper bounded element. By Zorn's lemma, $\mathscr{F}$ has a maximal element.

(i)$\Longleftrightarrow$(iii) This follows from Lemma 13.

(2) (i)$\Longrightarrow$(ii) This is trivial.

(ii)$\Longrightarrow$(i) Suppose that there exists a closed operator A in $\mathscr{H}$ such that $(T^{-1})^* \subset B$ for all $T \in \mathscr{F}$. Then we can similarly show that $\mathscr{F}_\psi$ has a maximal element, which implies by Lemma 13 that $\mathscr{F}$ has a minimal element.

(i)$\Longleftrightarrow$(iii) This follows from Lemma 13.

(3) This follows from (1) and (2). This completes the proof.

We remark that the closed operators A and B in Theorem 14 do not need any other conditions, for example, the existence of inverse.

By Theorem 14, we have the following.

Corollary 15. *Let $T_0 \in C_\varphi$ and put $\mathscr{F}_{T_0} = \{T \in C_\varphi;\ T_0 \subset T\}$. Then the following statements hold.*

(1) Suppose that there exists a closed operator A in $\mathscr{H}$ such that $T \subset A$ for all $T \in \mathscr{F}_{T_0}$. Then there exists a maximal element of C_φ which is an extension of T_0.

(2) Suppose that there exists a closed operator B in $\mathscr{H}$ such that $(T^{-1})^* \subset B$ for all $T \in \mathscr{F}_{T_0}$. Then there exists a minimal element of C_φ which is a restriction of T_0.

Proof. (1) By Theorem 14, $\mathscr{F}_{T_0}$ has a maximal element T_1. Here we show that T_1 is a maximal element of C_φ. Indeed, this follows since $T \in \mathscr{F}_{T_0}$ for any element T of C_φ satisfying $T_1 \subset T$. We can similarly show (2).

Next we investigate the existence of the smallest element and of the largest element of C_φ.

Theorem 16. *C_φ has the smallest element and the largest element if and only if there exist closed operators A and B in $\mathscr{H}$ such that $T \subset A$ and $(T^{-1})^* \subset B$ for all $T \in C_\varphi$.*

Proof. Suppose that there exist closed operators A and B in $\mathscr{H}$ such that $T \subset A$ and $(T^{-1})^* \subset B$ for all $T \in C_\varphi$. We define an operator T_0 as follows:

$$D(T_0) = \bigcap_{T\in C_\varphi} D(T),$$
$$T_0 x = Tx, \quad x \in D(T_0), \quad (46)$$

where T is an element of C_φ. Take an arbitrary $x \in D(T_0)$ and $T_1, T_2 \in C_\varphi$. Since $x \in D(T_1)$, $x \in D(T_2)$, $T_1 \subset A$ and $T_2 \subset A$, we have

$$T_1 x = T_2 x = Ax. \quad (47)$$

Thus, T_0 does not depend on the method of choosing $T \in C_\varphi$, and so T_0 is well defined. Since $\{e_n\} \subset \cap_{T\in C_\varphi} D(T) = D(T_0)$, T_0 is a densely defined closed operator in $\mathscr{H}$ such that $T_0 \subset T$ for all $T \in C_\varphi$. Since $(T^{-1})^* \subset B$ for all $T \in C_\varphi$, we have $D(B^*) \subset \bigcap_{T\in C_\varphi} R(T)$, which implies that $\bigcap_{T\in C_\varphi} R(T)$ is dense in $\mathscr{H}$. Hence, we can prove at the same way as the proof of Lemma 12 (2) that T_0 is the smallest element C_φ. Next we show that C_φ has the largest element. Take an arbitrary $T \in C_\varphi$. Then ψ_T is a generalized Riesz system with a constructing operator $(T^{-1})^*$ and $K \subset B$ and $(K^{-1})^* \subset A$ for all $K \in C^N_{\psi_T}$. Hence, as shown above there exists the smallest element K_1 of $C^N_{\psi_T}$, and so $K_1 = (T_1^{-1})^*$ for some $T_1 \in C_\varphi$ and $(T_1^{-1})^* \subset (T^{-1})^*$. Thus $T \subset T_1$ and T_1 is the largest element of C_φ. The converse is trivial. This completes the proof.

As seen in Section 2, for a biorthogonal pair $(\{\varphi_n\}, \{\psi_n\})$, the equality $(T^{-1})^* e_n = \psi_n$, $n = 0, 1, \cdots$ does not necessarily hold for all $T \in C_\varphi$. From this fact we define the notion of *natural* pair of generalized Riesz systems.

Definition 17. A biorthogonal pair $(\{\varphi_n\}, \{\psi_n\})$ of generalized Riesz systems is said to be natural, if $C_\varphi = C^N_\varphi$ and $C_\psi = C^N_\psi$, that is, $(T^{-1})^* e_n = \psi_n$ for all n and $T \in C_\varphi$ and $(K^{-1})^* e_n = \varphi_n$ for all n and $K \in C_\psi$.

Theorem 18. *Let $(\{\varphi_n\}, \{\psi_n\})$ be a natural pair of generalized Riesz systems. Then C_φ and C_ψ have the smallest element and the largest element, respectively, if and only if there exist closed operators A and B in $\mathscr{H}$ such that $T \subset A$ and $K \subset B$ for all $T \in C_\varphi$ and $K \in C_\psi$.*

Proof. This is shown using Theorem 16 for the generalized Riesz systems for $\{\varphi_n\}$ and $\{\psi_n\}$.

For a generalized Riesz system $\{\varphi_n\}$, suppose that there exist the largest element T_L of C_φ and the smallest element T_S of C_φ. Then every closed operator T in $\mathscr{H}$ satisfying $T_S \subset T \subset T_L$ is a constructing operator of $\{\varphi_n\}$, and so we can construct all kinds of non-self-adjoint Hamiltonians $TH^\alpha_e T^{-1}$, lowering operator $TA^\alpha_e T^{-1}$ and raising operator $TB^\alpha_e T^{-1}$ for $\{\varphi_n\}$. It may be possible to find constructing operators suitable for each of the physical models.

4. Conclusions

All the results presented in this paper are of pure mathematical nature, but we hope that they will be applied to more physical models in future. For example, we argue that cases like the CCR-algebras and their physical applications could probably studied by taking suitable constructing operators for convenient generalized Riesz systems.

Conflicts of Interest

The author declares that there are no conflicts of interest regarding the publication of this paper.

Acknowledgments

The author is supported by Daiichi University of Pharmacy.

References

[1] F. Bagarello, H. Inoue, and C. Trapani, "Biorthogonal vectors, sesquilinear forms, and some physical operators," *Journal of Mathematical Physics*, vol. 59, no. 3, 033506, 13 pages, 2018.

[2] H. Inoue and M. Takakura, "Regular biorthogonal pairs and pseudo-bosonic operators," *Journal of Mathematical Physics*, vol. 57, no. 8, 083503, 9 pages, 2016.

[3] H. Inoue, "Semi-regular biorthogonal pairs and generalized Riesz bases," *Journal of Mathematical Physics*, vol. 57, no. 11, 113502, 11 pages, 2016.

[4] O. Christensen, *An Introduction to Frames and Riesz Bases*, Birkhäuser, 2002.

[5] F. Bagarello and G. Bellomonte, "Hamiltonians defined by biorthogonal sets," *Journal of Physics A: Mathematical and General*, vol. 50, no. 14, 145203, 20 pages, 2017.

[6] G. Bellomonte and C. Trapani, "Riesz-like bases in Rigged Hilbert Spaces," *Journal of Analysis and its Applications*, vol. 35, no. 3, pp. 243–265, 2016.

[7] A. Mostafazadeh, "Pseudo-Hermitian representation of quantum mechanics," *International Journal of Geometric Methods in Modern Physics*, vol. 7, no. 7, pp. 1191–1306, 2010.

[8] F. Bagarello, C. Trapani, and S. Triolo, "Gibbs states defined by biorthogonal sequences," *Journal of Physics A: Mathematical and General*, vol. 49, no. 40, 405202, 15 pages, 2016.

[9] S. Di Bella and C. Trapani, "Some representation theorems for sesquilinear forms," *Journal of Mathematical Analysis and Applications*, vol. 451, no. 1, pp. 64–83, 2017.

[10] K. Schmüdgen, *Unbounded Operator Algebras and Representation Theory*, Birkhäuser, 1990.

[11] J. da Providência, N. Bebiano, and J. P. da Providência, "Non-Hermitian Hamiltonians with Real Spectrum in Quantum Mechanics," *Brazilian Journal of Physics*, vol. 41, no. 1, pp. 78–85, 2011.

[12] F. Bagarello, "Examples of pseudo-bosons in quantum mechanics," *Physics Letters A*, vol. 374, no. 37, pp. 3823–3827, 2010.

[13] M. S. Swanson, "Transition elements for a non-Hermitian quadratic HAMiltonian," *Journal of Mathematical Physics*, vol. 45, no. 2, pp. 585–601, 2004.

4

Convolutions of Harmonic Functions with Certain Dilatations

Om P. Ahuja and Jay M. Jahangiri

Mathematical Sciences, Kent State University, Burton, OH 44021-9500, USA

Correspondence should be addressed to Jay M. Jahangiri; jjahangi@kent.edu

Academic Editor: Teodor Bulboaca

The convolution of harmonic functions, unlike the analytic case, proved to be very challenging. In this paper, we introduce dilatation conditions that guarantee the convolution of two harmonic functions to be locally one-to-one, sense-preserving, and close-to-convex harmonic in the unit disk.

1. Introduction

Let $\mathscr{A}$ denote the class of functions that are analytic in the open unit disc $\mathbb{D} := \{z \in \mathbb{C} : |z| < 1\}$ and let $\mathscr{A}'$ be the subclass of $\mathscr{A}$ consisting of functions h with the normalization $h(0) = h'(0) - 1 = 0$. We let $\mathscr{K}(\alpha)$ denote the class of functions $h \in \mathscr{A}'$ so that

$$\operatorname{Re}\left\{1 + z\frac{h''(z)}{h'(z)}\right\} > \alpha; \quad -\frac{1}{2} \leq \alpha; \; z \in \mathbb{D}. \tag{1}$$

Consider the family of complex-valued harmonic functions $f = u + iv$ defined in $\mathbb{D}$, where u and v are real harmonic in $\mathbb{D}$. Such functions can be expressed as $f = h + \overline{g}$, where $h \in \mathscr{A}$ and $g \in \mathscr{A}$. Clunie and Sheil-Small in their remarkable paper [1] explored the functions of the form $f = h + \overline{g}$ that are locally one-to-one, sense-preserving, and harmonic in $\mathbb{D}$. By Lewy's Theorem (see [2] or [1]), a necessary and sufficient condition for the harmonic function $f = h + \overline{g}$ to be locally one-to-one and sense-preserving in $\mathbb{D}$ is that its Jacobian $J_f = |h'|^2 - |g'|^2$ is positive or equivalently, if and only if $h'(z) \neq 0$ in $\mathbb{D}$ and the second complex dilatation ω of f satisfies $|\omega| = |g'/h'| < 1$ in $\mathbb{D}$.

In an interesting article, Bshouty and Lyzzaik [3] proved the following.

Theorem 1. *Let $f = h + \overline{g}$ be a harmonic mapping of $\mathbb{D}$, with $h'(0) \neq 0$, that satisfies $g'(z) = zh'(z)$ and $h \in \mathscr{K}(-1/2)$ for all $z \in \mathbb{D}$. Then f is a univalent close-to-convex mapping.*

A simply connected proper subdomain of $\mathbb{C}$ is said to be close-to-convex if its complement in $\mathbb{C}$ is the union of closed half-lines with pairwise disjoint interiors. Consequently, a univalent analytic or harmonic function $f : \mathbb{D} \to \mathbb{C}$ is said to be close-to-convex if $f(\mathbb{D})$ is close-to-convex (e.g., see Clunie and Sheil-Small [1] or Bshouty and Lyzzaik [3]).

Ruscheweyh and Sheil-Small in a striking article [4] proved that the Hadamard product or convolution of two analytic convex functions is also convex analytic and that the convolution of an analytic convex function and an analytic close-to-convex function is close-to-convex analytic in the unit disk $\mathbb{D}$. Ironically, these results could not be extended to the harmonic case, since the convolution of harmonic functions, unlike the analytic case, proved to be very challenging. The purpose of the present paper is to introduce dilatation conditions that guarantee the convolution of two harmonic functions to be locally one-to-one, sense-preserving, and close-to-convex harmonic in the unit disk $\mathbb{D}$. In other words, we extend Theorem 1 to the convolution of two harmonic functions $f_1 = h_1 + \overline{g_1}$ and $f_2 = h_2 + \overline{g_2}$ with certain dilatations, where $h_1 * h_2 \in \mathbb{K}(\alpha)$.

The operator $*$ stands for the convolution or Hadamard product of two power series $h_1(z) = \sum_{n=1}^{\infty} a_n z^n$ and $h_2(z) = \sum_{n=1}^{\infty} b_n z^n$ given by $h_1(z) * h_2(z) = (h_1 * h_2)(z) = \sum_{n=1}^{\infty} a_n b_n z^n$. Similarly, the convolution of two harmonic functions $f_1 = h_1 + \overline{g_1}$ and $f_2 = h_2 + \overline{g_2}$ is given by $f_1 * f_2 = h_1 * h_2 + \overline{g_1 * g_2}$.

In regard to the convolution of harmonic univalent functions, Clunie and Sheil-Small [1] proved the following.

Theorem 2. *If $\phi \in \mathcal{K}(0)$ and if f is convex harmonic in $\mathbb{D}$, then their convolution $(\phi + \varepsilon\overline{\phi}) * f$ $(|\varepsilon| \leq 1)$ is close-to-convex harmonic in $\mathbb{D}$.*

A mapping $f : \mathbb{D} \to \mathbb{C}$ is called convex harmonic if $f(\mathbb{D})$ is a convex domain.

The convexity condition for the function ϕ in Theorem 2 cannot be compromised as it is demonstrated in the following.

Example 3. Set

$$f(z) = h(z) + \overline{g(z)} = \frac{z - (1/2) z^2}{(1 - z)^2} - \overline{\left(\frac{(1/2) z^2}{(1 - z)^2}\right)} \quad (2)$$

and consider the starlike analytic function $\phi(z) = z + z^n/n$; $n \geq 2$ in $\mathbb{D}$. Letting $\varepsilon = 0$ in Theorem 2, we observe that the harmonic convolution

$$\left(\phi(z) + (0) \cdot \overline{\phi(z)}\right) * \left(h(z) + \overline{g(z)}\right) = z + \frac{n+1}{2n} z^n \quad (3)$$

is not even univalent in $\mathbb{D}$.

In an attempt to investigate the possibilities of improving the required convexity condition for ϕ, the authors in [5] proved the following.

Theorem 4. *Let $\phi \in \mathcal{K}(0)$ and $h \in \mathcal{K}(0)$. Also let w be a Schwarz function. Then the convolution function*

$$\left(\phi(z) + \varepsilon\overline{\phi(z)}\right) * \left(h(z) + \overline{\int_0^z w(t) h'(t)\, dt}\right); \quad |\varepsilon| = 1, \quad (4)$$

is close-to-convex harmonic in $\mathbb{D}$.

Theorem 4 for $\phi(z) = z/(1 - z)$ and $\varepsilon = 1$ yields a theorem given by Bshouty et al. ([6], Theorem 2). From what is said above, especially Example 3, one wonders if there are other conditions that guarantee the close-to-convexity of the convolution of two harmonic functions. In the following theorem, we find such conditions.

Theorem 5. *Let $h_1 \in \mathcal{A}'$ and $h_2 \in \mathcal{A}'$ so that $h_1 * h_2 \in \mathcal{K}(\alpha)$, where $\mathcal{K}(\alpha)$ is given by inequality (1). If one of the following conditions hold*

(i) $\alpha \geq -1/2$; *$g_1(z) = zh_1(z)$ and $g_2'(z) = zh_2'(z)$,*

(ii) $\alpha \geq 0$; *$g_1'(z) = z^n h_1'(z)$ and $g_2'(z) = z^n h_2'(z)$, where $n \in \mathbb{N}$,*

then the convolution function $F(z) = h_1(z) * h_2(z) + \overline{g_1(z) * g_2(z)}$ is locally one-to-one, sense-preserving, and close-to-convex harmonic in $\mathbb{D}$.

Since the convolution of two convex analytic functions is also convex (see Ruscheweyh and Sheil-Small [4]), an obvious consequence of the above theorem would be as follows.

Corollary 6. *Let $h_1 \in \mathcal{K}(0)$ and $h_2 \in \mathcal{K}(0)$ and set $g_1(z) = z^n h_1'(z)$ and $g_2'(z) = z^n h_2'(z)$. Then the convolution function $F(z) = h_1(z) * h_2(z) + \overline{g_1(z) * g_2(z)}$ is locally one-to-one, sense-preserving, and close-to-convex harmonic in $\mathbb{D}$.*

2. Preliminary Lemmas and Proof of Theorem 5

To prove our Theorem 5, we shall need the following three lemmas, the first of which is a celebrated result by Clunie and Sheil-Small [1] and the second one is given by Kaplan [7]. The third lemma which is on subordination is a modification of a result given by Miller and Mocanu (e.g., see [8] Lemma 1 or [9]). For functions p and q, where $p(0) = q(0) = 0$, we write $p \prec q$ (i.e., p is subordinate to q) if there exists an analytic function ω with $\omega(0) = 0$ and $|\omega(z)| < 1$ so that $p(z) = q(\omega(z))$ in $\mathbb{D}$.

Lemma 7. *(i) If g and h are analytic in $\mathbb{D}$ so that $|g'(0)| < |h'(0)|$ and if $h + \varepsilon g$ is close-to-convex analytic in $\mathbb{D}$ for each ε $(|\varepsilon| = 1)$, then the function $f = h + \overline{g}$ is close-to-convex harmonic in $\mathbb{D}$.*

(ii) If h and g are analytic in $\mathbb{D}$ so that $h \in \mathcal{K}(0)$ and if $f = h + \overline{g}$ is locally univalent in $\mathbb{D}$, then the function $f = h + \overline{g}$ is close-to-convex harmonic in $\mathbb{D}$.

Lemma 8. *A necessary and sufficient condition for the analytic function $h : \mathbb{D} \to \mathbb{C}$ to be close-to-convex is that h' is nonvanishing in $\mathbb{D}$ and*

$$\int_{\theta_1}^{\theta_2} \operatorname{Re}\left\{1 + re^{i\theta} \frac{h''(re^{i\theta})}{h'(re^{i\theta})}\right\} d\theta > -\pi; \quad \theta_1 < \theta_2, \; 0 < r < 1. \quad (5)$$

Lemma 9. *If $\Re\, p(z) > 0$ and $q(z)$ is analytic in $\mathbb{D}$, then $q(z) + zp(z)q'(z) \prec z$ implies $q(z) \prec z$.*

Proof of Theorem 5.

Proof of Part (i). The convolution function $F(z) = h_1(z) * h_2(z) + \overline{g_1(z) * g_2(z)} = H(z) + \overline{G(z)}$ is locally univalent and sense-preserving since

$$\left|\frac{G'(z)}{H'(z)}\right| = \left|\frac{(1/z) g_1(z) * g_2'(z)}{(1/z) h_1(z) * h_2'(z)}\right| = \left|\frac{g_1(z) * zg_2'(z)}{h_1(z) * zh_2'(z)}\right| = \left|\frac{zh_1(z) * z^2 h_2'(z)}{h_1(z) * zh_2'(z)}\right| = |z| < 1. \quad (6)$$

Obviously $|G'(0)| < |H'(0)|$; therefore, in view of Lemma 7, it suffices to prove that $T_\lambda(z) = H(z) - \lambda G(z)$ for $|\lambda| = 1$ is close-to-convex analytic in $\mathbb{D}$.

We note that

$$T_\lambda'(z) = H'(z) - \lambda G'(z) = H'(z) - \lambda\left((1/z) g_1(z) * g_2'(z)\right)$$

$$= H'(z) - \lambda\left(h_1(z) * zh_2'(z)\right)$$
$$= H'(z) - \lambda z\left((1/z)h_1(z) * h_2'(z)\right)$$
$$= H'(z) - \lambda z\left(H'(z)\right) = (1-\lambda z)\cdot H'(z), \tag{7}$$

and $T_\lambda''(z) = -\lambda H'(z) + (1-\lambda z)H''(z)$.

We also observe that T_λ is nonvanishing in $\mathbb{D}$ since $H'(0) \neq 0$. Therefore,

$$\operatorname{Re}\left\{1 + z\frac{T_\lambda''(z)}{T_\lambda'(z)}\right\}$$
$$= \operatorname{Re}\left\{1 + z\frac{-\lambda H'(z) + (1-\lambda z)H''(z)}{(1-\lambda z)H'(z)}\right\} \tag{8}$$
$$= \operatorname{Re}\left\{1 + \frac{\lambda z}{\lambda z - 1} + z\frac{H''(z)}{H'(re^{i\theta})}\right\}.$$

Now, by Lemma 8 and inequality (5) for $\theta_1 < \theta_2 < \theta_1 + 2\pi$ and $0 < r < 1$, it suffices to show that

$$\int_{\theta_1}^{\theta_2} \operatorname{Re}\left\{1 + re^{i\theta}\frac{T_\lambda''(re^{i\theta})}{T_\lambda'(re^{i\theta})}\right\} d\theta > -\pi. \tag{9}$$

For $z \in \mathbb{D}$, one may verify (also see Bshouty and Lyzzaik [3] p. 770) that

$$\operatorname{Re}\left\{\frac{z}{z-1}\right\} = \frac{1}{2} - \frac{1}{2}\frac{1-|z|^2}{|1-z|^2}. \tag{10}$$

For $z = re^{i\theta}$, replacing z by λz and letting $\zeta = \overline{\lambda} r$ yield

$$\operatorname{Re}\left\{\frac{\lambda z}{\lambda z - 1}\right\} = \frac{1}{2} - \frac{1}{2}\frac{1-|\zeta|^2}{|e^{i\theta}-\zeta|^2} = \frac{1}{2} - \frac{1}{2}P_\zeta(\theta), \tag{11}$$

where $P_\zeta(\theta)$ is the Poisson Kernal. It then follows that

$$\int_{\theta_1}^{\theta_2} \operatorname{Re}\left\{\frac{\lambda re^{i\theta}}{\lambda re^{i\theta} - 1}\right\} d\theta = \int_{\theta_1}^{\theta_2} \operatorname{Re}\left\{\frac{1}{2} - \frac{1}{2}\cdot P_\zeta(\theta)\right\} d\theta$$
$$= \frac{\theta_2 - \theta_1}{2} - \frac{1}{2}\int_{\theta_1}^{\theta_2} P_\zeta(\theta)\, d\theta \tag{12}$$
$$> \frac{\theta_2 - \theta_1}{2} - \frac{1}{2}\int_0^{2\pi} P_\zeta(\theta)\, d\theta = \frac{\theta_2 - \theta_1}{2} - \pi.$$

On the other hand, since $H(z) = h_1(z) * h_2(z) \in \mathscr{K}(-1/2)$, we obtain

$$\int_{\theta_1}^{\theta_2} \operatorname{Re}\left\{1 + re^{i\theta}\frac{H''(re^{i\theta})}{H'(re^{i\theta})}\right\} d\theta > \frac{\theta_1 - \theta_2}{2}. \tag{13}$$

Therefore, in view of the required condition (9), we get

$$\int_{\theta_1}^{\theta_2} \operatorname{Re}\left\{1 + re^{i\theta}\frac{T_\lambda''(re^{i\theta})}{T_\lambda'(re^{i\theta})}\right\} d\theta$$
$$= \int_{\theta_1}^{\theta_2} \operatorname{Re}\left(1 + \frac{\lambda re^{i\theta}}{\lambda re^{i\theta} - 1}\right) d\theta$$
$$+ \int_{\theta_1}^{\theta_2} \operatorname{Re}\left(re^{i\theta}\frac{H''(re^{i\theta})}{H'(re^{i\theta})}\right) d\theta \tag{14}$$
$$> \frac{\theta_2 - \theta_1}{2} - \pi + \frac{\theta_1 - \theta_2}{2} = -\pi.$$

Proof of Part (ii). In view of Lemma 7, it suffices to show that the convolution function $F(z) = h_1(z) * h_2(z) + \overline{g_1(z) * g_2(z)} = H(z) + \overline{G(z)}$ is locally univalent and sense-preserving in $\mathbb{D}$. In other words, we need to show that

$$\left|\frac{G'(z)}{H'(z)}\right| = \left|\frac{[g_1(z) * g_2(z)]'}{[h_1(z) * h_2(z)]'}\right| < 1. \tag{15}$$

Using the Hadamard product properties of power series, we have

$$1 > |z^n| = \left|\frac{z^n(h_1' * h_2')}{h_1' * h_2'}\right| = \left|\frac{g_1' * g_2'}{h_1' * h_2'}\right| = \left|\frac{[g_1 * zg_2']'}{[h_1 * zh_2']'}\right| = \left|\frac{[z((1/z)g_1 * g_2')]'}{[z((1/z)h_1 * h_2')]'}\right| = \left|\frac{[z(g_1 * g_2)']'}{[z(h_1 * h_2)']'}\right|$$
$$= \left|\frac{(g_1 * g_2)' + z(g_1 * g_2)''}{[z(h_1 * h_2)']'}\right| = \left|\frac{(g_1 * g_2)'(h_1 * h_2)' + z(g_1 * g_2)''(h_1 * h_2)'}{(h_1 * h_2)'[z(h_1 * h_2)']'}\right|$$
$$= \left|\frac{(g_1 * g_2)'[(h_1 * h_2)' + z(h_1 * h_2)''] + z(g_1 * g_2)''(h_1 * h_2)' - z(g_1 * g_2)'(h_1 * h_2)''}{(h_1 * h_2)'[z(h_1 * h_2)']'}\right|$$
$$= \left|\frac{(g_1 * g_2)'[z(h_1 * h_2)']' + z(g_1 * g_2)''(h_1 * h_2)' - z(g_1 * g_2)'(h_1 * h_2)''}{(h_1 * h_2)'[z(h_1 * h_2)']'}\right|$$

$$= \left| \frac{(g_1 * g_2)'}{(h_1 * h_2)'} + \frac{z}{[z(h_1 * h_2)']'} \left(\frac{(g_1 * g_2)''(h_1 * h_2)' - (g_1 * g_2)'(h_1 * h_2)''}{(h_1 * h_2)'} \right) \right|$$

$$= \left| \frac{(g_1 * g_2)'}{(h_1 * h_2)'} + z \left(\frac{(h_1 * h_2)'}{[z(h_1 * h_2)']'} \right) \left(\frac{(g_1 * g_2)''(h_1 * h_2)' - (g_1 * g_2)'(h_1 * h_2)''}{[(h_1 * h_2)']^2} \right) \right|$$

$$= \left| \frac{(g_1 * g_2)'}{(h_1 * h_2)'} + z \frac{(h_1 * h_2)'}{[z(h_1 * h_2)']'} \left(\frac{(g_1 * g_2)'}{(h_1 * h_2)'} \right)' \right|. \quad (16)$$

Therefore,

$$\frac{(g_1 * g_2)'}{(h_1 * h_2)'} + z \frac{(h_1 * h_2)'}{[z(h_1 * h_2)']'} \left(\frac{(g_1 * g_2)'}{(h_1 * h_2)'} \right)' \prec z. \quad (17)$$

On the other hand, since $h_1 * h_2 \in \mathscr{K}(0)$, $z(h_1 * h_2)'$ is starlike or

$$\operatorname{Re} \left\{ \frac{z[z(h_1 * h_2)']'}{z(h_1 * h_2)'} \right\} = \operatorname{Re} \frac{[z(h_1 * h_2)']'}{(h_1 * h_2)'} > 0. \quad (18)$$

Thus, in view of Lemma 9, $(g_1(z) * g_2(z))'/(h_1(z) * h_2(z))' \prec z$ or $|G'(z)/H'(z)| = |[g_1(z) * g_2(z)]'/[h_1(z) * h_2(z)]'| < 1$.

Remark 10. It is left as an open problem whether Theorem 5(i) can be extended to the case $g_1(z) = z^n h_1(z)$ and $g_2'(z) = z^n h_2'(z)$ if $n > 1$.

Conflicts of Interest

The authors declare that there are no conflicts of interest regarding the publication of this paper.

References

[1] J. Clunie and T. Sheil-Small, "Harmonic univalent functions," *Annales Academiae Scientiarum Fennicae. Series A. I. Mathematica*, vol. 9, pp. 3–25, 1984.

[2] H. Lewy, "On the non-vanishing of the Jacobian in certain one-to-one mappings," *Bulletin (New Series) of the American Mathematical Society*, vol. 42, no. 10, pp. 689–692, 1936.

[3] D. Bshouty and A. Lyzzaik, "Close-to-convexity criteria for planar harmonic mappings," *Complex Analysis and Operator Theory*, vol. 5, no. 3, pp. 767–774, 2011.

[4] S. Ruscheweyh and T. Sheil-Small, "Hadamard products of SCHlicht functions and the Polya-SCHoenberg conjecture," *Commentarii Mathematici Helvetici*, vol. 48, pp. 119–135, 1973.

[5] O. P. Ahuja, J. M. Jahangiri, and H. Silverman, "Convolutions for special classes of harmonic univalent functions," *Applied Mathematics Letters*, vol. 16, no. 6, pp. 905–909, 2003.

[6] D. Bshouty, S. S. Joshi, and S. B. Joshi, "On close-to-convex harmonic mappings," *Complex Variables and Elliptic Equations*, vol. 58, no. 9, pp. 1195–1199, 2013.

[7] W. Kaplan, "Close-to-convex schlicht functions," *Michigan Mathematical Journal*, vol. 1, pp. 169–185, 1952.

[8] S. S. Miller and P. T. Mocanu, "Univalent solutions of Briot-Bouquet differential equations," *Journal of Differential Equations*, vol. 56, no. 3, pp. 297–309, 1985.

[9] S. S. Miller and P. T. Mocanu, *Differential Subordination—Theory and Applications*, vol. 225 of *Monorgraphs and Textbooks in Pure and Applied Mathematics*, Marcel Dekker, New York, NY, USA, 2000.

5

On the Differentiability of Weak Solutions of an Abstract Evolution Equation with a Scalar Type Spectral Operator on the Real Axis

Marat V. Markin

Department of Mathematics, California State University, Fresno, 5245 N. Backer Avenue, M/S PB 108, Fresno, CA 93740-8001, USA

Correspondence should be addressed to Marat V. Markin; mmarkin@csufresno.edu

Academic Editor: Seppo Hassi

Given the abstract evolution equation $y'(t) = Ay(t)$, $t \in \mathbb{R}$, with *scalar type spectral operator* A in a complex Banach space, found are conditions *necessary and sufficient* for all *weak solutions* of the equation, which a priori need not be strongly differentiable, to be strongly infinite differentiable on $\mathbb{R}$. The important case of the equation with a *normal operator* A in a complex Hilbert space is obtained immediately as a particular case. Also, proved is the following inherent smoothness improvement effect explaining why the case of the strong finite differentiability of the weak solutions is superfluous: if every weak solution of the equation is strongly differentiable at 0, then all of them are strongly infinite differentiable on $\mathbb{R}$.

"Curiosity is the lust of the mind."
Thomas Hobbes

1. Introduction

We find conditions on a *scalar type spectral* operator A in a complex Banach space necessary and sufficient for all *weak solutions* of the evolution equation

$$y'(t) = Ay(t), \quad t \in \mathbb{R}, \tag{1}$$

which a priori need not be strongly differentiable, to be strongly infinite differentiable on $\mathbb{R}$. The important case of the equation with a *normal operatorA* in a complex Hilbert space is obtained immediately as a particular case. We also prove the following inherent smoothness improvement effect explaining why the case of the strong finite differentiability of the weak solutions is superfluous: if every weak solution of the equation is strongly differentiable at 0, then all of them are strongly infinite differentiable on $\mathbb{R}$.

The found results develop those of paper [1], where similar consideration is given to the strong differentiability of the weak solutions of the equation

$$y'(t) = Ay(t), \quad t \geq 0, \tag{2}$$

on $[0, \infty)$ and $(0, \infty)$.

Definition 1 (weak solution). Let A be a densely defined closed linear operator in a Banach space X and I be an interval of the real axis $\mathbb{R}$. A strongly continuous vector function $y : I \longrightarrow X$ is called a *weak solution* of the evolution equation

$$y'(t) = Ay(t), \quad t \in I, \tag{3}$$

if, for any $g^* \in D(A^*)$,

$$\frac{d}{dt}\langle y(t), g^*\rangle = \langle y(t), A^* g^*\rangle, \quad t \in I, \tag{4}$$

where $D(\cdot)$ is the *domain* of an operator, A^* is the operator *adjoint* to A, and $\langle \cdot, \cdot \rangle$ is the *pairing* between the space X and its dual X^* (cf. [2]).

Remarks 2.

(i) Due to the *closedness* of A, a weak solution of (3) can be equivalently defined to be a strongly continuous vector function $y : I \longmapsto X$ such that, for all $t \in I$,

$$\int_{t_0}^{t} y(s)\,ds \in D(A) \quad \text{and} \qquad y(t) = y(t_0) + A\int_{t_0}^{t} y(s)\,ds, \tag{5}$$

where t_0 is an arbitrary fixed point of the interval I, and is also called a *mild solution* (cf. [3, Ch. II, Definition 6.3], see also [4, Preliminaries]).

(ii) Such a notion of *weak solution*, which need not be differentiable in the strong sense, generalizes that of *classical* one, strongly differentiable on I and satisfying the equation in the traditional plug-in sense, the classical solutions being precisely the weak ones strongly differentiable on I.

(iii) As is easily seen $y : \mathbb{R} \longrightarrow X$ is a weak solution of (1) *iff*

$$y_+(t) := y(t), \quad t \geq 0, \tag{6}$$

is a weak solution of (2) and

$$y_-(t) := y(-t), \quad t \geq 0, \tag{7}$$

is a weak solution of the equation

$$y'(t) = -Ay(t), \quad t \geq 0. \tag{8}$$

(iv) When a closed densely defined linear operator A in a complex Banach space X generates a strongly continuous group $\{T(t)\}_{t\in\mathbb{R}}$ of bounded linear operators (see, e.g., [3, 5]), i.e., the associated *abstract Cauchy problem* (*ACP*)

$$y'(t) = Ay(t), \quad t \in \mathbb{R}, \qquad y(0) = f \tag{9}$$

is *well-posed* (cf. [3, Ch. II, Definition 6.8]), the weak solutions of (1) are the orbits

$$y(t) = T(t)f, \quad t \in \mathbb{R}, \tag{10}$$

with $f \in X$ (cf. [3, Ch. II, Proposition 6.4], see also [2, Theorem]), whereas the classical ones are those with $f \in D(A)$ (see, e.g., [3, Ch. II, Proposition 6.3]).

(v) In our discourse, the associated *ACP* may be *ill-posed*, i.e., the scalar type spectral operator A need not generate a strongly continuous group of bounded linear operators (cf. [6]).

2. Preliminaries

Here, for the reader's convenience, we outline certain essential preliminaries.

Henceforth, unless specified otherwise, A is supposed to be a *scalar type spectral operator* in a complex Banach space $(X, \|\cdot\|)$ with strongly σ-additive *spectral measure* (the *resolution of the identity*) $E_A(\cdot)$ assigning to each Borel set δ of the complex plane $\mathbb{C}$ a projection operator $E_A(\delta)$ on X and having the operator's *spectrum* $\sigma(A)$ as its *support* [7, 8].

Observe that, in a complex finite-dimensional space, the scalar type spectral operators are all linear operators on the space, for which there is an *eigenbasis* (see, e.g., [7, 8]) and, in a complex Hilbert space, the scalar type spectral operators are precisely all those that are similar to the *normal* ones [9].

Associated with a scalar type spectral operator in a complex Banach space is the *Borel operational calculus* analogous to that for a *normal operator* in a complex Hilbert space [7, 8, 10, 11], which assigns to any Borel measurable function $F : \sigma(A) \longrightarrow \mathbb{C}$ a scalar type spectral operator

$$F(A) := \int_{\sigma(A)} F(\lambda)\,dE_A(\lambda) \tag{11}$$

(see [7, 8]).

In particular,

$$A^n = \int_{\sigma(A)} \lambda^n dE_A(\lambda), \quad n \in \mathbb{Z}_+, \tag{12}$$

($\mathbb{Z}_+ := \{0, 1, 2, \ldots\}$ is the set of *nonnegative integers*, $A^0 := I$, I is the *identity operator* on X), and

$$e^{zA} := \int_{\sigma(A)} e^{z\lambda} dE_A(\lambda), \quad z \in \mathbb{C}. \tag{13}$$

The properties of the *spectral measure* and *operational calculus*, exhaustively delineated in [7, 8], underlie the entire subsequent discourse. Here, we underline a few facts of particular importance.

Due to its *strong countable additivity*, the spectral measure $E_A(\cdot)$ is *bounded* [8, 12], i.e., there is such an $M > 0$ that, for any Borel set $\delta \subseteq \mathbb{C}$,

$$\|E_A(\delta)\| \leq M. \tag{14}$$

Observe that the notation $\|\cdot\|$ is used here to designate the norm in the space $L(X)$ of all bounded linear operators on X. We adhere to this rather conventional economy of symbols in what follows also adopting the same notation for the norm in the dual space X^*.

For any $f \in X$ and $g^* \in X^*$, the *total variation measure* $v(f, g^*, \cdot)$ of the complex-valued Borel measure $\langle E_A(\cdot)f, g^*\rangle$ is a *finite* positive Borel measure with

$$v(f, g^*, \mathbb{C}) = v(f, g^*, \sigma(A)) \leq 4M\|f\|\|g^*\| \tag{15}$$

(see, e.g., [13, 14]).

Also (Ibid.), for a Borel measurable function $F : \mathbb{C} \longrightarrow \mathbb{C}$, $f \in D(F(A))$, $g^* \in X^*$, and a Borel set $\delta \subseteq \mathbb{C}$,

$$\int_\delta |F(\lambda)|\,dv(f, g^*, \lambda) \leq 4M\|E_A(\delta)F(A)f\|\|g^*\|. \tag{16}$$

In particular, for $\delta = \sigma(A)$, $E_A(\sigma(A)) = I$ and

$$\int_{\sigma(A)} |F(\lambda)| \, dv(f, g^*, \lambda) \le 4M \|F(A) f\| \|g^*\|. \tag{17}$$

Observe that the constant $M > 0$ in (15)–(17) is from (14).

Further, for a Borel measurable function $F : \mathbb{C} \longrightarrow [0,\infty)$, a Borel set $\delta \subseteq \mathbb{C}$, a sequence $\{\Delta_n\}_{n=1}^{\infty}$ of pairwise disjoint Borel sets in $\mathbb{C}$, and $f \in X$, $g^* \in X^*$,

$$\begin{aligned} &\int_{\delta} F(\lambda) \, dv\left(E_A\left(\bigcup_{n=1}^{\infty} \Delta_n\right) f, g^*, \lambda\right) \\ &\quad = \sum_{n=1}^{\infty} \int_{\delta\cap\Delta_n} F(\lambda) \, dv(E_A(\Delta_n) f, g^*, \lambda). \end{aligned} \tag{18}$$

Indeed, since, for any Borel sets $\delta, \sigma \subseteq \mathbb{C}$,

$$E_A(\delta) E_A(\sigma) = E_A(\delta \cap \sigma) \tag{19}$$

[7, 8], for the total variation measure,

$$v(E_A(\delta) f, g^*, \sigma) = v(f, g^*, \delta \cap \sigma). \tag{20}$$

Whence, due to the *nonnegativity* of $F(\cdot)$ (see, e.g., [15]),

$$\begin{aligned} &\int_{\delta} F(\lambda) \, dv\left(E_A\left(\bigcup_{n=1}^{\infty} \Delta_n\right) f, g^*, \lambda\right) \\ &\quad = \int_{\delta\cap\bigcup_{n=1}^{\infty}\Delta_n} F(\lambda) \, dv(f, g^*, \lambda) \\ &\quad = \sum_{n=1}^{\infty} \int_{\delta\cap\Delta_n} F(\lambda) \, dv(f, g^*, \lambda) \\ &\quad = \sum_{n=1}^{\infty} \int_{\delta\cap\Delta_n} F(\lambda) \, dv(E_A(\Delta_n) f, g^*, \lambda). \end{aligned} \tag{21}$$

The following statement, allowing characterizing the domains of Borel measurable functions of a scalar type spectral operator in terms of positive Borel measures, is fundamental for our discourse.

Proposition 3 ([16, Proposition 3.1]). *Let A be a scalar type spectral operator in a complex Banach space $(X, \|\cdot\|)$ with spectral measure $E_A(\cdot)$ and $F : \sigma(A) \longrightarrow \mathbb{C}$ be a Borel measurable function. Then $f \in D(F(A))$ iff*

(i) *for each $g^* \in X^*$, $\int_{\sigma(A)} |F(\lambda)| dv(f, g^*, \lambda) < \infty$;*

(ii) $\sup_{\{g^*\in X^* \mid \|g^*\|=1\}} \int_{\{\lambda\in\sigma(A) \mid |F(\lambda)|>n\}} |F(\lambda)| dv(f, g^*, \lambda) \longrightarrow 0$, $n \longrightarrow \infty$,

where $v(f, g^*, \cdot)$ is the total variation measure of $\langle E_A(\cdot) f, g^*\rangle$.

The succeeding key theorem provides a description of the weak solutions of (2) with a scalar type spectral operator A in a complex Banach space.

Theorem 4 ([16, Theorem 4.2] with $T = \infty$). *Let A be a scalar type spectral operator in a complex Banach space $(X, \|\cdot\|)$. A vector function $y : [0,\infty) \longrightarrow X$ is a weak solution of (2) iff there is an $f \in \bigcap_{t\ge 0} D(e^{tA})$ such that*

$$y(t) = e^{tA} f, \quad t \ge 0, \tag{22}$$

the operator exponentials understood in the sense of the Borel operational calculus (see (13)).

Remark 5. Theorem 4 generalizes [17, Theorem 3.1], its counterpart for a normal operator A in a complex Hilbert space.

We also need the following characterizations of a particular weak solution's of (2) with a scalar type spectral operator A in a complex Banach space being strongly differentiable on a subinterval I of $[0,\infty)$.

Proposition 6 ([1, Proposition 3.1] with $T = \infty$). *Let $n \in \mathbb{N}$ and I be a subinterval of $[0,\infty)$. A weak solution $y(\cdot)$ of (2) is n times strongly differentiable on I iff*

$$y(t) \in D(A^n), \quad t \in I, \tag{23}$$

in which case

$$y^{(k)}(t) = A^k y(t), \quad k = 1, \ldots, n, \; t \in I. \tag{24}$$

Subsequently, the frequent terms *"spectral measure"* and *"operational calculus"* are abbreviated to *s.m.* and *o.c.*, respectively.

3. General Weak Solution

Theorem 7 (general weak solution). *Let A be a scalar type spectral operator in a complex Banach space $(X, \|\cdot\|)$. A vector function $y : \mathbb{R} \longrightarrow X$ is a weak solution of (1) iff there is an $f \in \bigcap_{t\in\mathbb{R}} D(e^{tA})$ such that*

$$y(t) = e^{tA} f, \quad t \in \mathbb{R}, \tag{25}$$

the operator exponentials understood in the sense of the Borel operational calculus (see (13)).

Proof. As is noted in the Introduction, $y : \mathbb{R} \longrightarrow X$ is a weak solution of (1) *iff*

$$y_+(t) := y(t), \quad t \ge 0, \tag{26}$$

is a weak solution of (2) and

$$y_-(t) := y(-t), \quad t \ge 0, \tag{27}$$

is a weak solution of (8).

Applying Theorem 4, to $y_+(\cdot)$ and $y_-(\cdot)$, we infer that this is equivalent to the fact

$$y(t) = e^{tA} f, \quad t \in \mathbb{R}, \text{ with some } f \in \bigcap_{t\in\mathbb{R}} D\left(e^{tA}\right). \tag{28}$$

Remarks 8.

(i) More generally, Theorem 4 and its proof can be easily modified to describe in the same manner all weak solution of (3) for an arbitrary interval I of the real axis $\mathbb{R}$.

(ii) Theorem 7 implies, in particular,

(a) that the subspace $\bigcap_{t\in\mathbb{R}} D(e^{tA})$ of all possible initial values of the weak solutions of (1) is the largest permissible for the exponential form given by (25), which highlights the naturalness of the notion of weak solution;

(b) that associated *ACP* (9), whenever solvable, is solvable *uniquely*.

(iii) Observe that the initial-value subspace $\bigcap_{t\in\mathbb{R}} D(e^{tA})$ of (1), containing the dense in X subspace $\bigcup_{\alpha>0} E_A(\Delta_\alpha)X$, where

$$\Delta_\alpha := \{\lambda \in \mathbb{C} \mid |\lambda| \le \alpha\}, \quad \alpha > 0, \tag{29}$$

which coincides with the class $\mathscr{E}^{\{0\}}(A)$ of *entire* vectors of A of *exponential type* [18], is *dense* in X as well.

(iv) When a scalar type spectral operator A in a complex Banach space generates a strongly continuous group $\{T(t)\}_{t\in\mathbb{R}}$ of bounded linear operators,

$$T(t) = e^{tA} \text{ and } D\left(e^{tA}\right) = X, \quad t \in \mathbb{R}, \tag{30}$$

[6], and hence, Theorem 7 is consistent with the well-known description of the weak solutions for this setup (see (10)).

(v) Clearly, the initial-value subspace $\bigcap_{t\in\mathbb{R}} D(e^{tA})$ of (1) is narrower than the initial-value subspace $\bigcap_{t\ge 0} D(e^{tA})$ of (2) and the initial-value subspace $\bigcap_{t\ge 0} D(e^{t(-A)}) = \bigcap_{t\le 0} D(e^{tA})$ of (8); in fact it is the intersection of the latter two.

4. Differentiability of a Particular Weak Solution

Here, we characterize a particular weak solution's of (1) with a scalar type spectral operator A in a complex Banach space being strongly differentiable on a subinterval I of $\mathbb{R}$.

Proposition 9 (differentiability of a particular weak solution). *Let $n \in \mathbb{N}$ and I be a subinterval of $\mathbb{R}$. A weak solution $y(\cdot)$ of (1) is n times strongly differentiable on I iff*

$$y(t) \in D(A^n), \quad t \in I, \tag{31}$$

in which case,

$$y^{(k)}(t) = A^k y(t), \quad k = 1, \dots, n, \; t \in I. \tag{32}$$

Proof. The statement immediately follows from the prior theorem and Proposition 6 applied to

$$y_+(t) := y(t) \text{ and } y_-(t) := y(-t), \quad t \ge 0, \tag{33}$$

for an arbitrary weak solution $y(\cdot)$ of (1).

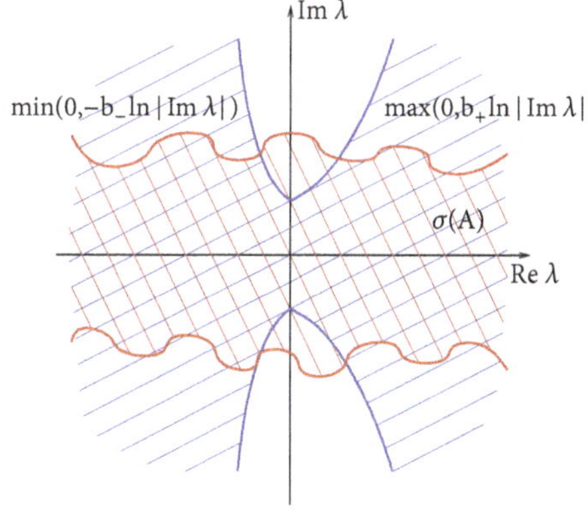

FIGURE 1

Remark 10. Observe that, as well as for Proposition 6, for $n = 1$, the subinterval I can degenerate into a singleton.

Inductively, we immediately obtain the following analog of [1, Corollary 3.2].

Corollary 11 (infinite differentiability of a particular weak solution). *Let A be a scalar type spectral operator in a complex Banach space $(X, \|\cdot\|)$ and I be a subinterval of $\mathbb{R}$. A weak solution $y(\cdot)$ of (1) is strongly infinite differentiable on I ($y(\cdot) \in C^\infty(I, X)$) iff, for each $t \in I$,*

$$y(t) \in C^\infty(A) := \bigcap_{n=1}^{\infty} D(A^n), \tag{34}$$

in which case

$$y^{(n)}(t) = A^n y(t), \quad n \in \mathbb{N}, \; t \in I. \tag{35}$$

5. Infinite Differentiability of Weak Solutions

In this section, we characterize the strong infinite differentiability on $\mathbb{R}$ of all weak solutions of (1) with a scalar type spectral operator A in a complex Banach space.

Theorem 12 (infinite differentiability of weak solutions). *Let A be a scalar type spectral operator in a complex Banach space $(X, \|\cdot\|)$ with spectral measure $E_A(\cdot)$. Every weak solution of (1) is strongly infinite differentiable on $\mathbb{R}$ iff there exist $b_+ > 0$ and $b_- > 0$ such that the set $\sigma(A)\backslash\mathscr{L}_{b_-,b_+}$, where*

$$\mathscr{L}_{b_-,b_+} := \{\lambda \in \mathbb{C} \mid \operatorname{Re}\lambda \le \min(0, -b_- \ln|\operatorname{Im}\lambda|) \text{ or } \operatorname{Re}\lambda \ge \max(0, b_+ \ln|\operatorname{Im}\lambda|)\}, \tag{36}$$

is bounded (see Figure 1).

Proof. *"If" part*: suppose that there exist $b_+ > 0$ and $b_- > 0$ such that the set $\sigma(A)\backslash\mathscr{L}_{b_-,b_+}$ is *bounded* and let $y(\cdot)$ be an arbitrary weak solution of (1).

By Theorem 7,

$$y(t) = e^{tA}f, \quad t \in \mathbb{R}, \text{ with some } f \in \bigcap_{t\in\mathbb{R}} D\left(e^{tA}\right). \tag{37}$$

Our purpose is to show that $y(\cdot) \in C^\infty(\mathbb{R}, X)$, which, by Corollary 11, is attained by showing that, for each $t \in \mathbb{R}$,

$$y(t) \in C^\infty(A) := \bigcap_{n=1}^{\infty} D(A^n). \tag{38}$$

Let us proceed by proving that, for any $t \in \mathbb{R}$ and $m \in \mathbb{N}$

$$y(t) \in D(A^m) \tag{39}$$

via Proposition 3.

For any $t \in \mathbb{R}$, $m \in \mathbb{N}$, and an arbitrary $g^* \in X^*$,

$$\begin{aligned}
&\int_{\sigma(A)} |\lambda|^m e^{t\operatorname{Re}\lambda}\, dv(f, g^*, \lambda) \\
&= \int_{\sigma(A)\backslash\mathscr{L}_{b_-,b_+}} |\lambda|^m e^{t\operatorname{Re}\lambda}\, dv(f, g^*, \lambda) \\
&+ \int_{\{\lambda\in\sigma(A)\cap\mathscr{L}_{b_-,b_+} \mid -1<\operatorname{Re}\lambda<1\}} |\lambda|^m e^{t\operatorname{Re}\lambda}\, dv(f, g^*, \lambda) \\
&+ \int_{\{\lambda\in\sigma(A)\cap\mathscr{L}_{b_-,b_+} \mid \operatorname{Re}\lambda\geq 1\}} |\lambda|^m e^{t\operatorname{Re}\lambda}\, dv(f, g^*, \lambda) \\
&+ \int_{\{\lambda\in\sigma(A)\cap\mathscr{L}_{b_-,b_+} \mid \operatorname{Re}\lambda\leq -1\}} |\lambda|^m e^{t\operatorname{Re}\lambda}\, dv(f, g^*, \lambda) \\
&< \infty.
\end{aligned} \tag{40}$$

Indeed,

$$\int_{\sigma(A)\backslash\mathscr{L}_{b_-,b_+}} |\lambda|^m e^{t\operatorname{Re}\lambda}\, dv(f, g^*, \lambda) < \infty \tag{41}$$

and

$$\int_{\{\lambda\in\sigma(A)\cap\mathscr{L}_{b_-,b_+} \mid -1<\operatorname{Re}\lambda<1\}} |\lambda|^m e^{t\operatorname{Re}\lambda}\, dv(f, g^*, \lambda) < \infty \tag{42}$$

due to the boundedness of the sets

$$\sigma(A)\backslash\mathscr{L}_{b_-,b_+} \text{ and } \left\{\lambda \in \sigma(A)\cap\mathscr{L}_{b_-,b_+} \mid -1 < \operatorname{Re}\lambda < 1\right\}, \tag{43}$$

the continuity of the integrated function on $\mathbb{C}$, and the finiteness of the measure $v(f, g^*, \cdot)$.

Further, for any $t \in \mathbb{R}$, $m \in \mathbb{N}$, and an arbitrary $g^* \in X^*$,

$$\begin{aligned}
&\int_{\{\lambda\in\sigma(A)\cap\mathscr{L}_{b_-,b_+} \mid \operatorname{Re}\lambda\geq 1\}} |\lambda|^m e^{t\operatorname{Re}\lambda}\, dv(f, g^*, \lambda) \\
&\leq \int_{\{\lambda\in\sigma(A)\cap\mathscr{L}_{b_-,b_+} \mid \operatorname{Re}\lambda\geq 1\}} [|\operatorname{Re}\lambda| + |\operatorname{Im}\lambda|]^m e^{t\operatorname{Re}\lambda}\, dv(f, g^*, \lambda) \\
&\qquad \text{since, for } \lambda \in \sigma(A)\cap\mathscr{L}_{b_-,b_+} \text{ with } \operatorname{Re}\lambda \geq 1,\ e^{b_+^{-1}\operatorname{Re}\lambda} \geq |\operatorname{Im}\lambda|; \\
&\leq \int_{\{\lambda\in\sigma(A)\cap\mathscr{L}_{b_-,b_+} \mid \operatorname{Re}\lambda\geq 1\}} \left[\operatorname{Re}\lambda + e^{b_+^{-1}\operatorname{Re}\lambda}\right]^m e^{t\operatorname{Re}\lambda}\, dv(f, g^*, \lambda) \\
&\qquad \text{since, in view of } \operatorname{Re}\lambda \geq 1,\ b_+ e^{b_+^{-1}\operatorname{Re}\lambda} \geq \operatorname{Re}\lambda; \\
&\leq \int_{\{\lambda\in\sigma(A)\cap\mathscr{L}_{b_-,b_+} \mid \operatorname{Re}\lambda\geq 1\}} \left[b_+ e^{b_+^{-1}\operatorname{Re}\lambda} + e^{b_+^{-1}\operatorname{Re}\lambda}\right]^m \cdot e^{t\operatorname{Re}\lambda}\, dv(f, g^*, \lambda) \\
&= [b_+ + 1]^m \int_{\{\lambda\in\sigma(A)\cap\mathscr{L}_{b_-,b_+} \mid \operatorname{Re}\lambda\geq 1\}} e^{[mb_+^{-1}+t]\operatorname{Re}\lambda}\, dv(f, g^*, \lambda) \\
&\qquad \text{since } f \in \bigcap_{t\in\mathbb{R}} D\left(e^{tA}\right), \text{ by Proposition 3;} \\
&< \infty.
\end{aligned} \tag{44}$$

Finally, for any $t \in \mathbb{R}$, $m \in \mathbb{N}$ and an arbitrary $g^* \in X^*$,

$$\begin{aligned}
&\int_{\{\lambda\in\sigma(A)\cap\mathscr{L}_{b_-,b_+} \mid \operatorname{Re}\lambda\leq -1\}} |\lambda|^m e^{t\operatorname{Re}\lambda}\, dv(f, g^*, \lambda) \\
&\leq \int_{\{\lambda\in\sigma(A)\cap\mathscr{L}_{b_-,b_+} \mid \operatorname{Re}\lambda\leq -1\}} [|\operatorname{Re}\lambda| + |\operatorname{Im}\lambda|]^m e^{t\operatorname{Re}\lambda}\, dv(f, g^*, \lambda) \\
&\qquad \text{since, for } \lambda \in \sigma(A)\cap\mathscr{L}_{b_-,b_+} \text{ with } \operatorname{Re}\lambda \leq -1,\ e^{b_-^{-1}(-\operatorname{Re}\lambda)} \geq |\operatorname{Im}\lambda|; \\
&\leq \int_{\{\lambda\in\sigma(A)\cap\mathscr{L}_{b_-,b_+} \mid \operatorname{Re}\lambda\leq -1\}} \left[-\operatorname{Re}\lambda + e^{b_-^{-1}(-\operatorname{Re}\lambda)}\right]^m e^{t\operatorname{Re}\lambda}\, dv(f, g^*, \lambda) \quad \text{since, in view of } -\operatorname{Re}\lambda \geq 1,\ b_- e^{b_-^{-1}(-\operatorname{Re}\lambda)} \geq -\operatorname{Re}\lambda; \\
&\leq \int_{\{\lambda\in\sigma(A)\cap\mathscr{L}_{b_-,b_+} \mid \operatorname{Re}\lambda\leq -1\}} \left[b_- e^{b_-^{-1}(-\operatorname{Re}\lambda)} + e^{b_-^{-1}(-\operatorname{Re}\lambda)}\right]^m e^{t\operatorname{Re}\lambda}\, dv(f, g^*, \lambda) \\
&= [b_- + 1]^m \int_{\{\lambda\in\sigma(A)\cap\mathscr{L}_{b_-,b_+} \mid \operatorname{Re}\lambda\leq -1\}} e^{[t-mb_-^{-1}]\operatorname{Re}\lambda}\, dv(f, g^*, \lambda) \quad \text{since } f \in \bigcap_{t\in\mathbb{R}} D\left(e^{tA}\right), \text{ by Proposition 3;} \\
&< \infty.
\end{aligned} \tag{45}$$

Also, for any $t \in \mathbb{R}$, $m \in \mathbb{N}$, and an arbitrary $n \in \mathbb{N}$,

$$\begin{aligned} &\sup_{\{g^* \in X^* \mid \|g^*\|=1\}} \int_{\{\lambda \in \sigma(A) \mid |\lambda|^m e^{t \operatorname{Re} \lambda} > n\}} |\lambda|^m e^{t \operatorname{Re} \lambda} \, dv(f, g^*, \lambda) \\ &\le \sup_{\{g^* \in X^* \mid \|g^*\|=1\}} \int_{\{\lambda \in \sigma(A) \setminus \mathscr{L}_{b_-,b_+} \mid |\lambda|^m e^{t \operatorname{Re} \lambda} > n\}} |\lambda|^m \\ &\quad \cdot e^{t \operatorname{Re} \lambda} \, dv(f, g^*, \lambda) \\ &+ \sup_{\{g^* \in X^* \mid \|g^*\|=1\}} \int_{\{\lambda \in \sigma(A) \cap \mathscr{L}_{b_-,b_+} \mid -1 < \operatorname{Re} \lambda < 1, |\lambda|^m e^{t \operatorname{Re} \lambda} > n\}} |\lambda|^m \\ &\quad \cdot e^{t \operatorname{Re} \lambda} \, dv(f, g^*, \lambda) \\ &+ \sup_{\{g^* \in X^* \mid \|g^*\|=1\}} \int_{\{\lambda \in \sigma(A) \cap \mathscr{L}_{b_-,b_+} \mid \operatorname{Re} \lambda \ge 1, |\lambda|^m e^{t \operatorname{Re} \lambda} > n\}} |\lambda|^m \\ &\quad \cdot e^{t \operatorname{Re} \lambda} \, dv(f, g^*, \lambda) \\ &+ \sup_{\{g^* \in X^* \mid \|g^*\|=1\}} \int_{\{\lambda \in \sigma(A) \cap \mathscr{L}_{b_-,b_+} \mid \operatorname{Re} \lambda \le -1, |\lambda|^m e^{t \operatorname{Re} \lambda} > n\}} |\lambda|^m \\ &\quad \cdot e^{t \operatorname{Re} \lambda} \, dv(f, g^*, \lambda) \\ &\longrightarrow 0, \quad n \longrightarrow \infty. \end{aligned} \tag{46}$$

Indeed, since, due to the boundedness of the sets

$$\sigma(A) \setminus \mathscr{L}_{b_-,b_+} \text{ and } \{\lambda \in \sigma(A) \cap \mathscr{L}_{b_-,b_+} \mid -1 < \operatorname{Re} \lambda < 1\} \tag{47}$$

and the continuity of the integrated function on $\mathbb{C}$, the sets

$$\{\lambda \in \sigma(A) \setminus \mathscr{L}_{b_-,b_+} \mid |\lambda|^m e^{t \operatorname{Re} \lambda} > n\} \tag{48}$$

and

$$\{\lambda \in \sigma(A) \cap \mathscr{L}_{b_-,b_+} \mid -1 < \operatorname{Re} \lambda < 1, \ |\lambda|^m e^{t \operatorname{Re} \lambda} > n\} \tag{49}$$

are *empty* for all sufficiently large $n \in \mathbb{N}$, we immediately infer that, for any $t \in \mathbb{R}$ and $m \in \mathbb{N}$,

$$\begin{aligned} \lim_{n \to \infty} &\sup_{\{g^* \in X^* \mid \|g^*\|=1\}} \int_{\{\lambda \in \sigma(A) \setminus \mathscr{L}_{b_-,b_+} \mid |\lambda|^m e^{t \operatorname{Re} \lambda} > n\}} |\lambda|^m \\ &\cdot e^{t \operatorname{Re} \lambda} \, dv(f, g^*, \lambda) = 0 \end{aligned} \tag{50}$$

and

$$\begin{aligned} \lim_{n \to \infty} &\sup_{\{g^* \in X^* \mid \|g^*\|=1\}} \int_{\{\lambda \in \sigma(A) \cap \mathscr{L}_{b_-,b_+} \mid -1 < \operatorname{Re} \lambda < 1, |\lambda|^m e^{t \operatorname{Re} \lambda} > n\}} |\lambda|^m \\ &\cdot e^{t \operatorname{Re} \lambda} \, dv(f, g^*, \lambda) = 0. \end{aligned} \tag{51}$$

Further, for any $t \in \mathbb{R}$, $m \in \mathbb{N}$, and an arbitrary $n \in \mathbb{N}$,

$$\begin{aligned} &\sup_{\{g^* \in X^* \mid \|g^*\|=1\}} \int_{\{\lambda \in \sigma(A) \cap \mathscr{L}_{b_-,b_+} \mid \operatorname{Re} \lambda \ge 1, |\lambda|^m e^{t \operatorname{Re} \lambda} > n\}} |\lambda|^m \\ &\quad \cdot e^{t \operatorname{Re} \lambda} \, dv(f, g^*, \lambda) \quad \text{as in (44);} \\ &\le \sup_{\{g^* \in X^* \mid \|g^*\|=1\}} [b_+ + 1]^m \\ &\quad \cdot \int_{\{\lambda \in \sigma(A) \cap \mathscr{L}_{b_-,b_+} \mid \operatorname{Re} \lambda \ge 1, |\lambda|^m e^{t \operatorname{Re} \lambda} > n\}} e^{[m b_+^{-1} + t] \operatorname{Re} \lambda} \, dv(f, g^*, \lambda) \\ &\qquad \text{since } f \in \bigcap_{t \in \mathbb{R}} D(e^{tA}), \text{ by (16);} \\ &\le \sup_{\{g^* \in X^* \mid \|g^*\|=1\}} [b_+ + 1]^m \\ &\quad \cdot 4M \left\| E_A(\{\lambda \in \sigma(A) \cap \mathscr{L}_{b_-,b_+} \mid \operatorname{Re} \lambda \ge 1, \ |\lambda|^m e^{t \operatorname{Re} \lambda} > n\}) \right. \\ &\quad \left. \cdot e^{[m b_+^{-1} + t] A} f \right\| \|g^*\| \\ &\le [b_+ + 1]^m \\ &\quad \cdot 4M \left\| E_A(\{\lambda \in \sigma(A) \cap \mathscr{L}_{b_-,b_+} \mid \operatorname{Re} \lambda \ge 1, \ |\lambda|^m e^{t \operatorname{Re} \lambda} > n\}) \right. \\ &\quad \left. \cdot e^{[m b_+^{-1} + t] A} f \right\| \quad \text{by the strong continuity of the } \textit{s.m.}; \\ &\longrightarrow [b_+ + 1]^m 4M \left\| E_A(\emptyset) e^{[m b_+^{-1} + t] A} f \right\| = 0, \quad n \longrightarrow \infty. \end{aligned} \tag{52}$$

Finally, for any $t \in \mathbb{R}$, $m \in \mathbb{N}$, and an arbitrary $n \in \mathbb{N}$,

$$\begin{aligned} &\sup_{\{g^* \in X^* \mid \|g^*\|=1\}} \int_{\{\lambda \in \sigma(A) \cap \mathscr{L}_{b_-,b_+} \mid \operatorname{Re} \lambda \le -1, |\lambda|^m e^{t \operatorname{Re} \lambda} > n\}} |\lambda|^m \\ &\quad \cdot e^{t \operatorname{Re} \lambda} \, dv(f, g^*, \lambda) \quad \text{as in (45);} \\ &\le \sup_{\{g^* \in X^* \mid \|g^*\|=1\}} [b_- + 1]^m \\ &\quad \cdot \int_{\{\lambda \in \sigma(A) \cap \mathscr{L}_{b_-,b_+} \mid \operatorname{Re} \lambda \le -1, |\lambda|^m e^{t \operatorname{Re} \lambda} > n\}} e^{[t - m b_-^{-1}] \operatorname{Re} \lambda} \, dv(f, g^*, \lambda) \\ &\qquad \text{since } f \in \bigcap_{t \in \mathbb{R}} D(e^{tA}), \text{ by (16);} \\ &\le \sup_{\{g^* \in X^* \mid \|g^*\|=1\}} [b_- + 1]^m \\ &\quad \cdot 4M \left\| E_A(\{\lambda \in \sigma(A) \cap \mathscr{L}_{b_-,b_+} \mid \operatorname{Re} \lambda \le -1, \ |\lambda|^m e^{t \operatorname{Re} \lambda} > n\}) \right. \\ &\quad \left. \cdot e^{[t - m b_-^{-1}] A} f \right\| \|g^*\| \\ &\le [b_- + 1]^m \\ &\quad \cdot 4M \left\| E_A(\{\lambda \in \sigma(A) \cap \mathscr{L}_{b_-,b_+} \mid \operatorname{Re} \lambda \le -1, \ |\lambda|^m e^{t \operatorname{Re} \lambda} > n\}) \right. \\ &\quad \left. \cdot e^{[t - m b_-^{-1}] A} f \right\| \quad \text{by the strong continuity of the } \textit{s.m.}; \\ &\longrightarrow [b_+ + 1]^m 4M \left\| E_A(\emptyset) e^{[t - m b_-^{-1}] A} f \right\| = 0, \quad n \longrightarrow \infty. \end{aligned} \tag{53}$$

By Proposition 3 and the properties of the *o.c.* (see [8, Theorem XVIII.2.11 (f)]), (40) and (46) jointly imply that, for any $t \in \mathbb{R}$ and $m \in \mathbb{N}$,

$$f \in D(A^m e^{tA}), \tag{54}$$

which further implies that, for each $t \in \mathbb{R}$,

$$y(t) = e^{tA} f \in \bigcap_{n=1}^{\infty} D(A^n) =: C^{\infty}(A). \quad (55)$$

Whence, by Corollary 11, we infer that

$$y(\cdot) \in C^{\infty}(\mathbb{R}, X), \quad (56)$$

which completes the proof of the *"if"* part.

"Only if" part: let us prove this part *by contrapositive* assuming that, for any $b_+ > 0$ and $b_- > 0$, the set $\sigma(A) \backslash \mathscr{L}_{b_-,b_+}$ is *unbounded*. In particular, this means that, for any $n \in \mathbb{N}$, unbounded is the set

$$\sigma(A) \backslash \mathscr{L}_{(2n)^{-1},(2n)^{-1}} = \{\lambda \in \sigma(A) \mid -(2n)^{-1} \ln |\operatorname{Im} \lambda| < \operatorname{Re} \lambda < (2n)^{-1} \ln |\operatorname{Im} \lambda|\}. \quad (57)$$

Hence, we can choose a sequence of points $\{\lambda_n\}_{n=1}^{\infty}$ in the complex plane as follows:

$$\lambda_n \in \sigma(A), \quad n \in \mathbb{N},$$
$$-(2n)^{-1} \ln |\operatorname{Im} \lambda_n| < \operatorname{Re} \lambda_n < (2n)^{-1} \ln |\operatorname{Im} \lambda_n|, \quad n \in \mathbb{N}, \quad (58)$$
$$\lambda_0 := 0,$$
$$|\lambda_n| > \max [n^4, |\lambda_{n-1}|], \quad n \in \mathbb{N}.$$

The latter implies, in particular, that the points λ_n, $n \in \mathbb{N}$, are *distinct* ($\lambda_i \neq \lambda_j$, $i \neq j$).

Since, for each $n \in \mathbb{N}$, the set

$$\{\lambda \in \mathbb{C} \mid -(2n)^{-1} \ln |\operatorname{Im} \lambda| < \operatorname{Re} \lambda < (2n)^{-1} \ln |\operatorname{Im} \lambda|, \ |\lambda| > \max [n^4, |\lambda_{n-1}|]\} \quad (59)$$

is *open* in $\mathbb{C}$; along with the point λ_n, it contains an *open disk*

$$\Delta_n := \{\lambda \in \mathbb{C} \mid |\lambda - \lambda_n| < \varepsilon_n\} \quad (60)$$

centered at λ_n of some radius $\varepsilon_n > 0$, i.e., for each $\lambda \in \Delta_n$,

$$-(2n)^{-1} \ln |\operatorname{Im} \lambda| < \operatorname{Re} \lambda < (2n)^{-1} \ln |\operatorname{Im} \lambda| \ \text{and} \ |\lambda| > \max [n^4, |\lambda_{n-1}|]. \quad (61)$$

Furthermore, we can regard the radii of the disks to be small enough so that

$$0 < \varepsilon_n < \frac{1}{n}, \quad n \in \mathbb{N}, \ \text{and}$$
$$\Delta_i \cap \Delta_j = \emptyset, \quad i \neq j \ (\text{i.e., the disks are } \textit{pairwise disjoint}). \quad (62)$$

Whence, by the properties of the *s.m.*,

$$E_A(\Delta_i) E_A(\Delta_j) = 0, \quad i \neq j, \quad (63)$$

where 0 stands for the *zero operator* on X.

Observe also, that the subspaces $E_A(\Delta_n)X$, $n \in \mathbb{N}$, are *nontrivial* since

$$\Delta_n \cap \sigma(A) \neq \emptyset, \quad n \in \mathbb{N}, \quad (64)$$

with Δ_n being an *open set* in $\mathbb{C}$.

By choosing a unit vector $e_n \in E_A(\Delta_n)X$ for each $n \in \mathbb{N}$, we obtain a sequence $\{e_n\}_{n=1}^{\infty}$ in X such that

$$\|e_n\| = 1, \quad n \in \mathbb{N}, \ \text{and}$$
$$E_A(\Delta_i) e_j = \delta_{ij} e_j, \quad i, j \in \mathbb{N}, \quad (65)$$

where δ_{ij} is the *Kronecker delta*.

As is easily seen, (65) implies that the vectors e_n, $n \in \mathbb{N}$, are *linearly independent*.

Furthermore, there is an $\varepsilon > 0$ such that

$$d_n := \operatorname{dist}(e_n, \operatorname{span}(\{e_i \mid i \in \mathbb{N}, \ i \neq n\})) \geq \varepsilon, \quad n \in \mathbb{N}. \quad (66)$$

Indeed, the opposite implies the existence of a subsequence $\{d_{n(k)}\}_{k=1}^{\infty}$ such that

$$d_{n(k)} \longrightarrow 0, \quad k \longrightarrow \infty. \quad (67)$$

Then, by selecting a vector

$$f_{n(k)} \in \operatorname{span}(\{e_i \mid i \in \mathbb{N}, \ i \neq n(k)\}), \quad k \in \mathbb{N}, \quad (68)$$

such that

$$\|e_{n(k)} - f_{n(k)}\| < d_{n(k)} + \frac{1}{k}, \quad k \in \mathbb{N}, \quad (69)$$

we arrive at

$$\begin{aligned} 1 &= \|e_{n(k)}\| \quad \text{since, by (65)}, E_A(\Delta_{n(k)}) f_{n(k)} = 0; \\ &= \|E_A(\Delta_{n(k)})(e_{n(k)} - f_{n(k)})\| \\ &\leq \|E_A(\Delta_{n(k)})\| \|e_{n(k)} - f_{n(k)}\| \quad \text{by (14)}; \\ &\leq M \|e_{n(k)} - f_{n(k)}\| \leq M \left[d_{n(k)} + \frac{1}{k}\right] \longrightarrow 0, \quad k \longrightarrow \infty, \end{aligned} \quad (70)$$

which is a *contradiction* proving (66).

As follows from the *Hahn-Banach Theorem*, for any $n \in \mathbb{N}$, there is an $e_n^* \in X^*$ such that

$$\|e_n^*\| = 1, \quad n \in \mathbb{N}, \ \text{and}$$
$$\langle e_i, e_j^* \rangle = \delta_{ij} d_i, \quad i, j \in \mathbb{N}. \quad (71)$$

Let us consider separately the two possibilities concerning the sequence of the real parts $\{\operatorname{Re} \lambda_n\}_{n=1}^{\infty}$: its being *bounded* or *unbounded*.

First, suppose that the sequence $\{\operatorname{Re} \lambda_n\}_{n=1}^{\infty}$ is *bounded*, i.e., there is such an $\omega > 0$ that

$$|\operatorname{Re}\lambda_n| \le \omega, \quad n \in \mathbb{N}, \tag{72}$$

and consider the element

$$f := \sum_{k=1}^{\infty} k^{-2} e_k \in X, \tag{73}$$

which is well defined since $\{k^{-2}\}_{k=1}^{\infty} \in l_1$ (l_1 is the space of absolutely summable sequences) and $\|e_k\| = 1$, $k \in \mathbb{N}$ (see (65)).

In view of (65), by the properties of the *s.m.*,

$$E_A\left(\bigcup_{k=1}^{\infty}\Delta_k\right) f = f \text{ and } E_A(\Delta_k) f = k^{-2} e_k, \quad k \in \mathbb{N}. \tag{74}$$

For any $t \ge 0$ and an arbitrary $g^* \in X^*$,

$$\begin{aligned}
&\int_{\sigma(A)} e^{t\operatorname{Re}\lambda}\, dv(f, g^*, \lambda) \quad \text{by (74);} \\
&= \int_{\sigma(A)} e^{t\operatorname{Re}\lambda}\, dv\left(E_A\left(\bigcup_{k=1}^{\infty}\Delta_k\right) f, g^*, \lambda\right) \quad \text{by (18);} \\
&= \sum_{k=1}^{\infty} \int_{\sigma(A)\cap\Delta_k} e^{t\operatorname{Re}\lambda}\, dv(E_A(\Delta_k) f, g^*, \lambda) \quad \text{by (74);} \\
&= \sum_{k=1}^{\infty} k^{-2} \int_{\sigma(A)\cap\Delta_k} e^{t\operatorname{Re}\lambda}\, dv(e_k, g^*, \lambda) \\
&\text{since, for } \lambda \in \Delta_k, \text{ by (72) and (62)}, \operatorname{Re}\lambda = \operatorname{Re}\lambda_k + (\operatorname{Re}\lambda - \operatorname{Re}\lambda_k) \le \operatorname{Re}\lambda_k + |\lambda - \lambda_k| \le \omega + \varepsilon_k \le \omega + 1; \\
&\le e^{t(\omega+1)} \sum_{k=1}^{\infty} k^{-2} \int_{\sigma(A)\cap\Delta_k} 1\, dv(e_k, g^*, \lambda) = e^{t(\omega+1)} \sum_{k=1}^{\infty} k^{-2} v(e_k, g^*, \Delta_k) \quad \text{by (15);} \\
&\le e^{t(\omega+1)} \sum_{k=1}^{\infty} k^{-2} 4M \|e_k\| \|g^*\| = 4M e^{t(\omega+1)} \|g^*\| \sum_{k=1}^{\infty} k^{-2} < \infty.
\end{aligned} \tag{75}$$

Also, for any $t < 0$ and an arbitrary $g^* \in X^*$,

$$\begin{aligned}
&\int_{\sigma(A)} e^{t\operatorname{Re}\lambda}\, dv(f, g^*, \lambda) \quad \text{by (74);} \\
&= \int_{\sigma(A)} e^{t\operatorname{Re}\lambda}\, dv\left(E_A\left(\bigcup_{k=1}^{\infty}\Delta_k\right) f, g^*, \lambda\right) \quad \text{by (18);} \\
&= \sum_{k=1}^{\infty} \int_{\sigma(A)\cap\Delta_k} e^{t\operatorname{Re}\lambda}\, dv(E_A(\Delta_k) f, g^*, \lambda) \quad \text{by (74);} \\
&= \sum_{k=1}^{\infty} k^{-2} \int_{\sigma(A)\cap\Delta_k} e^{t\operatorname{Re}\lambda}\, dv(e_k, g^*, \lambda) \\
&\text{since, for } \lambda \in \Delta_k, \text{ by (72) and (62)}, \operatorname{Re}\lambda = \operatorname{Re}\lambda_k - (\operatorname{Re}\lambda_k - \operatorname{Re}\lambda) \ge \operatorname{Re}\lambda_k - |\operatorname{Re}\lambda_k - \operatorname{Re}\lambda| \ge -\omega - \varepsilon_k \ge -\omega - 1; \\
&\le e^{-t(\omega+1)} \sum_{k=1}^{\infty} k^{-2} \int_{\sigma(A)\cap\Delta_k} 1\, dv(e_k, g^*, \lambda) = e^{-t(\omega+1)} \sum_{k=1}^{\infty} k^{-2} v(e_k, g^*, \Delta_k) \quad \text{by (15);} \\
&\le e^{-t(\omega+1)} \sum_{k=1}^{\infty} k^{-2} 4M \|e_k\| \|g^*\| = 4M e^{-t(\omega+1)} \|g^*\| \sum_{k=1}^{\infty} k^{-2} < \infty.
\end{aligned} \tag{76}$$

Similarly to (75), for any $t \geq 0$ and an arbitrary $n \in \mathbb{N}$,

$$\sup_{\{g^* \in X^* \mid \|g^*\|=1\}} \int_{\{\lambda \in \sigma(A) \mid e^{t \operatorname{Re}\lambda} > n\}} e^{t \operatorname{Re}\lambda}\, dv(f, g^*, \lambda)$$
$$\leq \sup_{\{g^* \in X^* \mid \|g^*\|=1\}} e^{t(\omega+1)} \sum_{k=1}^{\infty} k^{-2} \int_{\{\lambda \in \sigma(A) \mid e^{t \operatorname{Re}\lambda} > n\} \cap \Delta_k} 1\, dv(e_k, g^*, \lambda) \quad \text{by (74);}$$
$$= e^{t(\omega+1)} \cdot \sup_{\{g^* \in X^* \mid \|g^*\|=1\}} \sum_{k=1}^{\infty} \int_{\{\lambda \in \sigma(A) \mid e^{t \operatorname{Re}\lambda} > n\} \cap \Delta_k} 1\, dv(E_A(\Delta_k) f, g^*, \lambda) \quad \text{by (18);}$$
$$= e^{t(\omega+1)} \sup_{\{g^* \in X^* \mid \|g^*\|=1\}} \int_{\{\lambda \in \sigma(A) \mid e^{t \operatorname{Re}\lambda} > n\}} 1\, dv\left(E_A\left(\bigcup_{k=1}^{\infty} \Delta_k\right) f, g^*, \lambda\right) \quad \text{by (74);}$$
$$= e^{t(\omega+1)} \sup_{\{g^* \in X^* \mid \|g^*\|=1\}} \int_{\{\lambda \in \sigma(A) \mid e^{t \operatorname{Re}\lambda} > n\}} 1\, dv(f, g^*, \lambda) \quad \text{by (16);}$$
$$\leq e^{t(\omega+1)} \sup_{\{g^* \in X^* \mid \|g^*\|=1\}} 4M \left\| E_A\left(\left\{\lambda \in \sigma(A) \mid e^{t \operatorname{Re}\lambda} > n\right\}\right) f \right\| \cdot \|g^*\|$$
$$\leq 4Me^{t(\omega+1)} \left\| E_A\left(\left\{\lambda \in \sigma(A) \mid e^{t \operatorname{Re}\lambda} > n\right\}\right) f \right\| \quad \text{by the strong continuity of the } s.m.;$$
$$\longrightarrow 4Me^{t(\omega+1)} \left\| E_A(\emptyset) f \right\| = 0, \quad n \longrightarrow \infty. \tag{77}$$

Similarly to (76), for any $t < 0$ and an arbitrary $n \in \mathbb{N}$,

$$\sup_{\{g^* \in X^* \mid \|g^*\|=1\}} \int_{\{\lambda \in \sigma(A) \mid e^{t \operatorname{Re}\lambda} > n\}} e^{t \operatorname{Re}\lambda}\, dv(f, g^*, \lambda)$$
$$\leq \sup_{\{g^* \in X^* \mid \|g^*\|=1\}} e^{-t(\omega+1)} \sum_{k=1}^{\infty} k^{-2} \int_{\{\lambda \in \sigma(A) \mid e^{t \operatorname{Re}\lambda} > n\} \cap \Delta_k} 1\, dv(e_k, g^*, \lambda) \quad \text{by (74);}$$
$$= e^{-t(\omega+1)} \cdot \sup_{\{g^* \in X^* \mid \|g^*\|=1\}} \sum_{k=1}^{\infty} \int_{\{\lambda \in \sigma(A) \mid e^{t \operatorname{Re}\lambda} > n\} \cap \Delta_k} 1\, dv(E_A(\Delta_k) f, g^*, \lambda) \quad \text{by (18);}$$
$$= e^{-t(\omega+1)} \sup_{\{g^* \in X^* \mid \|g^*\|=1\}} \int_{\{\lambda \in \sigma(A) \mid e^{t \operatorname{Re}\lambda} > n\}} 1\, dv\left(E_A\left(\bigcup_{k=1}^{\infty} \Delta_k\right) f, g^*, \lambda\right) \quad \text{by (74);}$$
$$= e^{-t(\omega+1)} \sup_{\{g^* \in X^* \mid \|g^*\|=1\}} \int_{\{\lambda \in \sigma(A) \mid e^{t \operatorname{Re}\lambda} > n\}} 1\, dv(f, g^*, \lambda) \quad \text{by (16);}$$
$$\leq e^{-t(\omega+1)} \sup_{\{g^* \in X^* \mid \|g^*\|=1\}} 4M \left\| E_A\left(\left\{\lambda \in \sigma(A) \mid e^{t \operatorname{Re}\lambda} > n\right\}\right) f \right\| \cdot \|g^*\|$$
$$\leq 4Me^{-t(\omega+1)} \left\| E_A\left(\left\{\lambda \in \sigma(A) \mid e^{t \operatorname{Re}\lambda} > n\right\}\right) f \right\| \quad \text{by the strong continuity of the } s.m.;$$
$$\longrightarrow 4Me^{-t(\omega+1)} \left\| E_A(\emptyset) f \right\| = 0, \quad n \longrightarrow \infty. \tag{78}$$

By Proposition 3, (75), (76), (77), and (78) jointly imply that

$$f \in \bigcap_{t \in \mathbb{R}} D\left(e^{tA}\right), \tag{79}$$

and hence, by Theorem 7,

$$y(t) := e^{tA} f, \quad t \in \mathbb{R}, \tag{80}$$

is a weak solution of (1).

Let

$$h^* := \sum_{k=1}^{\infty} k^{-2} e_k^* \in X^*, \tag{81}$$

the functional being well defined since $\{k^{-2}\}_{k=1}^{\infty} \in l_1$ and $\|e_k^*\| = 1$, $k \in \mathbb{N}$ (see (71)).

In view of (71) and (66), we have

$$\langle e_n, h^* \rangle = \left\langle e_k, k^{-2} e_k^* \right\rangle = d_k k^{-2} > \varepsilon k^{-2}, \quad k \in \mathbb{N}. \tag{82}$$

Hence,

$$\int_{\sigma(A)} |\lambda|\, dv(f, h^*, \lambda) \quad \text{by (18) as in (75);}$$
$$= \sum_{k=1}^{\infty} k^{-2} \int_{\sigma(A) \cap \Delta_k} |\lambda|\, dv(e_k, h^*, \lambda) \quad \text{since, for } \lambda \in \Delta_k, \text{ by (61), } |\lambda| \geq k^4;$$
$$\geq \sum_{k=1}^{\infty} k^{-2} k^4 v(e_k, h^*, \Delta_k) \geq \sum_{k=1}^{\infty} k^2 \left|\langle E_A(\Delta_k) e_k, h^* \rangle\right| \quad \text{by (65) and (82);}$$
$$\geq \sum_{k=1}^{\infty} k^2 \varepsilon k^{-2} = \infty. \tag{83}$$

By Proposition 3, (83) implies that

$$y(0) = f \notin D(A), \tag{84}$$

which, by Proposition 9 ($n = 1$, $I = \{0\}$) further implies that the weak solution $y(t) = e^{tA} f$, $t \in \mathbb{R}$, of (1) is not strongly differentiable at 0.

Now, suppose that the sequence $\{\operatorname{Re}\lambda_n\}_{n=1}^{\infty}$ is *unbounded*. Therefore, there is a subsequence $\{\operatorname{Re}\lambda_{n(k)}\}_{k=1}^{\infty}$ such that

$$\operatorname{Re}\lambda_{n(k)} \longrightarrow \infty \text{ or}$$
$$\operatorname{Re}\lambda_{n(k)} \longrightarrow -\infty, \tag{85}$$
$$k \longrightarrow \infty.$$

Let us consider separately each of the two cases.

First, suppose that

$$\operatorname{Re}\lambda_{n(k)} \longrightarrow \infty, \quad k \longrightarrow \infty \tag{86}$$

Then, without loss of generality, we can regard that

$$\operatorname{Re}\lambda_{n(k)} \geq k, \quad k \in \mathbb{N}. \tag{87}$$

Consider the elements

$$\begin{aligned} f &:= \sum_{k=1}^{\infty} e^{-n(k)\operatorname{Re}\lambda_{n(k)}} e_{n(k)} \in X \text{ and} \\ h &:= \sum_{k=1}^{\infty} e^{-(n(k)/2)\operatorname{Re}\lambda_{n(k)}} e_{n(k)} \in X, \end{aligned} \tag{88}$$

well defined since, by (87),

$$\left\{e^{-n(k)\operatorname{Re}\lambda_{n(k)}}\right\}_{k=1}^{\infty}, \left\{e^{-(n(k)/2)\operatorname{Re}\lambda_{n(k)}}\right\}_{k=1}^{\infty} \in l_1 \tag{89}$$

and $\|e_{n(k)}\| = 1$, $k \in \mathbb{N}$ (see (65)).

By (65),

$$\begin{aligned} &E_A\left(\bigcup_{k=1}^{\infty}\Delta_{n(k)}\right) f = f \text{ and} \\ &E_A\left(\Delta_{n(k)}\right) f = e^{-n(k)\operatorname{Re}\lambda_{n(k)}} e_{n(k)}, \\ &k \in \mathbb{N}, \end{aligned} \tag{90}$$

and

$$\begin{aligned} &E_A\left(\bigcup_{k=1}^{\infty}\Delta_{n(k)}\right) h = h \text{ and} \\ &E_A\left(\Delta_{n(k)}\right) h = e^{-(n(k)/2)\operatorname{Re}\lambda_{n(k)}} e_{n(k)}, \\ &k \in \mathbb{N}. \end{aligned} \tag{91}$$

For any $t \geq 0$ and an arbitrary $g^* \in X^*$,

$$\begin{aligned} &\int_{\sigma(A)} e^{t\operatorname{Re}\lambda}\, dv(f, g^*, \lambda) \quad \text{by (18) as in (75);} \\ &= \sum_{k=1}^{\infty} e^{-n(k)\operatorname{Re}\lambda_{n(k)}} \int_{\sigma(A)\cap\Delta_{n(k)}} e^{t\operatorname{Re}\lambda}\, dv(e_{n(k)}, g^*, \lambda) \\ &\text{since, for } \lambda \in \Delta_{n(k)}, \text{ by (62), } \operatorname{Re}\lambda = \operatorname{Re}\lambda_{n(k)} + (\operatorname{Re}\lambda - \operatorname{Re}\lambda_{n(k)}) \leq \operatorname{Re}\lambda_{n(k)} + |\lambda - \lambda_{n(k)}| \leq \operatorname{Re}\lambda_{n(k)} + 1; \\ &\leq \sum_{k=1}^{\infty} e^{-n(k)\operatorname{Re}\lambda_{n(k)}} e^{t(\operatorname{Re}\lambda_{n(k)}+1)} \int_{\sigma(A)\cap\Delta_{n(k)}} 1\, dv(e_{n(k)}, g^*, \lambda) \\ &= e^t \sum_{k=1}^{\infty} e^{-[n(k)-t]\operatorname{Re}\lambda_{n(k)}} v(e_{n(k)}, g^*, \Delta_{n(k)}) \quad \text{by (15);} \\ &\leq e^t \sum_{k=1}^{\infty} e^{-[n(k)-t]\operatorname{Re}\lambda_{n(k)}} 4M \|e_{n(k)}\| \|g^*\| = 4Me^t \|g^*\| \sum_{k=1}^{\infty} e^{-[n(k)-t]\operatorname{Re}\lambda_{n(k)}} \\ &< \infty. \end{aligned} \tag{92}$$

Indeed, for all $k \in \mathbb{N}$ sufficiently large so that

$$n(k) \geq t + 1, \tag{93}$$

in view of (87),

$$e^{-[n(k)-t]\operatorname{Re}\lambda_{n(k)}} \leq e^{-k}. \tag{94}$$

For any $t < 0$ and an arbitrary $g^* \in X^*$,

$$\begin{aligned} &\int_{\sigma(A)} e^{t\operatorname{Re}\lambda}\, dv(f, g^*, \lambda) \quad \text{by (18) as in (75);} \\ &= \sum_{k=1}^{\infty} e^{-n(k)\operatorname{Re}\lambda_{n(k)}} \int_{\sigma(A)\cap\Delta_{n(k)}} e^{t\operatorname{Re}\lambda}\, dv(e_{n(k)}, g^*, \lambda) \end{aligned}$$

since, for $\lambda \in \Delta_{n(k)}$, by (62), $\operatorname{Re}\lambda = \operatorname{Re}\lambda_{n(k)} - (\operatorname{Re}\lambda_{n(k)} - \operatorname{Re}\lambda) \geq \operatorname{Re}\lambda_{n(k)} - |\operatorname{Re}\lambda_{n(k)} - \operatorname{Re}\lambda| \geq \operatorname{Re}\lambda_{n(k)} - 1$;

$$\leq \sum_{k=1}^{\infty} e^{-n(k)\operatorname{Re}\lambda_{n(k)}} e^{t(\operatorname{Re}\lambda_{n(k)}-1)} \int_{\sigma(A)\cap\Delta_{n(k)}} 1\, dv(e_{n(k)}, g^*, \lambda)$$

$$= e^{-t} \sum_{k=1}^{\infty} e^{-[n(k)-t]\operatorname{Re}\lambda_{n(k)}} v(e_{n(k)}, g^*, \Delta_{n(k)}) \quad \text{by (15)};$$

$$\leq e^{-t} \sum_{k=1}^{\infty} e^{-[n(k)-t]\operatorname{Re}\lambda_{n(k)}} 4M \|e_{n(k)}\| \|g^*\| = 4Me^{-t} \|g^*\| \sum_{k=1}^{\infty} e^{-[n(k)-t]\operatorname{Re}\lambda_{n(k)}}$$

$$< \infty. \tag{95}$$

Indeed, for all $k \in \mathbb{N}$, in view of $t < 0$,

$$n(k) - t \geq n(k) \geq 1, \tag{96}$$

and hence, in view of (87),

$$e^{-[n(k)-t]\operatorname{Re}\lambda_{n(k)}} \leq e^{-k}. \tag{97}$$

Similarly to (92), for any $t \geq 0$ and an arbitrary $n \in \mathbb{N}$,

$$\sup_{\{g^*\in X^* \mid \|g^*\|=1\}} \int_{\{\lambda\in\sigma(A) \mid e^{t\operatorname{Re}\lambda}>n\}} e^{t\operatorname{Re}\lambda}\, dv(f, g^*, \lambda)$$

$$\leq \sup_{\{g^*\in X^* \mid \|g^*\|=1\}} e^t \sum_{k=1}^{\infty} e^{-[n(k)-t]\operatorname{Re}\lambda_{n(k)}} \int_{\{\lambda\in\sigma(A) \mid e^{t\operatorname{Re}\lambda}>n\}\cap\Delta_{n(k)}} 1\, dv(e_{n(k)}, g^*, \lambda)$$

$$= e^t \sup_{\{g^*\in X^* \mid \|g^*\|=1\}} \sum_{k=1}^{\infty} e^{-[(n(k)/2)-t]\operatorname{Re}\lambda_{n(k)}} e^{-(n(k)/2)\operatorname{Re}\lambda_{(k)}}$$

$$\int_{\{\lambda\in\sigma(A) \mid e^{t\operatorname{Re}\lambda}>n\}\cap\Delta_{n(k)}} 1\, dv(e_{n(k)}, g^*, \lambda)$$

since, by (87), there is an $L > 0$ such that $e^{-[(n(k)/2)-t]\operatorname{Re}\lambda_{n(k)}} \leq L$, $k \in \mathbb{N}$;

$$\leq Le^t \sup_{\{g^*\in X^* \mid \|g^*\|=1\}} \sum_{k=1}^{\infty} e^{-(n(k)/2)\operatorname{Re}\lambda_{n(k)}} \int_{\{\lambda\in\sigma(A) \mid e^{t\operatorname{Re}\lambda}>n\}\cap\Delta_{n(k)}} 1\, dv(e_{n(k)}, g^*, \lambda) \quad \text{by (91)};$$

$$= Le^t \sup_{\{g^*\in X^* \mid \|g^*\|=1\}} \sum_{k=1}^{\infty} \int_{\{\lambda\in\sigma(A) \mid e^{t\operatorname{Re}\lambda}>n\}\cap\Delta_{n(k)}} 1\, dv(E_A(\Delta_{n(k)})h, g^*, \lambda) \quad \text{by (18)};$$

$$= Le^t \sup_{\{g^*\in X^* \mid \|g^*\|=1\}} \int_{\{\lambda\in\sigma(A) \mid e^{t\operatorname{Re}\lambda}>n\}} 1\, dv\left(E_A\left(\bigcup_{k=1}^{\infty}\Delta_{n(k)}\right)h, g^*, \lambda\right) \quad \text{by (91)};$$

$$= Le^t \sup_{\{g^*\in X^* \mid \|g^*\|=1\}} \int_{\{\lambda\in\sigma(A) \mid e^{t\operatorname{Re}\lambda}>n\}} 1\, dv(h, g^*, \lambda) \quad \text{by (16)};$$

$$\leq Le^t \sup_{\{g^*\in X^* \mid \|g^*\|=1\}} 4M \left\|E_A\left(\{\lambda \in \sigma(A) \mid e^{t\operatorname{Re}\lambda} > n\}\right)h\right\| \|g^*\|$$

$$\le 4LMe^{t}\left\|E_A\left(\left\{\lambda\in\sigma(A)\mid e^{t\operatorname{Re}\lambda}>n\right\}\right)h\right\|$$

by the strong continuity of the *s.m.*;

$$\longrightarrow 4LMe^{t}\left\|E_A(\emptyset)h\right\|=0,\quad n\longrightarrow\infty. \tag{98}$$

Similar to 5, for any $t<0$ and an arbitrary $n\in\mathbb{N}$,

$$\sup_{\{g^*\in X^*\mid\|g^*\|=1\}}\int_{\{\lambda\in\sigma(A)\mid e^{t\operatorname{Re}\lambda}>n\}}e^{t\operatorname{Re}\lambda}\,dv(f,g^*,\lambda)$$

$$\le \sup_{\{g^*\in X^*\mid\|g^*\|=1\}} e^{-t}\sum_{k=1}^{\infty}e^{-[n(k)-t]\operatorname{Re}\lambda_{n(k)}}\int_{\{\lambda\in\sigma(A)\mid e^{t\operatorname{Re}\lambda}>n\}\cap\Delta_{n(k)}}1\,dv(e_{n(k)},g^*,\lambda)$$

$$=e^{-t}\sup_{\{g^*\in X^*\mid\|g^*\|=1\}}\sum_{k=1}^{\infty}e^{-[(n(k)/2)-t]\operatorname{Re}\lambda_{n(k)}}e^{-(n(k)/2)\operatorname{Re}\lambda_{(k)}}$$

$$\int_{\{\lambda\in\sigma(A)\mid e^{t\operatorname{Re}\lambda}>n\}\cap\Delta_{n(k)}}1\,dv(e_{n(k)},g^*,\lambda)$$

since, by (87), there is an $L>0$ such that $e^{-[(n(k)/2)-t]\operatorname{Re}\lambda_{n(k)}}\le L,\ k\in\mathbb{N}$;

$$\le Le^{-t}\sup_{\{g^*\in X^*\mid\|g^*\|=1\}}\sum_{k=1}^{\infty}e^{-(n(k)/2)\operatorname{Re}\lambda_{n(k)}}\int_{\{\lambda\in\sigma(A)\mid e^{t\operatorname{Re}\lambda}>n\}\cap\Delta_{n(k)}}1\,dv(e_{n(k)},g^*,\lambda)$$

by (91);

$$=Le^{-t}\sup_{\{g^*\in X^*\mid\|g^*\|=1\}}\sum_{k=1}^{\infty}\int_{\{\lambda\in\sigma(A)\mid e^{t\operatorname{Re}\lambda}>n\}\cap\Delta_{n(k)}}1\,dv(E_A(\Delta_{n(k)})h,g^*,\lambda) \tag{99}$$

by (18);

$$=Le^{-t}\sup_{\{g^*\in X^*\mid\|g^*\|=1\}}\int_{\{\lambda\in\sigma(A)\mid e^{t\operatorname{Re}\lambda}>n\}}1\,dv\left(E_A\left(\bigcup_{k=1}^{\infty}\Delta_{n(k)}\right)h,g^*,\lambda\right)$$

by (91);

$$=Le^{-t}\sup_{\{g^*\in X^*\mid\|g^*\|=1\}}\int_{\{\lambda\in\sigma(A)\mid e^{t\operatorname{Re}\lambda}>n\}}1\,dv(h,g^*,\lambda)\quad\text{by (16)};$$

$$\le Le^{-t}\sup_{\{g^*\in X^*\mid\|g^*\|=1\}}4M\left\|E_A\left(\left\{\lambda\in\sigma(A)\mid e^{t\operatorname{Re}\lambda}>n\right\}\right)h\right\|\|g^*\|$$

$$\le 4LMe^{-t}\left\|E_A\left(\left\{\lambda\in\sigma(A)\mid e^{t\operatorname{Re}\lambda}>n\right\}\right)h\right\|$$

by the strong continuity of the *s.m.*;

$$\longrightarrow 4LMe^{-t}\left\|E_A(\emptyset)h\right\|=0,\quad n\longrightarrow\infty.$$

By Proposition 3, (92), 5, 5, and (99) jointly imply that

$$f\in\bigcap_{t\in\mathbb{R}}D\left(e^{tA}\right), \tag{100}$$

and hence, by Theorem 7,

$$y(t):=e^{tA}f,\quad t\in\mathbb{R}, \tag{101}$$

is a weak solution of (1).

Since, for any $\lambda \in \Delta_{n(k)}$, $k \in \mathbb{N}$, by (62), (87),

$$\begin{aligned}\operatorname{Re}\lambda &= \operatorname{Re}\lambda_{n(k)} - \left(\operatorname{Re}\lambda_{n(k)} - \operatorname{Re}\lambda\right) \\ &\geq \operatorname{Re}\lambda_{n(k)} - \left|\operatorname{Re}\lambda_{n(k)} - \operatorname{Re}\lambda\right| \geq \operatorname{Re}\lambda_{n(k)} - \varepsilon_{n(k)} \\ &\geq \operatorname{Re}\lambda_{n(k)} - \frac{1}{n(k)} \geq k - 1 \geq 0\end{aligned} \tag{102}$$

and, by (61),

$$\operatorname{Re}\lambda < (2n(k))^{-1}\ln|\operatorname{Im}\lambda|, \tag{103}$$

we infer that, for any $\lambda \in \Delta_{n(k)}$, $k \in \mathbb{N}$,

$$|\lambda| \geq |\operatorname{Im}\lambda| \geq e^{2n(k)\operatorname{Re}\lambda} \geq e^{2n(k)(\operatorname{Re}\lambda_{n(k)} - 1/n(k))}. \tag{104}$$

Using this estimate, for the functional $h^* \in X^*$ defined by (81), we have

$$\begin{aligned}&\int_{\sigma(A)} |\lambda|\, dv(f, h^*, \lambda) \quad \text{by (18) as in (75);} \\ &= \sum_{k=1}^{\infty} e^{-n(k)\operatorname{Re}\lambda_{n(k)}} \int_{\Delta_{n(k)}} |\lambda|\, dv(e_{n(k)}, h^*, \lambda) \\ &\geq \sum_{k=1}^{\infty} e^{-n(k)\operatorname{Re}\lambda_{n(k)}} e^{2n(k)(\operatorname{Re}\lambda_{n(k)} - 1/n(k))} v(e_{n(k)}, h^*, \Delta_{n(k)}) \\ &= \sum_{k=1}^{\infty} e^{-2} e^{n(k)\operatorname{Re}\lambda_{n(k)}} \left|\langle E_A(\Delta_{n(k)}) e_{n(k)}, h^*\rangle\right| \\ &\qquad \text{by (87), (65), and (82);} \\ &\geq \sum_{k=1}^{\infty} e^{-2}\varepsilon \frac{e^{n(k)}}{n(k)^2} = \infty.\end{aligned} \tag{105}$$

By Proposition 3, (83) implies that

$$y(0) = f \notin D(A), \tag{106}$$

which, by Proposition 9 ($n = 1$, $I = \{0\}$), further implies that the weak solution $y(t) = e^{tA}f$, $t \in \mathbb{R}$, of (1) is not strongly differentiable at 0.

The remaining case of

$$\operatorname{Re}\lambda_{n(k)} \longrightarrow -\infty, \quad k \longrightarrow \infty, \tag{107}$$

is symmetric to the case of

$$\operatorname{Re}\lambda_{n(k)} \longrightarrow \infty, \quad k \longrightarrow \infty, \tag{108}$$

and is considered in absolutely the same manner, which furnishes a weak solution $y(\cdot)$ of (1) such that

$$y(0) \notin D(A), \tag{109}$$

and hence, by Proposition 9 ($n = 1$, $I = \{0\}$), not strongly differentiable at 0.

With every possibility concerning $\{\operatorname{Re}\lambda_n\}_{n=1}^{\infty}$ considered, we infer that assuming the opposite to the "*if*" part's premise allows to find a weak solution of (1) on $[0,\infty)$ that is not strongly differentiable at 0, much less strongly infinite differentiable on $\mathbb{R}$.

Thus, the proof by contrapositive of the "*only if*" *part* is complete and so is the proof of the entire statement.

From Theorem 12 and [1, Theorem 4.2], the latter characterizing the strong infinite differentiability of all weak solution of (2) on $(0,\infty)$, we also obtain the following.

Corollary 13. *Let A be a scalar type spectral operator in a complex Banach space. If all weak solutions of (2) are strongly infinite differentiable on $(0,\infty)$, then all weak solutions of (1) are strongly infinite differentiable on $\mathbb{R}$.*

Remark 14. As follows from Theorem 12, all weak solutions of (1) with a scalar type spectral operator A in a complex Banach space can be *strongly infinite differentiable* while the operator A is *unbounded*, e.g., when A is an unbounded *self-adjoint* operator in a complex Hilbert space (cf. [17, Theorem 7.1]). This fact contrasts the situation when a closed densely defined linear operator A in a complex Banach space generates a strongly continuous group $\{T(t)\}_{t\in\mathbb{R}}$ of bounded linear operators, i.e., the associated abstract Cauchy problem is *well-posed* (see Remarks 1.1), in which case even the (left or right) strong differentiability of all weak solutions of (1) at 0 immediately implies *boundedness* for A (cf. [3]).

6. The Cases of Normal and Self-Adjoint Operators

As an important particular case of Theorem 12, we obtain

Corollary 15 (the case of a normal operator). *Let A be a normal operator in a complex Hilbert space. Every weak solution of (1) is strongly infinite differentiable on $\mathbb{R}$ iff there exist $b_+ > 0$ and $b_- > 0$ such that the set $\sigma(A)\setminus\mathscr{L}_{b_-,b_+}$, where*

$$\begin{aligned}\mathscr{L}_{b_-,b_+} := \{\lambda \in \mathbb{C} \mid \operatorname{Re}\lambda &\leq \min(0, -b_-\ln|\operatorname{Im}\lambda|) \text{ or } \operatorname{Re}\lambda \\ &\geq \max(0, b_+\ln|\operatorname{Im}\lambda|)\},\end{aligned} \tag{110}$$

is bounded (see Figure 1).

Remark 16. Corollary 15 develops the results of paper [17], where similar consideration is given to the strong differentiability of the weak solutions of (2) with a normal operator A in a complex Hilbert space on $[0,\infty)$ and $(0,\infty)$.

From Corollary 13, we immediately obtain the following.

Corollary 17. *Let A be a normal operator in a complex Hilbert space. If all weak solutions of (2) are strongly infinite differentiable on $(0,\infty)$ (cf. [17, Theorem 5.2]), then all weak solutions of (1) are strongly infinite differentiable on $\mathbb{R}$.*

Considering that, for a self-adjoint operator A in a complex Hilbert space X,

$$\sigma(A) \subseteq \mathbb{R} \tag{111}$$

(see, e.g., [10, 11]), we further arrive at the following.

Corollary 18 (the case of a self-adjoint operator). *Every weak solution of (1) with a self-adjoint operator A in a complex Hilbert space is strongly infinite differentiable on $\mathbb{R}$.*

Cf. [17, Theorem 7.1].

7. Inherent Smoothness Improvement Effect

As is observed in the proof of the *"only if"* part of Theorem 12, the opposite to the *"if"* part's premise implies that there is a weak solution of (1), which is not strongly differentiable at 0. This renders the case of finite strong differentiability of the weak solutions superfluous and we arrive at the following inherent effect of smoothness improvement.

Proposition 19. *Let A be a scalar type spectral operator in a complex Banach space $(X, \|\cdot\|)$. If every weak solution of (1) is strongly differentiable at 0, then all of them are strongly infinite differentiable on $\mathbb{R}$.*

Cf. [1, Proposition 5.1].

8. Concluding Remark

Due to the *scalar type spectrality* of the operator A, Theorem 12 is stated exclusively in terms of the location of its *spectrum* in the complex plane, similarly to the celebrated *Lyapunov stability theorem* [19] (cf. [3, Ch. I, Theorem 2.10]), and thus is an intrinsically qualitative statement (cf. [1, 20]).

Conflicts of Interest

The author declares that there are no conflicts of interest regarding the publication of this paper.

Acknowledgments

The author extends sincere appreciation to his colleague, Dr. Maria Nogin of the Department of Mathematics, California State University, Fresno, for her kind assistance with the graphics.

References

[1] M. V. Markin, "On the differentiability of weak solutions of an abstract evolution equation with a scalar type spectral operator," *International Journal of Mathematics and Mathematical Sciences*, vol. 2011, Article ID 825951, 27 pages, 2011.

[2] J. M. Ball, "Strongly continuous semigroups, weak solutions, and the variation of constants formula," *Proceedings of the American Mathematical Society*, vol. 63, no. 2, pp. 370–373, 1977.

[3] K.-J. Engel and R. Nagel, *One-Parameter Semigroups for Linear Evolution Equations*, vol. 194 of *Graduate Texts in Mathematics*, Springer, New York, NY, USA, 2000.

[4] M. V. Markin, "On the mean ergodicity of weak solutions of an abstract evolution equation," *Methods of Functional Analysis and Topology*, vol. 24, no. 1, pp. 53–70, 2018.

[5] E. Hille and R. S. Phillips, *Functional Analysis and Semi-Groups*, vol. 31 of *American Mathematical Society Colloquium Publications*, American Mathematical Society, Providence, RI, USA, 1957.

[6] M. V. Markin, "A note on the spectral operators of scalar type and semigroups of bounded linear operators," *International Journal of Mathematics and Mathematical Sciences*, vol. 32, no. 10, pp. 635–640, 2002.

[7] N. Dunford, "A survey of the theory of spectral operators," *Bulletin of the American Mathematical Society*, vol. 64, pp. 217–274, 1958.

[8] N. Dunford and J. T. Schwartz, *Linear Operators. Part III: Spectral Operators*, Interscience Publishers, New York, NY, USA, 1971.

[9] J. Wermer, "Commuting spectral measures on Hilbert space," *Pacific Journal of Mathematics*, vol. 4, no. 3, pp. 355–361, 1954.

[10] N. Dunford and J. T. Schwartz, *Linear Operators. Part II: Spectral Theory. Self Adjoint Operators in Hilbert Space*, Interscience Publishers, New York, NY, USA, 1963.

[11] A. I. Plesner, *Spectral Theory of Linear Operators*, Nauka, Moscow, 1965 (Russian).

[12] N. Dunford and J. T. Schwartz, *Linear Operators. Part I: General Theory*, with the assistance of W. G. Bade and R. G. Bartle, Interscience Publishers, New York, NY, USA, 1958.

[13] M. V. Markin, "On scalar type spectral operators, infinite differentiable and Gevrey ultradifferentiable C_0-semigroups," *International Journal of Mathematics and Mathematical Sciences*, vol. 2004, no. 45, pp. 2401–2422, 2004.

[14] M. V. Markin, "On the Carleman classes of vectors of a scalar type spectral operator," *International Journal of Mathematics and Mathematical Sciences*, vol. 2004, no. 60, pp. 3219–3235, 2004.

[15] P. R. Halmos, *Measure Theory*, vol. 18 of *Graduate Texts in Mathematics*, Springer, New York, NY, USA, 1974.

[16] M. V. Markin, "On an abstract evolution equation with a spectral operator of scalar type," *International Journal of Mathematics and Mathematical Sciences*, vol. 32, no. 9, pp. 555–563, 2002.

[17] M. V. Markin, "On the strong smoothness of weak solutions of an abstract evolution equation. I. Differentiability," *Applicable Analysis*, vol. 73, no. 3-4, pp. 573–606, 1999.

[18] M. V. Markin, "On the Carleman ultradifferentiable vectors of a scalar type spectral operator," *Methods of Functional Analysis and Topology*, vol. 21, no. 4, pp. 361–369, 2015.

[19] A. M. Lyapunov, *Stability of motion [Ph.D. thesis]*, University of Kharkov, 1892, English translation, Academic Press, New York, NY, USA, 1966.

[20] A. Pazy, "On the differentiability and compactness of semigroups of linear operators," *Journal of Mathematics and Mechanics*, vol. 17, no. 12, pp. 1131–1141, 1968.

Extension of Kirk-Saliga Fixed Point Theorem in a Metric Space with a Reflexive Digraph

Karim Chaira,[1] Abderrahim Eladraoui,[2] Mustapha Kabil,[2] and Samih Lazaiz[1]

[1]*Laboratory of Algebra, Analysis and Applications, Faculty of Sciences Ben M'sik, University of Hassan II Casablanca, Casablanca, Morocco*
[2]*Laboratory of Mathematics and Applications, Faculty of Sciences and Technologies, Mohammedia, University of Hassan II Casablanca, Casablanca, Morocco*

Correspondence should be addressed to Abderrahim Eladraoui; a.adraoui@live.fr

Academic Editor: Nawab Hussain

We extend the result of Kirk-Saliga and we generalize Alfuraidan and Khamsi theorem for reflexive graphs. As a consequence, we obtain the ordered version of Caristi's fixed point theorem. Some concrete examples are given to support the obtained results.

1. Introduction

Fixed point theory is one of the most useful tools in mathematics; it is used to solve many existence problems such as differential equations, control theory, optimization, and several other branches (for the literature see [1]). The most well-known fixed point result is Banach contraction principle [2]; it is famous for its applications, proving the existence of solution of integral equations by converting the problem to fixed point problem (see [3]). Recall that a point $x \in X$ is called a fixed point for a map $T : X \to X$ if $Tx = x$. Due to its importance, this theorem found a number of generalizations and extensions in many directions; for more details see [4] and the references therein. In 1976, Caristi (see [5]) gave an elegant generalization of Banach contraction principle, where the assumption that "$T : X \to X$ is continuous" is dropped and replaced by a weak assumption. Since then, various proofs, extensions, and generalizations are given by many authors (see [6–8]). It is worth mentioning that Caristi's fixed point theorem is equivalent to the Ekeland variational principle [8]. Also, it characterizes the completeness of the metric space as showed by Kirk in [9]. Among those generalizations, there is Kirk-Saliga fixed point theorem (see [10]) which states that any map $T : X \to X$ has a fixed point provided that X is complete metric space and there exist an integer $p \in \mathbb{N}$ and a lower semicontinuous function $\varphi : X \to [0, \infty)$ such that

$$d(x, Tx) \le \varphi(x) - \varphi(T^p x) \tag{1}$$

and $\varphi(Tx) \le \varphi(x)$ for any $x \in X$. For more on the latter result, one can consult [11].

Recently, Ran and Reurings [12] extend the Banach contraction principle in the context of partially ordered set where the contraction is restricted to the comparable elements which allowed them to give a meaningful application to linear and nonlinear matrix equations. Moreover, Nieto and Rodríguez-López in [13] have weakened the continuity assumption using a more suitable condition where the order is combined with the topological properties. For more details, one can consult [14, 15]. Also, in [16] Alfuraidan and Khamsi gave an analogue version of Caristi's fixed point theorem in the setting of partially ordered metric space where the inequality holds only for comparable elements. However, the new approach in their work is mixing the concept of the reflexive acyclic digraph with fixed point results. In this article, we discuss an extension of Kirk-Saliga result and we generalize Alfuraidan and Khamsi theorem for reflexive graphs. As a corollary, we obtain the ordered version of Caristi's fixed point theorem. Some concrete examples are given to support the obtained results. Throughout this paper we denote by $\mathbb{N}$ the set of all integers and by $\mathbb{N}^*$ the set of all positive integers.

2. Preliminaries

We start by recalling some basic notions on graphs borrowed from [17].

Definition 1. Let V be an arbitrary set.

(i) A directed graph, or digraph, is a pair $G = (V, E)$, where E is a subset of the Cartesian product $V \times V$. The elements of V are called vertices or nodes of G and the elements of E are the edges also called oriented edges or arcs of G. An edge of the form (v, v) is a loop on v. Another way to express that E is a subset of $V \times V$ is to say that E is a binary relation over V. Given a digraph G, the set of vertices (of edges) of G is denoted by $V(G)$ $(E(G))$.

(ii) The digraph $G = (V, E)$ is said to be transitive if whenever $(x, y) \in E$ and $(y, z) \in E$, $(x, z) \in E$.

Definition 2. A digraph $G = (V, E)$ is said to be reflexive if $\Delta := \{(v, v) \mid v \in V\}$ is a subset of E. Otherwise, every vertex has a loop.

Definition 3. Let $G = (V, E)$ be a digraph.

(i) A vertex x is said to be isolated if for all vertex $y \neq x$, we have neither $(x, y) \in E$ nor $(y, x) \in E$.

(ii) Two vertices $x, y \in V$. A path in G, from (or joining) x to y, is a sequence of vertices $p = \{a_i\}_{0 \le i \le n}$, $n \in \mathbb{N}^*$ such that $a_0 = x$, $a_n = y$ and $(a_i, a_{i+1}) \in E$, for all $i \in \{0, 1, \ldots, n-1\}$. The integer n is the length of the path p. If $x = y$ and $n > 1$, the path p is called a directed cycle. An acyclic digraph is a digraph which has no directed cycle.

(iii) We denote by $y \in [x]_G$ the fact that y can be reached from x by means of a path in G.

A metric space (X, d) endowed with a digraph G such that $V(G) = X$ is denoted by (X, d, G). The following notion of regularity is borrowed from Alfuraidan and Khamsi in [16] that considered it for posets.

Definition 4. Let $(X, d, \preceq)$ be a partially ordered metric space. We say that X satisfies the condition (OSC) if for any decreasing sequence $\{x_n\} \subseteq X$ that is convergent to $x \in X$, $x = \inf\{x_n : n \in \mathbb{N}\}$.

In the setting of digraphs, the analogue of the infimum of chain may be stated as follows.

Definition 5. Let (X, d, G) be metric space endowed with a digraph. We say that X satisfies the condition (OSCL) if for any sequence $\{x_n\} \subseteq X$ that is convergent to $x \in X$ and for all $n \in \mathbb{N}$, $x_{n+1} \in [x_n]_G$, $x \in [x_n]_G$ for all $n \in \mathbb{N}$ and if there exists $y \in X$ such that $y \in [x_n]_G$, for all $n \in \mathbb{N}$, then $y \in [x]_G$.

Remark 6. Let $(X, d, \preceq)$ be a partially ordered metric space. Let $G_{\preceq}$ be the digraph associated with the order $\preceq$ (see [16]). One can see that

$$x \preceq y \Longleftrightarrow (y, x) \in E(G_{\preceq}) \Longleftrightarrow x \in [y]_{G_{\preceq}}. \tag{2}$$

Under the above observations, the (OSCL) property is reduced to the (OSC) condition.

Let ω be the first transfinite ordinal and let Ω be the first uncountable transfinite ordinal. ω is the order type of $\mathbb{N}$ "the set of integers" and Ω is the order type of $\mathbb{R}$ the set of real numbers. Note that, for each $\xi < \Omega$, ξ is countable.

Proposition 7 (see [11]). *The following is valid:*

(i) *The ordinal Ω cannot be attained via sequential limits of countable ordinals. That is if $\{\alpha_n\}$ is an ascending sequence of countable ordinals, then the ordinal*

$$\alpha = \sup\{\alpha_n\} = \lim \alpha_n \tag{3}$$

is countable too.

(ii) *Each second kind countable ordinal is attainable via such sequences. In other words: if $\alpha < \Omega$ is of second kind (ordinal limit), then there exists a strictly ascending sequence $\{\alpha_n\}$ of countable ordinals with property (3).*

The following result is needed throughout this work; for the proof see [18, Proposition A.6, pp. 284].

Proposition 8. *Suppose that a sequence $\{x_\alpha\}_{\alpha \in \Omega} \subseteq \mathbb{R}$ is bounded and either nonincreasing or nondecreasing. Then there exists $\beta \in \Omega$ such that $x_\alpha = x_\beta$ for all $\Omega > \alpha \ge \beta$.*

We conclude this section by the following useful definitions.

Definition 9. Let (X, d, G) be metric space endowed with a digraph, $p \in \mathbb{N}$ and $\varphi : X \to [0, +\infty[$ a lower semicontinuous function. Let $T : X \to X$ be a self-mapping. We say the following:

(1) T is a G-monotone if for all $(x, y) \in X^2$,

$$x \in [y]_G \Longrightarrow Tx \in [Ty]_G. \tag{4}$$

(2) T is a G-Caristi mapping if for all $x \in X$,

$$Tx \in [x]_G \Longrightarrow d(Tx, x) \le \varphi(x) - \varphi(Tx). \tag{5}$$

(3) T is a G-Kirk-Saliga mapping if for all $x \in X$,

$$Tx \in [x]_G \Longrightarrow \begin{cases} (\text{KS1}) : d(Tx, x) \le \varphi(x) - \varphi(T^p x); \\ (\text{KS2}) : \varphi(Tx) \le \varphi(x). \end{cases} \tag{6}$$

3. Main Results

Theorem 10. *Let (X, d, G) be a complete metric space endowed with a reflexive digraph satisfying the (OSCL) condition. Let $T : X \to X$ be a G-monotone and G-Kirk-Saliga mapping. If there exists an element $x_0 \in X$ such that $Tx_0 \in [x_0]_G$, then T admits a fixed point in X.*

Proof. If $p = 0$ then $Tx = x$, for all $x \in X$ such that $Tx \in [x]_G$. Assume that $p \geq 1$ and consider the function ϕ defined from X into $[0, +\infty[$ by

$$\phi(x) = \sum_{i=0}^{p-1} \varphi(T^i x), \quad \forall x \in X. \tag{7}$$

The idea of the proof is to construct a transfinite orbit $(x_\alpha)_{\alpha\in\Omega}$, where Ω is the first uncountable ordinal satisfying, for each $\alpha \in \Omega$,

$A(\alpha)$: $Tx_\alpha = x_{\alpha+1}$;

$B(\alpha)$: $x_\alpha = \lim_{\lambda\to\alpha^-} x_\lambda$, whenever α is an ordinal limit;

$C(\alpha)$: $x_\alpha \in [x_\mu]_G$, whenever $\mu \prec \alpha$;

$D(\alpha)$: $d(x_\alpha, x_\mu) \leq \phi(x_\mu) - \phi(x_\alpha)$, whenever $\mu \preceq \alpha$.

Consider the sequence $\{x_n\}$ defined for each $n \in \mathbb{N}$ by $x_{n+1} = Tx_n$. Since $Tx_0 \in [x_0]_G$ and using the monotony of T, we obtain $x_{n+1} \in [x_n]_G$ for each $n \in \mathbb{N}$. According to (KS2), the nonnegative sequence $\{\phi(x_n)\}$ is decreasing and then converges. From (KS1), we get that for all integers $n > m$

$$\begin{aligned} d(x_m, x_n) &\leq \sum_{i=m}^{n-1} d(x_i, x_{i+1}) \leq \sum_{i=m}^{n-1} \phi(x_i) - \phi(x_{i+1}) \\ &\leq \phi(x_m) - \phi(x_n). \end{aligned} \tag{8}$$

Hence, $\{x_n\}$ is a Cauchy sequence and then converges to $x_\omega \in X$. Let us put $x_{\omega+1} = Tx_\omega$. Clearly the properties $A(\alpha)$–$D(\alpha)$ are satisfied for each $\alpha \leq \omega$. Let $\beta \in \Omega$. Assume that the orbit $\{x_\alpha\}_{\alpha<\beta}$ has been defined. We need to define x_β and show that the four properties $A(\beta)$–$D(\beta)$ hold. For that, we have to distinguish two cases, when β is an immediate successor or β is an ordinal limit. Clearly $A(\beta)$ and $B(\beta)$ are satisfied; let us focus on $C(\beta)$ and $D(\beta)$.

Claim 1 ($C(\beta)$ holds)

Case 1. Assume that β is an ordinal limit; that is, there exists a strictly ascending sequence $(\beta_n)_n$ of ordinals in Ω such that $\beta = \sup\{\beta_n; n \in \mathbb{N}\}$ and $\beta_m \preceq \beta_n \prec \beta$ whenever $m \leq n$. Since $D(\alpha)$ holds for all $\alpha \prec \beta$, we get

$$d(x_{\beta_n}, x_{\beta_m}) \leq \phi(x_{\beta_m}) - \phi(x_{\beta_n}), \tag{9}$$

which implies that $(\phi(x_{\beta_n}))_n$ is decreasing sequence in $[0, \infty)$ and hence it is convergent. Then (x_{β_n}) is Cauchy sequence, so it converges in X. Set $x_\beta = \lim_{n\to\infty} x_{\beta_n}$. By (OSCL) property, we obtain $x_\beta \in [x_{\beta_n}]_G$ for all $n \in \mathbb{N}$. Let $\alpha \prec \beta$. There exists $n_0 \in \mathbb{N}$ such that for each $n \geq n_0$ we have

$$\alpha \preceq \beta_n \prec \beta, \tag{10}$$

and thus for each $n \geq n_0$,

$$x_{\beta_n} \in [x_\alpha]_G, x_\beta \in [x_{\beta_n}]_G \Longrightarrow x_\beta \in [x_\alpha]_G. \tag{11}$$

Since α is taken arbitrary, we obtain $C(\beta)$.

Case 2. Assume that β is an immediate successor; there exists $\alpha \prec \beta$ such that $\beta = \alpha + 1$.

(i) If α is an immediate successor, there exists an ordinal μ such $\alpha = \mu + 1$. From $C(\alpha)$, we have $x_\alpha \in [x_\mu]_G$ and using the G-monotonicity of T it follows that $x_\beta \in [x_\alpha]_G$ and so $C(\beta)$ holds.

(ii) If α is an ordinal limit, from Proposition 7, there exists an ascending sequence $\{\alpha_n\} \subset \Omega$ such that $\alpha = \sup\{\alpha_n : n \in \mathbb{N}\}$. From $B(\alpha)$ we have $x_\alpha = \lim_{n\to+\infty} x_{\alpha_n}$. Using the (OSCL) condition, we have $x_\alpha \in [x_{\alpha_n}]_G$. Since T is G-monotone, $x_\beta \in [x_{\alpha_n+1}]_G$ and as $x_{\alpha_n+1} \in [x_{\alpha_n}]_G$, we get $x_\beta \in [x_{\alpha_n}]_G$. Again, (OSCL) insures that $x_\beta \in [x_\alpha]_G$. Then $C(\beta)$ holds.

Claim 2 ($D(\beta)$ holds)

Case 1. Assume that β is ordinal limit. Let $\alpha \prec \beta$. There exists $n_0 \in \mathbb{N}$ such that for each $n \geq n_0$ we have

$$\alpha \preceq \beta_n \prec \beta. \tag{12}$$

Then we get for each $n \geq n_0$ that

$$d(x_\alpha, x_{\beta_n}) \leq \phi(x_\alpha) - \phi(x_{\beta_n}), \tag{13}$$

and for all $i \in \{0, 1, \ldots, p-1\}$

$$\lim_{n\to\infty} T^i x_{\beta_n} = \lim_{n\to\infty} x_{\beta_n+i} = x_\beta. \tag{14}$$

Since φ is lower semicontinuous, we get

$$\varphi(x_\beta) \leq \liminf_{n\to\infty} \varphi(T^i x_{\beta_n}). \tag{15}$$

From $C(\beta)$, we have $x_\beta \in [x_{\beta_n}]_G$ for all $n \in \mathbb{N}$. Using the same argument as above, we get $Tx_\beta \in [x_{\beta_n}]_G$ and (OSCL) insures that $Tx_\beta \in [x_\beta]_G$. Hence, for all $i \in \{0, 1, \ldots, p-1\}$, $T^{i+1}x_\beta \in [T^i x_\beta]_G$. This implies that

$$\varphi(T^p x_\beta) \leq \varphi(T^{p-1} x_\beta) \leq \cdots \leq \varphi(Tx_\beta) \leq \varphi(x_\beta). \tag{16}$$

By passing to limit superior in inequality (13), it follows that

$$\begin{aligned} d(x_\alpha, x_\beta) &\leq \phi(x_\alpha) - \liminf_{n\to+\infty} \phi(x_{\beta_n}) \\ &\leq \phi(x_\alpha) - \sum_{i=0}^{p-1} \liminf_{n\to+\infty} \varphi(T^i x_{\beta_n}) \\ &\leq \phi(x_\alpha) - p\varphi(x_\beta) \leq \phi(x_\alpha) - \phi(x_\beta). \end{aligned} \tag{17}$$

Hence, $D(\beta)$ holds.

Case 2. Assume that $\beta = \alpha + 1$ is an immediate successor; we have shown above that $C(\beta)$ holds. Then $Tx_\alpha = x_\beta \in [x_\alpha]_G$ and by assumption we get

$$d(x_\alpha, x_\beta) \leq \phi(x_\alpha) - \phi(x_\beta), \tag{18}$$

and for all $\gamma \preceq \alpha$, we have

$$d\left(x_\gamma, x_\alpha\right) \leq \phi\left(x_\gamma\right) - \phi\left(x_\alpha\right). \quad (19)$$

The triangle inequality implies that

$$d\left(x_\gamma, x_\beta\right) \leq \phi\left(x_\gamma\right) - \phi\left(x_\beta\right), \quad (20)$$

for each $\gamma \preceq \beta$, which completes the proof of $D(\beta)$ in both cases.

Thus, the orbit $(x_\alpha)_{\alpha\in\Omega}$ is well constructed. Since $\{\phi(x_\alpha)\}$ is nonincreasing on $\{x_\alpha\}$ and Ω is uncountable, there must exist $\alpha_0 \in \Omega$ such that $\phi(x_\alpha)$ is constant for all $\alpha \geq \alpha_0$. From $D(\alpha_0 + 1)$, we get

$$d\left(x_{\alpha_0+1}, x_{\alpha_0}\right) \leq \phi\left(x_{\alpha_0}\right) - \phi\left(x_{\alpha_0+1}\right) = 0. \quad (21)$$

Hence, $Tx_{\alpha_0} = x_{\alpha_0+1} = x_{\alpha_0}$.

We support our result by giving an example of a mapping which is G-Kirk-Saliga mapping, for some integer $p > 1$, but not G-Caristi.

Example 11. Consider the metric space (X, d), where $X = [0,1]$ and $d(x,y) = |x-y|$, for all $x, y \in X$. Endow X with the directed graph $G = (X, E)$ represented in Figure 1, where

$$E = \Delta \cup \left\{\left(\sqrt{\frac{1}{2}}, 0\right), (1,0), \left(\frac{1}{2^n}, 0\right), \left(\frac{1}{2^n}, \frac{1}{2^{n+1}}\right) : n \in \mathbb{N}^*\right\}. \quad (22)$$

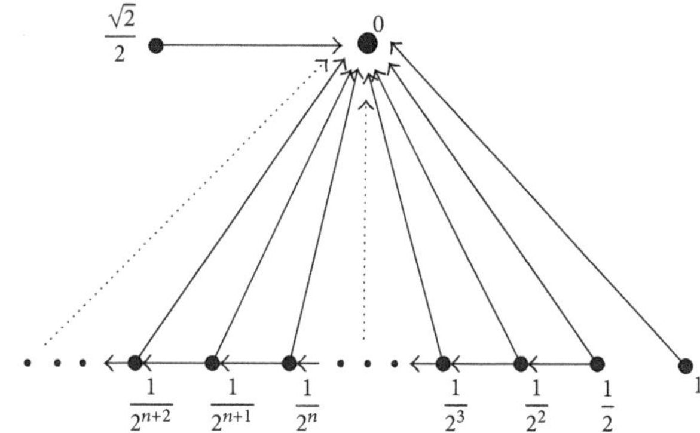

FIGURE 1: Graph G (the loops and the isolated vertices are not represented).

Consider the function $\varphi : X \to [0, +\infty[$ defined by

$$\varphi(x) = \begin{cases} \sqrt{x}, & \text{if } x \in [0,1[; \\ 0, & \text{if } x = 1, \end{cases} \quad (23)$$

and the mapping $T : X \to X$ defined by $Tx = x^2$, if $x \in [0,1[$; $T1 = \sqrt{1/2}$.

One can see that $T1 \notin [1]_G$, $T\sqrt{1/2} \notin [\sqrt{1/2}]_G$ and

$$X_G := \{x \in X : Tx \in [x]_G\} = \left\{0, \frac{1}{2^n} : n \in \mathbb{N}^*\right\}. \quad (24)$$

We verify the following assertions:

(i) (X, d) is complete and T is G-monotone obviously.

(ii) G satisfies the (OSCL) property. Indeed, let $\{x_n\}$ be a sequence in X such that $\{x_n\}$ converges to some $x \in X$ and $x_{n+1} \in [x_n]_G$, for all $n \in \mathbb{N}$. Two cases to distinguish are as follows:

(1) There exists $n_0 \in \mathbb{N}$ such that $x_n = x_{n_0}$, for all $n \geq n_0$. Then for all $n \geq n_0$, $x_n = x$. If x is an isolated vertex, the (OSCL) is obviously satisfied. If not, $y \in [x_n]_G$ for all $n \in \mathbb{N}$ implies $y \in [x]_G$. Thus, (OSCL) is satisfied.

(2) For all $k \in \mathbb{N}$, there exists $m_k \in \mathbb{N}$ such $x_{m_k} \neq x_k$. Then $\{x_n\} \subseteq \{1/2^n : n \in \mathbb{N}\}$; that is, there exists a nondecreasing function $\psi : \mathbb{N} \to \mathbb{N}$ such that $x_n = 1/2^{\psi(n)}$ for all $n \in \mathbb{N}$, and $x = 0$. If $y \in [x_n]_G$ for all $n \in \mathbb{N}$, then $y = 0$. Thus, (OSCL) is satisfied.

(iii) T is G-kirk-Saliga mapping in X with $p = 3$. Indeed, for all $x \in X_G$,

$$\begin{gathered} x - x^2 \leq \sqrt{x} - x^4 \Longleftrightarrow \\ d(x, Tx) \leq \varphi(x) - \varphi\left(T^3 x\right), \end{gathered} \quad (25)$$

but T is not G-Caristi mapping, since

$$d\left(\frac{1}{2}, T\frac{1}{2}\right) > \varphi\left(\frac{1}{2}\right) - \varphi\left(T\frac{1}{2}\right). \quad (26)$$

(iv) $T0 \in [0]_G$.

and T admits a fixed point in X which is 0.

If we remove the (OSCL) property, we are not certain that the fixed point will be obtained. Let us illustrate that by this counterexample.

Example 12. Replace in the above example the digraph G by the digraph G' represented in Figure 2, where

$$E' = \Delta \cup \left\{\left(\frac{1}{2^n}, 0\right), \left(\frac{1}{2^n}, 1\right), \left(\frac{1}{2^n}, \frac{1}{2^{n+1}}\right) : n \in \mathbb{N}^*\right\}, \quad (27)$$

and we consider the mapping $T : X \to X$ defined as follows:

$$\begin{gathered} T0 = 1; \\ Tx = x^2, \quad \text{if } x \in]0,1[; \\ T1 = 0. \end{gathered} \quad (28)$$

One can see that G' satisfies (OSC) property but does not satisfy the (OSCL), since $1/2^n \to 0$ and for all $n \in \mathbb{N}$, $0 \in [1/2^n]_{G'}$ and $1 \in [1/2^n]_{G'}$ but $1 \notin [0]_{G'}$. The mapping T satisfies all others conditions of Theorem 10 but has no fixed point in X.

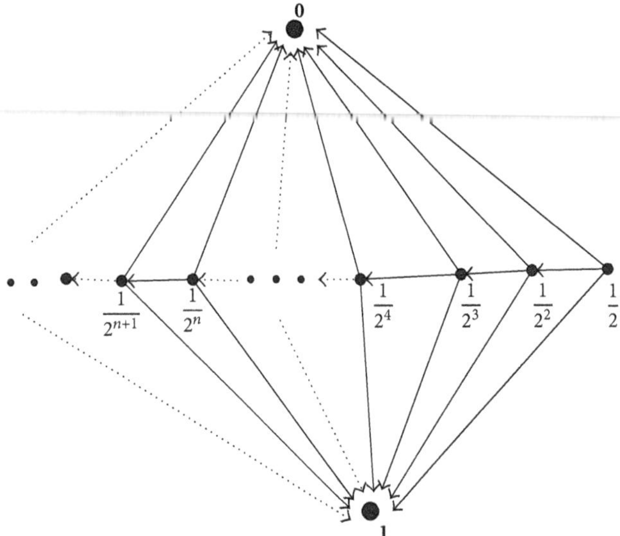

FIGURE 2: Graph G' (the loops and the isolated vertices are not represented).

Corollary 13. *Let (X, d, G) be a complete metric space endowed with a reflexive digraph which satisfies the (OSCL) property. Let $n = \prod_{i=0}^{k} p_i^{\alpha_i}$, where p_i is prime integer and $(k, \alpha_i) \in \mathbb{N}\times\mathbb{N}^*$, for each $i \in \{0, 1, \ldots, k\}$, and let $T : X \to X$ be a G-monotone mapping such that there exists $j \in \{0, 1, \ldots, k\}$, for all $x \in X$,*

$$T^{p_j}x \in [x]_G \Longrightarrow \begin{cases} \max\{d(x, T^{p_i}x) : i \in \{0, 1, \ldots, k\}\} \leq \varphi(x) - \varphi(T^n x), \\ \varphi(T^{p_j}x) \leq \varphi(x). \end{cases} \quad (29)$$

Then T admits a fixed point in X provided that there exists an element $x_0 \in X$ such that $Tx_0 \in [x_0]_G$.

Proof. Clearly, T^{p_j} satisfies all conditions of Theorem 10; then there exists $\overline{x} \in X$ such that $T^{p_j}\overline{x} = \overline{x}$ and so, $T^n\overline{x} = \overline{x}$. Since $T^{p_j}\overline{x} \in [\overline{x}]_G$, we get from (29) that

$$d(\overline{x}, T^{p_i}\overline{x}) \leq \varphi(\overline{x}) - \varphi(T^n\overline{x}) = 0, \quad \forall i \in \{0, 1, \ldots, k\} \setminus \{j\}. \quad (30)$$

Now, let $i \in \{0, 1, \ldots, k\} \setminus \{j\}$; then $T^{p_i}\overline{x} = \overline{x}$. By Bezout identity, there exists $(u, v) \in \mathbb{Z}^2$ such that $uv \leq 0$ and $p_j u + p_i v = 1$. Without loss of generality, we suppose that $u < 0$ and $v > 0$. Since $T^{p_j}\overline{x} = \overline{x}$ and $T^{p_i}\overline{x} = \overline{x}$, then $T^{1-p_j u}\overline{x} = T\overline{x}$ and $T^{p_i v}\overline{x} = \overline{x}$. Since $p_i v = 1 - p_j u$, then $T\overline{x} = \overline{x}$.

We conclude this work by a discussion about preordered sets.

Let $(X, \leqslant)$ be a preordered set; that is, the binary relation "$\leqslant$" is reflexive and transitive.

Given a reflexive digraph $G = (X, E)$, we can always define a preorder $\leqslant_G$ on X as follows:

$$x \leqslant_G y \Longleftrightarrow x \in [y]_G. \quad (31)$$

Conversely, if $(X, \leqslant)$ is a preordered set, we define the reflexive digraph $G_\leqslant$ as follows: two vertices $x, y \in X$ are connected by an arc from x to y if $x \leqslant y$. Note that $G_\leqslant$ is transitive (i.e., if $(x, y) \in E(G_\leqslant)$ and $(y, z) \in E(G_\leqslant)$; then $(x, z) \in E(G_\leqslant)$), so $x \in [y]_{G_\leqslant} \Leftrightarrow (x, y) \in E(G_\leqslant)$. These remarks lead to the following definition.

Definition 14. Let $(X, \leqslant)$ be a preordered set. We say that $(X, \leqslant)$ satisfies the (OSCL) condition if and only if $G_\leqslant$ satisfies the (OSCL) condition.

We shall say that $T : X \to X$ is $\leqslant$-monotone (resp., $\leqslant$-Kirk-Saliga) mapping if $G_\leqslant$-monotone (resp., $G_\leqslant$-Kirk-Saliga) mapping.

An analogue version of Theorem 10 in the setting of the preordered metric spaces may be stated as follows.

Theorem 15. *Let $(X, d, \leqslant)$ be a preordered complete metric space satisfying the (OSCL) condition. Let $T : X \to X$ be a $\leqslant$-monotone and $\leqslant$-Kirk-Saliga mapping. If there exists an element $x_0 \in X$ such that $Tx_0 \leqslant x_0$, then T admits a fixed point in X.*

Remark 16. If moreover the above binary relation $\leqslant$ is antisymmetric (i.e., ($x \leqslant y$ and $y \leqslant x$) imply $x = y$), we obtain, from Remark 6, the following result established by Alfuraidan and Khamsi.

Corollary 17 (see [16, Theorem 5]). *Let $(X, d, \leqslant)$ be a complete partially ordered metric space satisfying the property (OSC). Let $T : X \to X$ be a $\leqslant$-monotone and $\leqslant$-Kirk-Saliga mapping with $p = 1$. If there exists an element $x_0 \in X$ such that $Tx_0 \leqslant x_0$, then T admits a fixed point in X.*

Conflicts of Interest

The authors declare that they have no conflicts of interest.

References

[1] B. Lemmens and R. Nussbaum, *Nonlinear Perron-Frobenius theory*, vol. 189 of *Cambridge Tracts in Mathematics*, Cambridge University Press, Cambridge, UK, 2012.

[2] S. Banach, "Sur les opérations dans les ensembles abstraits et leur application aux équations intégrales," *Fundamenta Mathematicae*, vol. 3, pp. 133–181, 1922.

[3] G. Teschl, *Ordinary Differential Equations and Dynamical Systems*, American Mathematical Society, Providence, RI, USA, 2012.

[4] M. Edelstein, "An extension of Banach's contraction principle," *Proceedings of the American Mathematical Society*, vol. 12, pp. 7–10, 1961.

[5] J. Caristi, "Fixed point theorems for mappings satisfying inwardness conditions," *Transactions of the American Mathematical Society*, vol. 215, pp. 241–251, 1976.

[6] H. Brézis and F. E. Browder, "A general principle on ordered sets in nonlinear functional analysis," *Advances in Mathematics*, vol. 21, no. 3, pp. 355–364, 1976.

[7] A. Brøndsted, "On a lemma of Bishop and Phelps," *Pacific Journal of Mathematics*, vol. 55, no. 2, pp. 335–341, 1974.

[8] I. Ekeland, "On the variational principle," *Journal of Mathematical Analysis and Applications*, vol. 47, pp. 324–353, 1974.

[9] W. A. Kirk, "Caristi's fixed point theorem and metric convexity," *Colloquium Mathematicum*, vol. 36, no. 1, pp. 81–86, 1976.

[10] W. A. Kirk and L. M. Saliga, "The Brézis-Browder order principle and extensions of Caristi's theorem," *Nonlinear Analysis: Theory, Methods & Applications*, vol. 47, no. 4, pp. 2765–2778, 2001.

[11] M. Turinici, "Extension of the Kirk-Saliga fixed point theorem," *Analele stiintifice ale Universitatii Ovidius Constanta*, vol. 15, no. 2, pp. 91–102, 2007.

[12] A. C. Ran and M. C. Reurings, "A fixed point theorem in partially ordered sets and some applications to matrix equations," *Proceedings of the American Mathematical Society*, vol. 132, no. 5, pp. 1435–1443, 2004.

[13] J. J. Nieto and R. Rodríguez-López, "Contractive mapping theorems in partially ordered sets and applications to ordinary differential equations," *Order*, vol. 22, no. 3, pp. 223–239, 2005.

[14] R. P. Agarwal, M. A. El-Gebeily, and D. O'Regan, "Generalized contractions in partially ordered metric spaces," *Applicable Analysis: An International Journal*, vol. 87, no. 1, pp. 109–116, 2008.

[15] J. J. Nieto and R. Rodríguez-López, "Existence and uniqueness of fixed point in partially ordered sets and applications to ordinary differential equations," *Acta Mathematica Sinica*, vol. 23, no. 12, pp. 2205–2212, 2007.

[16] M. R. Alfuraidan and M. A. Khamsi, "Caristi fixed point theorem in metric spaces with a graph," *Abstract and Applied Analysis*, vol. 2014, Article ID 303484, 5 pages, 2014.

[17] M. Rigo, "Advanced Graph Theory and Combinatorics," *John Wiley & Sons*, 2016.

[18] M. A. Khamsi and W. A. Kirk, *An Introdunction toMetric Spaces and Fixed Point Theory*, Wiley, New York, NY, USA, 2001.

Greedy Expansions with Prescribed Coefficients in Hilbert Spaces

Artur R. Valiullin, Albert R. Valiullin, and Vladimir V. Galatenko

Faculty of Mechanics and Mathematics, Lomonosov Moscow State University, Russia

Correspondence should be addressed to Albert R. Valiullin; albert.valiullin1997@gmail.com

Academic Editor: Brigitte Forster-Heinlein

Greedy expansions with prescribed coefficients, which have been studied by V. N. Temlyakov in Banach spaces, are considered here in a narrower case of Hilbert spaces. We show that in this case the positive result on the convergence does not require monotonicity of coefficient sequence $\mathscr{C}$. Furthermore, we show that the condition sufficient for the convergence, namely, the inclusion $\mathscr{C} \in \ell^2 \setminus \ell^1$, can not be relaxed at least in the power scale. At the same time, in finite-dimensional spaces, the condition $\mathscr{C} \in \ell^2$ can be replaced by convergence of $\mathscr{C}$ to zero.

1. Introduction

Expansion in Fourier series [1] is a classical and comprehensively studied tool of theoretical and applied mathematics which takes an expanded function as an input and constructs a sequence of its expansion coefficients. Greedy expansions [2, 3], which are equivalent in the simplest case to Fourier series reordered by decreasing norms of terms and known in statistics and signal processing as Projection Pursuit Regression [4, 5] and Matching Pursuit [6], respectively, perform parallel computation of expansion coefficients and selection of expansion elements from a predefined dictionary. V. N. Temlyakov [3, 7] (see also [8]) proposed a type of a greedy expansion that performs only selection of expansion elements, while coefficients are prescribed in advance. The definition proposed by V. N. Temlyakov for the case of Banach spaces, in the case of Hilbert spaces, takes the following form.

Definition 1. Let H be a Hilbert space over $\mathbb{R}$ with a scalar product $(\cdot,\cdot)$, D be a symmetric unit-normed dictionary in H (i.e., span $D = H$, all elements in D have a unit norm, and if $g \in D$, then $-g$ also belongs to D). In addition, let $f \in H$, $t \in (0;1]$, and $\mathscr{C} = \{c_n\}_{n=1}^{\infty}$ be a sequence of positive numbers. We define inductively a sequence of remainders $\{r_n\}_{n=0}^{\infty} \subset H$ and a sequence of expanding elements $\{e_n\}_{n=1}^{\infty} \subset D$. First, we set $r_0 = f$. Then, if $r_{n-1} \in H$ $(n \in 1,2,3,\ldots)$ has already been defined, we select e_n as an (arbitrary) element which satisfies the condition

$$(r_{n-1}, e_n) \geqslant t \sup_{e\in D} (r_{n-1}, e), \tag{1}$$

and set $r_n = r_{n-1} - c_n e_n$.

The series $\sum_{n=1}^{\infty} c_n e_n(f)$ is called a greedy expansion of f in the dictionary D with the prescribed coefficients $\mathscr{C}$ and the weakness parameter t.

It immediately follows from the definition of a greedy expansion that $r_N = f - \sum_{n=1}^{N} c_n e_n(f)$ $(N \in \mathbb{N})$, and hence the convergence of the expansion to an expanded element is equivalent to the convergence of remainders r_n to zero as $n \longrightarrow \infty$.

As a selection of an expanding element e_n is potentially not unique, there may exist different realizations of a greedy expansion for a given dictionary D, weakness parameter t and sequence of coefficients $\mathscr{C}$. Furthermore, for $t = 1$ greedy expansion may turn out to be nonrealizable due to the absence of an element $e \in D$ which provides $\sup_{e\in D}(r_{n-1}, e)$.

V. N. Temlyakov showed [3, Theorem 2.1] that if a number of conditions hold which are equivalent in the case of Hilbert spaces to the divergence of the series $\sum_{n=1}^{\infty} c_n$ and the convergence of the series $\sum_{n=1}^{\infty} c_n^2$, a greedy expansion with prescribed coefficients $\mathscr{C} = \{c_n\}_{n=1}^{\infty}$ converges to an expanded element at least for a subsequence of indexes, i.e., $\liminf_{n\to\infty}\|r_n\| = 0$. Later V. N. Temlyakov proved the standard convergence (i.e., $\lim_{n\to\infty} r_n = 0$) under the additional condition of monotonicity of $\mathscr{C}$ [7, Theorem 4]. Yet, it remained unknown whether the condition $\mathscr{C} \in \ell^2$ and the monotonicity condition could be essentially relaxed without violating the guaranteed convergence to an expanded element.

2. Main Results

We start with a positive result which states that in Hilbert spaces the monotonicity is not required for the standard convergence. Namely, the following theorem holds.

Theorem 2. *Let H be a Hilbert space, D be a symmetric unit-normed dictionary in H, $t \in (0,1]$, $\mathscr{C} = \{c_n\}_{n=1}^{\infty}$ be a sequence of positive numbers which satisfies the conditions*

$$\sum_{n=1}^{\infty} c_n = \infty; \qquad \sum_{n=1}^{\infty} c_n^2 < \infty \tag{2}$$

(i.e., $\mathscr{C} \in \ell^2 \setminus \ell^1$). Then for every element $f \in H$ all realizations of its greedy expansion in the dictionary D with the prescribed coefficients $\mathscr{C}$ and the weakness parameter t converge to f.

It is clear that if the first condition on $\mathscr{C}$ is violated, then there is no convergence to an expanded element for all f with the norm exceeding the sum $\sum_{n=1}^{\infty} c_n$. The significance of the second condition on $\mathscr{C}$ follows from the following theorem.

Theorem 3. *There exist a Hilbert space H, a symmetric unit-normed dictionary $D \subset H$, an element $f \in H$ and a sequence of positive numbers $\mathscr{C} = \{c_n\}_{n=1}^{\infty}$ such that*

$$\sum_{n=1}^{\infty} c_n = \infty, \qquad c_n \leqslant \frac{1}{\sqrt{n}} \quad \forall n \in \{1,2,3\ldots\}, \tag{3}$$

but a greedy expansion of f in the dictionary D with the prescribed coefficients $\mathscr{C}$ and the weakness parameter $t = 1$ does not converge to f.

As for the second condition of Theorem 2 a boundary in the power scale is $1/\sqrt{n}$, Theorem 3 in fact shows that this condition in Theorem 2 can not be relaxed at least in the power scale.

However, the question about a possibility of a more delicate relaxation of the condition $\mathscr{C} \in \ell^2$ remains open. This question can be stated as follows: is it true that for every sequence $\mathscr{C}$ of positive numbers that converges to zero but does not belong to ℓ^2 there exist a Hilbert space H, a symmetric unit-normed dictionary $D \subset H$ and an expanded element $f \in H$ such that at least one realization of greedy expansion of f in D with the prescribed coefficients $\mathscr{C}$ (and, e.g., the weakness parameter $t = 1$) does not converge to f?

We note that assertions similar to Theorems 2 and 3 have been announced by O. Rassudova in her conference talk [9], but the proofs have not been published. To the best of our knowledge, in her proof of an analogue of Theorem 3 O. Rassudova used a modification of the construction [10, Theorem 3] which is based on analytical estimates and does not have a clear geometric interpretation. The construction presented in our work is geometrically demonstrative.

We also note that at least for $t = 1$ for the natural class of monotonic coefficients in case of finite-dimensional Hilbert spaces the condition $\sum_{n=1}^{\infty} c_n^2 < \infty$ in Theorem 2 can be replaced by an essentially weaker condition $c_n \to 0$ $(n \to \infty)$. The proof of this fact is presented in section *The case of finite-dimensional spaces.*

3. Proof of Theorem 2

The theorem can be easily derived from the equality $\liminf_{n\to\infty}\|r_n\| = 0$, which holds due to the aforementioned result by V. N. Temlyakov [3, Theorem 2.1]. From the definition of the greedy expansion it immediately follows that

$$\begin{aligned}\|r_m\|^2 &= (r_{m-1} - c_m e_m, r_{m-1} - c_m e_m) \\ &= \|r_{m-1}\|^2 - 2c_m (r_{m-1}, e_m) + c_m^2.\end{aligned} \tag{4}$$

As coefficients c_n are positive and the dictionary is symmetric, $2c_m(r_{m-1}, e_m) \geqslant 0$. Hence

$$\|r_m\|^2 \leqslant \|r_{m-1}\|^2 + c_m^2, \tag{5}$$

and thus

$$\|r_{m+k}\|^2 \leqslant \|r_m\|^2 + \sum_{j=m+1}^{m+k} c_j^2. \tag{6}$$

The condition $\mathscr{C} = \{c_n\}_{n=1}^{\infty} \in \ell^2$ implies that

$$\forall \varepsilon > 0 \ \exists N_1 > 0: \ \forall m > N_1 \ \sum_{j=m+1}^{\infty} c_j^2 < \frac{\varepsilon}{2}. \tag{7}$$

Due to the equality $\liminf_{n\to\infty}\|r_n\| = 0$ we have that

$$\forall \varepsilon > 0 \ \forall N > 0 \ \exists m > N: \ \|r_m\|^2 < \frac{\varepsilon}{2}. \tag{8}$$

From two last assertions we obtain that for every $\varepsilon > 0$ there exists $m > 0$ such that the following two conditions simultaneously hold:

$$\sum_{j=m+1}^{\infty} c_j^2 < \frac{\varepsilon}{2}, \qquad \|r_m\|^2 < \frac{\varepsilon}{2}. \tag{9}$$

Thus using estimate (6) we get that $\|r_{m+k}\|^2 < \varepsilon$ for all $k \in \mathbb{N}$. But according to the definition of the limit it directly means that $\lim_{n\to\infty}\|r_n\| = 0$. The proof of Theorem 2 is complete.

We note that for monotonic coefficients and the weakness parameter $t = 1$ the statement of Theorem 2, which is covered in this case by [7, Theorem 4], can be also derived as a corollary of the same result by V. N. Temlyakov about the convergence for a subsequence of indexes [3, Theorem 2.1] and the following lemma. We find this lemma to be interesting on its own.

Lemma 4. *Let $t = 1$ and c_n monotonically converge to zero as $n \longrightarrow \infty$. Then for every realization of greedy expansion with the prescribed coefficients $\mathscr{C} = \{c_n\}_{n=1}^{\infty}$ the sequence of norms of its remainders $\{\|r_n\|\}_{n=0}^{\infty}$ converges.*

We begin our proof of this lemma with an estimate of a possible increase of the remainder norms. Let M denote the set of all indexes n for which $\|r_{n+1}\| > \|r_n\|$. If the set M is finite, then starting from a certain index the sequence $\{\|r_n\|\}_{n=0}^{\infty}$ is monotonic and thus convergent. Hence it remains to consider the case of countably infinite M.

For the sake of brevity we denote the scalar products (r_n, e_{n+1}) by x_{n+1} $(n = 0, 1, 2, \ldots)$. It follows from the definition of the greedy expansion and the symmetric property of the dictionary that all x_n are nonnegative.

The Pythagorean theorem implies that for all indexes n the equality

$$\|r_{n+1}\|^2 = \|r_n\|^2 - x_{n+1}^2 + (c_{n+1} - x_{n+1})^2 \tag{10}$$

holds. If $n \in M$, then $c_{n+1} > 2x_{n+1}$, $\|r_{n+1}\|^2 \leqslant \|r_n\|^2 + c_{n+1}^2$ and thus

$$x_{n+2} = \sup_{e\in D}(r_{n+1}, e) \geqslant -(r_{n+1}, e_{n+1}) = c_{n+1} - x_{n+1}. \tag{11}$$

Consequently

$$\begin{aligned}
\|r_{n+2}\|^2 &= \|r_{n+1}\|^2 - x_{n+2}^2 + (c_{n+2} - x_{n+2})^2 \\
&= \|r_n\|^2 - x_{n+1}^2 - x_{n+2}^2 + (c_{n+1} - x_{n+1})^2 \\
&\quad + (c_{n+2} - x_{n+2})^2 \\
&= \|r_n\|^2 + c_{n+1}^2 + c_{n+2}^2 - 2c_{n+1}x_{n+1} - 2c_{n+2}x_{n+2} \\
&\leqslant \|r_n\|^2 + c_{n+1}^2 + c_{n+2}^2 - 2c_{n+1}x_{n+1} \\
&\quad - 2c_{n+2}(c_{n+1} - x_{n+1}) \\
&= \|r_n\|^2 + (c_{n+1} - c_{n+2})^2 - 2x_{n+1}(c_{n+1} - c_{n+2}) \\
&\leqslant \|r_n\|^2 + (c_{n+1} - c_{n+2})^2.
\end{aligned} \tag{12}$$

At the same time

$$\begin{aligned}
\|r_{n+2}\|^2 &= \|r_{n+1}\|^2 + c_{n+2}^2 - 2c_{n+2}x_{n+2} \\
&= \|r_{n+1}\|^2 + c_{n+2}(c_{n+2} - 2x_{n+2}) \\
&\leqslant \|r_{n+1}\|^2 + c_{n+2}(c_{n+2} - 2(c_{n+1} - x_{n+1})) \\
&= \|r_{n+1}\|^2 \\
&\quad + c_{n+2}((c_{n+2} - c_{n+1}) + (2x_{n+1} - c_{n+1})) \\
&\leqslant \|r_{n+1}\|^2.
\end{aligned} \tag{13}$$

It means that if the remainder norm increased at the transition from r_n to r_{n+1}, then the increase of the norm square does not exceed c_{n+1}^2, for the next expansion step the increase is impossible, and the joint increase of the square of remainder norm for two steps of the expansion does not exceed $(c_{n+1} - c_{n+2})^2$ and hence does not exceed $C(c_{n+1} - c_{n+2})$, where C can be set to $2c_1$.

Having this estimate, let us complete the proof. We note that the series $\sum_{n\in M}(c_{n+1} - c_{n+2})$ converges: it can be easily derived either from the Leibniz's alternating series test or from the inequality

$$\begin{aligned}
\sum_{n\in M, n\leqslant K}(c_{n+1} - c_{n+2}) &\leqslant \sum_{n=0}^{K}(c_{n+1} - c_{n+2}) = c_1 - c_{K+2} \\
&\leqslant c_1.
\end{aligned} \tag{14}$$

Let us fix an arbitrary positive ε and find an index N_0 such that $\sum_{n\in M, n>N_0}(c_{n+1} - c_{n+2}) < \varepsilon/(4C)$ and simultaneously $\sup_{n>N_0} c_n^2 < \varepsilon/4$. Next we find an index $N_1 > N_0$ such that $\|r_{N_1}\|^2 < r^2 + \varepsilon/4$, where r denotes the infimum of the remainder norms $\{\|r_n\|\}_{n>N_0}$. Then for every $n > N_1$ we have that

$$\begin{aligned}
r^2 \leqslant \|r_n\|^2 &\leqslant \|r_{N_1}\|^2 + \sum_{n\in M, n\geqslant N_1} C(c_{n+1} - c_{n+2}) + c_n^2 \\
&< r^2 + \varepsilon.
\end{aligned} \tag{15}$$

Hence $\|r_n\|^2 \longrightarrow r^2$ $(n \longrightarrow \infty)$ and consequently $\|r_n\| \longrightarrow r$. The proof of Lemma 4 is complete.

4. Proof of Theorem 3

Our proof of Theorem 3 includes the following blocks: construction of a dictionary with simultaneous construction of coefficients $\mathscr{C} = \{c_n\}_{n=1}^{\infty}$; description of realization of greedy expansion; proof of the absence of convergence to the expanded element; obtaining the required estimate of c_n. As a Hilbert space H we take an arbitrary infinite-dimensional separable space, e.g., ℓ^2.

Figure 1 illustrates certain steps of the proof.

4.1. Description of the Construction. We first present the structure of the example; i.e., we describe the construction of dictionary elements $\{e_n\}$ and coefficients $\{c_n\}$. As a part of this construction we also define the sequence of vectors (remainders) $\{r_n\}$, including the expanded element $f = r_0$.

Let f be an arbitrary non-zero element of H with $\|f\| \leqslant 1/2$, $r_0 = f$. We define dictionary elements e_{-1} and e_0 as arbitrary (unequal) unit vectors such that e_{-1}, e_0 and r_0 lie in one plane and the angle α_0 between e_0 and r_0 equals

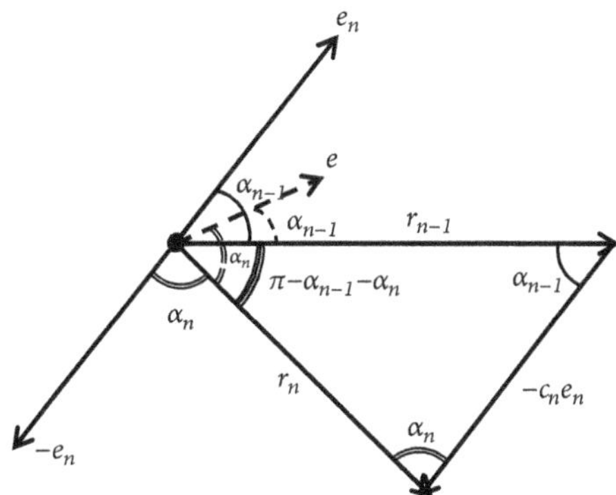

FIGURE 1: An illustration for certain steps of the proof of Theorem 3. All vectors except e lie in the vertical plane; e (as well as r_{n-1}) lies in the horizontal plane; e represents any vector from the set $\{e_{-1}, e_0, -e_1, -e_2, \dots, -e_{n-1}\}$. The spherical law of cosines is applied to angles formed by the vectors r_{n-1}, r_n, e. The law of sines is applied to the triangle formed by vectors $r_{n-1}, r_n, -c_ne_n$.

the angle between e_{-1} and r_0 and belongs to the interval $(\pi/4, \pi/2)$.

Next, we set dictionary element e_1 to an arbitrary unit vector with the following two properties: its orthogonal projection on the plane $< e_{-1}, e_0 >$ lies on the line with directing vector r_0, and the angle between this vector and r_0 also equals α_0. We also find the coefficient c_1 such that for $r_1 = r_0 - c_1e_1$ scalar products $(r_1, -e_1), (r_1, e_0), (r_1, e_{-1})$ are equal or, equivalently, angles between r_1 and the vectors e_{-1}, e_0, $-e_1$ are equal. We denote this angle by α_1. Note that $\alpha_1 \in (\alpha_0, \pi/2)$: the formal justification of this fact can be based, e.g., on the equality $\cos\alpha_1 = \cos\alpha_0\cos(\pi - \alpha_0 - \alpha_1)$, which directly follows from the spherical law of cosines.

Similarly, we set e_2 to an arbitrary unit vector with an orthogonal projection on the subspace $< e_{-1}, e_0, e_1 >$ lying on the line with directing vector r_1 and the angle between this vector and r_1 equal α_1, and select the coefficient c_2 in such a way that for $r_2 = r_1 - c_2e_2$ scalar products $(r_2, -e_2)$, $(r_2, -e_1)$, (r_2, e_0), (r_2, e_{-1}) are equal or, equivalently, angles between r_2 and the vectors e_{-1}, e_0, $-e_1$, $-e_2$ are equal. We denote this angle by α_2. Again it is easy to see that $\alpha_2 \in (\alpha_1, \pi/2)$, as due to the spherical law of cosines $\cos\alpha_2 = \cos\alpha_1\cos(\pi-\alpha_1-\alpha_2)$.

We continue the construction inductively. Namely, after constructing $\{r_j\}_{j=0}^{n-1}$, $\{e_j\}_{j=-1}^{n-1}$, $\{c_j\}_{j=1}^{n-1}$ and $\{\alpha_j\}_{j=0}^{n-1}$ we set e_n to an arbitrary unit vector with an orthogonal projection on the subspace $< e_{-1}, e_0, \dots, e_{n-1} >$ lying on the line with directing vector r_{n-1} and angle between this vector and r_{n-1} equal α_{n-1}, and select the coefficient c_n in such a way that for $r_n = r_{n-1} - c_ne_n$ scalar products $(r_n, -e_n), (r_n, -e_{n-1}), \dots, (r_n, -e_1), (r_n, e_0)$, (r_n, e_{-1}) are equal or, equivalently, angles between r_n and the vectors e_{-1}, e_0, $-e_1$, $-e_2$, $\dots$, $-e_{n-1}$, $-e_n$ are equal. We denote this angle by α_n and note that as $\cos\alpha_n = \cos\alpha_{n-1}\cos(\pi - \alpha_{n-1}-\alpha_n)$ due to the spherical law of cosines, $\alpha_n \in (\alpha_{n-1}, \pi/2)$.

Let us justify formally that it is possible to find e_n and c_n with the required properties. Let θ_n denote an arbitrary unit vector orthogonal to the subspace $< e_{-1}, e_0, \dots, e_{n-1} >$ (which also contains vector r_{n-1}), and E_n denote the set of vectors $\{e_{-1}, e_0, -e_1, -e_2, \dots, -e_{n-1}\}$. As e_n we can take any of two unit vectors from the plane $< r_{n-1}, \theta_n >$ that have an angle with r_{n-1} equal to α_{n-1}. Let $e_n = a_nr_{n-1} + b_n\theta_n$; note that $a_n > 0$ as the angle α_{n-1} is acute. By construction for all $e \in E_n$ the scalar product (r_{n-1}, e) equals $\|r_{n-1}\|\cos\alpha_{n-1}$. For the sake of brevity we denote $\|r_{n-1}\|\cos\alpha_{n-1}$ by β_n. Hence for all $e \in E$, as $\theta_n \perp e$, scalar products of e_n and e are the same and equal $a_n\beta_n$. Consequently, for an arbitrary positive c and every $e \in E_n$ we have equalities $(r_{n-1} - ce_n, -e_n) = -\beta_n + c$, $(r_{n-1} - ce_n, e) = \beta_n - ca_n\beta_n$. Thus, it remains to set c_n to the solution of the linear equation $-\beta_n + c = \beta_n - ca_n\beta_n$, i.e., to $2\beta_n/(1 + a_n\beta_n)$.

4.2. Realization of Greedy Expansion. We note that a possible realization of greedy expansion in the dictionary

$$D = \{\pm e_{-1}, \pm e_0, \pm e_1, \pm e_2, \dots\} \tag{16}$$

with the prescribed coefficients $\mathscr{C} = \{c_n\}_{n=1}^{\infty}$ (and the weakness parameter $t = 1$) is a realization in which e_n is selected as an expanding element at the n-th step, and hence n-th remainder coincides with r_n. Indeed, an angle between a vector and its orthogonal projection on a subspace does not exceed an angle between this vector and any non-zero vector from the subspace. Thus while for every $n \in \{1, 2, 3, \dots\}$ and every $e \in \{\pm e_{-1}, \pm e_0, \pm e_1, \pm e_2, \dots, \pm e_n\}$ we have an equality

$$|(r_{n-1}, e)| = \|r_{n-1}\|\cos\alpha_{n-1} \tag{17}$$

and for $e = e_n$ the scalar product (r_{n-1}, e) is positive, for $e = \pm e_{n+j}$ $(j \in \{1, 2, 3, \dots\})$ we have an inequality

$$|(r_{n-1}, e)| \le \|r_{n-1}\|\cos\alpha_{n+j-1} \le \|r_{n-1}\|\cos\alpha_{n-1}. \tag{18}$$

4.3. Absence of Convergence. In this subsection we show that the greedy expansion $\sum_{n=1}^{\infty} c_ne_n$ does not converge to f or, equivalently, remainders r_n do not converge to zero.

First we find the limit of α_n. It exists as $\{\alpha_n\}_{n=1}^{\infty}$ is a nondecreasing sequence with all values belonging to the interval $(\pi/4, \pi/2)$. As noted above, for all $n \in \{0, 1, 2, \dots\}$ the equality

$$\cos\alpha_{n+1} = \cos\alpha_n\cos(\pi - \alpha_n - \alpha_{n+1}) \tag{19}$$

holds. Consequently, if α denotes the limit $\lim_{n\to\infty}\alpha_n$, then $\cos(\pi - 2\alpha)\cos\alpha = \cos\alpha$. It implies that $\alpha = \pi/2$.

Now we show that $\sum_{n=0}^{\infty}(1/\tan\alpha_n) = \infty$. Indeed,

$$\begin{aligned}
&\cos(\pi - \alpha_n - \alpha_{n+1})\cos\alpha_n = \cos\alpha_{n+1} \\
&\quad\Longrightarrow \cos(\alpha_n + \alpha_{n+1})\cos\alpha_n + \cos\alpha_{n+1} = 0 \\
&\quad\Longrightarrow (\cos\alpha_n\cos\alpha_{n+1} - \sin\alpha_n\sin\alpha_{n+1})\cos\alpha_n \\
&\qquad + \cos\alpha_{n+1} = 0 \\
&\quad\Longrightarrow (\cos\alpha_n - \sin\alpha_n\tan\alpha_{n+1})\cos\alpha_n + 1 = 0 \\
&\quad\Longrightarrow \cos^2\alpha_n - \sin\alpha_n\tan\alpha_{n+1}\cos\alpha_n + 1 = 0 \\
&\quad\Longrightarrow \tan\alpha_{n+1} = \frac{\cos^2\alpha_n + 1}{\sin\alpha_n\cos\alpha_n} = \frac{(\cos 2\alpha_n + 1)/2 + 1}{(\sin 2\alpha_n)/2} \\
&= \frac{\cos 2\alpha_n + 3}{\sin 2\alpha_n}
\end{aligned}$$

$$= \frac{\left(1 - \tan^2\alpha_n\right) / \left(1 + \tan^2\alpha_n\right) + 3}{2\tan\alpha_n / \left(1 + \tan^2\alpha_n\right)} = \frac{2\tan^2\alpha_n + 4}{2\tan\alpha_n}$$

$$= \tan\alpha_n + \frac{2}{\tan\alpha_n}. \tag{20}$$

Thus

$$\tan\alpha_{n+1} = \tan\alpha_0 + 2\left(\frac{1}{\tan\alpha_0} + \frac{1}{\tan\alpha_1} + \dots + \frac{1}{\tan\alpha_n}\right). \tag{21}$$

As $\alpha_n \longrightarrow \pi/2$, $\tan\alpha_n \longrightarrow \infty$ and hence $\sum_{n=0}^{\infty}(1/\tan\alpha_n) = \infty$.

The law of sines gives the equalities

$$\frac{\|r_0\|}{\sin\alpha_1} = \frac{\|r_1\|}{\sin\alpha_0} \Longrightarrow$$

$$\|r_1\| = \frac{\sin\alpha_0}{\sin\alpha_1}\|r_0\|;$$

$$\frac{\|r_1\|}{\sin\alpha_2} = \frac{\|r_2\|}{\sin\alpha_1} \Longrightarrow$$

$$\|r_2\| = \frac{\sin\alpha_1}{\sin\alpha_2}\|r_1\|; \tag{22}$$

$$\vdots$$

$$\frac{\|r_{n-1}\|}{\sin\alpha_n} = \frac{\|r_n\|}{\sin\alpha_{n-1}} \Longrightarrow$$

$$\|r_n\| = \frac{\sin\alpha_{n-1}}{\sin\alpha_n}\|r_{n-1}\|.$$

Consequently

$$\|r_n\| = \frac{\sin\alpha_{n-1}}{\sin\alpha_n}\frac{\sin\alpha_{n-2}}{\sin\alpha_{n-1}}\dots\frac{\sin\alpha_0}{\sin\alpha_1}\|r_0\| = \frac{\|r_0\|\sin\alpha_0}{\sin\alpha_n}. \tag{23}$$

Taking into consideration the convergence $\alpha_n \longrightarrow \pi/2$ $(n \longrightarrow \infty)$, we derive from this equality that

$$\|r_n\| \underset{n\longrightarrow\infty}{\not\longrightarrow} 0, \tag{24}$$

so the absence of convergence to the expanded element is proved.

4.4. Estimate of c_n. Here we first show that $\sum_{n=1}^{\infty} c_n = \infty$. Applying again the law of sines, but this time for the other pairs of angles, we get equalities

$$\frac{\|r_0\|}{\sin\alpha_1} = \frac{c_1}{\sin\left(\pi - \alpha_0 - \alpha_1\right)} \Longrightarrow$$

$$c_1 = \|r_0\|\frac{\sin\left(\alpha_0 + \alpha_1\right)}{\sin\alpha_1};$$

$$\vdots$$

$$\frac{\|r_n\|}{\sin\alpha_{n+1}} = \frac{c_{n+1}}{\sin\left(\pi - \alpha_n - \alpha_{n+1}\right)} \Longrightarrow$$

$$c_{n+1} = \|r_n\|\frac{\sin\left(\alpha_n + \alpha_{n+1}\right)}{\sin\alpha_{n+1}}. \tag{25}$$

Let us use equality (23):

$$c_{n+1} = \|r_0\|\sin\alpha_0\frac{\sin\left(\alpha_n + \alpha_{n+1}\right)}{\sin\alpha_{n+1}\sin\alpha_n}$$

$$= \|r_0\|\sin\alpha_0\frac{\sin\alpha_n\cos\alpha_{n+1} + \cos\alpha_n\sin\alpha_{n+1}}{\sin\alpha_{n+1}\sin\alpha_n} \tag{26}$$

$$= \|r_0\|\sin\alpha_0\left(\frac{1}{\tan\alpha_n} + \frac{1}{\tan\alpha_{n+1}}\right).$$

Taking into account the monotonicity of $\{\alpha_n\}$, it implies the inequality

$$c_{n+1} > \frac{2\|r_0\|\sin\alpha_0}{\tan\alpha_{n+1}}. \tag{27}$$

Consequently, as the series $\sum_{n=0}^{\infty}(1/\tan\alpha_n)$ diverges, the series $\sum_{n=1}^{\infty} c_n$ also diverges.

Then we show that $c_n \leqslant 1/\sqrt{n}$. The equality (26) and monotonicity of $\{\alpha_n\}$ imply that

$$c_{n+1} < \frac{2\|r_0\|\sin\alpha_0}{\tan\alpha_n} \quad (n \in \{0, 1, 2, \dots\}) \tag{28}$$

It remains to establish the inequality $\tan\alpha_n > \sqrt{n+1}$: as $\|r_0\| \leqslant 1/2$, it directly gives the required upper estimate of c_n. For $n = 0$ this inequality holds because $\alpha_0 \in (\pi/4, \pi/2)$. Next, from equality (20), taking into account that the minimum value of the function $f(x) = x + 2/x$ on the positive semi-axis equals $2\sqrt{2}$, we obtain that $\tan\alpha_1 \geqslant 2\sqrt{2} > \sqrt{2}$. Besides, $f(x)$ is an increasing function on $[\sqrt{2}, +\infty)$, and the justification of the inequality can be completed by induction:

$$\tan\alpha_{n+1} = f\left(\tan\alpha_n\right) \geqslant f\left(\sqrt{n+1}\right)$$

$$= \sqrt{n+1} + \frac{2}{\sqrt{n+1}} > \sqrt{n+2}. \tag{29}$$

Justification of this inequality completes the proof of the upper estimate of c_n and in total the proof Theorem 3.

We note that the construction described in the proof of Theorem 3 can be straightforwardly adapted from the case of greedy expansions with prescribed coefficients to the generalized Approximate Weak Greedy Algorithm (see [10, Theorem 3]).

5. The Case of Finite-Dimensional Spaces

In this section we prove the following theorem.

Theorem 5. *Let the space H be finite-dimensional, and let the coefficients $\mathscr{C} = \{c_n\}_{n=1}^{\infty}$ converge to zero monotonically and* *satisfy the condition* $\sum_{n=1}^{\infty} c_n = \infty$. *Then for every symmetric unit-normed dictionary* $D \subset H$ *and any* $f \in H$ *all realizations of greedy expansion of* f *in* D *with the prescribed coefficients* $\mathscr{C}$ *and the weakness parameter* $t = 1$ *converge to* f.

Thus in the finite-dimensional case at least for $t = 1$ in the natural class of monotonic coefficient sequences conditions sufficient for the convergence of a greedy expansion to an expanded element include only the convergence of coefficients to zero and infinity of their sum. Clearly, convergence of coefficients to zero is also a necessary condition for the convergence of greedy expansion.

As the first step of the proof we note that due to Lemma 4 there exists a limit of norms of remainders of a greedy expansion with prescribed coefficients $\lim_{n\to\infty}\|r_n\|$. Let us suppose that this limit is non-zero. Then starting from a certain index remainder norms are separated from zero. Due to the compactness of a sphere in a finite-dimensional space and the completeness and the symmetric property of the dictionary it implies that scalar products

$$x_{n+1} = (r_n, e_{n+1}) = \sup_{e\in D} (r_n, e) \tag{30}$$

are also separated from zero. In other words, there exist a positive number γ and an index N_0 such that for all $n > N_0$ the inequality $x_{n+1} > \gamma$ holds. In addition there exists such an index $N_1 > N_0$ that for all $n > N_1$ coefficient c_n does not exceed γ. But consequently for $n \geqslant N_1$ it follows from equality (10) that

$$\begin{aligned} \|r_{n+1}\|^2 &= \|r_n\|^2 - x_{n+1}^2 + (c_{n+1} - x_{n+1})^2 \\ &= \|r_n\|^2 - c_{n+1}(2x_{n+1} - c_{n+1}) \\ &\leqslant \|r_n\|^2 - \gamma c_{n+1}. \end{aligned} \tag{31}$$

Hence

$$\|r_{N_1+K}\|^2 \leqslant \|r_{N_1}\|^2 - \gamma \sum_{n=N_1+1}^{N_1+K} c_n \longrightarrow -\infty \quad (K \longrightarrow \infty). \tag{32}$$

This contradiction completes the proof of Theorem 5.

6. Conclusion

The main results of the paper state that in Hilbert spaces a greedy expansion with prescribed coefficients converges to an expanded element if coefficients satisfy certain relatively weak conditions that do not include monotonicity, and these conditions can not be essentially relaxed. At the same time we showed that for the finite-dimensional case the relaxation is possible.

Conflicts of Interest

The authors declare that there are no conflicts of interest regarding the publication of this article.

Acknowledgments

We thank Prof. T.P. Lukashenko and Dr. E.D. Livshits for valuable comments and discussions. The research was supported by the Russian Federation Government Grant No. 14.W03.31.0031.

References

[1] B. S. Kashin and A. A. Saakyan, *Orthogonal series*, vol. 75 of *Translations of Mathematical Monographs*, American Mathematical Society, Providence, Rhode Island, RI, USA, 1989.

[2] V. N. Temlyakov, "Weak greedy algorithms," *Advances in Computational Mathematics*, vol. 12, no. 2-3, pp. 213–227, 2000.

[3] V. N. Temlyakov, "Greedy expansions in Banach spaces," *Advances in Computational Mathematics*, vol. 26, no. 4, pp. 431–449, 2007.

[4] J. H. Friedman and W. Stuetzle, "Projection pursuit regression," *Journal of the American Statistical Association*, vol. 76, no. 376, pp. 817–823, 1981.

[5] L. K. Jones, "A simple lemma on greedy approximation in Hilbert space and convergence rates for projection pursuit regression and neural network training," *The Annals of Statistics*, vol. 20, no. 1, pp. 608–613, 1992.

[6] S. G. Mallat and Z. Zhang, "Matching pursuits with time-frequency dictionaries," *IEEE Transactions on Signal Processing*, vol. 41, no. 12, pp. 3397–3415, 1993.

[7] V. Temlyakov, "Greedy algorithms with prescribed coefficients," *Journal of Fourier Analysis and Applications*, vol. 13, no. 1, pp. 71–86, 2007.

[8] V. V. Galatenko, "Expansion in a system of functions with fixed coefficients," *Moscow University Mathematics Bulletin*, vol. 53, no. 1, pp. 12–15, 2003.

[9] O. Rassudova, Zhadnye razlozhenija v gil'bertovyh prostranstvah [Greedy expansions in Hilbert spaces; in Russian], Materialy Voronezhskoj zimnej matematicheskoj shkoly "Sovremennye metody teorii funkcij i smezhnye problemy" [Proc. Voronezh Winter Mathematical School, Contemporary Methods in Theory of Functions and Adjacent Problems], Voronezh: VSU Publishing House, pp. 151-152, 2009.

[10] V. V. Galatenko and E. D. Livshits, "Generalized approximate weak greedy algorithms," *Mathematical Notes*, vol. 78, no. 1-2, pp. 170–184, 2005.

8

Coincidence Point Theorems for (α, β, γ)-Contraction Mappings in Generalized Metric Spaces

Chaiporn Thangthong and Anchalee Khemphet

Center of Excellence in Mathematics and Applied Mathematics, Department of Mathematics, Faculty of Science, Chiang Mai University, Chiang Mai 50200, Thailand

Correspondence should be addressed to Anchalee Khemphet; anchalee.k@cmu.ac.th

Academic Editor: Nawab Hussain

The result of our study is that a coincidence point of two mappings P and Q can be achieved when the ordered pair (P, Q) is an (α, β, γ)-contraction with respect to a generalized metric space. Moreover, with some additional condition, a common fixed point can be obtained as a consequence of our main theorems. Further, we apply our findings to some examples and integral equation problems.

1. Introduction

There has been a wide range of research in discovering fixed points, or the only fixed point, of certain types of mappings that are contractions in the past. Many aspects have been used to accomplish the result. At the very beginning, Geraghty [1] generally developed the Banach contraction principle by considering the class Θ whose elements are functions $\theta : [0, \infty) \to [0, 1)$ such that

$$\begin{aligned} \theta(t_n) &\longrightarrow 1 \\ &\Longrightarrow t_n \to 0. \end{aligned} \tag{1}$$

In 2012, Samet et al. [2] studied the existing results for α-ψ-contractions. His concept was given in the following definition. Suppose that $X \neq \emptyset$ and α is a real-valued function on $X \times X$.

Definition 1 (see [2]). Let A be a self-mapping on X and $u, v \in X$. If $\alpha(Au, Av) \geq 1$ whenever $\alpha(u, v) \geq 1$, then we say that A is α-admissible.

Later, Karapinar [3] added more conditions to Definition 1.

Definition 2 (see [3]). Let A be an α-admissible self-mapping on X and $u, v, w \in X$. If $\alpha(u, w) \geq 1$ and $\alpha(w, v) \geq 1$ imply $\alpha(u, v) \geq 1$, then we say that A is triangular α-admissible.

Furthermore, another essential part in this topic is a metric space. There were a large number of literatures that worked not only on a metric space, but also on other topological spaces; for examples, see [4–6]. Three years ago, Jleli and Samet [7] defined a generalized metric, known as a JS-metric. The advantage of their idea is that many topological spaces are covered by the JS-metric space. With this reason, results of fixed point theorems on JS-metric spaces have been recently interesting (e.g., see [8]).

Let $D : X \times X \to [0, \infty]$ be a function and, for $u \in X$, denote a set of sequences $\{u_n\}$ in X such that $\lim_{n\to\infty} D(u_n, u) = 0$ by $C(D, X, u)$.

Definition 3 (see [7]). Suppose that, for any $u, v \in X$,

(D_1) if $D(u, v) = 0$, then $u = v$;

(D_2) $D(u, v) = D(v, u)$;

(D_3) there is a $K > 0$ so that, for each $w, z \in X$, if $\{u_n\} \in C(D, X, w)$, then $D(w, z) \leq K \limsup_{n\to\infty} D(u_n, z)$.

Then, we call D a generalized metric, or a JS-metric on X, and also (X, D) a generalized metric space, or a JS-metric space.

Definition 4 (see [7]). Suppose that (X, D) is a JS-metric space and $\{u_n\} \subseteq X$.

(1) If $\{u_n\} \in C(D, X, u)$ for some $u \in X$, then we say that $\{u_n\}$ D-converges to u.

(2) If $\lim_{m,n\to\infty} D(u_n, u_m) = 0$, then we say that $\{u_n\}$ is D-Cauchy.

(3) If any D-Cauchy sequence $\{u_n\}$ in X D-converges to some u in X, then we say that (X, D) is D-complete.

Proposition 5 (see [7]). *Suppose that (X, D) is a JS-metric space. Let $\{u_n\} \subseteq X$ and $u, v \in X$. If $\{u_n\} \in C(D, X, u)$ and $\{u_n\} \in C(D, X, v)$, then $u = v$.*

Definition 6 (see [7]). Suppose that (X, D) is a JS-metric space. Let P be a self-mapping on X and $\{u_n\} \subseteq X$. If $\{u_n\} \in C(D, X, u_0)$ implies $\{Pu_n\} \in C(D, X, Pu_0)$ for some $u_0 \in X$, then we say that P is continuous at u_0. Moreover, if P is continuous at every $u \in X$, then we say P is continuous.

Next, Martínez-Moreno et al. [9] had a new perspective to obtain common fixed points of particular contractive mappings on a space with two metrics.

Inspired by the above, we consider some existence results for a coincidence point of two functions when the ordered pair of these functions is an (α, β, γ)-contraction on a space with two JS-metrics. In addition, some examples and an application of an integral equation are presented.

2. Main Results

First, we assume throughout this section that all functions P and Q are self-mappings defined on X.

Definition 7. For $u, v, w \in X$, if

(1) $\alpha(Qu, Qv) \geq 1$ implies $\alpha(Pu, Pv) \geq 1$,

(2) $\alpha(u, w) \geq 1$ and $\alpha(w, v) \geq 1$ imply $\alpha(u, v) \geq 1$,

then we say that P is triangular α-admissible with respect to Q.

Next, we define $\mathscr{B}$ as the class of mappings $\beta : [0, \infty] \to [0, 1)$ such that

$$\beta(t_n) \longrightarrow 1 \Longrightarrow t_n \to 0. \tag{2}$$

Note that $\beta(t)$ is defined at $t = \infty$. Also, let $\gamma : [0, \infty] \to [0, \infty]$ be a nondecreasing continuous function satisfying

$$\gamma(t) = 0 \Longleftrightarrow t = 0. \tag{3}$$

Denote the class of all such functions γ by Υ.

Definition 8. Suppose that (X, D) is a JS-metric space. If

(1) P is triangular α-admissible with respect to Q,

(2) there exist $\beta \in \mathscr{B}$ and $\gamma \in \Upsilon$,

$$\alpha(Qu, Qv)\gamma(D(Pu, Pv)) \leq \beta(D(Qu, Qv))\gamma(D(Qu, Qv)) \tag{4}$$

for all $u, v \in X$,

then we say that the pair (P, Q) is an (α, β, γ)-contraction with respect to D.

Here, we are interested in the existence of a coincidence point of P and Q, where (P, Q) is an (α, β, γ)-contraction with respect to some generalized metric on X. This can be done under suitable relations between P and Q.

Definition 9. Suppose that (X, D) and (Y, D') are two JS-metric spaces and $\{u_n\} \subseteq X$. If $A : X \to Y$ and $B : X \to X$ are functions such that $\{Bu_n\}$ being D-Cauchy in (X, D) implies $\{Au_n\}$ is D'-Cauchy in (Y, D'), then we say that A is B-Cauchy on X.

Last but not least, we need the comparison notations for any two generalized metrics. If D and D' are two generalized metrics on X, the notation $D \geq D'$ represents $D(u, v) \geq D'(u, v)$ for every $u, v \in X$. If the inequality fails for some $u, v \in X$, we use the notation $D \ngeq D'$. All other inequality signs can be defined in the same fashion.

Theorem 10. *Suppose that (X, D') is a D'-complete JS-metric space and D is a generalized metric on X. If*

(i) (P, Q) is an (α, β, γ)-contraction with respect to D,

(ii) $P(X)$ is a subspace of $Q(X)$,

(iii) there is a $u_0 \in X$, $\alpha(Qu_0, Pu_0) \geq 1$, and $\sup\{D(Qu_0, Pv) : v \in X\} < \infty$,

(iv) $P : (X, D) \to (X, D')$ is Q-Cauchy on X whenever $D \ngeq D'$,

(v) P and Q commute,

(vi) $P, Q : (X, D') \to (X, D')$ are continuous,

then there exists a coincidence point of P and Q.

Proof. From assumption *(iii)*, let $u_0 \in X$ such that $\alpha(Qu_0, Pu_0) \geq 1$ and $\sup\{D(Qu_0, Pv) : v \in X\} < \infty$. Since $P(X) \subseteq Q(X)$ and $Pu_0 \in X$, a sequence $\{u_n\}$ in X can be constructed such that

$$Qu_n = Pu_{n-1} \tag{5}$$

for all $n \in \mathbb{N}$. Observe that if $Qu_{n_0} = Qu_{n_0-1}$ for some $n_0 \in \mathbb{N}$, then $Qu_{n_0-1} = Pu_{n_0-1}$, and so we are done. Assume that $Qu_n \neq Qu_{n-1}$ for each $n \in \mathbb{N}$.

Since $\alpha(Qu_0, Qu_1) = \alpha(Qu_0, Pu_0) \geq 1$ and P is triangular α-admissible with respect to Q, $\alpha(Qu_1, Qu_2) = \alpha(Pu_0, Pu_1) \geq 1$. Repeating this process inductively, we obtain that

$$\alpha(Qu_n, Qu_{n+1}) \geq 1 \tag{6}$$

for each $n \in \mathbb{N}$.

Our task is now to prove that $\{Qu_n\}$ is D-Cauchy.

Assume that this is not true. Equivalently, there is an $\epsilon > 0$ so that, for each $k \in \mathbb{N}$,

$$D(Qu_{n_k}, Qu_{m_k}) \geq \epsilon \tag{7}$$

for some $m_k \geq n_k \geq k$. By inequality (6), together with the assumption that P is triangular α-admissible with respect to Q, we have that

$$\alpha(Qu_{n_k}, Qu_{m_k}) \geq 1 \tag{8}$$

for all $k \in \mathbb{N}$. Then, by assumption *(i)*,

$$\begin{aligned}\gamma\left(D\left(Qu_{n_k+1}, Qu_{m_k+1}\right)\right) &= \gamma\left(D\left(Pu_{n_k}, Pu_{m_k}\right)\right) \\ &\leq \alpha\left(Qu_{n_k}, Qu_{m_k}\right)\gamma\left(D\left(Pu_{n_k}, Pu_{m_k}\right)\right) \\ &\leq \beta\left(D\left(Qu_{n_k}, Qu_{m_k}\right)\right)\gamma\left(D\left(Qu_{n_k}, Qu_{m_k}\right)\right).\end{aligned} \quad (9)$$

Continuing to apply this concept totally n_k+1 times, we finally get the equality

$$\begin{aligned}&\gamma\left(D\left(Qu_{n_k+1}, Qu_{m_k+1}\right)\right) \\ &\leq \prod_{i=0}^{n_k}\beta\left(D\left(Qu_{n_k-i}, Qu_{m_k-i}\right)\right)\gamma\left(D\left(Qu_0, Qu_{m_k-n_k}\right)\right).\end{aligned} \quad (10)$$

Let $0 \leq i_k \leq n_k$ such that

$$\begin{aligned}&\beta\left(D\left(Qu_{n_k-i_k}, Qu_{m_k-i_k}\right)\right) \\ &\quad = \max\left\{\beta\left(D\left(Qu_{n_k-i}, Qu_{m_k-i}\right)\right) : 0 \leq i \leq n_k\right\}.\end{aligned} \quad (11)$$

Denote $\delta = \limsup_{k\to\infty}\beta(D(Qu_{n_k-i_k}, Qu_{m_k-i_k}))$.

Notice that if $\delta < 1$, then $\lim_{k\to\infty}\gamma(D(Qu_{n_k+1}, Qu_{m_k+1})) = 0$, and so

$$\lim_{k\to\infty} D\left(Qu_{n_k+1}, Qu_{m_k+1}\right) = 0 \quad (12)$$

which contradicts inequality (7).

Thus, $\delta = 1$. Then, there is a subsequence of $\{\beta(D(Qu_{n_k-i_k}, Qu_{m_k-i_k}))\}$ which converges to 1. Without loss of generality, assume that

$$\lim_{k\to\infty}\beta\left(D\left(Qu_{n_k-i_k}, Qu_{m_k-i_k}\right)\right) = 1. \quad (13)$$

By the definition of β,

$$\lim_{k\to\infty} D\left(Qu_{n_k-i_k}, Qu_{m_k-i_k}\right) = 0. \quad (14)$$

Therefore, there exists a $k_0 \in N$ such that

$$D\left(Qu_{n_{k_0}-i_{k_0}}, Qu_{m_{k_0}-i_{k_0}}\right) < \frac{\epsilon}{2}. \quad (15)$$

It follows that, also from inequality (7),

$$\begin{aligned}\gamma(\epsilon) &\leq \gamma\left(D\left(Qu_{n_{k_0}}, Qu_{m_{k_0}}\right)\right) \\ &\leq \prod_{j=1}^{i_{k_0}}\beta\left(D\left(Qu_{n_{k_0}-j}, Qu_{m_{k_0}-j}\right)\right) \\ &\quad\cdot\gamma\left(D\left(Qu_{n_{k_0}-i_{k_0}}, Qu_{m_{k_0}-i_{k_0}}\right)\right) \\ &< \gamma\left(D\left(Qu_{n_{k_0}-i_{k_0}}, Qu_{m_{k_0}-i_{k_0}}\right)\right) \leq \gamma\left(\frac{\epsilon}{2}\right).\end{aligned} \quad (16)$$

This is a contradiction since γ is nondecreasing. Thus, $\{Qu_n\}$ is D-Cauchy.

The next goal is to show that $\{Qu_n\}$ is also D'-Cauchy. It can be observed that if $D \geq D'$, we are done. Assume that $D \ngeq D'$. Since P is Q-Cauchy on X, $\{Pu_n\}$ is D'-Cauchy. Consequently,

$$\begin{aligned}\lim_{n,m\to\infty} D'\left(Qu_{n+1}, Qu_{m+1}\right) &= \lim_{n,m\to\infty} D'\left(Pu_n, Pu_m\right) \\ &= 0,\end{aligned} \quad (17)$$

and so $\{Qu_n\}$ is D'-Cauchy.

Since (X, D') is D'-complete, one can find a $c \in X$ satisfying

$$\lim_{n\to\infty} D'\left(Qu_n, c\right) = \lim_{n\to\infty} D'\left(Pu_n, c\right) = 0. \quad (18)$$

That is,

$$\{Qu_n\}, \{Pu_n\} \in C\left(D', X, c\right). \quad (19)$$

By assumption *(vi)*,

$$\begin{aligned}&\{PQu_n\} \in C\left(D', X, Pc\right) \\ &\text{and } \{QPu_n\} \in C\left(D', X, Qc\right).\end{aligned} \quad (20)$$

Since P and Q commute, $Pc = Qc$. This completes the proof.

As a consequence of Theorem 10, if $D = D'$, then we have the following theorem. Besides, we can replace properties (v) and (vi) by other conditions as stated in the theorem below.

Theorem 11. *Suppose that (X, D) is a D-complete JS-metric space. Assume that*

(i) (P, Q) is an (α, β, γ)-contraction with respect to D;

(ii) $P(X)$ is a subspace of $Q(X)$;

(iii) there is a $u_0 \in X$, $\alpha(Qu_0, Pu_0) \geq 1$, and $\sup\{D(Qu_0, Pv) : v \in X\} < \infty$;

(iv) either (a) or (b) holds:

(a) P and Q are continuous mappings that commute.

(b) $Q(X)$ is D-complete and, for any $\{u_n\} \subseteq X$ satisfying $\alpha(u_n, u_{n+1}) \geq 1$ for each $n \in \mathbb{N}$, if there is a $c \in X$ such that $\{u_n\} \in C(D, X, c)$, then $\alpha(u_n, c) \geq 1$ for all $n \in \mathbb{N}$.

Then, there exists a coincidence point of P and Q.

Proof. It is easy to see that if condition *(a)* is true, then applying Theorem 10 to the case $D = D'$ yields the desired result. Assume that statement *(a)* does not hold. Thus, *(b)* must be valid. Let $\{u_n\} \subseteq X$ and $Qu_n = Pu_{n-1}$ for each $n \in \mathbb{N}$. An argument similar to the one used in Theorem 10 shows that $\{Qu_n\}$ is D-Cauchy and $\alpha(Qu_n, Qu_{n+1}) \geq 1$ for all $n \in \mathbb{N}$. From *(b)*,

$$\lim_{n\to\infty} D\left(Qu_n, Qc\right) = \lim_{n\to\infty} D\left(Pu_n, Qc\right) = 0 \quad (21)$$

for some $c \in X$. That is,

$$\{Qu_n\}, \{Pu_n\} \in C(D, X, Qc). \quad (22)$$

Again, since condition *(b)* holds, for any $n \in \mathbb{N}$, $\alpha(Qu_n, Qc) \geq 1$. From (4) and assumption *(i)*, it follows that

$$\begin{aligned}\gamma(D(Pu_n, Pc)) &\leq \alpha(Qu_n, Qc)\gamma(D(Pu_n, Pc)) \\ &\leq \beta(D(Qu_n, Qc))\gamma(D(Qu_n, Qc)) \quad (23) \\ &< \gamma(D(Qu_n, Qc)).\end{aligned}$$

Since γ is nondecreasing,

$$D(Pu_n, Pc)) < D(Qu_n, Qc), \quad (24)$$

and so

$$\lim_{n\to\infty} D(Pu_n, Pc)) = 0. \quad (25)$$

Consider

$$D(Qc, Pc) \leq K \limsup_{n\to\infty} D(Pu_n, Pc) \quad (26)$$

for some $K > 0$. Therefore, $D(Qc, Pc) = 0$. Hence, $Qc = Pc$, completing the proof.

Adding some extra condition to Theorem 10, the coincidence point is actually a common fixed point. This can be shown in the following theorem. Denote

$$C(P, Q) = \{u \in X : Pu = Qu\}. \quad (27)$$

Theorem 12. *Suppose that (X, D') is a D'-complete JS-metric space and D is a generalized metric on X. If all assumptions (i)-(vi) in Theorem 10 are satisfied and $\alpha(Qu, Qv) \geq 1$ whenever $Qu \neq Qv$, where $u, v \in C(P, Q)$, then there exists a common fixed point of P and Q.*

Proof. According to Theorem 10, $C(P, Q) \neq \emptyset$. Then, we can let $u, v \in X$ so that $Pu = Qu$ and $Pv = Qv$.

Suppose that $Qu \neq Qv$. By the assumption, $\alpha(Qu, Qv) \geq 1$. From the fact that (P, Q) is an (α, β, γ)-contraction with respect to D, we have that

$$\begin{aligned}\gamma(D(Pu, Pv)) &\leq \alpha(Qu, Qv)\gamma(D(Pu, Pv)) \\ &\leq \beta(D(Qu, Qv))\gamma(D(Qu, Qv)) \quad (28) \\ &< \gamma(D(Qu, Qv)) = \gamma(D(Pu, Pv))\end{aligned}$$

which leads to a contradiction. Therefore, $Qu = Qv$.

Next, let $c = Qu = Pu$. Since P and Q commute, $Qc = QPu = PQu = Pc$. Thus, $c \in C(P, Q)$. Referring to the proof above, we can conclude that $Pc = Qc = Qu = c$. Hence, the proof is complete.

We give examples to illustrate Theorems 10 and 11, respectively.

Example 13. Suppose that $X = [0, 1]$. Given the generalized metrics D and D' on X defined by

$$D(u, v) = \begin{cases} u + v, & u \neq 0 \text{ and } v \neq 0, \\ \dfrac{u}{2}, & v = 0, \\ \dfrac{v}{2}, & u = 0, \end{cases} \quad (29)$$

and

$$D'(u, v) = \begin{cases} L(u + v), & u \neq 0 \text{ and } v \neq 0, \\ \dfrac{Lu}{2}, & v = 0, \\ \dfrac{Lv}{2}, & u = 0, \end{cases} \quad (30)$$

where $u, v \in X$ and L is a real number such that $L > 1$, we have that (X, D') is D'-complete. Let α be a function defined by

$$\alpha(u, v) = \begin{cases} 1, & u, v \in \left[0, \dfrac{1}{4}\right] \text{ with } u \neq 0 \text{ or } v = 0, \\ 0, & \text{otherwise.} \end{cases} \quad (31)$$

Given the self-mappings P and Q on X defined by

$$P(u) = u^4 \text{ and } Q(u) = u^2, \quad (32)$$

some tedious manipulation yields assumptions (ii), (v), and (vi) in Theorem 10. Further, notice that $1/2 \in X$ such that $\alpha(Q(1/2), P(1/2)) = \alpha(1/4, 1/16) \geq 1$ and $\sup\{D(Q(1/2), Pv) : v \in X\} < \infty$.

Claim 1. P is triangular α-admissible with respect to Q.

Let $u, v, w \in X$. Assume that $\alpha(Qu, Qv) \geq 1$. Then, $u^2, v^2 \in [0, 1/4]$ and $Qu = u^2 \neq 0$ or $Qv = v^2 = 0$. Accordingly, $u, v \in [0, 1/2]$, and $u \neq 0$ or $v = 0$. It follows that $u^4, v^4 \in [0, 1/16]$, and $Pu = u^4 \neq 0$ or $Pv = v^4 = 0$. Therefore, $\alpha(Pu, Pv) \geq 1$.

Next, assume that $\alpha(u, w) \geq 1$ and $\alpha(w, v) \geq 1$. It can be observed that if $w = 0$, then $v = 0$, and if $w \neq 0$, then $u \neq 0$. That is, $u \neq 0$ or $v = 0$. Therefore, $\alpha(u, v) \geq 1$. Thus, we have Claim 1.

Claim 2. (P, Q) is an (α, β, γ)-contraction with respect to D, where $\beta \in \mathscr{B}$ and $\gamma \in \Upsilon$ are as follows:

$$\gamma(t) = \frac{t}{2}$$
$$\text{and } \beta(t) = \begin{cases} \dfrac{1}{4}, & 0 \leq t < 1, \\ \dfrac{1}{t^2 + 2}, & t \geq 1 \end{cases} \quad (33)$$

for $t \in [0, \infty]$.

Given $u, v \in X$, if $\alpha(Qu, Qv) = 0$, inequality (4) holds. Assume that $\alpha(Qu, Qv) \geq 1$. Similar as above, $u, v \in [0, 1/2]$. Consider the following cases.

Case 1. $Qv = 0$. We have that

$$\alpha(Qu, Qv)\gamma(D(Pu, Pv)) = \frac{Pu/2}{2} = \frac{u^4}{4} \leq \frac{1}{4}\left(\frac{u^2}{4}\right) \\ = \beta(D(Qu, Qv))\gamma(D(Qu, Qv)). \tag{34}$$

Case 2. $Qv \neq 0$. Then, $Qu \neq 0$. Consider

$$\alpha(Qu, Qv)\gamma(D(Pu, Pv)) = \frac{Pu + Pv}{2} = \frac{u^4 + v^4}{2} \\ \leq \frac{1}{4}\left(\frac{u^2 + v^2}{2}\right) = \beta(D(Qu, Qv))\gamma(D(Qu, Qv)). \tag{35}$$

Therefore, (P, Q) is a (α, β, γ)-contraction with respect to D.

Claim 3. $P : (X, D) \to (X, D')$ is Q-Cauchy on X.

Notice that $D \leq D'$. Suppose that $\{u_n\} \subseteq X$ such that $\{Qu_n\}$ is D-Cauchy. Let $\epsilon > 0$. There is a $k \in \mathbb{N}$ so that, for all $m, n \geq k$, we have that

$$u_n^2 + u_m^2 = D(Qu_n, Qu_m) < \frac{\epsilon}{L}. \tag{36}$$

With this k,

$$D'(Pu_n, Pu_m) \leq L\left(u_n^4 + u_m^4\right) \leq L\left(u_n^2 + u_m^2\right) < \epsilon \tag{37}$$

for any $m, n \geq k$. This completes Claim 3.

Thus, by Theorem 10, P and Q have a coincidence point, precisely, 0.

Example 14. Let $X = [0, 1]$. Therefore, (X, D) is D-complete, where D is the generalized metric as defined in Example 13. Suppose that α is a function as follows:

$$\alpha(u, v) = \begin{cases} 1, & u \neq 0 \text{ or } v = 0, \\ 0, & \text{otherwise.} \end{cases} \tag{38}$$

Define self-mappings P and Q on X by

$$P(u) = \frac{u}{u + 6} \\ \text{and } Q(u) = \frac{u}{3}. \tag{39}$$

Note that $P(X) \subseteq Q(X)$ and $Q(X)$ is D-complete. Moreover, we have $1 \in X$ such that $\alpha(Q1, P1) \geq 1$ and $\sup\{D(Q1, Pv) : v \in X\} < \infty$.

Claim 1. P is triangular α-admissible with respect to Q.

Let $u, v, w \in X$. Assume that $\alpha(Qu, Qv) \geq 1$. Then, $Qu \neq 0$ or $Qu = 0$. That is, $u \neq 0$ or $v = 0$. Thus, $Pu \neq 0$ or $Pv = 0$. Therefore, $\alpha(Pu, Pv) \geq 1$.

Similar as in the proof of the previous example, if $\alpha(u, w) \geq 1$ and $\alpha(w, v) \geq 1$, then $\alpha(u, v) \geq 1$. Therefore, Claim 1 is obtained.

Claim 2. (P, Q) is an (α, β, γ)-contraction with respect to D, where $\beta \in \mathscr{B}$ and $\gamma \in \Upsilon$ defined by $\beta(t) = 1/2$ and $\gamma(t) = t/2$ for $t \in [0, \infty]$, respectively.

Suppose $u, v \in X$. If $\alpha(Qu, Qv) = 0$, then inequality (4) holds. Assume that $\alpha(Qu, Qv) \geq 1$. Consider the following cases.

Case 1. $Qv = 0$. We have that

$$\alpha(Qu, Qv)\gamma(D(Pu, Pv)) = \frac{Pu/2}{2} = \frac{1}{2}\left(\frac{u}{2(u + 6)}\right) \\ \leq \frac{1}{2}\left(\frac{u}{6}\right) = \beta(D(Qu, Qv))\gamma(D(Qu, Qv)). \tag{40}$$

Case 2. $Qv \neq 0$. Then, $Qu \neq 0$. Consider

$$\alpha(Qu, Qv)\gamma(D(Pu, Pv)) = \frac{Pu + Pv}{2} \\ = \frac{1}{2}\left(\frac{u}{u + 6} + \frac{v}{v + 6}\right) \leq \frac{1}{2}\left(\frac{u}{6} + \frac{v}{6}\right) \\ = \beta(D(Qu, Qv))\gamma(D(Qu, Qv)). \tag{41}$$

Therefore, we have Claim 2.

Next, let $\{u_n\} \subseteq X$ such that, for any $n \in \mathbb{N}$, $\alpha(u_n, u_{n+1}) \geq 1$. Assume that $\{u_n\} \in C(D, X, c)$ for some $c \in X$. By the definition of α, for each $n \in \mathbb{N}$, $u_n \neq 0$ or $u_{n+1} = 0$. Fix $n \in \mathbb{N}$. If $u_n \neq 0$, then $\alpha(u_n, c) \geq 1$. Assume that $u_n = 0$. Suppose that $c \neq 0$. Then, $D(u_n, c) = D(0, c) = c/2 \neq 0$. This is a contradiction since $\{u_n\} \in C(D, X, c)$. Thus, $c = 0$. Therefore, we get that $\alpha(u_n, c) \geq 1$. Since n is arbitrary, this is true for every $n \in \mathbb{N}$. Hence, by Theorem 11, $c = 0$ is a coincidence point of P and Q.

3. Application

We wish to apply our finding to the existence problem of a solution to the integral equation. This is one of the crucial uses of fixed point theorems that can be found in the literatures (see [10–13]).

$$u(t) = \int_0^T p(t, s, u(s))\,ds + b(t) \tag{42}$$

for $t \in [0, T]$, where T is a real number such that $T > 0$.

Suppose that $X = C([0, T], \mathbb{R})$ and

$$D(u, v) = \max_{t \in [0,T]} |u(t)| + \max_{t \in [0,T]} |v(t)| \tag{43}$$

for $u, v \in C([0, T], \mathbb{R})$. We have that (X, D) is a D-complete JS-metric space. The following theorem shows when the equation (42) has a solution if the integral equation is homogeneous.

Theorem 15. *Consider equation (42). Suppose that*

(i) $p : [0, T] \times [0, T] \times \mathbb{R} \to \mathbb{R}$ *is continuous;*

(ii) for any $u, v \in \mathbb{R}$, if $u \leq v$, then $p(t,s,u) \leq p(t,s,v)$ and

$$|p(t,s,u)| + |p(t,s,v)| \leq \frac{\ln(1+|u|+|v|)}{T}, \quad (44)$$

where $s, t \in [0,T]$;

(iii) there is a $u_0 \in X$, $u_0(t) \leq \int_0^T p(t,s,u_0(s))ds$, where $t \in [0,T]$;

(iv) $\sup\{\max_{t\in[0,T]}|\int_0^T p(t,s,v(s))ds| : v \in X\} < \infty$.

Then, the integral equation (42) has a solution.

Proof. Define the self-mappings P and Q on X as follows:

$$Pu(t) = \int_0^T p(t,s,u(s))\,ds, \quad (45)$$

and $Qu(t) = u(t)$ for $u \in X$ and $t \in [0,T]$. Suppose that $\alpha : X \times X \to \mathbb{R}$ is a function defined by

$$\alpha(u,v) = \begin{cases} 1, & u(t) \leq v(t) \text{ for any } t \in [0,T], \\ 0, & \text{otherwise.} \end{cases} \quad (46)$$

Clearly, $P(X) \subseteq Q(X)$, and P and Q are continuous mappings that commute. Moreover, by assumptions *(iii)* and *(iv)*, it is straightforward to show that condition *(iii)* of Theorem 11 is satisfied. Our problem reduces to show that the pair (P,Q) is an (α,β,γ)-contraction with respect to D for some $\beta \in \mathscr{B}$ and $\gamma \in \Upsilon$.

First, we establish that P is triangular-α-admissible with respect to Q.

Assume that $\alpha(Qu, Qv) \geq 1$. Then, $Qu(t) \leq Qv(t)$; that is, $u(t) \leq v(t)$ for any $t \in [0,T]$. By assumption (ii), $p(t,s,u) \leq p(t,s,v)$. Therefore,

$$Pu(t) = \int_0^T p(t,s,u(s))\,ds \leq \int_0^T p(t,s,v(s))\,ds = Pv(t). \quad (47)$$

That is, $\alpha(Pu,Pv) \geq 1$. It is simple to show that the second condition of Definition 7 holds. Thus, P is triangular-α-admissible with respect to Q.

Finally, it remains to prove inequality (4).

If $u(t) > v(t)$ for some $t \in [0,T]$, we are done. Suppose that $u(t) \leq v(t)$ for all $t \in [0,T]$. From assumption *(ii)*, consider

$$\begin{aligned} &|Pu(t)| + |Pv(t)| \\ &\leq \int_0^T |p(t,s,u(s))| + |p(t,s,v(s))|\,ds \\ &\leq \frac{1}{T}\int_0^T \ln(1+|u(s)|+|v(s)|)\,ds \\ &\leq \ln\left(1+\max_{t\in[0,T]}|Qu(t)| + \max_{t\in[0,T]}|Qv(t)|\right), \end{aligned} \quad (48)$$

and, then,

$$\begin{aligned} &\ln(|Pu(t)|+|Pv(t)|+1) \\ &\leq \ln\left(\int_0^T |p(t,s,u(s))| + |p(t,s,v(s))|\,ds + 1\right) \\ &\leq \ln\left(\frac{1}{T}\int_0^T \ln(1+|u(s)|+|v(s)|)\,ds + 1\right) \\ &\leq \ln\left(\ln\left(1+\max_{t\in[0,T]}|Qu(t)| + \max_{t\in[0,T]}|Qv(t)|\right)+1\right) \\ &= \frac{\ln(\ln(1+\max_{t\in[0,T]}|Qu(t)|+\max_{t\in[0,T]}|Qv(t)|)+1)}{\ln(1+\max_{t\in[0,T]}|Qu(t)|+\max_{t\in[0,T]}|Qv(t)|)} \\ &\cdot \ln\left(1+\max_{t\in[0,T]}|Qu(t)|+|Qv(t)|\right). \end{aligned} \quad (49)$$

This gives us the desired inequality for $\gamma(t) = \ln(t+1)$ and

$$\beta(t) = \begin{cases} \dfrac{\ln(\ln(1+t)+1)}{\ln(1+t)}, & t > 0, \\ k, & t = 0 \end{cases} \quad (50)$$

for some $k \in [0,1)$, where $t \in [0,\infty)$. Thus, (P,Q) is an (α,β,γ)-contraction with respect to D.

Hence, there exists a coincidence point of P and Q which is a solution to the integral equation (42).

Conflicts of Interest

The authors declare that there are no conflicts of interest regarding the publication of this paper.

Acknowledgments

I would like to thank the editor and the referees for their comments and suggestions on the manuscript. This research is supported by Chiang Mai University, Thailand.

References

[1] M. A. Geraghty, "On contractive mappings," *Proceedings of the American Mathematical Society*, vol. 40, pp. 604–608, 1973.

[2] B. Samet, C. Vetro, and P. Vetro, "Fixed point theorems for α-ψ-contractive type mappings," *Nonlinear Analysis: Theory, Methods & Applications*, vol. 75, no. 4, pp. 2154–2165, 2012.

[3] E. Karapinar, "α-ψ-Geraghty contraction type mappings and some related fixed point results," *Filomat*, vol. 28, no. 1, pp. 37–48, 2014.

[4] S. Czerwik, "Contraction mappings in *b*-metric spaces," *Communications in Mathematics*, vol. 1, pp. 5–11, 1993.

[5] P. Hitzler and A. K. Seda, "Dislocated topologies," *Journal of Electrical Engineering*, vol. 51, no. 12, pp. 3–7, 2000.

[6] W. M. Kozlowski, *Modular function spaces, Monographs and Textbooks in Pure and Applied Mathematics*, vol. 122, Dekker, New York, NY, USA, 1988.

[7] M. Jleli and B. Samet, "A generalized metric space and related fixed point theorems," *Fixed Point Theory and Applications*, 2015:61, 14 pages, 2015.

[8] I. Altun, N. Al Arifi, M. Jleli, A. Lashin, and B. Samet, "Feng-Liu type fixed point results for multivalued mappings on JS-metric spaces," *Journal of Nonlinear Sciences and Applications. JNSA*, vol. 9, no. 6, pp. 3892–3897, 2016.

[9] J. Martínez-Moreno, W. Sintunavarat, and Y. J. Cho, "Common fixed point theorems for Geraghty's type contraction mappings using the monotone property with two metrics," *Fixed Point Theory and Applications*, vol. 2015, no. 1, article no. 174, 2015.

[10] J. Ahmad, N. Hussain, A. Azam, and M. Arshad, "Common fixed point results in complex valued metric space with applications to system of integral equations," *Journal of Nonlinear and Convex Analysis. An International Journal*, vol. 16, no. 5, pp. 855–871, 2015.

[11] N. Hussain, A. Azam, J. Ahmad, and M. Arshad, "Common fixed point results in complex valued metric spaces with application to integral equations," *Filomat*, vol. 28, no. 7, pp. 1363–1380, 2014.

[12] J. R. Roshan, V. Parvaneh, and I. Altun, "Some coincidence point results in ordered b-metric spaces and applications in a system of integral equations," *Applied Mathematics and Computation*, vol. 226, pp. 725–737, 2014.

[13] X. Wu and L. Zhao, "Fixed point theorems for generalized α-ψ type contractive mappings in b-metric spaces and applications," *Journal of Mathematics and Computer Science*, vol. 18, pp. 49–62, 2018.

9

Application of ADM Using Laplace Transform to Approximate Solutions of Nonlinear Deformation for Cantilever Beam

Ratchata Theinchai, Siriwan Chankan, and Weera Yukunthorn

Faculty of Science and Technology, Kanchanaburi Rajabhat University, Kanchanaburi 71000, Thailand

Correspondence should be addressed to Weera Yukunthorn; weera.mec@gmail.com

Academic Editor: Birendra Nath Mandal

We investigate semianalytical solutions of Euler-Bernoulli beam equation by using Laplace transform and Adomian decomposition method (LADM). The deformation of a uniform flexible cantilever beam is formulated to initial value problems. We separate the problems into 2 cases: integer order for small deformation and fractional order for large deformation. The numerical results show the approximated solutions of deflection curve, moment diagram, and shear diagram of the presented method.

1. Introduction

The Euler-Bernoulli beam theory states that the action load produces the bending moment $M(x) \in C([0, L])$ which is proportional to deflection characteristics of the beam. The equation of this law can be written as follows:

$$\frac{y''}{\left[1+(y')^2\right]^{3/2}} = -\frac{M(x)}{EI}, \tag{1}$$

where $y \in C^2([0, L])$ is deflection curve of a uniform beam, the modulus of elasticity E, and the moment of inertia I. We note that E and I are both constant and the product of E and I is called beam stiffness. In a case of small deformation, we assume that $y'(x)$ is infinitesimal. Equation (1) is reduced to the well-known fourth-order linear differential equation:

$$EI\frac{d^4y}{dx^4} = \frac{d^2M}{dx^2}. \tag{2}$$

In this study we consider the uniform flexible of cantilever beam; see Figure 1. The parameters L and Δ are undeformed length and horizontal displacement, respectively. The deformed length of beam is verified by the integral $\int_0^l [1 + (y')^2]^{1/2}dx$, where $l = L - \Delta$. It was shown in [1] that the slope of deflection curve represents the following equation $y'(x) = G(x)/[1 - G^2(x)]^{1/2}$, where $G(x) = (\int_x^l M(s)ds)/(EI)$ for the known function M. The deflection curve in Figure 1 corresponds to initial value problem of the geometric problem:

$$\begin{gathered} \frac{y''(x)}{\left[1+(y'(x))^2\right]^{3/2}} = -\frac{M(x)}{EI}, \quad x \in [0, l] \\ y(0) = 0, \\ y'(0) = \frac{G(0)}{\left[1-G^2(0)\right]^{1/2}}, \end{gathered} \tag{3}$$

where $M(x) = Px$. If the slope is very small, the linear Euler-Bernoulli beam theory [2] governs the problem

$$\begin{gathered} EI\frac{d^4y}{dx^4} = 0, \quad x \in [0, L], \\ EIy''(0) = -PL, \\ EIy^{(3)}(0) = -P, \\ y(L) = \frac{PL^3}{3EI}, \\ y'(L) = 0. \end{gathered} \tag{4}$$

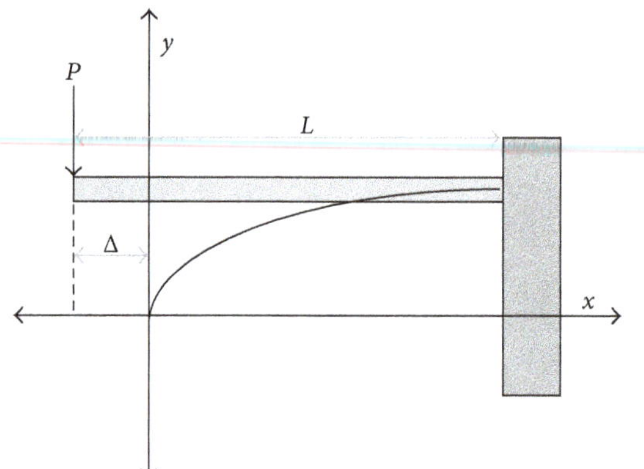

FIGURE 1: Cantilever beam with a concentrated load at the free end.

The solution of the problem (4) is

$$y(x) = \frac{Px^3}{6EI} - \frac{PL^2x}{2EI}. \quad (5)$$

Our method proposes to reform problem (3) in a sense of fractional calculus without linearization. Because fractional calculus is the great idea to describe behavior in nature, for example, force response of viscoelastic material [3, 4], fluid flow [5], and fitting experimental data [6]. For convenience, we set $k = -P/EI$ and $c = G(0)/[1-G^2(0)]^{1/2}$. We develop (1) to deal with fractional order $\alpha \in (1,2]$; system (3) becomes

$$\begin{aligned} D_0^\alpha y(x) &= kx\left[1+\left(y'(x)\right)^2\right]^{3/2}, \quad x \in [0,l], \\ y(0) &= 0, \\ y'(0) &= c. \end{aligned} \quad (6)$$

We use Adomian polynomial to approximate a nonlinear term and derive a semianalytical solution by use of Laplace transform for the initial value problem (6).

2. Preliminaries

In this section we introduce some definitions and theory of fractional calculus and Laplace transform which are used in our method. See [7] for more details.

Definition 1. The Riemann-Liouville fractional integral operator of order $\alpha > 0$ is defined as

$$I_0^\alpha y(x) = \frac{1}{\Gamma(\alpha)} \int_0^t (t-\tau)^{\alpha-1} f(\tau)\, d\tau. \quad (7)$$

Definition 2. The Caputo fractional derivative of order $\alpha > 0$ is defined as

$$D_0^\alpha f(t) = \frac{1}{\Gamma(n-\alpha)} \int_0^t (t-\tau)^{n-\alpha-1} f^{(n)}(\tau)\, d\tau, \quad n = \lceil \alpha \rceil. \quad (8)$$

Theorem 3. *If $p > 0$ and $q > 0$, then $(D_0^p D_0^q f)(x) = D^{p+q} f(x)$.*

Theorem 4. *The Laplace transform of the Caputo fractional derivative is given by*

$$L\{D_0^\alpha y(x)\} = s^\alpha L\{y(x)\} - \sum_{k=0}^{n} s^{\alpha-k-1} y^{(k)}(0), \quad (9)$$

where $n = \lfloor \alpha \rfloor$ and k is a nonnegative integer. The inverse Laplace transform of power function is given by

$$L^{-1}\left\{\frac{\Gamma(p+1)}{s^{p+1}}\right\} = t^p, \quad p > -1. \quad (10)$$

3. Semianalytical Solution via LADM

To formulate the general solution of problem (6), we replace y' with u in the equation of problem (6) and applied Theorem 3. We obtain

$$D_0^{\alpha-1} u = kx\left[1+u^2\right]^{3/2}. \quad (11)$$

Taking Laplace transform of (11) gives

$$L\left\{D_0^{\alpha-1} u\right\} = kL\left\{x\left[1+u^2\right]^{3/2}\right\}. \quad (12)$$

Using Theorem 4 and replacing $y'(0)$ by c, we can rewrite as

$$L\{u\} = \frac{c}{s} + ks^{1-\alpha} L\left\{x\left[1+u^2\right]^{3/2}\right\}. \quad (13)$$

We take inverse Laplace transform of (13), and the following operator equation is obtained:

$$u = c + kL^{-1}\left\{s^{1-\alpha} L\left\{x\left[1+u^2\right]^{3/2}\right\}\right\}. \quad (14)$$

We apply Adomian polynomial [8, 9] to approximate a nonlinear term of (14), by setting

$$\begin{aligned} Nu &= \left[1+u^2\right]^{3/2} = \sum_{n=0}^{\infty} A_n(u); \\ A_n(u) &= \frac{1}{n!}\left(\sum_{i=0}^{\infty} c(v,n) \frac{d^v Nu}{du^v}\right). \end{aligned} \quad (15)$$

Equation (14) becomes

$$u = c + kL^{-1}\left\{s^{1-\alpha} L\left\{x \sum_{n=0}^{\infty} A_n(u)\right\}\right\}. \quad (16)$$

Assume that the solution of (11) can be written as $u = \sum_{n=1}^{\infty} u_n$. From the combination of Laplace transform and Adomian decomposition method, we obtain the formulation of LADM for fractional order of cantilever beam deflection equation:

$$\sum_{n=0}^{\infty} u_n = c + kL^{-1}\left\{s^{1-\alpha} L\left\{x \sum_{n=0}^{\infty} A_n(u)\right\}\right\}. \quad (17)$$

In this work, we apply the first four terms of LADM which are provided from following recurrence:

$$u_0 = c,$$
$$u_{i+1} = kL^{-1}\left\{s^{1-\alpha}L\left\{xA_i\right\}\right\}, \quad \text{for } i = 0, 1, 2, \tag{18}$$

where

$$A_0 = Nu_0,$$
$$A_1 = u_1\frac{d(Nu_0)}{du_0}, \tag{19}$$
$$A_2 = u_2\frac{d(Nu_0)}{du_0} + \frac{u_1^2}{2!}\frac{d^2(Nu_0)}{du_0^2}. \tag{20}$$

The components of adomian polynomial are given by

$$A_0 = \left(1+u_0^2\right)^{3/2},$$
$$A_1 = 3u_0u_1\left(1+u_0^2\right)^{1/2},$$
$$A_2 = 3u_0u_2\left(1+u_0^2\right)^{1/2} + \frac{3}{2}u_1^2\left(1+u_0^2\right)^{1/2} + \frac{3}{2}\left(u_0u_1\right)^2\left(1+u_0^2\right)^{-1/2}. \tag{21}$$

The iteration results of recurrence (18) are shown as follows:

$$u_0 = c,$$
$$A_0 = \left(1+c^2\right)^{3/2},$$
$$u_1 = -kL^{-1}\left\{s^{1-\alpha}L\left\{xA_0\right\}\right\} = -\frac{k\left(1+c^2\right)^{3/2}}{\Gamma(1+\alpha)}x^{\alpha},$$
$$A_1 = -\frac{3ck\left(1+c^2\right)^2}{\Gamma(2+\alpha-1)}x^{\alpha},$$
$$u_2 = -kL^{-1}\left\{s^{1-\alpha}L\left\{xA_1\right\}\right\} = \frac{3ck^2\left(1+c^2\right)^2\Gamma(2+\alpha)}{\Gamma(1+\alpha)\Gamma(1+2\alpha)} \cdot x^{2\alpha},$$
$$A_2 = \left(\frac{9c^2\Gamma(2+\alpha)}{\Gamma(1+2\alpha)} + \frac{3\left(1+2c^2\right)}{2\Gamma(1+\alpha)}\right)\frac{k^2\left(1+c^2\right)^{5/2}}{\Gamma(1+\alpha)} \cdot x^{2\alpha},$$
$$u_3 = -kL^{-1}\left\{s^{1-\alpha}L\left\{xA_2\right\}\right\} = -\left(\frac{9c^2\Gamma(2+\alpha)}{\Gamma(1+2\alpha)} + \frac{3\left(1+2c^2\right)}{2\Gamma(1+\alpha)}\right) \cdot \frac{\Gamma(2+2\alpha)k^3\left(1+c^2\right)^{5/2}}{\Gamma(1+\alpha)\Gamma(1+3\alpha)}x^{3\alpha}. \tag{22}$$

The first four terms' approximate solution is $u = u_0+u_1+u_2+u_3$:

$$u(x) = c - \frac{k\left(1+c^2\right)^{3/2}}{\Gamma(1+\alpha)}x^{\alpha} + \frac{3ck^2\left(1+c^2\right)^2\Gamma(2+\alpha)}{\Gamma(1+\alpha)\Gamma(1+2\alpha)}x^{2\alpha} - \left(\frac{9c^2\Gamma(2+\alpha)}{\Gamma(1+2\alpha)} + \frac{3\left(1+2c^2\right)}{2\Gamma(1+\alpha)}\right) \cdot \frac{\Gamma(2+2\alpha)k^3\left(1+c^2\right)^{5/2}}{\Gamma(1+\alpha)\Gamma(1+3\alpha)}x^{3\alpha}. \tag{23}$$

Since $u = y'$ and $y(0) = 0$, thus $y(x) = \int_0^x u(s)ds$. This yields

$$y(x) = cx - \frac{k\left(1+c^2\right)^{3/2}}{\Gamma(2+\alpha)}x^{1+\alpha} + \frac{3ck^2\left(1+c^2\right)^2\Gamma(2+\alpha)}{\Gamma(1+\alpha)\Gamma(2+2\alpha)}x^{1+2\alpha} - \left(\frac{9c^2\Gamma(2+\alpha)}{\Gamma(1+2\alpha)} + \frac{3\left(1+2c^2\right)}{2\Gamma(1+\alpha)}\right) \cdot \frac{\Gamma(2+2\alpha)k^3\left(1+c^2\right)^{5/2}}{\Gamma(1+\alpha)\Gamma(2+3\alpha)}x^{1+3\alpha}. \tag{24}$$

3.1. Integer Order for Small Deformation. We simulate some deflection curve of the results in (24) and (5) which are solution of LADM and classical method, respectively. The parameters are $L = 100$ in., $\Delta = 0$ in., $EI = 1.8 \times 10^5$ kip in.^{2}, $\alpha = 2$, and $P = 1$ kip. Figure 2 shows the deflection of cantilever beam, obtained by using classical method and LADM. LADM has accurate slope around the free end of the beam as well as classical method. Moreover LADM shows that the deflection curve remains nearly straight line in the case of small deformation.

The effects of various loads influencing maximum deflection at the free end and length of the deformed beam

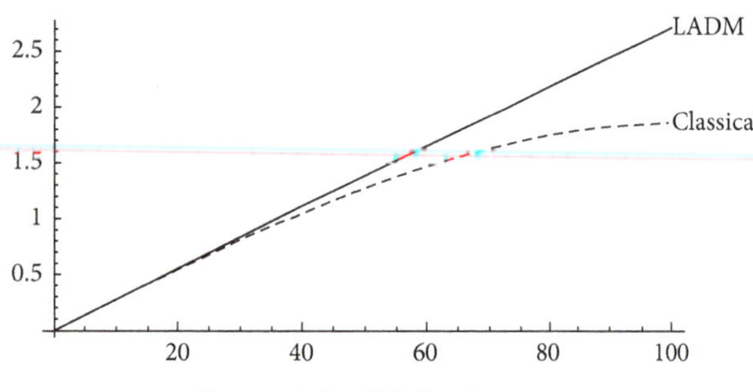

FIGURE 2: Small deflection curve.

TABLE 1: Numerical results under various load testing.

Load (kip)	Linear $y_{\max}$ (in.)	Linear Curve length (in.)	LADM $y_{\max}$ (in.)	LADM Curve length (in.)
0.1	0.1852	100.0002	0.2770	100.0004
0.2	0.3704	100.0008	0.5525	100.0015
0.3	0.5556	100.0019	0.8264	100.0034
0.4	0.7407	100.0033	1.0989	100.0060
0.5	0.9259	100.0051	1.3698	100.0094
1	1.8519	100.0206	2.7020	100.0365
5	9.2593	100.5125	12.0619	100.7396
10	18.519	102.0283	20.0049	102.2721
50	92.5926	140.3049	—	—
100	185.1852	216.3719	—	—

are presented in Table 1. We observe that the calculation of nonlinear integer order deformation via LADM fails at $P = 50$ kips and $P = 100$ kips. Since $G(0) > 1$ thus integral of bending moment is greater than stiffness of the beam. This yields large deformation, large rotation, and large strain, namely, some concept of large deformation. Not only does LADM fails, but also classical method does, because it contradicts the assumption that the lengths of deformed and undeformed beam are identical. This is the motivation for reform in fractional order model.

3.2. Fractional Order for Large Deformation. There are many methods for describing a large deformation of a beam, for example, the Laplace-Padé coupling with NDHPM and HPM [10], VIM [11], and pseudolinear system [1]. In this section, we introduce the process of investigating semianalytical solution via LADM and some numerical results of pseudolinear system in [1].

To illustrate the numerical results we suppose that horizontal displacement $\Delta = 40$. The beam parameters are $L = 1000$ in. and $EI = 1.8 \times 10^5$ kip in.^{2}. The fact from our assumption is

$$y'(840) = 0. \tag{25}$$

In this case we fixed the concentrate load $P = 0.4$ kip. We get the deformed length $l = 840$ and initial condition

TABLE 2: Values of slope at 840 in. and maximum vertical displacement for various fractional orders.

Fractional order	Slope at 840 in.	Deformed length (in.)
1.56	−0.185702	768.244392
1.55	−0.102380	798.176219
1.54	−0.023820	829.711010
1.53	0.050256	862.954991
1.52	0.120111	898.022706

TABLE 3: Numerical results of slope deflection.

x (in.)	Pseudolinear (in.)	LADM (in.)
0	1.1049	1.2630
100.0	1.0698	1.2132
200.0	0.9720	1.1186
300.0	0.8360	0.9943
400.0	0.6822	0.8460
500.0	0.5232	0.6773
600.0	0.3632	0.4903
700.0	0.2009	0.2862
800.0	0.0304	0.0679
816.9	0	0.0294

$c = y'(0) - 1.2626$. Fractional order is determined by simulating the results of slope at 840 and deformed length. Table 2 shows that order $\alpha = 1.54$ is given, $y'(840) = -0.0238$ and $y'(829.711) = 0$, which is the most accurate for the fact in (25). We design order $\alpha = 1.54$ for this problem. Using the first four terms' approximation, the deflection is shown in Figure 3(a). The slope deflection of psudolinear system and LADM are presented in Table 3. Moment diagram (M_e) can be computed from the deflection curve as follows: $M_e(x) = EIy''(x)/[1 + (y'(x))^2]^{3/2}$ and $V(x) = M_e'(x)$ for shear diagram. The results of both moment and shear diagram are shown in Figures 3(b) and 3(c), respectively.

4. Conclusion

In this study, we use Euler-Bernoulli beam equation for describing the uniform flexible cantilever beam with a concentrated load. The initial conditions are given by calculating slope of the beam. We use LADM to determine the semianalytical solution. LADM with integer order system can approximate solution without cancellation a nonlinear term in the case of small deformation. For the large deformation, we reform the problem to fractional order system and estimate fractional order α which conserves the fact from the assumption. Finally, we show that LADM gives the solution as a polynomial expression which is the advantage for analyzing moment and shear diagrams. LADM may be a powerful and successive method for solving nonlinear science and engineering problem.

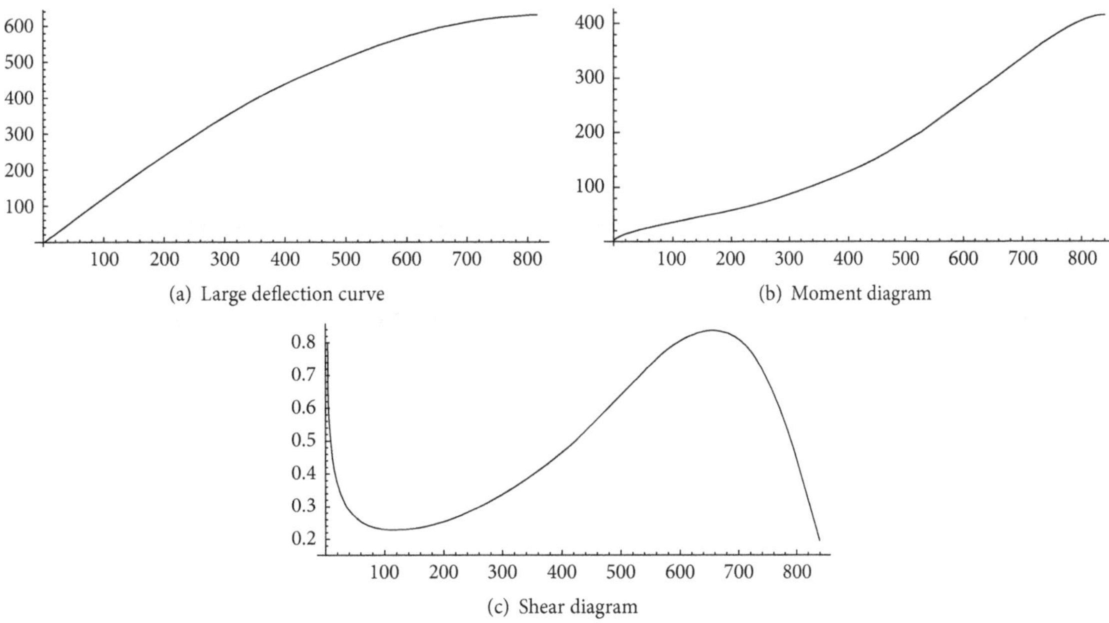

(a) Large deflection curve

(b) Moment diagram

(c) Shear diagram

FIGURE 3: Numerical results of deformation via LADM (4 terms).

Competing Interests

The authors declare that there is no conflict of interests regarding the publication of this paper.

References

[1] D. G. Fertis, *Nonlinear Structural Engineering: With Unique Theories and Methods to Solve Effectively Complex Nonlinear Problems*, Springer, New York, NY, USA, 2006.

[2] A. F. Bower, *Applied Mechanics of Solids*, CRC Press, Boca Raton, Fla, USA, 2010.

[3] M. Fukunaga, N. Shimizu, and H. Nasuno, "A nonlinear fractional derivative model of impulse motion for viscoelastic materials," *Physica Scripta*, vol. 136, Article ID 014010, 6 pages, 2009.

[4] F. Mainardi, *Fractional Calculus and Waves in Linear Viscoelasticity*, Imperial College Press, London, UK, 2010.

[5] P. Lino, G. Maione, and F. Saponaro, "Fractional-order modeling of high-pressure fluid-dynamic flows: an automotive application," *IFAC-PapersOnLine*, vol. 48, no. 1, pp. 382–387, 2015.

[6] M. Di Paola, A. Pirrotta, and A. Valenza, "Visco-elastic behavior through fractional calculus: an easier method for best fitting experimental results," *Mechanics of Materials*, vol. 43, no. 12, pp. 799–806, 2011.

[7] A. A. Kilbas, H. M. Srivastava, and J. J. Truillo, *Theory and Applications of Fractional Differential Equations*, Elsevier, Amsterdam, The Netherlands, 2006.

[8] G. Adomian, "A review of the decomposition method in applied mathematics," *Journal of Mathematical Analysis and Applications*, vol. 135, no. 2, pp. 501–544, 1988.

[9] A. M. WazWaz and M. S. Mehanna, "The combined Laplace-Adomian method for handing singular integral equation of heat transfer," *International Journal of Nonlinear Sciences*, vol. 10, no. 2, pp. 248–252, 2010.

[10] H. Vázquez-Leal, Y. Khan, A. L. Herrera-May et al., "Approximations for large deflection of a cantilever beam under a terminal follower force and nonlinear pendulum," *Mathematical Problems in Engineering*, vol. 2013, Article ID 148537, 12 pages, 2013.

[11] H. Ghaffarzadeh and A. Nikkar, "Explicit solution to the large deformation of a cantilever beam under point load at the free tip using the variational iteration method-II," *Journal of Mechanical Science and Technology*, vol. 27, no. 11, pp. 3433–3438, 2013.

On Surface Completion and Image Inpainting by Biharmonic Functions: Numerical Aspects

S. B. Damelin [1] **and N. S. Hoang** [2]

[1] *Mathematical Reviews, The American Mathematical Society, 416 Fourth Street, Ann Arbor, MI 48104, USA*
[2] *Department of Mathematics, University of Oklahoma, Norman, OK 73019-3103, USA*

Correspondence should be addressed to S. B. Damelin; damelin@umich.edu

Academic Editor: Irena Lasiecka

Numerical experiments with smooth surface extension and image inpainting using harmonic and biharmonic functions are carried out. The boundary data used for constructing biharmonic functions are the values of the Laplacian and normal derivatives of the functions on the boundary. Finite difference schemes for solving these harmonic functions are discussed in detail.

1. Introduction

The smooth function extension problem is a classical problem that has been studied extensively in the literature from various viewpoints. Some of the well-known results include Urysohn's Lemma, the Tietze Extension, and Whitney's Extension Theorem (see, e.g., [1–9]).

Inpainting was first introduced in [10] and then has been studied extensively by several authors (see, e.g., [11, 12]). Although smooth image inpainting is a smooth function extension problem, the most common approach in image inpainting so far is to use the solution to some PDE which are obtained from minimum-energy models as the recovered image. The most commonly used density function for these energy models is total derivation.

In [12], by considering smooth inpainting as a smooth surface extension problem, the author studied methods for linear inpainting and cubic inpainting. Error bounds for these inpainting methods are derived in [12]. In [13], several surface completion methods have been studied. An optimal bound for the errors of the cubic inpainting method in [12] is given. Applications to smooth inpainting have also been discussed in [13]. There, error bounds of completion methods are derived in terms of the radius of the domains on which the functions are completed. In one of the methods in [13], the author proposed to use p-harmonic functions for smooth surface completion and smooth surface inpainting. Later in [14], p-harmonic functions are also studied for smooth surface completion and inpainting. The differences of the method using p-harmonic functions in [13, 14] are as follows: the method in [13] uses $\Delta^i u|_S$, $i = 0, 1, \ldots, p-1$, as boundary data while the method in [14] uses $(\partial^i/\partial N^i)u|_S$, $i = 0, 1, \ldots, p-1$, as boundary data to solve for a p-harmonic function. Here, u is the function to be inpainted and S is the boundary of the inpainted region.

The goals of this paper are to implement and compare the performance of the two surface completion schemes in [13, 14]. In particular, we focus our study on smooth surface completion and smooth surface inpainting by biharmonic functions.

2. Surface Completion by Biharmonic Functions

Let D be a simply connected region in $\mathbb{R}^2$ with C^1-boundary $S = \partial D$ and d be the diameter of D. Let u_0 be a smooth function on any region containing D. Assume that u_0 is known on a neighbourhood outside D. The surface completion problem consists of finding a function u on a region containing D such that

$$u = u_0 \quad \text{outside } D. \tag{1}$$

There are several ways to construct the function u so that (1) holds. For smooth surface completion, one is often interested in finding a sufficiently smooth function u satisfying (1).

An application of smooth surface completion is in smooth image inpainting. In smooth image inpainting, one has a smooth image u_0 which is known in a neighbourhood outside of a region D while the data inside D is missing. The goal of image inpainting is to extend the function u over the region D in such a way that the extension over the missing region is not noticeable with human eyes.

In image inpainting, an inpainting scheme is said to be of linear order, or simply linear inpainting, if, for any smooth test image u_0, as the diameter d of the inpainting region D shrinks to 0, one has

$$\|u - u_0\|_D = O\left(d^2\right), \tag{2}$$

where u is the image obtained from the inpainting scheme. Here, $\|\cdot\|_D$ denotes the $L^\infty(D)$ norm. Here and throughout, $f = O(g)$ if $|f|/|g|$ is bounded uniformly by some constant $C > 0$.

Note that harmonic inpainting, that is, the extension found from the equation

$$\begin{aligned} \Delta u &= 0 \quad \text{in } D, \\ u|_S &= u_0|_S, \end{aligned} \tag{3}$$

is a linear inpainting scheme [12].

In [12], the following result for cubic inpainting is proved.

Theorem 1 (cubic inpainting, Theorem 6.5 [12]). *Let u_1 be the harmonic inpainting of u_0. Let u_ℓ be a linear inpainting of Δu_0 on D (not necessarily harmonic), and let $u(x)$ be defined by*

$$u(x) = u_1(x) + u_2(x), \quad x \in D, \tag{4}$$

where u_2 solves Poisson's equation

$$\begin{aligned} \Delta u_2 &= u_\ell, \quad \text{in } D, \\ u_2|_S &= 0. \end{aligned} \tag{5}$$

Then, u defines a cubic inpainting of u_0; that is,

$$\|u - u_0\|_D = O\left(d^4\right). \tag{6}$$

Remark 2. If u_ℓ is the harmonic inpainting of Δu_0 in D, that is, u_ℓ solves the equation

$$\begin{aligned} \Delta u_\ell &= 0 \quad \text{in } D, \\ u_\ell|_S &= \Delta u_0|_S, \end{aligned} \tag{7}$$

then the element u defined by (4) is a biharmonic function which solves the following problem:

$$\begin{aligned} \Delta^2 u &= 0, \\ \Delta u|_S &= \Delta u_0|_S, \\ u|_S &= u_0|_S. \end{aligned} \tag{8}$$

In [13], this result is generalized to a multiresolution approximation extension scheme in which the Laplacian is replaced by more general lagged diffusivity anisotropic operators. It is proved in [13] that if u solves the equation

$$\begin{aligned} \Delta^n u &= 0 \quad \text{in } D, \\ \Delta^i u\big|_S &= \Delta^i u_0\big|_S, \quad i = 0, 1, \dots, n-1, \end{aligned} \tag{9}$$

then

$$\|u - u_0\|_D = O\left(d^{2n}\right). \tag{10}$$

A sharper error bound than (10) is obtained in [13].

Equation (9) can be written as a system of Poisson's equations as follows:

$$\begin{aligned} v_0 &= 0, \quad \text{in } D, \\ \Delta v_i &= v_{i-1} \quad \text{in } D, \\ v_i|_S &= \Delta^{n-i} u_0\big|_S, \\ &\qquad i = 1, 2, \dots, n \\ u &= v_n. \end{aligned} \tag{11}$$

Thus, the problem of solving (9) is reduced to the problem of solving Poisson's equations of the form

$$\begin{aligned} \Delta u &= f \quad \text{in } D, \\ u|_S &= g. \end{aligned} \tag{12}$$

Numerical methods for solving (12) have been extensively studied in the literature.

Note that the normal derivatives $(\partial^i/\partial N^i)u_0|_S$ are not presented in (9). Thus, extension using (9) may not be differentiable across the boundary S. To improve the smoothness of the extension across S, it is suggested to find u from the equation

$$\begin{aligned} \Delta^n u &= 0 \quad \text{in } D, \\ \frac{\partial^i}{\partial N^i} u &= \frac{\partial^i}{\partial N^i} u_0, \quad i = 0, 1, \dots, n-1. \end{aligned} \tag{13}$$

It is proved in [14] that if u is the solution to (13), then (10) also holds.

Equation (13) cannot be reduced to a system of Poisson's equations as (9). In fact, to solve (13), one often uses a finite difference approach which consists of finding discrete approximations to the operators Δ^n and $\partial^i/\partial N^i$, $i = 1, \dots, n-1$. For "large" n, it is quite complicated, though possible, to obtain these approximations.

As we can see from the above discussions, (9) is easier to solve numerically than (13). However, scheme (9) does not use any information about the normal derivatives of the surface on S. Thus, the extension surface, obtained from scheme (9), may not be smooth across the boundary S. On the contrary, (13) uses normal derivatives as boundary data and, therefore, is expected to yield better results than scheme (9) does.

In the next section, we will implement and compare the two surface completion schemes using (9) and (13). In particular, we focus our study on biharmonic functions which are solutions to (9) and (13) for $n = 2$.

3. Implementation

Let us discuss a numerical method for solving the equation

$$\Delta^2 u = 0 \quad \text{in } D,$$
$$\Delta u|_S = f, \tag{14}$$
$$u|_S = g.$$

To solve this equation, one often defines $v = \Delta u$ and solves for u from the following system:

$$\Delta v = 0,$$
$$v|_S = f,$$
$$\Delta u = v, \tag{15}$$
$$u|_S = g.$$

Thus, the problem of solving (15) is reduced to the problem of solving the following Poisson's equation:

$$\Delta u = f_1 \quad \text{in } D,$$
$$u|_S = g_1. \tag{16}$$

To solve (16), we use a 5-point finite difference scheme to approximate the Laplacian operator. This 5-point scheme is based on the following well-known formula:

$$\Delta_5 u := \frac{1}{h^2}\begin{pmatrix} & 1 & \\ 1 & -4 & 1 \\ & 1 & \end{pmatrix} u = \Delta u + O\left(h^2\right). \tag{17}$$

Here, h is the discretization step size. This scheme is well defined at points P, whose nearest neighbours are interior points of D. If P has a neighbour $Q \in \partial D$, then we use a stencil of the form

$$\Delta_5 u(P) := \frac{1}{h^2}\begin{pmatrix} & 1 & \\ & -4 & 1 \\ & 1 & \end{pmatrix} u(P) + \frac{u(Q)}{h^2}$$
$$= \Delta u(P) + O\left(h^2\right). \tag{18}$$

In the above formula, Q is the nearest neighbour to the left of P. Similar formulae when Q is the nearest neighbour on the top, in the bottom, and to the right of P can be obtained easily.

In our experiments, we choose D as a square and the solution u on the computation grid is presented as a vector. Using the 5-point finite difference scheme, (16) is reduced to the following algebraic system:

$$Au = \left(f_1 - \frac{\bar{g}_1}{h^2}\right), \tag{19}$$

where A is the 5-point finite difference approximation to the Laplacian and $\bar{g}_1$ is a vector containing boundary values of u on S at suitable entries. Matrix A is a tridiagonal matrix; that is, all nonzero elements of A lie on the main diagonal and the first diagonals above and below the main diagonal. Matrix $-h^2 A$ can be obtained by the function delsq available in Matlab.

Let us discuss a numerical method for solving the equation

$$\Delta^2 u = 0,$$
$$u_N|_S = f, \tag{20}$$
$$u|_S = g.$$

For a discrete approximation to the bi-Laplacian, we use a 13-point finite difference scheme which is based on the following formula (see [15]):

$$\Delta_{13}^2 u = \frac{1}{h^4}\begin{pmatrix} & & 1 & & \\ & 2 & -8 & 2 & \\ 1 & -8 & 20 & -8 & 1 \\ & 2 & -8 & 2 & \\ & & 1 & & \end{pmatrix} u$$
$$= \Delta^2 u + O\left(h^2\right). \tag{21}$$

This stencil is well defined for a grid point P if all its nearest neighbours are in the interior of the domain. If P has a neighbour $Q \in \partial D$ and Q is on the left of P, then we use the following formula:

$$\Delta_{13}^2 u(P) := \frac{1}{h^4}\begin{pmatrix} & 1 & & \\ 2 & -8 & 2 & \\ -8 & 20 & -8 & 1 \\ 2 & -8 & 2 & \\ & 1 & & \end{pmatrix} u(P) + \frac{u_N(Q)}{h^3}$$
$$+ \frac{u(Q)}{h^4} = \Delta^2 u + O\left(h^{-1}\right). \tag{22}$$

Using the above finite difference scheme, (20) is reduced to a linear algebraic system of the form $Au = b$, where A is a five-diagonal matrix. Numerical solutions to u on the grid can be obtained by solving this linear algebraic system.

Before we proceed with numerical experiments, we need the following.

3.1. Quantitative Comparisons. It is constructive to provide quantitative correlations between original and processed images and in particular code to compare figures such as those below. In order to calculate these required correlations (and many are provided), we refer the reader to a free access code for our method in scikit-image processing in python unit completely at the disposal of the reader which readily provides quantitative correlations—in particular for the figures. The scikit-image processing package

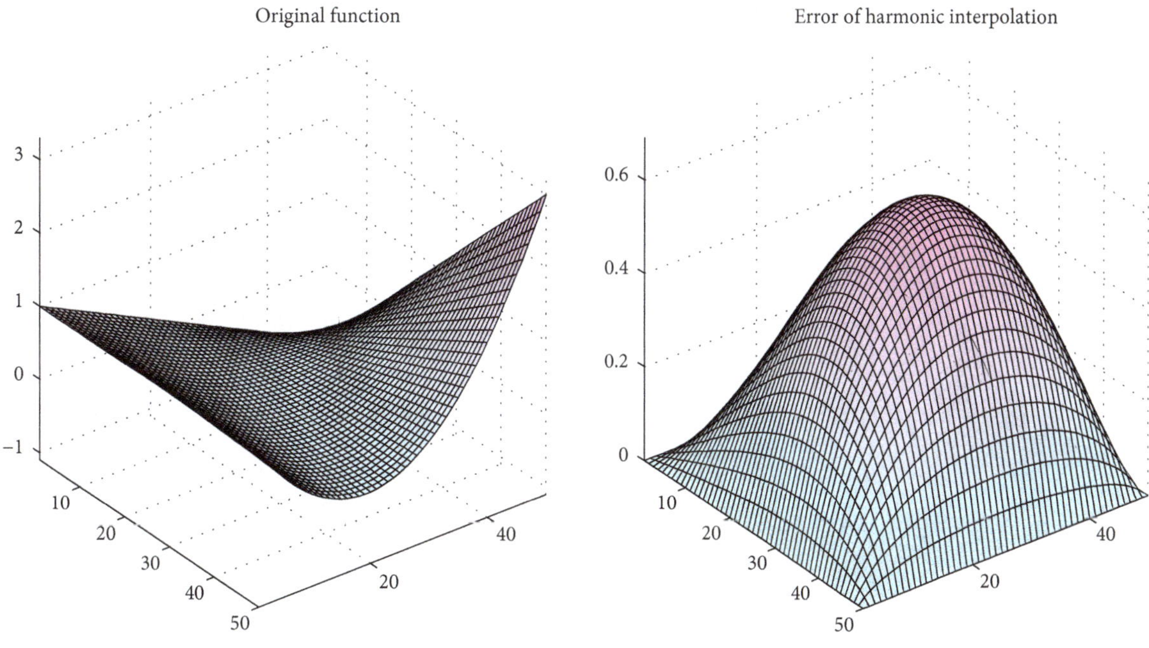

Figure 1

is at http://scikit-image.org/ and is a collection of open access algorithms for image processing with peer-reviewed code.

For our method, see http://scikit-image.org/docs/stable/api/skimage.restoration.html?highlight=biharmonicskimage.restoration.inpaint-biharmonic.

Moreover, for the benefit of the reader, many comparison methods with associated code are given in this unit; see some below, but see the full package for a longer list with references.

(i) Denoise-bilateral (see [16]).

(ii) Denoise-nl-means (see [17–19]).

(iii) Denoise-tv-bregman, denoise-tv-chambolle, denoise-wavelet, estimate-sigma, inpaint-biharmonic (the present paper), nl-means-denoising, richardson-lucy, unsupervised-wiener, unwrap-phase, Wiener-Hunt deconvolution.

4. Numerical Experiments

4.1. Smooth Surface Completion. Let us first do some numerical experiments with smooth surface completion. In our experiments, we compare numerical solutions from the three surface completion methods: the method with harmonic functions, the method with biharmonic functions in [12, 13], and the method with biharmonic functions in [14].

The function u to be completed in our first experiment is

$$u(x, y) = xy + x^2 (y + 1), \quad x, y \in D = [-1, 1]. \tag{23}$$

Note that this function is a biharmonic function. The domain D is discretized by a grid of size $(n+1) \times (n+1)$ points. In the first experiment, we used $n = 50$.

In our numerical experiments, we denote by u_H the extension by a harmonic function, by u_L the biharmonic extension from [13], and by u_N the biharmonic extension from [14].

Figure 1 plots the original function and the error of reconstruction by a harmonic function. Figure 2 plots the errors of surface reconstructions by biharmonic functions from [13, 14].

From Figures 1 and 2, one can see that the biharmonic reconstructions from [13, 14] are much better than the reconstruction by harmonic functions. The method in [13] in this experiment yields numerical results with accuracy a bit higher than the method in [14]. However, this does not imply that the method in [13] is better in terms of accuracy than the method in [14]. In the condition number of A, the finite difference approximation to the bi-Laplacian in this experiment is larger than that of the finite difference approximation to the Laplacian. Due to these condition numbers, the algorithm using the method in [13] yields results with higher accuracy than the algorithm using the method in [14]. As we can see in later experiments, the method in [14] often gives better results than the method in [13]. The conclusion from this example is that both methods in [13, 14] yield numerical solutions at very high accuracy. The harmonic reconstruction in this experiment is not very good. This stems from the fact that the function to be reconstructed is not harmonic.

In the next experiment, the function to be reconstructed is chosen by

$$u(x, y) = \frac{(1 + \cos(x))(1 + \cos(y))}{4}, \quad x, y \in [-1, 1]. \tag{24}$$

This function $u(x, y)$ is not a biharmonic function.

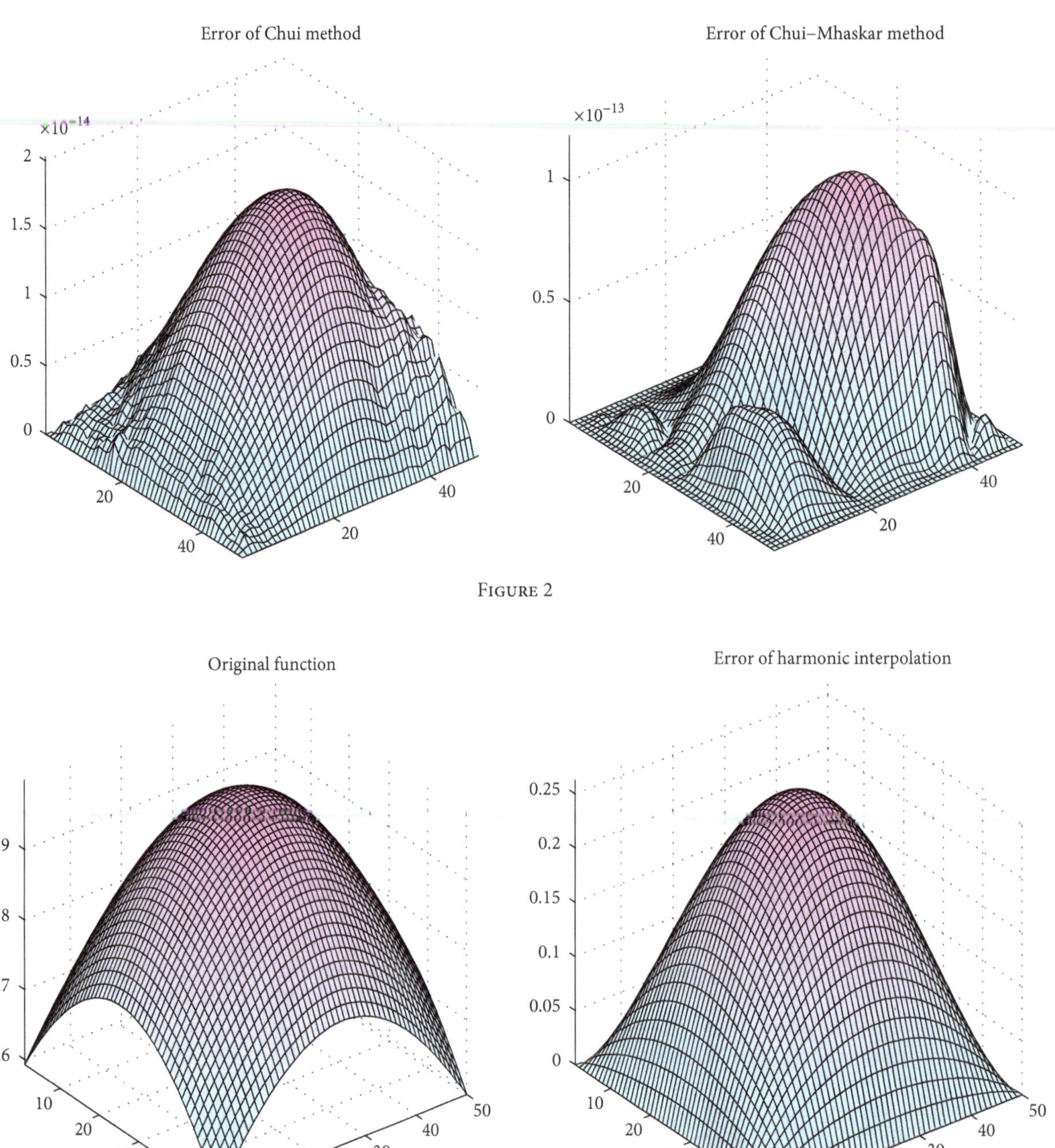

Figure 2

Figure 3

Figures 3 and 4 plot the errors of harmonic and biharmonic reconstructions. From these figures, it is clear that the method in [14] yields the best approximation. The harmonic reconstruction is the worst amongst the 3 methods in this experiment.

4.2. Image Inpainting. Let us do some numerical experiments with image inpainting.

Figure 5 plots a damaged image and a reconstructed image by harmonic functions. Figure 6 plots restored images by biharmonic functions following the methods from [13, 14].

It can be seen from Figures 5 and 6 that the biharmonic extension method from [14] yields the best reconstruction. Although the biharmonic extension method from [13] is better than harmonic extension in our experiments with smooth surface completion, it is not as good as the harmonic extension in this experiment. This is understandable since our image contains edges and is not a smooth function. It can be seen from the restored image by the method in [13] in Figure 6 that the reconstruction may not be smooth, or even differentiable, across the boundary.

Error of Chui method

Error of Chui–Mhaskar method

Figure 4

Damaged image

Harmonic inpainting

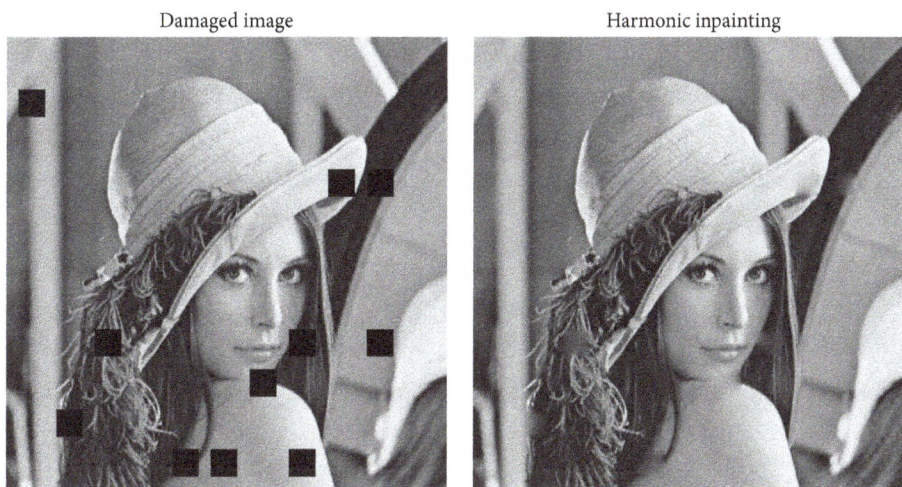

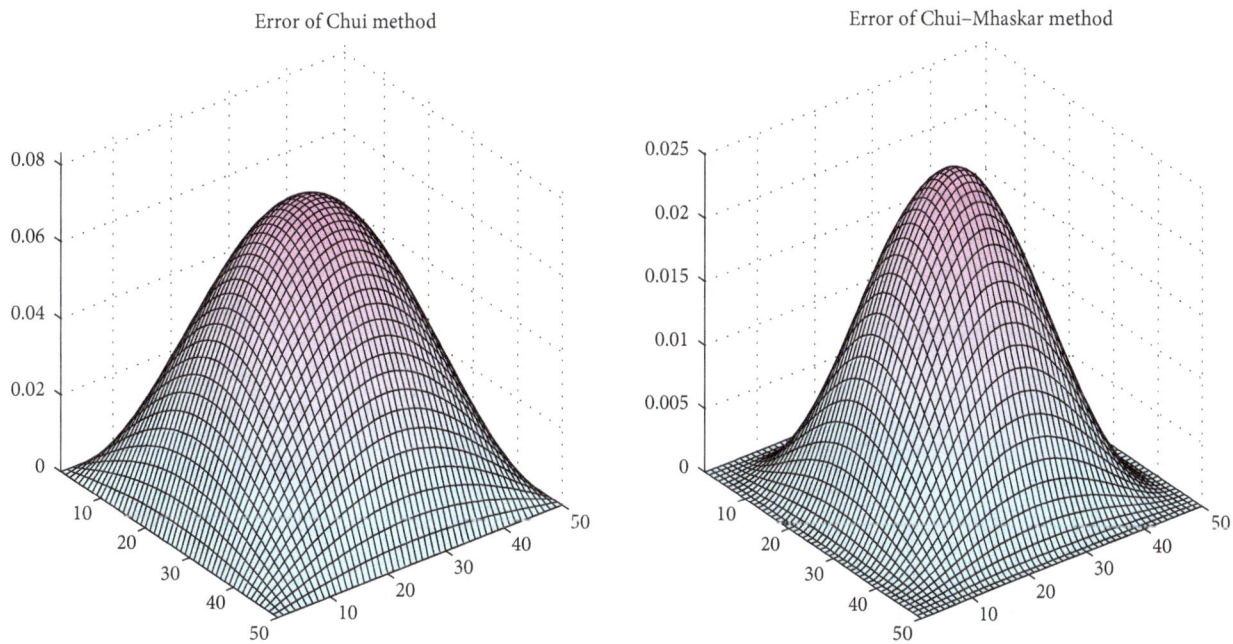

Figure 5: Damaged image and restored image by harmonic functions.

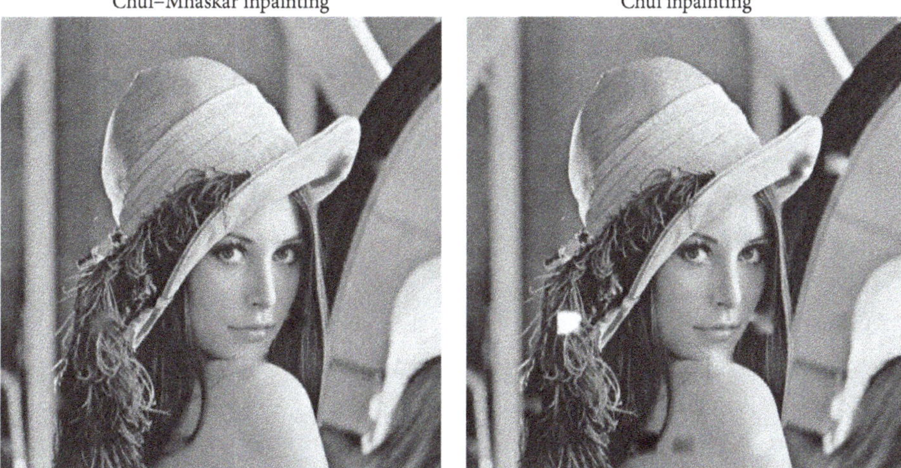

Figure 6: Restored image by biharmonic functions.

FIGURE 7: Damaged image and restored image by harmonic functions.

FIGURE 8: Restored image by biharmonic functions.

TABLE 1: Results for $D_i = [-2^{-i}, 2^{-i}] \times [-2^{-i}, 2^{-i}]$ for $i = \overline{0,9}$.

i	$\log_2 \|u_H - u\|_\infty$	$\log_2 \|u_L - u\|_\infty$	$\log_2 \|u_N - u\|_\infty$
0	2.36	0.37	−1.46
1	0.50	−3.48	−5.34
2	−1.46	−7.44	−9.32
3	−3.45	−11.44	−13.31
4	−5.45	−15.43	−17.31
5	−7.45	−19.43	−21.31
6	−9.45	−23.43	−25.31
7	−11.45	−27.43	−29.31
8	−13.45	−31.43	−33.34
9	−15.45	−35.46	−38.54

Appendix

Table 1 presents numerical results for the function u defined by (24) and on the domain $D_i = [-2^{-i}, 2^{-i}] \times [-2^{-i}, 2^{-i}]$, $i = \overline{0,9}$. The diameter of D_i is $d_i = 2^{1-i}$. From Table 1, one can see that the harmonic reconstruction has an order of accuracy of 2 while the biharmonic reconstruction methods have an order of accuracy of 4. This agrees with the theoretical estimates in [12–14].

Figures 7 and 8 plot a damaged picture of peppers and reconstructed images by the 3 methods. From these figures, one gets the same conclusion as in the previous experiment. The biharmonic reconstruction in [14] yields the best restoration while the biharmonic reconstruction in [13] yields the worst reconstruction.

Conflicts of Interest

The authors declare that there are no conflicts of interest regarding the publication of this paper.

References

[1] P. Shvartsman, "The Whitney extension problem and Lipschitz selections of set-valued mappings in jet-spaces," *Transactions of the American Mathematical Society*, vol. 360, no. 10, pp. 5529–5550, 2008.

[2] P. Shvartsman, "Whitney-type extension theorems for jets generated by Sobolev functions," *Advances in Mathematics*, vol. 313, pp. 379–469, 2017.

[3] Y. Brudnyi and P. Shvartsman, "Whitney's extension problem for multivariate C1,ω-functions," *Transactions of the American Mathematical Society*, vol. 353, no. 6, pp. 2487–2512, 2001.

[4] S. B. Damelin and C. Fefferman, *On Smooth Whitney Extensions of almost isometries with small distortion, Interpolation and Alignment in RD-Part 1*, arxiv: 1411.2451.

[5] S. B. Damelin and C. Fefferman, *On Smooth Whitney extensions of almost isometries with small distortion in Cm(Rn)*, arxiv: 1505.06950.

[6] C. Fefferman, S. B. Damelin, and W. Glover, "A BMO theorem for ε and an application to comparing manifolds of speech and sound Involve," *A Journal of Mathematics*, vol. 5, no. 2, pp. 159–172, 2012.

[7] C. Fefferman, "Whitney's extension problems and interpolation of data," *Bulletin (New Series) of the American Mathematical Society*, vol. 46, no. 2, pp. 207–220, 2009.

[8] E. J. McShane, "Extension of range of functions," *Bulletin (New Series) of the American Mathematical Society*, vol. 40, no. 12, pp. 837–842, 1934.

[9] H. Whitney, "Analytic extensions of differentiable functions defined in closed sets," *Transactions of the American Mathematical Society*, vol. 36, no. 1, pp. 63–89, 1934.

[10] M. Bertalmio, G. Sapiro, C. Ballester, and V. Caselles, "Image Inpainting," in *Proceedings of the 27th annual conference on Computer graphics and interactive techniques (SIGGRAPH '00)*, pp. 417–424, July 2000.

[11] M. Bertalmio, A. Bertozzi, and G. Sapiro, "Navier-stokes, fluid dynamics, and image and video inpainting," in *Proceedings of the 2001 IEEE Computer Society Conference on Computer Vision and Pattern Recognition. CVPR 2001*, pp. I-355–I-362, Kauai, HI, USA.

[12] T. F. Chan and J. Shen, *Image Processing and Analysis*, SIAM, Philadelphia, Pa, USA, 2005.

[13] C. K. Chui, "An MRA approach to surface completion and image inpainting," *Applied and Computational Harmonic Analysis* , vol. 26, no. 2, pp. 270–276, 2009.

[14] C. K. Chui and H. N. Mhaskar, "MRA contextual-recovery extension of smooth functions on manifolds," *Applied and Computational Harmonic Analysis* , vol. 28, no. 1, pp. 104–113, 2010.

[15] P. Bjorstad, "Fast numerical solution of the biharmonic Dirichlet problem on rectangles," *SIAM Journal on Numerical Analysis*, vol. 20, no. 1, pp. 59–71, 1983.

[16] C. Tomasi and R. Manduchi, "Bilateral filtering for gray and color images," in *Proceedings of the 6th International Conference on Computer Vision (ICCV '98)*, pp. 839–846, Bombay, India, January 1998.

[17] A. Buades, B. Coll, and J.-M. Morel, "A non-local algorithm for image denoising," in *Proceedings of the 2005 IEEE Computer Society Conference on Computer Vision and Pattern Recognition, CVPR 2005*, pp. 60–65, usa, June 2005.

[18] J. Darbon, A. Cunha, T. F. Chan, S. Osher, and G. J. Jensen, "Fast nonlocal filtering applied to electron cryomicroscopy," in *Proceedings of the 5th IEEE International Symposium on Biomedical Imaging: From Nano to Macro (ISBI '08)*, pp. 1331–1334, IEEE, Paris, France, May 2008.

[19] J. Froment, "Parameter-Free Fast Pixelwise Non-Local Means Denoising," *Image Processing On Line*, vol. 4, pp. 300–326, 2014.

Hopf Bifurcation Analysis of a New SEIRS Epidemic Model with Nonlinear Incidence Rate and Nonpermanent Immunity

M. P. Markakis and P. S. Douris

Department of Electrical & Computer Engineering, University of Patras, 26504 Patras, Greece

Correspondence should be addressed to M. P. Markakis; markakis@upatras.gr

Academic Editor: Hans Engler

A new SEIRS epidemic model with nonlinear incidence rate and nonpermanent immunity is presented in the present paper. The fact that the incidence rate per infective individual is given by a nonlinear function and product of rational powers of two state variables, as well as the introduction of an epidemic-induced death rate, leads to a more realistic modeling of the physical problem itself. A stability analysis is performed and the features of Hopf bifurcation are investigated. Both the corresponding critical regions in the parameter space and their stability characteristics are presented. Furthermore, by using algorithms based on a new symbolic form as regards the restriction of an n-dimensional nonlinear parametric system to the center manifold and the normal forms of the corresponding Hopf bifurcation, as well, the associated bifurcation diagram is derived, and finally various emerging limit cycles are numerically obtained by appropriate implemented methods.

1. Introduction

The realistic modeling of epidemic models constitutes an important issue of modern research as it can contribute to both a better understanding and more accurate modeling of the actual dynamics and the interrelation of the populations involved. Nonpermanent immunity leads to SEIRS or SIRS models which have been studied with respect to the effects of the epidemiological parameters, with bilinear (see, e.g., [1]) or nonlinear incidence rate (see [2, 3] and the references therein). In particular, in [3] the writers investigate the stability of both the disease-free equilibrium and the endemic one. Also, they determine conditions regarding the existence and stability of Hopf-bifurcated limit cycles with respect to the latter, concerning both SEIRS and SIRS models. The nonlinear incidence rate offers a deeper insight into the actual relation between the populations of susceptible and infective individuals. Furthermore, we introduce an additional death rate solely due to the disease, and hence the SEIRS model is enriched with new nonlinear terms.

Let us now present the specifics of the aforementioned model, described by the following 4D differential system:

$$\begin{aligned} \dot{S} &= rN - dS + \varepsilon R - h(S,I)\,I, \\ \dot{E} &= -(d+\sigma)\,E + h(S,I)\,I, \\ \dot{I} &= -(\gamma + d + \alpha)\,I + \sigma E, \\ \dot{R} &= \gamma I - (d+\varepsilon)\,R, \end{aligned} \tag{1}$$

$$\dot{N} = rN - dN - \alpha I, \tag{2}$$

where S, E, I, R, N denote the number of susceptible, exposed (incubating), infective, and recovered individuals and the total population, respectively, while $h(S,I)$ represents the incidence rate per infective individual and $r, d, \varepsilon, \sigma, \alpha, \gamma$ stand for the system parameters. As regards their physical meaning, r denotes the birth rate, d denotes the physical death rate, ε denotes the rate of loss of immunity, σ denotes the rate of incubation, α is the additional death rate due to the epidemic, and γ denotes the recovery rate. Then by normalizing with respect to the total population $N = S + E + I + R$ which is considered constant and taking into account (2), the system becomes

$$\begin{aligned} \dot{x} &= r - rx + \varepsilon z + \alpha xy - \overline{h}(x, y)\, y, \\ \dot{w} &= -(r + \sigma)\, w + \alpha wy + \overline{h}(x, y)\, y, \\ \dot{y} &= \sigma w - (r + \gamma + \alpha)\, y + \alpha y^2, \\ \dot{z} &= \gamma y - (r + \varepsilon)\, z + \alpha yz \end{aligned} \tag{3}$$

with

$$\begin{aligned} x &= \frac{S}{N}, \\ w &= \frac{E}{N}, \\ y &= \frac{I}{N}, \\ z &= \frac{R}{N}, \\ x + w &+ y + z = 1. \end{aligned} \tag{4}$$

Now, by eliminating $z = 1 - x - w - y$, system (3) is reduced to the following three-dimensional one:

$$\begin{aligned} \dot{x} &= r + \varepsilon - (r + \varepsilon)\, x - \varepsilon w - \varepsilon y + \alpha xy - \overline{h}(x, y)\, y, \\ \dot{w} &= -(r + \sigma)\, w + \alpha wy + \overline{h}(x, y)\, y, \\ \dot{y} &= \sigma w - (r + \gamma + \alpha)\, y + \alpha y^2. \end{aligned} \tag{5}$$

By setting

$$\overline{h}(x, y) = \beta x^m y^{s-1}, \tag{6}$$

where β, m are positive constants and $s > 1$, (5) takes the final form

$$\begin{aligned} \dot{x} &= r + \varepsilon - (r + \varepsilon)\, x - \varepsilon w - \varepsilon y + \alpha xy - \beta x^m y^s, \\ \dot{w} &= -(r + \sigma)\, w + \alpha wy + \beta x^m y^s, \\ \dot{y} &= \sigma w - (r + \gamma + \alpha)\, y + \alpha y^2. \end{aligned} \tag{7}$$

The analysis is multiparametric in that the parameter space of the system is structured by three varying parameters. Thus in Section 2 a stability analysis of the system is performed, where the active parameters are determined and various graphical representations are obtained concerning the critical (with respect to Hopf bifurcations) values of the varying parameters, as well as the critical, noncritical, and stability regions in the parameter space considered. Then, based on a new proper symbolic form as regards the center manifold analysis (see [4]), the basic features and steps of which are generally presented in Section 3, effective algorithms are implemented, by using symbolic computational software, which result in the associated bifurcation portraits throughout the regions of the parameter space under consideration. These portraits are presented in Section 4, together with limit cycles corresponding to the cases resulting from the respective analysis and obtained by use of a custom orthogonal collocation method on finite elements. Finally, Appendices A and B include algebraic manipulations and formulae related to the analysis carried out throughout this work.

2. Stability Analysis-Hopf Bifurcation

Final reduced system (7) possesses two types of equilibria: a disease-free one, namely,

$$\Sigma_0 = (1, 0, 0), \tag{8}$$

and an endemic one of the form $\Sigma_1(x^0, y^0, z^0)$ with $y^0 \neq 0$ obtained after some tedious algebraic manipulations (see Appendix A), as

$$\begin{aligned} x^0 &= \frac{1}{\sigma(\alpha y^0 - \kappa_0)} \left[\alpha^2 \left(y^0\right)^3 - \alpha\kappa_4 \left(y^0\right)^2 \right. \\ &\quad \left. + (\kappa_3 + \sigma\varepsilon)\, y^0 - \sigma\kappa_0\right], \end{aligned} \tag{9}$$

$$\begin{aligned} \beta\sigma \left(x^0\right)^m \left(y^0\right)^{s-1} &= \alpha^2 \left(y^0\right)^2 - \alpha(\kappa_4 - \varepsilon)\, y^0 + \kappa_3 \\ &\quad - \varepsilon\kappa_2, \end{aligned} \tag{10}$$

$$w^0 = \frac{1}{\sigma}\left[-\alpha \left(y^0\right)^2 + \kappa_2 y^0\right], \tag{11}$$

where $\kappa_0 = r + \varepsilon$, $\kappa_1 = r + \varepsilon + \sigma$, $\kappa_2 = r + \gamma + \alpha$, $\kappa_3 = \kappa_1\kappa_2$, and $\kappa_4 = \kappa_1 + \kappa_2$. We focus on the endemic equilibrium $\Sigma_1 = (x^0, y^0, w^0)$ with x^0, y^0, w^0 given in (9)–(11), since it corresponds to persistence of the disease.

As regards the local stability of Σ_1, taking into account (7), (10), and (11), the Jacobian matrix evaluated at Σ_1 becomes

$$J_0 = \begin{pmatrix} \alpha y^0 - \kappa_0 - \dfrac{m}{\sigma} A \dfrac{y^0}{x^0} & -\varepsilon & -\varepsilon + \alpha x^0 - \dfrac{s}{\sigma} A \\ \dfrac{m}{\sigma} A \dfrac{y^0}{x^0} & \alpha y^0 - \kappa_1 + \varepsilon & \dfrac{\alpha}{\sigma}\left[-\alpha \left(y^0\right)^2 + \kappa_2 y^0\right] + \dfrac{s}{\sigma} A \\ 0 & \sigma & 2\alpha y^0 - \kappa_2 \end{pmatrix}, \tag{12}$$

where

$$\begin{aligned} y^0 &= y^0(\gamma, r, \alpha, \varepsilon, \sigma, \beta, m, s), \\ x^0 &= x^0\left(y^0; \gamma, r, \alpha, \varepsilon, \sigma\right), \\ A &= A\left(y^0; \gamma, r, \alpha, \varepsilon, \sigma\right) \\ &= \alpha^2\left(y^0\right)^2 - \alpha\left(\kappa_4 - \varepsilon\right) y^0 + \kappa_3 - \varepsilon\kappa_2. \end{aligned} \tag{13}$$

The associated characteristic equation is

$$\lambda^3 + B_2\lambda^2 + B_1\lambda + B_0 = 0. \tag{14}$$

Now, by considering the well-known Routh-Hurwitz necessary and sufficient stability conditions, namely,

$$\begin{aligned} B_0 &> 0, \\ B_1 &> 0, \\ B_1B_2 - B_0 &> 0, \end{aligned} \tag{15}$$

related to the equilibrium Σ_1, we conclude with the formulae

$$\begin{aligned} B_0 &= \frac{1}{\sigma x^0}\left(P_{00} + P_{01}x^0\right), \\ B_1 &= \frac{1}{\sigma x^0}\left(P_{10} + P_{11}x^0\right), \\ B_3 &= B_1B_2 - B_0 \\ &= \frac{1}{\sigma^2\left(x^0\right)^2}\left(P_{30} + P_{31}x^0 + P_{32}\left(x^0\right)^2\right), \end{aligned} \tag{16}$$

where $P_{ij} = f_{ij}(y^0; \gamma, r, \alpha, \varepsilon, \sigma, m, s)$, $i = 0, 1, 3$, $j = 0, 1, 2$, are polynomials with respect to y^0.

Moreover, by solving the equation $B_3 = 0$ (resulting from (14) for $\lambda = i\omega$) with respect to y^0 (after substitution of the right-hand side of (9) for x^0, we finally evaluate the real roots of a 9th-degree polynomial numerically), we further evaluate x^0 and w^0 by substituting the obtained root of B_3 into (9) and (11), respectively. Then, taking into account the fact that $\kappa_2 > \alpha$, we arrive at $w^0 > 0$ by means of (11) in the case where $0 < y^0 < 1$. Thus if

$$\begin{aligned} 0 < x^0 &< 1, \\ 0 < y^0 &< 1, \\ w^0 &< 1, \\ B_0 &> 0, \\ B_1 &> 0 \end{aligned} \tag{17}$$

then (x^0, y^0, w^0) represent the critical values $(x^0_{cr}, y^0_{cr}, w^0_{cr})$.

Then, by solving (10) with respect to β and considering (γ, σ, β) varying parameters of the problem we obtain the critical value

$$\beta_{cr} = \frac{\alpha^2\left(y^0_{cr}\right)^2 - \alpha\left(\kappa_4 - \varepsilon\right) y^0_{cr} + \kappa_3 - \varepsilon\kappa_2}{\sigma\left(x^0_{cr}\right)^m\left(y^0_{cr}\right)^{s-1}} \tag{18}$$

provided that $0 < \beta_{cr} < 1$, with x^0_{cr}, y^0_{cr} being the aforementioned critical equilibrium of the system (7). Thus a *critical surface* $\beta_{cr} = \beta_{cr}(\gamma, \sigma)$ is generated in the parameter space (γ, σ, β) (we have $y^0_{cr} = y^0_{cr}(\gamma, \sigma)$ and due to (9) we also have $x^0_{cr} = x^0_{cr}(\gamma, \sigma, y^0_{cr}(\gamma, \sigma))$, with $r, \alpha, \varepsilon, m, s$ being fixed), defined over the area of the parameter plane (γ, σ), where the critical values of β are obtained via (17) and (18); we call this area *critical region*. Moreover, we differentiate B_3, given by (16), with respect to the active parameters p $(p = \gamma, \sigma, \beta)$, namely,

$$\begin{aligned} B_{3,p} &= \frac{1}{\sigma^2\left(x^0\right)^2}\left(P_{30,p} + P_{31,p}x^0 + P_{32,p}\left(x^0\right)^2\right. \\ &\quad \left. + \left(P_{31} + 2P_{32}x^0\right) x^0{}_p\right) - \frac{2B_3}{\sigma x^0}\left(\delta x^0 + \sigma x^0{}_p\right), \\ \delta &= \begin{cases} 1, & p = \sigma \\ 0, & p = \gamma, \beta \end{cases} \\ P_{3j,p} &= f_{3j,p} + f_{3j,y^0}y^0{}_p, \quad j = 0, 1, 2, \end{aligned} \tag{19}$$

where $\square_{,p} = \partial\square/\partial p$ and $\square_{,y^0} = \partial\square/\partial y^0$, and also $x^0{}_p$ and $y^0{}_p$ denote the partial derivatives of x^0 and y^0 with respect to p, provided in Appendix A (A.2). Then by introducing the critical equilibrium $(x^0_{cr}(y^0_{cr}), y^0_{cr})$ and taking into account the fact that y^0_{cr} is a numerically obtained root of a high degree polynomial, we conclude that by setting the right-hand side of (19) equal to zero, no explicit relation can be extracted involving the parameters of the system. Furthermore, for any fixed values of $r, \alpha, \varepsilon, m, s$, we numerically compute

$$B_{3,p}\left(\gamma, \sigma, \beta_{cr}, x^0_{cr}, y^0_{cr}\right) \neq 0, \tag{20}$$

everywhere on the *critical surface*. Thus, considering (17) and (20), according to Liu criterion [5] J_0 has a pair of purely imaginary eigenvalues together with a negative real one on this surface; the *transversality condition* (see Appendix B, (C1-2)) holds, as well. Hence, a Hopf bifurcation occurs at the critical equilibrium. Graphical representations of β_{cr} (evaluated by using (18)) versus γ (for different values of σ, with $r, \alpha, \varepsilon, m, s$ being fixed), versus σ (for different values of γ, with $r, \alpha, \varepsilon, m, s$ being fixed), and versus s (for different values of γ, with $r, \alpha, \varepsilon, \sigma, m$ being fixed) are presented in Figures 1(a), 1(b), and 1(c), respectively, while *critical regions* are obtained in the parameter plane (γ, σ) for fixed values of $r, \alpha, \varepsilon, m, s$ in Figures 2(a) and 2(b).

We should note that variation of the values of fixed parameters does not affect the number of critical values as regards β. Thus, in any case, the expression (17), under restrictions (18), yields zero or at most one critical value $(0 < \beta_{cr} < 1)$. Additionally, we note that an increase in s or m gives rise to an expansion of the *zero region*, namely, the area of the parameter plane (γ, σ) where no critical values of β exist.

Furthermore, the status of the equilibrium points corresponding to the values of β in the range $(0, 1)$ is shown in Figures 3(a) and 3(b). More precisely, after the right-hand side of (9) has been substituted for x^0, by using standard

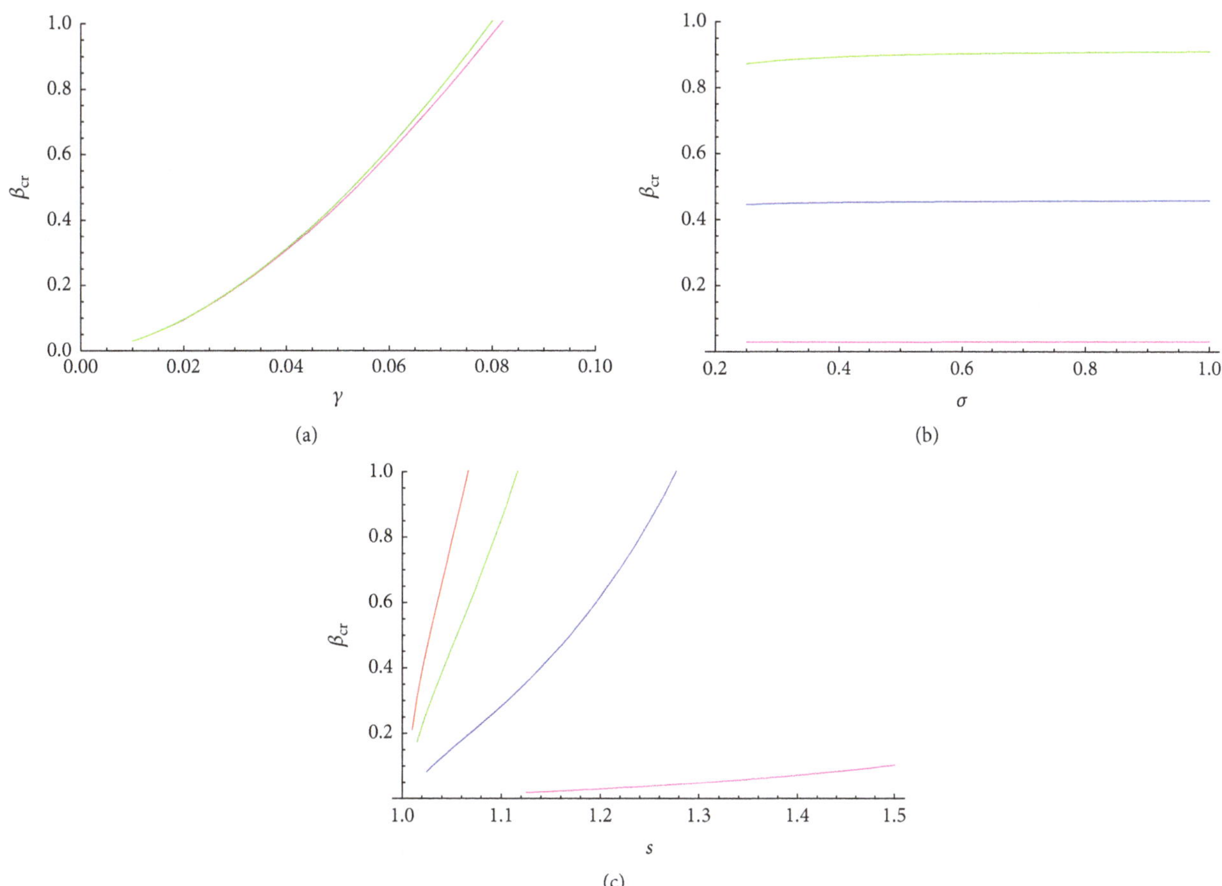

FIGURE 1: The critical value, β_{cr}, as a function of the parameters: (a) γ for $\sigma = 0.25$ (magenta) and $\sigma = 0.75$ (green), (b) σ for $\gamma = 0.01$ (magenta), $\gamma = 0.05$ (blue), and $\gamma = 0.075$ (green), and (c) s for $\sigma = 0.5$ and $\gamma = 0.01$ (magenta), $\gamma = 0.06$ (blue), $\gamma = 0.12$ (green), and $\gamma = 0.17$ (red).

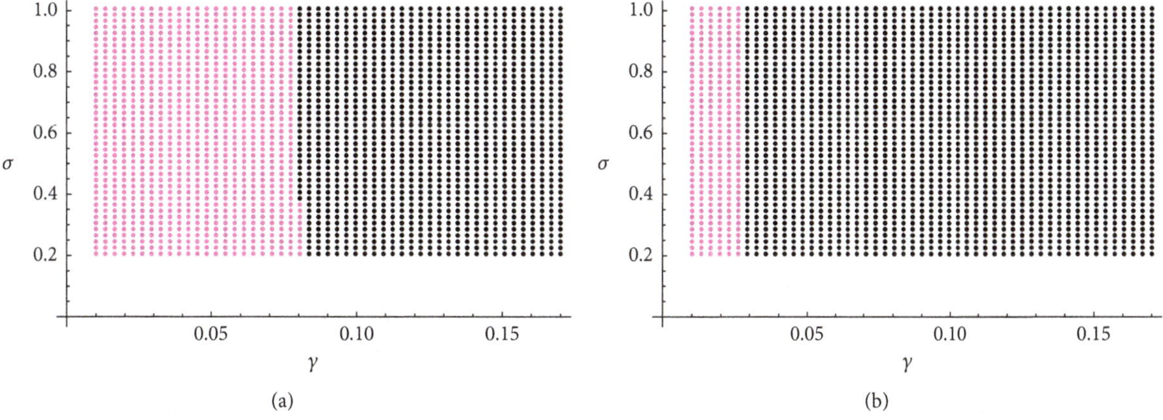

FIGURE 2: *Critical regions* (magenta – one critical value for β) and *zero regions* (black – no critical value for β), for $r = 10^{-4}$, $\alpha = 10^{-4}$, $\varepsilon = 10^{-3}$, and $m = 0.5$ and (a) $s = 1.2$ and (b) $s = 1.6$.

numerical computation routines, (10) is solved with respect to y^0 for any given value of β and fixed values for the other parameters. Then x^0 and w^0 can be obtained by means of (9) and (11), respectively, and finally the coefficients B_i, $i = 0, 1, 2$ of the characteristic equation (14) are determined. Thus, concerning the *critical* pairs (γ, σ) (the points of the *critical region*), we focus on the slope of the real part of the complex conjugate eigenvalues of the Jacobian as a function of β, inside an interval $(\beta_{cr} - d_1, \beta_{cr} + d_2)$, $d_1, d_2 > 0$ (where a pair of complex eigenvalues exist), in order to determine the

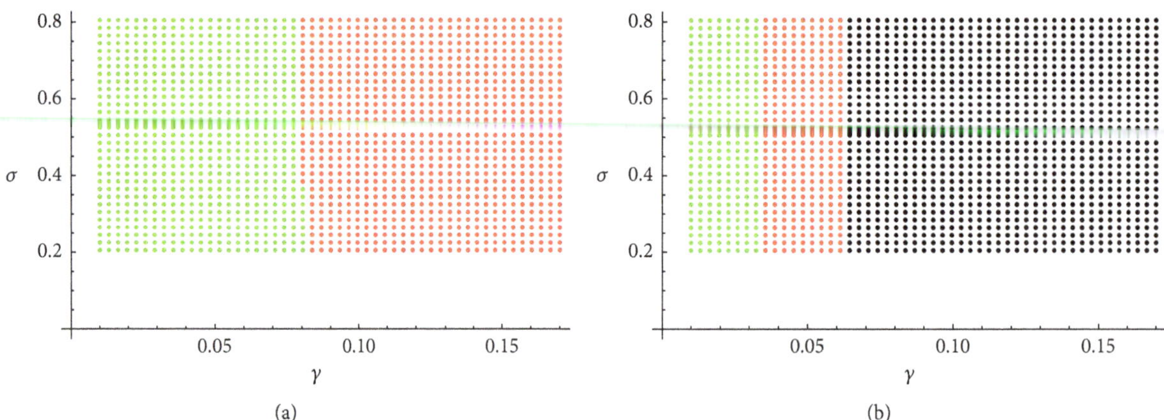

FIGURE 3: Stability features regarding the *critical* (green) and the *zero* (black and red) *regions*. Green denotes the negative slope of the real part of the complex eigenvalues versus β, around β_{cr}, red indicates the existence of unstable endemic equilibria for $\beta \in (\delta, 1)$, and black outlines the pairs (γ, σ) for which no endemic equilibria arise in the whole range $(0, 1)$ for β. The parameters are defined as $r = 10^{-4}$, $\alpha = 10^{-4}$, $\varepsilon = 10^{-3}$, and $m = 0.5$ and (a) $s = 1.2$ and (b) $s = 1.6$.

direction and stability of the occurring codimension 1 Hopf bifurcation (see Appendix B). As regards the *zero regions* (no β_{cr}), one investigates the existence of equilibrium points, as well as their stability, as β varies within the range $(0, 1)$.

As a result, we conclude that regardless of the parameter values, the real part of the complex eigenvalues transverses the β-axis (at β_{cr}) with negative slope, all over the *critical surface* (for β lying into the aforementioned "complex" interval, the sign of the real eigenvalue is always negative). Moreover, we encounter an unstable endemic equilibrium depending on β in the *zero regions*, for this active parameter lying inside an interval $(\delta, 1)$, $\delta = \delta(\gamma, r, \alpha, \varepsilon, \sigma, m, s) > 0$ (we have no equilibria at $(0, \delta)$), with δ notably sensitive to γ-variation, in the sense that as γ increases, δ shifts rapidly towards unity, shrinking the (unstable) "equilibrium" β- interval. Additionally, by increasing s or m, a "0- equilibrium" region emerged inside the *zero region* (there exist no equilibria in the whole range $(0, 1)$), getting larger as these two rational exponents (especially s) increase.

3. Analytical Formulae for a Parametric n-Dimensional System

3.1. Reduction to an $(n + 2)$-Dimensional Coordinate Space. Now, we briefly discuss the basic features of a new formulation regarding a parametric n-dimensional nonlinear system analysis. The procedure adopted is presented in detail in [4, Section 2], resulting in the derivation of the two-dimensional restriction to the center manifold of the system, as well as in the fast numerical computation of the Lyapunov coefficients associated with simple or degenerate Hopf bifurcations of the system. Thus considering a smooth continuous-time three-parameter system with smooth dependence on the parameters, namely,

$$\frac{dx}{dt} = f(x; a),$$
$$x \in \mathbb{R}^n, \; a = (a_1, a_2, a_3)^T \in \mathbb{R}^3, \; f : \mathbb{R}^{n+3} \longrightarrow \mathbb{R}^n, \; f \in C^\infty \quad (21)$$

(for SEIRS system (7) we have that $(a_1, a_2, a_3) = (\gamma, \sigma, \beta)$), then expansion around the equilibrium path yields

$$\frac{d\xi}{dt} = J_0(a)\xi + F(\xi; a), \quad \xi = x - x^0(a), \quad (22)$$

with $J_0(a) = D_x f(x^0(a); a)$ being the Jacobian matrix evaluated at this equilibrium path. The smooth vector function $F : \mathbb{R}^{n+3} \rightarrow \mathbb{R}^n$ represents the nonlinear terms; that is, $F = O(\|x\|^2)$ and

$$F(\xi; a) = \sum_{m=2}^{\infty} \frac{1}{m!} F^{(m)}\left(\underbrace{\xi, \ldots, \xi}_{m}; a\right), \quad (23)$$

where

$$F^{(m)}\left({}^1u, \ldots, {}^mu; a\right) = \sum_{j_1, \ldots, j_m = 1}^{n} \left.\frac{\partial^m f(x; a)}{\partial x_{j_1} \cdots \partial x_{j_m}}\right|_{x = x^0} {}^1u_{j_1} \cdots {}^mu_{j_m}, \quad (24)$$
$$m = 2, 3, \ldots$$

with ${}^iu = ({}^iu_1, \ldots, {}^iu_n)^T \in \mathbb{C}^n$, $i = 1, \ldots, m$. Additionally, if a critical triplet $a_0 = (a_{10}, a_{20}, a_{30})$ exists, where J_0 has a pair of pure imaginary eigenvalues $\lambda_{1,2} = \pm i\omega_0$, $\omega_0 > 0$, while the real part of the rest of the eigenvalues is negative, then a Hopf bifurcation occurs at a_0, leading to a family of limit cycles. This implies that J_0 has a pair of complex conjugate eigenvalues $\lambda(a)$, $\overline{\lambda}(a)$ in a region around a_0, with

$$\lambda(a) = \mu(a) + i\omega(a),$$
$$\mu(a_0) = 0, \quad (25)$$
$$\omega(a_0) = \omega_0 > 0.$$

We also note that the classic Hopf theory additionally demands that, at a_0, $\lambda(a)$ cross the imaginary axis with nonzero speed (*transversality* condition). Moreover we consider the normalized complex eigenvectors $q(a) \in \mathbb{C}^n$ and $p(a) \in \mathbb{C}^n$ of J_0 and $J_0{}^T$, respectively, having the properties

$$\begin{aligned} J_0 q &= \lambda q, \\ J_0{}^T p &= \overline{\lambda} p, \\ \langle p, q \rangle &= \sum_{i=1}^{n} \overline{p}_i q_i = 1, \\ \langle p, \overline{q} \rangle &= 0. \end{aligned} \tag{26}$$

Also, the two-dimensional parameter-dependent real eigenspace (corresponding to $\lambda, \overline{\lambda}$) (spanned by $\{\operatorname{Re} q(a), \operatorname{Im} q(a)\}$), is denoted by $T(a)$ (which becomes critical at a_0: $T^c = T(a_0)$), and the $(n-2)$-dimensional $(n > 2)$ real eigenspace, corresponding to all eigenvalues of J_0 other than $\lambda, \overline{\lambda}$, is denoted by $T^{su}(a)$. Finally, the following Lemma ([6], Lemma 5.4) holds.

Lemma 1. *$y \in T^{su}$ if and only if $\langle p, y \rangle = 0$.*

Then, based on the decomposition

$$\begin{gathered} x - x^0 = \xi = zq + \overline{zq} + y, \\ x \in \mathbb{R}^n,\ z \in \mathbb{C}^1,\ y \in \mathbb{R}^n \cap T^{su},\ zq + \overline{z}\overline{q} \in T, \end{gathered} \tag{27}$$

by means of the third and fourth relation of (26), the last equation yields

$$\begin{aligned} z &= \langle p, \xi \rangle, \\ y &= \xi - \langle p, \xi \rangle q - \langle \overline{p}, \xi \rangle \overline{q}. \end{aligned} \tag{28}$$

Thus taking into account (22) and the first and second equation of (26), as well as (27), differentiation of (28) results in the reduced form of system (22) in the $(n+2)$-dimensional coordinate space (z, y):

$$\dot{z} = \lambda(a) z + g(z, \overline{z}, y; a), \tag{29}$$

$$\begin{aligned} \dot{y} = J_0(a) y + F(zq + \overline{z}\overline{q} + y; a) - g(z, \overline{z}, y; a) q \\ - \overline{g}(z, \overline{z}, y; a) \overline{q}, \end{aligned} \tag{30}$$

where the dot $(\cdot)$ denotes differentiation with respect to t and

$$g(z, \overline{z}, y; a) = \langle p, F(zq + \overline{z}\overline{q} + y; a) \rangle. \tag{31}$$

Systems (22) and (29)-(30) are in fact dimensionally equivalent, due to the two real constraints imposed on y by means of Lemma 1.

3.2. A New Symbolic Representation of the System. By expressing the complex function $g(z, \overline{z}, y)$ in the form

$$\begin{aligned} g = \frac{1}{2} g_{00}{}^{(2)} + \frac{1}{6} g_{00}{}^{(3)} + \cdots \\ + \left(g_{10}{}^{(1)} + \frac{1}{2} g_{10}{}^{(2)} + \frac{1}{6} g_{10}{}^{(3)} + \cdots \right) z \\ + \left(g_{01}{}^{(1)} + \frac{1}{2} g_{01}{}^{(2)} + \frac{1}{6} g_{01}{}^{(3)} + \cdots \right) \overline{z} + g^{(2)} \\ + g^{(3)} + \cdots, \end{aligned} \tag{32}$$

where

$$\begin{aligned} & g^{(m)}(z, \overline{z}, y) \\ & \quad = \sum_{k+l=m} \frac{1}{k!l!} \left(g_{kl} + g_{kl}{}^{(1)} + \frac{1}{2} g_{kl}{}^{(2)} + \frac{1}{6} g_{kl}{}^{(3)} + \cdots \right) \\ & \quad \cdot z^k \overline{z}^l, \\ & g_{ij}{}^{(m^+)}(y) \\ & \quad = \left\langle p, F^{(i+j+m^+)} \left(\underbrace{q, \ldots, q}_{i}, \underbrace{\overline{q}, \ldots, \overline{q}}_{j}, \underbrace{y, \ldots, y}_{m^+} \right) \right\rangle, \\ & \qquad g_{kl} = g_{kl}{}^{(0)} \end{aligned} \tag{33}$$

with $g_{ij}{}^{(m^+)} \in \mathbb{C}$, $m = 2, 3, \ldots, i + j = 0, 1, 2, \ldots, m^+ = 0, 1, 2, \ldots$, then (29) and (30) become

$$\begin{aligned} \dot{z} &= \frac{1}{2} g_{00}{}^{(2)} + \frac{1}{6} g_{00}{}^{(3)} + \cdots + \left(\lambda + g_{10}{}^{(1)} + \frac{1}{2} g_{10}{}^{(2)} \right. \\ & \quad \left. + \frac{1}{6} g_{10}{}^{(3)} + \cdots \right) z + \left(g_{01}{}^{(1)} + \frac{1}{2} g_{01}{}^{(2)} + \frac{1}{6} g_{01}{}^{(3)} \right. \\ & \quad \left. + \cdots \right) \overline{z} + g^{(2)} + g^{(3)} + \cdots, \\ \dot{y} &= J_0(a) y + \frac{1}{2} r_{00}{}^{(2)} + \frac{1}{6} r_{00}{}^{(3)} + \cdots - \left(\frac{1}{2} g_{00}{}^{(2)} \right. \\ & \quad \left. + \frac{1}{6} g_{00}{}^{(3)} + \cdots \right) q - \left(\frac{1}{2} \overline{g}_{00}{}^{(2)} + \frac{1}{6} \overline{g}_{00}{}^{(3)} + \cdots \right) \overline{q} \\ & \quad + \left[r_{10}{}^{(1)} + \frac{1}{2} r_{10}{}^{(2)} + \frac{1}{6} r_{10}{}^{(3)} + \cdots \right. \\ & \quad - \left(g_{10}{}^{(1)} + \frac{1}{2} g_{10}{}^{(2)} + \frac{1}{6} g_{10}{}^{(3)} + \cdots \right) q \end{aligned} \tag{34}$$

$$- \left(\overline{g}_{01}{}^{(1)} + \frac{1}{2}\overline{g}_{01}{}^{(2)} + \frac{1}{6}\overline{g}_{01}{}^{(3)} + \cdots \right) \overline{q} \Bigg] z + \Bigg[r_{01}{}^{(1)}$$

$$+ \frac{1}{2} r_{01}{}^{(2)} + \frac{1}{6} r_{01}{}^{(3)} + \cdots$$

$$- \left(g_{01}{}^{(1)} + \frac{1}{2} g_{01}{}^{(2)} + \frac{1}{6} g_{01}{}^{(3)} + \cdots \right) q$$

$$- \left(\overline{g}_{10}{}^{(1)} + \frac{1}{2}\overline{g}_{10}{}^{(2)} + \frac{1}{6}\overline{g}_{10}{}^{(3)} + \cdots \right) \overline{q} \Bigg] \overline{z} + r^{(2)}$$

$$+ r^{(3)} + \cdots, \tag{35}$$

where

$$r^{(m)}(z, \overline{z}, y) = \sum_{k+l=m} \frac{1}{k!l!} \Bigg[r_{kl} + r_{kl}{}^{(1)} + \frac{1}{2} r_{kl}{}^{(2)}$$

$$+ \frac{1}{6} r_{kl}{}^{(3)} + \cdots$$

$$- \left(g_{kl} + g_{kl}{}^{(1)} + \frac{1}{2} g_{kl}{}^{(2)} + \frac{1}{6} g_{kl}{}^{(3)} + \cdots \right) q \tag{36}$$

$$- \left(\overline{g}_{lk} + \overline{g}_{lk}{}^{(1)} + \frac{1}{2}\overline{g}_{lk}{}^{(2)} + \frac{1}{6}\overline{g}_{lk}{}^{(3)} + \cdots \right) \overline{q} \Bigg] z^k \overline{z}^l,$$

$$r_{ij}{}^{(m^+)}(y) = F^{(i+j+m^+)} \Bigg(\underbrace{q, \ldots, q}_{i}, \underbrace{\overline{q}, \ldots, \overline{q}}_{j}, \underbrace{y, \ldots, y}_{m^+} \Bigg),$$

$$r_{kl} = r_{kl}{}^{(0)}$$

with $g_{ij}{}^{(m^+)}$ as in (33), $r_{ij}{}^{(m^+)} \in \mathbb{C}^n$, $m = 2, 3, \ldots$, $i + j = 0, 1, 2, \ldots$, $m^+ = 0, 1, 2, \ldots$.

Now, by introducing the center manifold expression

$$y = V(z, \overline{z}) = \sum_{m=2}^{\infty} \sum_{k+l=m} \frac{1}{k!l!} w_{kl} z^k \overline{z}^l, \tag{37}$$

with $w_{kl} \in \mathbb{C}^n, k+l = 2, 3, \ldots$ and $\langle p, w_{kl} \rangle = 0$, and combining (37), as well as the invariance relation $\dot{y} = V_z \dot{z} + V_{\overline{z}} \dot{\overline{z}}$, with (34) and (35), the analysis proceeds according to the steps figured out in [4, Subsection 2.3]. Thus after extensive algebraic manipulations combined with computer assisted calculations carried out by use of the symbolic mathematical computational software Mathematica 7, the necessary (depending on the considered bifurcation codimension) center manifold coefficients, as well as the coefficients of the system restricted to the center manifold, are derived as explicit and implicit expressions of the parameters involved. Finally, the corresponding Lyapunov coefficients are numerically evaluated fast, not only at the critical parameter values, but throughout the whole parameter space, as well, and consequently the respective bifurcation portraits can be constructed.

4. Bifurcation Results-Discussion

After setting $r = 10^{-4}$, $\alpha = 10^{-4}$, $\varepsilon = 10^{-3}$, $m = 0.5$, $s = 1.2$ regarding the values of the fixed parameters, by following the procedure developed in [4] (briefly discussed in Section 3 of the present paper), we arrive at the bifurcation portrait of the system with respect to the Σ_1 equilibrium path, presented in Figure 4. Regarding the sign of the Lyapunov coefficient, it remains strictly negative on the whole *critical surface* $\beta_{cr} = \beta_{cr}(\gamma, \sigma)$. Thus stable limit cycles are generated through supercritical Hopf bifurcations, arising for (γ, σ) taking values in the *critical region* and $\beta < \beta_{cr}(\gamma, \sigma)$ (the real part of the complex conjugate eigenvalues is a decreasing function of β around β_{cr}; see Section 2). Since the critical Lyapunov coefficient l_1 never becomes zero, the system undergoes solely a *codimension 1* Hopf bifurcation, for any critical triplet of the active parameters (hence, in Appendix B, we only refer (after a general inspection) to the features of the third-order normal form of the planar equation).

The fact that the limit cycles bifurcated are stable means that the phenomenon is persistent; that is, the flows in a nearby neighbourhood of the limit cycle are attracted by the cycle itself, leading to the corresponding disequilibrium fluctuations defined by the periodic trajectory, as expected in a considerable number of epidemics. The bifurcation results are verified by the computation and presentation of one cycle for specific values of the parameters by use of a custom algorithm of orthogonal collocation on finite elements, shown in Figures 5(a), 5(b), and 5(c) and a family of limit cycles obtained for different values of the epidemic-induced parameter α, shown in Figures 6(a), 6(b), and 6(c), where the corresponding β_{cr} and also the period T of the periodic orbits increase with α. The stability of the obtained cycles is additionally verified by numerical computation of the respective Floquet-multipliers and exponents. For the limit cycles presented in Figure 5, let the Floquet-multipliers be $\mu_i = e^{\lambda_i \cdot T}$, $i = 1, 2, 3$, with $\lambda_i = (1/T) \cdot \ln(\mu_i)$ being the respective exponents and T being the fundamental period of the cycle, computed as follows: $\mu_i = \{1, 0.82, 1.1 \times 10^{-17}\}$ and $\lambda_i = \{0, -4.7 \times 10^{-4}, -0.095\}$, respectively. Moreover we note that the same cycles are generated by the variable-step, variable-order Adams-Bashforth-Moulton predictor-corrector method of orders 1 to 12, which is the standard integrated Matlab routine used to solve nonstiff ODEs, noted as "ode113" in the graphs illustrated in Figures 5 and 6.

As regards the role and significance of the introduction of the additional epidemic-induced death rate, a, apart from the fact that it contributes to a more accurate and realistic description of the occurrence of the epidemic, which leads to the de facto increase of the mortality rate of infected individuals, it also offers the opportunity for a more detailed and richer parameterisation as well as the effect of the epidemic on the dynamics and interrelation between the populations involved, especially in aggressive diseases. Evidently, the introduction of a gives rise to new nonlinear terms involved in all four equations of the original system, as shown in (3), and has an important effect on the numeric value of the fundamental period of the bifurcating cycles, which would have been overseen, otherwise. Moreover, the introduction of the above-mentioned parameter could also contribute to making estimations of the additional resources needed, based on the changes in the duration of critical phases and stages

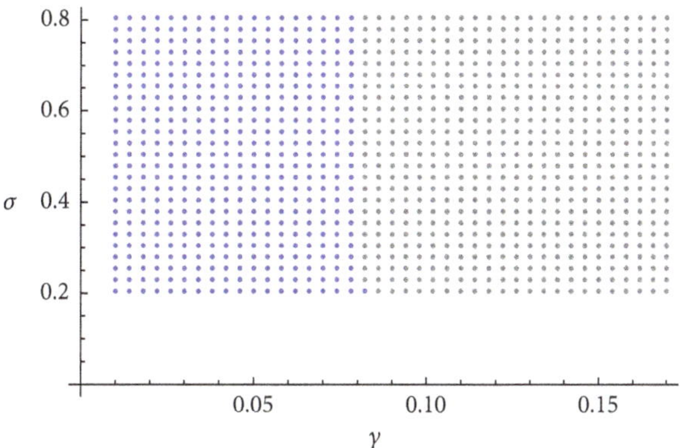

FIGURE 4: Bifurcation portrait as regards the supercritical Hopf bifurcation for $r = 10^{-4}$, $\alpha = 10^{-4}$, $\varepsilon = 10^{-3}$, $m = 0.5$, and $s = 1.2$. Negative sign (blue region) of the first Lyapunov coefficient at the *critical surface* $\beta_{cr} = \beta_{cr}(\gamma, \sigma)$, in the parameter plane (γ, σ) (no β_{cr} in the gray region).

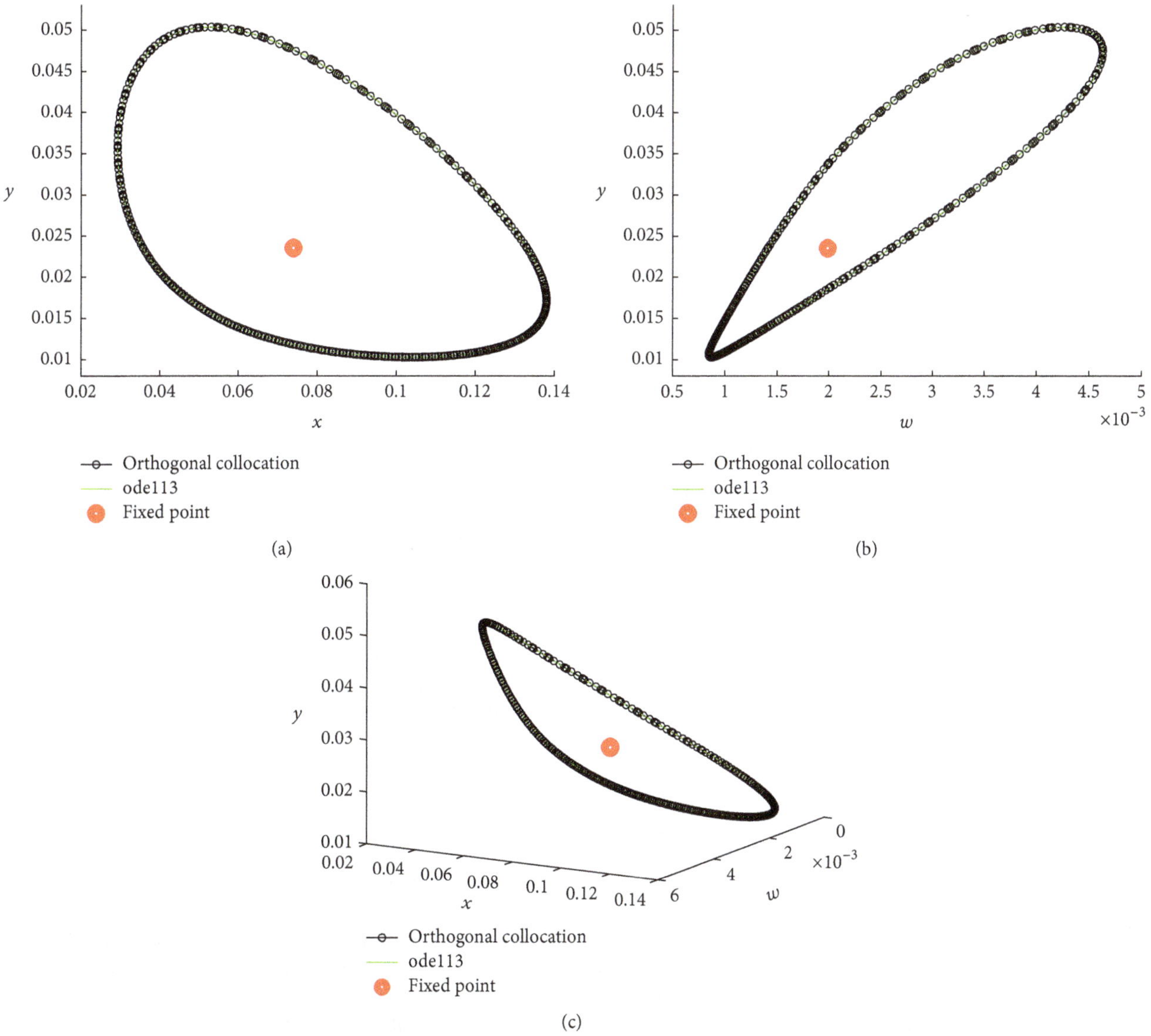

(a)

(b)

(c)

FIGURE 5: Stable cycle generated by supercritical Hopf bifurcation in xy (*SI*), wy (*EI*) planes and in 3D (*SEI*), respectively, for $(r, \alpha, \varepsilon, m, s, \gamma, \sigma, \beta) = (10^{-4}, 10^{-4}, 10^{-3}, 0.5, 1.2, 0.042, 0.5, 0.328525)$ ($\beta_{cr} = 0.338525$), determined by a custom orthogonal collocation on finite elements algorithm. Unstable endemic equilibrium (red marker): $(x^0, w^0, y^0) = (0.073948, 0.001986, 0.023535)$. Period: $T = 410.572533$ days. ode113: Adams-Bashforth-Moulton PECE solver of orders 1 to 12 (Matlab).

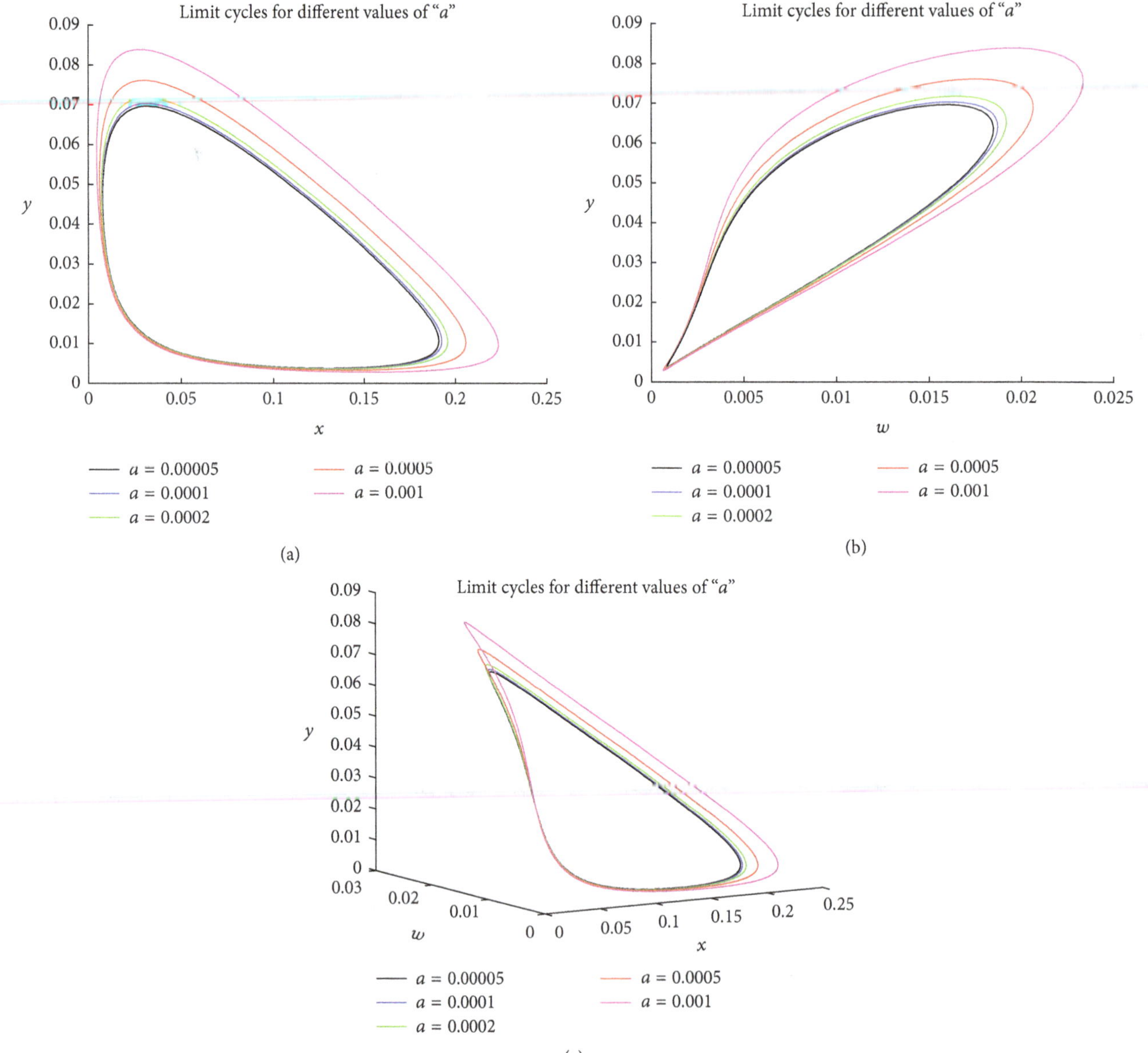

FIGURE 6: Stable cycles generated by supercritical Hopf bifurcation in xy (SI), wy (EI) planes and in 3D (SEI), respectively, for $(r, \varepsilon, m, s, \gamma, \sigma, \beta) = (10^{-4}, 10^{-3}, 0.5, 1.2, 0.0548, 0.24, 0.46825)$, $\alpha \in [5 \cdot 10^{-5}, 10^{-3}]$, determined by a custom orthogonal collocation on finite elements algorithm. min $\beta_{cr} = 0.51825$ ($\alpha = 5 \cdot 10^{-5}$). Periods: $T_{min} = 410.414990$ days ($\alpha = 5 \cdot 10^{-5}$); $T_{max} = 441.550487$ days ($\alpha = 10^{-3}$). ode113: Adams-Bashforth-Moulton PECE solver of orders 1 to 12 (Matlab).

during an epidemic cycle and the maximum number of infected individuals (i.e., estimating the additional cost and health resources needed), thus making it possible to manage and control the epidemics more effectively, efficiently, and with a better allocation of available resources.

Concerning the results presented in [3] with respect to SEIRS model, especially the ones related to Hopf bifurcation, firstly we would like to note that in the system investigated therein an equilibrium status between the birth and the death rate is considered, without taking into account the aforementioned pure epidemic-induced death rate introduced herein. Furthermore, the bifurcation analysis in [3] is performed in a two-dimensional parameter space. More precisely the respective diagrams are obtained with respect to the rational power s and the parameter $c = \beta\sigma/[(\sigma + r)(\gamma + r)]$ (called "contact number"; see [7]). We believe that the analysis and hence the results which are established in a higher dimension parameter space, like the three-dimensional one structured in the present work (with γ, σ, β considered as the varying parameters of the system), enhance the parametric "resolution," that is, the physical insight into the dynamics associated with the interaction of the parameters involved in the physical background. Therefore, taking into account these essential differences, we see that, in [3], depending on the parameter values and hence the sign of the *first Lyapunov coefficient*, either only a subcritical bifurcation (positive sign) or an alternation (due to multiple sign changes) of subcritical and supercritical (negative sign) bifurcations occurs, located by the corresponding *critical curves* in the parameter plane (s, c).

5. Conclusion

In the present paper a new SEIRS epidemic model with nonpermanent immunity, nonlinear incidence rate, and an additional disease-induced death rate is presented. Then as regards the characteristic equilibrium equation, by means of the Liu criterion we conclude with the determination of critical loci in the parameter space, where a Hopf bifurcation associated with the endemic equilibrium occurs. In addition, a three-dimensional parameter space structured by the recovery, incubation, and transmission rate forms the basis of the dynamic analysis, where we proceed by using a new symbolic form introduced by the writers in a previous work, applicable in any multidimensional and multiparameter system.

Therefore we believe that the adopted general form of the system combined with the dimension of the active parameter space satisfies the demand for a more realistic modeling and offers a deeper insight into the dynamic behaviour of the epidemic, with respect to the interaction of the populations involved. Additionally, due to the lengthy expressions derived in the treatment of the dynamics concerning the restriction of the system to the center manifold and the associated normal forms, as well, the algebraic platform mentioned above is proved to be appropriate in manipulating a large amount of analytical data. Thus constructing algorithms of an implicit structure, by use of chain numerical computations, all the quantities associated with the bifurcation analysis are evaluated fast and the bifurcation diagrams of the system can be obtained throughout the whole parameter space under consideration, for any values of the fixed parameters.

We should further note that variations of the fixed parameters change the shape and the size of the *critical* and *zero* regions, without affecting the qualitative profile of the results. In particular, increase in the rational exponents involved in the incidence rate, especially in that concerning the infective individual, results in a shrinkage of the *critical region*, while the *zero region*, as well as the nonequilibrium area inside that region (as regards the endemic one), expands towards lower values of the recovery rate. Finally, the algorithm of orthogonal collocation on finite elements with Legendre orthogonal polynomials is proved to be excellent in the fast and precise numerical computation of the bifurcated stable limit cycles.

Appendix

A. Algebraic Treatment of System (7)

A.1. Algebraic Manipulations Leading to the Final Form of System (7). System (7), with $y^0 \neq 0$, results in

$$\begin{aligned}
&r + \varepsilon - (r+\varepsilon)x^0 - \varepsilon w^0 - \varepsilon y^0 + a x^0 y^0 \\
&\quad - \beta\left(x^0\right)^m\left(y^0\right)^s = 0 \\
&- (r+\sigma)w^0 + a w^0 y^0 + \beta\left(x^0\right)^m\left(y^0\right)^s = 0 \\
&\sigma w^0 - (r+\gamma+\alpha)y^0 + a\left(y^0\right)^2 = 0.
\end{aligned} \tag{A.1}$$

By replacing the second equation of (A.1) with the sum of the last two ones, we get

$$\begin{aligned}
&r + \varepsilon - (r+\varepsilon)x^0 - \varepsilon w^0 - \varepsilon y^0 + a x^0 y^0 \\
&\quad - \beta\left(x^0\right)^m\left(y^0\right)^s = 0 \\
&- r w^0 - (r+\gamma+\alpha)y^0 + a w^0 y^0 + a\left(y^0\right)^2 \\
&\quad + \beta\left(x^0\right)^m\left(y^0\right)^s = 0 \\
&\sigma w^0 - (r+\gamma+\alpha)y^0 + a\left(y^0\right)^2 = 0.
\end{aligned} \tag{A.2}$$

Then, by multiplying the second relation of (A.2) by σ and the last one by r and adding them up we obtain

$$\begin{aligned}
&r + \varepsilon - (r+\varepsilon)x^0 - \varepsilon w^0 - \varepsilon y^0 + a x^0 y^0 \\
&\quad - \beta\left(x^0\right)^m\left(y^0\right)^s = 0 \\
&- (r+\sigma)(r+\gamma+\alpha) + a\sigma w^0 + a(r+\sigma)y^0 \\
&\quad + \beta\sigma\left(x^0\right)^m\left(y^0\right)^{s-1} = 0 \\
&\sigma w^0 - (r+\gamma+\alpha)y^0 + a\left(y^0\right)^2 = 0.
\end{aligned} \tag{A.3}$$

By further multiplying the last relation of (A.3) by $-a$ and adding it to the second one, we obtain

$$\begin{aligned}
&r + \varepsilon - (r+\varepsilon)x^0 - \varepsilon w^0 - \varepsilon y^0 + a x^0 y^0 \\
&\quad - \beta\left(x^0\right)^m\left(y^0\right)^s = 0 \\
&- (r+\sigma)(r+\gamma+\alpha) + a(2r+\sigma+\gamma+\alpha)y^0 \\
&\quad - a^2\left(y^0\right)^2 + \beta\sigma\left(x^0\right)^m\left(y^0\right)^{s-1} = 0 \\
&\sigma w^0 - (r+\gamma+\alpha)y^0 + a\left(y^0\right)^2 = 0.
\end{aligned} \tag{A.4}$$

Moreover, by multiplying the first relation of (A.4) by σ and the third one by ε and adding them up we get

$$\begin{aligned}
&\sigma(r+\varepsilon) - \sigma(r+\varepsilon)x^0 - \sigma\varepsilon y^0 + \sigma a x^0 y^0 \\
&\quad - \varepsilon(r+\gamma+\alpha)y^0 + \varepsilon a\left(y^0\right)^2 \\
&\quad - \beta\sigma\left(x^0\right)^m\left(y^0\right)^s = 0 \\
&\beta\sigma\left(x^0\right)^m\left(y^0\right)^{s-1} - a^2\left(y^0\right)^2 + a(2r+\sigma+\gamma+\alpha)y^0 \\
&\quad - (r+\sigma)(r+\gamma+\alpha) = 0 \\
&\sigma w^0 - (r+\gamma+\alpha)y^0 + a\left(y^0\right)^2 = 0.
\end{aligned} \tag{A.5}$$

Then by multiplying the second relation of (A.5) by y^0 and by adding the first two relations we finally obtain

$$\begin{aligned}
&\sigma(r+\varepsilon) - \sigma(r+\varepsilon)x^0 - \sigma\varepsilon y^0 + \sigma a x^0 y^0 - a^2\left(y^0\right)^2 \\
&\quad + a(2r+\sigma+\gamma+\alpha+\varepsilon)\left(y^0\right)^2 \\
&\quad - (r+\sigma+\varepsilon)(r+\gamma+\alpha)y^0 = 0 \\
&\beta\sigma\left(x^0\right)^m\left(y^0\right)^{s-1} - a^2\left(y^0\right)^2 + a(2r+\sigma+\gamma+\alpha)y^0 \\
&\quad - (r+\sigma)(r+\gamma+\alpha) = 0 \\
&\sigma w^0 - (r+\gamma+\alpha)y^0 + a\left(y^0\right)^2 = 0.
\end{aligned} \tag{A.6}$$

The latter system yields (9)–(11).

A.2. Derivatives of x^0, y^0 with respect to the Varying Parameters. The derivatives of x^0, y^0, with respect to the varying parameters of the problem (γ, σ, β), are involved in the *transversality condition* associated with the emerged Hopf bifurcation (see relation (C1-2) in Appendix B), as well as in the equivalent (according to Liu criterion) derivatives of $B_3 = B_1B_2 - B_0$ (see (19)). Thus taking into account the fact that y^0 and x^0 are obtained from (10) and (9), respectively, as

$$\begin{aligned}
y^0 &= y^0(\gamma, r, \alpha, \varepsilon, \sigma, \beta), \\
x^0 &= x^0\left(y^0; \gamma, r, \alpha, \varepsilon, \sigma\right),
\end{aligned} \tag{A.7}$$

then by differentiating (9) and (10), and by denoting the partial derivative of x^0 with respect to y^0, $(x^0)_{y^0}$ and the partial derivatives of x^0 and y^0 with respect to the parameters p $(p = \gamma, \sigma, \beta)$, $x^0{}_p$, and $y^0{}_p$, respectively, the following formulae are obtained:

$$\begin{aligned}
y^0{}_\gamma &= \frac{1}{Y_d}\left\{\kappa_1 - \varepsilon - \alpha y^0\right. \\
&\quad \left. - \frac{mA}{x^0 S_2}\left[-\alpha\left(y^0\right)^2 + \kappa_1 y^0\right]\right\} \\
x^0{}_\gamma &= \left(x^0\right)_{y^0} y^0{}_\gamma + \frac{-\alpha\left(y^0\right)^2 + \kappa_1 y^0}{S_2} \\
y^0{}_\sigma &= \frac{1}{Y_d}\left\{\kappa_2 - \alpha y^0 + \frac{(m-1)A}{\sigma}\right. \\
&\quad \left. - \frac{mA}{x^0 S_2}\left[-\alpha\left(y^0\right)^2 + (\kappa_2+\varepsilon)y^0 - \kappa_0\right]\right\}, \\
x^0{}_\sigma &= \left(x^0\right)_{y^0} y^0{}_\sigma - \frac{x^0}{\sigma} \\
&\quad + \frac{-\alpha\left(y^0\right)^2 + (\kappa_2+\varepsilon)y^0 - \kappa_0}{S_2} \\
y^0{}_\beta &= -\frac{1}{Y_d}\frac{A}{\beta}, \\
x^0{}_\beta &= \left(x^0\right)_{y^0} y^0{}_\beta,
\end{aligned} \tag{A.8}$$

where

$$\begin{aligned}
Y_d &= \frac{mA\left(x^0\right)_{y^0}}{x^0} + \frac{(s-1)A}{y^0} - 2\alpha^2 y^0 \\
&\quad + \alpha(\kappa_4 - \varepsilon),
\end{aligned} \tag{A.9}$$

$$\left(x^0\right)_{y^0} = \frac{S_0}{S_2} - \frac{\alpha\sigma S_1}{S_2{}^2}.$$

$A(y^0; \gamma, r, \alpha, \varepsilon, \sigma)$ is given by (13) and

$$\begin{aligned}
S_0\left(y^0; \gamma, r, \alpha, \varepsilon, \sigma\right) &= 3\alpha^2\left(y^0\right)^2 - 2\alpha\kappa_4 y^0 + \kappa_3 + \sigma\varepsilon \\
S_1\left(y^0; \gamma, r, \alpha, \varepsilon, \sigma\right) &= \alpha^2\left(y^0\right)^3 - \alpha\kappa_4\left(y^0\right)^2 \\
&\quad + (\kappa_3 + \sigma\varepsilon)y^0 - \sigma\kappa_0 \\
S_2\left(y^0; \gamma, r, \alpha, \varepsilon, \sigma\right) &= \sigma\left(\alpha y^0 - \kappa_0\right).
\end{aligned} \tag{A.10}$$

B. Normal Forms

B.1. Poincaré Normal Forms for the Planar Case ($n = 2$). In this case the terms $g_{ij}{}^{(m^+)}$ with $i+j = 0, 1, 2, \ldots, m^+ = 1, 2, \ldots$ in the first part of (33) and in (34) vanish. Moreover, (35) does not exist. Now, in order to obtain the desired normal form, we first introduce the transformation

$$\begin{aligned}
z &= w + h^{(2)}(w, \overline{w}) + \cdots + h^{(2n_r+1)}(w, \overline{w}), \\
&\qquad n_r \in \mathbb{N}, \\
h^{(m)}(w, \overline{w}) &= \sum_{k+l=m} \frac{1}{k!l!} h_{kl} w^k \overline{w}^l, \\
&\qquad m = 2, \ldots, 2n_r + 1,
\end{aligned} \tag{B.1}$$

the inversion of which is given by (see [6, Section 3.8])

$$\begin{aligned}
w &= z - h^{(2)}(z, \overline{z}) - \cdots - h^{(2n_r+1)}(z, \overline{z}) \\
&\quad + O\left(|z|^{2n_r+2}\right).
\end{aligned} \tag{B.2}$$

Then by differentiating (B.2) with respect to t and substituting the right-hand side of (29) for $\dot{z}$ (with $g = g^{(2)} + g^{(3)} + \cdots$), as well as the corresponding conjugate equation for $\dot{\overline{z}}$, by keeping terms up to $2n_r+1$ order with respect to $|z|$, we obtain

$$\begin{aligned}\dot{w} &= \lambda z + g^{(2)} + \cdots + g^{(2n_r+1)} \\ &- \sum_{m=1}^{2n_r-1} \left[\left(\lambda z + g^{(2)} + \cdots + g^{(2n_r+1-m)}\right) H^{(m)}\right. \\ &- \left.\left(\overline{\lambda}\overline{z} + \overline{g}^{(2)} + \cdots + \overline{g}^{(2n_r+1-m)}\right) \widetilde{H}^{(m)}\right] - \lambda z H^{(2n_r)} \\ &- \overline{\lambda}\overline{z}\widetilde{H}^{(2n_r)} + O\left(|z|^{2n_r+2}\right), \\ H^{(m)}(z,\overline{z}) &= \frac{\partial h^{(m+1)}(z,\overline{z})}{\partial z}, \\ \widetilde{H}^{(m)}(z,\overline{z}) &= \frac{\partial h^{(m+1)}(z,\overline{z})}{\partial \overline{z}}, \\ & m = 1, \ldots, 2n_r. \end{aligned} \tag{B.3}$$

Finally substitution of the transformation (B.1) for z (as well as the respective conjugate relation for $\overline{z}$) in (B.3) yields

$$\begin{aligned}\dot{w} &= \lambda(a) w + K^{(2)} + \cdots + K^{(2n_r+1)} + O\left(|w|^{2n_r+2}\right), \\ K^{(m)}&(w,\overline{w};a) \\ &= \sum_{k+l=m} K_{kl}\left(\lambda,\overline{\lambda},h_{kl},g_{kl},h_{ij},\overline{h}_{ij},g_{ij},\overline{g}_{ij}\right) w^k \overline{w}^l, \\ & m = 2,\ldots,2n_r+1,\ i+j = 2,\ldots,m-1, \end{aligned} \tag{B.4}$$

where terms up to $2n_r+1$ order with respect to $|w|$ have been evaluated. Now, by setting K_{kl}, $k+l = 2,\ldots,2n_r+1$ equal to zero we define the coefficients h_{kl} of transformation (B.1), while the rest of the *resonant* terms of the odd order, $c_i = K_{i+1,i}$, $i = 1,\ldots,n_r$, constitute the desirable normal form of (29). Furthermore, by substituting the obtained expressions of $h_{kl}, \overline{h}_{kl}$ involved in c_i $(k+l = 2,\ldots,2i)$ and then by means of computer calculations we arrive at the analytic formulae of the normal coefficients

$$\begin{aligned} c_i &= c_i\left(\lambda,\overline{\lambda},g_{i+1,i},g_{st},\overline{g}_{st}\right), \\ & s+t = 2,\ldots,2i,\ i = 1,\ldots,n_r, \end{aligned} \tag{B.5}$$

which are lengthy for $i \geq 2$.

B.2. Normal Form of the Third Order ($n_r = 1$). In this case, the above procedure results in the third-order *normal form*:

$$\dot{w} = \lambda(a) w + c_1(a) w |w|^2 + O\left(|w|^4\right), \tag{B.6}$$

where $c_1(a)w^2\overline{w}$ represents the cubic *resonant* term, due to the term $(\lambda+\overline{\lambda})h_{21}$ included in K_{21} (we set $h_{21} = 0$). Then by means of the linear time scaling

$$\theta = \omega(a) t \tag{B.7}$$

and the nonlinear time reparameterisation

$$d\theta = \left[1 - \frac{\operatorname{Im} c_1(a)}{\omega(a)} |w|^2\right] d\tau, \tag{B.8}$$

(B.6) becomes

$$\begin{aligned}\frac{dw}{d\tau} &= (b+i) w + l_1(a) w |w|^2 + O\left(|w|^4\right), \\ b(a) &= \frac{\mu(a)}{\omega(a)}, \\ l_1(a) &= \frac{\operatorname{Re} c_1(a)}{\omega(a)} - b(a) \frac{\operatorname{Im} c_1(a)}{\omega(a)}, \end{aligned} \tag{B.9}$$

where the parameter function $l_1(a)$ represents the *first Lyapunov coefficient.* If there exist *critical* values a_0 of the parameters, where

$$\begin{aligned} b(a_0) &= \frac{\mu(a_0)}{\omega_0} = 0, \\ l_1(a_0) &= \frac{\operatorname{Re} c_1(a_0)}{\omega_0} \neq 0, \end{aligned} \tag{C1-1}$$

then by the linear local (around a_0) transformation $w = u/|l_1(a)|^{1/2}$, (B.9) yields

$$\frac{du}{d\tau} = (b+i) u + s u |u|^2 + O\left(|u|^4\right), \tag{B.10}$$

with $s = \pm 1 = \operatorname{sign} l_1(a_0) = \operatorname{signRe}[c_1(a_0)]$. According to [6] (Lemma 3.2, valid for both supercritical and subcritical case) the terms which are higher than the third order in (B.10) do not affect the bifurcation behaviour of the system near x^0, and thus we consider the *locally topologically equivalent* system (B.10), where the $O(|u|^4)$ terms have been dropped. Moreover, if the *transversality* condition holds at the critical values of the parameters, that is,

$$\left.\frac{\partial b}{\partial a_i}\right|_{a=a_0} = \frac{1}{\omega_0} \left.\frac{\partial \mu}{\partial a_i}\right|_{a=a_0} \neq 0, \quad i = 1,2,3, \tag{C1-2}$$

then one limit cycle is bifurcated from the equilibrium x^0 at a_0. By using polar coordinates $u = \rho e^{i\varphi}$, the equivalent system (B.10) (without the $O(|u|^4)$ terms) gives $d\rho/d\tau = \rho(b \mp \rho^2)$, $s = \mp 1$, corresponding to the well-known supercritical ($s = -1, b > 0 \to \mu > 0$) and subcritical ($s = 1, b < 0 \to \mu < 0$) cases of the Hopf bifurcation. We call $x^0(a)$ a Hopf point of *codimension 1 (H1 point).*

In the case where $l_1(a_0) = 0$, we proceed to higher order normal forms.

Conflicts of Interest

The authors declare that there are no conflicts of interest regarding the publication of this paper.

Acknowledgments

P. S. Douris is pleased to acknowledge his financial support from "Andreas Mentzelopoulos Scholarships for the University of Patras."

References

[1] H. M. Yang and A. S. B. Silveira, "The loss of immunity in directly transmitted infections modeling: Effects on the epidemiological parameters," *Bulletin of Mathematical Biology*, vol. 60, no. 2, pp. 355–372, 1998.

[2] W. M. Liu, S. A. Levin, and Y. Iwasa, "Influence of nonlinear incidence rates upon the behavior of SIRS epidemiological models," *Journal of Mathematical Biology*, vol. 23, no. 2, pp. 187–204, 1986.

[3] W.-M. Liu, H. W. Hethcote, and S. A. Levin, "Dynamical behavior of epidemiological models with nonlinear incidence rates," *Mathematical and Computer Modelling*, vol. 51, pp. 810–822, 2010.

[4] M. P. Markakis and P. S. Douris, "On the computation of degenerate hopf bifurcations for n-dimensional multiparameter vector fields," *International Journal of Mathematics and Mathematical Sciences*, vol. 2016, Article ID 7658364, 12 pages, 2016.

[5] W. M. Liu, "Criterion of Hopf bifurcations without using eigenvalues," *Journal of Mathematical Analysis and Applications*, vol. 182, no. 1, pp. 250–256, 1994.

[6] Y. A. Kuznetsov, *Elements of Applied Bifurcation Theory*, Springer, New York, NY, USA, 3rd edition, 2004.

[7] H. W. Hethcote, "Qualitative analyses of communicable disease models," *Mathematical Biosciences*, vol. 28, no. 3/4, pp. 335–356, 1976.

A New Approach to Approximate Solutions for Nonlinear Differential Equation

Safia Meftah

Operators Theory and DPE Foundations and Applications Laboratory, Science Exact Faculty, Echahid Hamma Lakhdar University, P.O. Box 789, El Oued 39000, Algeria

Correspondence should be addressed to Safia Meftah; safia-meftah@univ-eloued.dz

Academic Editor: Theodore E. Simos

The question discussed in this study concerns one of the most helpful approximation methods, namely, the expansion of a solution of a differential equation in a series in powers of a small parameter. We used the Lindstedt-Poincaré perturbation method to construct a solution closer to uniformly valid asymptotic expansions for periodic solutions of second-order nonlinear differential equations.

1. Introduction

In the last few years, the study of approximations methods for systems of differential equations has been extensively developed; see, for example, [1]. This technique, known as the perturbation method (see [2]), has many applications in the theory of fractional differentiation operators (see [3]), in reaction-diffusion equations, stochastic stability, and asymptotic stability (see [4–9]), and for some numerical considerations (see, for example, [10–12]).

In current applications, some considerations require only the use of a small number of terms in the perturbation expansion, but the simple application of the perturbation is problematic if we want to calculate a uniformly valid solution.

Therefore, to structure a uniformly valid solution, one must look for an approximation that eliminates the terms causing the problem (secular terms). A technique to avoid the presence of these terms has been developed by Lindstedt. The principle of the Lindstedt method is to find approximations for periodic solutions, by convergent series using the expansion theorem and the periodicity of the solution [13, 14]. This method has various applications and properties; see, for example, [15]. Later, Poincaré proved that the expansion obtained by the Lindstedt technique is both asymptotic and uniformly valid.

The aim of this work is to present an analytical approximation study of periodic solutions for systems of second-order nonlinear differential equations. Although our analysis is based on the Lindstedt method, nevertheless the chosen development is according to a different approach from the one usually used. Thus, we recover an improvement in the process of the approximation.

Our paper consists of three sections. In the first section we present the general framework of our study. In the second, we recall most of the preliminary notions and the necessary definitions, and we prove the third approximation in the general case. Finally, in Section 3 we define and study the approximations of a new nonclassical equation.

2. Preliminaries and Definitions

In this section, we present an approximation method, based on the expansion of a solution of a differential equation in a series in a small parameter. It is used to construct uniformly valid periodic solutions to second-order nonlinear differential equations in the form

$$\frac{d^2y}{dt^2}(t,\epsilon)+y(t,\epsilon)=\epsilon F\left(y(t,\epsilon),\frac{dy}{dt}(t,\epsilon)\right),\quad 0<\epsilon<<1, \tag{1}$$

with $y(0,\epsilon)=A,(dy/dt)(0,\epsilon)(0)=0$, where $0<\epsilon<<1$ means that the positive parameter ϵ is small enough to be close to zero and F is supposed to be an analytical function of $y(t,\epsilon)$ and $(dy/dt)(t,\epsilon)$.

If $\epsilon = 0$ we obtain the following nonperturbed problem:

$$\frac{d^2 y}{dt^2}(t,0) + y(t,0) = 0. \quad (2)$$

Before we discuss our subject, we present some basic concepts concerning the perturbation theory. Then we introduce the Lindstedt method, which we use to determine uniformly valid solutions, in order to find a closer approximate solution for (2) (y^{**} is closer to y than y^* means that $|y - y^{**}| < |y - y^*|$). For further developments concerning the Lindstedt method see [16, 17].

2.1. Approximation Technique. We assume that the $(n+1)$th approximate solution of (1) can be written as

$$y(t,\epsilon) = \sum_{m=0}^{n} \epsilon^m y_m(t,0) + O\left(\epsilon^{n+1}\right). \quad (3)$$

The general procedure of the simple approximation is to substitute (3) into (1), develop in powers of ϵ, and put all coefficients of the powers of ϵ equal to zero. This gives a system of linear nonhomogeneous differential equations that we can solve recursively.

But the simple approximation takes us on a problem, if we need to calculate an analytical approximations of periodic solutions of nonlinear differential equations in the form given by (1). We illustrate this type of difficulty in the following example.

2.1.1. Example. We apply the simple approximation to the following equation:

$$\frac{d^2 y}{dt^2}(t,\epsilon) + \epsilon\left(\frac{dy}{dt}(t,\epsilon)\right)^2 + y(t,\epsilon) = 0, \quad 0 < \epsilon << 1, \quad (4)$$

with the initial values $y(0,\epsilon) = A$, $(dy/dt)(0,\epsilon) = 0$.

The fourth approximate solution of (4) is $y(t,\epsilon) = y_0(t,0) + \epsilon y_1(t,0) + \epsilon^2 y_2(t,0) + \epsilon^3 y_3(t,0) + O(\epsilon^4)$. After substituting and calculating, we find

$$\begin{aligned} y(t,\epsilon) &= A\cos t + \epsilon\frac{A^2}{6}(-3 + 4\cos t - \cos 2t) + \epsilon^2 \\ &\cdot \frac{A^3}{72}(-48 + 61\cos t - 16\cos 2t + 12t\sin t \\ &- 3\cos 3t) + \epsilon^3 A^4\left(-\frac{23}{24} + \frac{659}{540}\cos t - \frac{1}{3}\cos 2t\right. \\ &\left. + \frac{1}{12}\cos 3t - \frac{13}{1080}\cos 4t + \frac{1}{3}t\sin t - \frac{1}{18}t\sin 2t\right) \\ &+ O\left(\epsilon^4\right). \end{aligned} \quad (5)$$

We remark that the terms $y_2(t,0)$ and $y_3(t,0)$ are nonperiodic and unbounded as $t \longrightarrow +\infty$. This leads to the notion of secular terms.

2.2. Secular Terms. The conservation of a finite numbers of terms on the right-side of expansion (5) determines a function that is not only nonperiodic, but also unbounded as $t \longrightarrow +\infty$.

Definition 1. Terms such as $t^m\cos(pt)$ or $t^m\sin(nt)$ where $m, n \in \mathbb{N}^*$, $p \in \mathbb{N}$ are called secular terms.

These expressions appear because expansion (5) is not uniformly valid. The existence of such expressions destroys the periodicity of expansion (5) when only a finite number of terms is conserved. Therefore, to obtain a uniformly valid solution, we must look for an approximation that eliminates secular terms. A technique to avoid the presence of secular terms and allows for an approximation that is valid for all time has been developed by Lindstedt-Poincaré as described above in what follows.

2.3. Lindstedt-Poincaré Method. The substance of this method is to introduce a new independent variable linearly linked to the old independent variable. This transformation completely eliminates the secular terms. The basic idea came from the astronomer Lindstedt, based on the change of variable $\theta = \omega(\epsilon)t$ with $\omega_0 = w(0) = 1, \omega(\epsilon) \neq 1$, and both $y(\theta,\epsilon)$ and $\omega(\epsilon)$ are expanded in powers of ϵ as follows:

$$\begin{aligned} y(\theta,\epsilon) &= y_0(\theta,0) + \epsilon y_1(\theta,0) + \ldots + \epsilon^n y_n(\theta,0) + \ldots, \\ \omega(\epsilon) &= 1 + \epsilon\omega_1 + \ldots + \epsilon^n\omega_n + \ldots, \end{aligned} \quad (6)$$

and we note that, in this step, ω_j are unknowns; we obtain them by elimination of the secular terms.

First, we introduce the following notations:

$$\begin{aligned} \dot{y} &\equiv \frac{dy}{d\theta}(\theta,\epsilon), \\ \ddot{y} &\equiv \frac{d^2 y}{d\theta^2}(\theta,\epsilon), \\ F_y(y,\omega\dot{y}) &\equiv \frac{\partial F(y(\theta,\epsilon),\dot{y})}{\partial y(\theta,\epsilon)}, \\ F_{\dot{y}}(y,\dot{y}) &\equiv \frac{\partial F(y(\theta,\epsilon),\dot{y})}{\partial \dot{y}}, \end{aligned} \quad (7)$$

and (1) becomes

$$\omega^2\ddot{y} + y = \epsilon F(y,\omega\dot{y}), \quad 0 < \epsilon << 1, \quad (8)$$

with $y(0,\epsilon) = A$, $\dot{y}(0,\epsilon) = 0$. When we substitute expansion (6) into (8) we have

$$\begin{aligned} &\left(1 + \epsilon\omega_1 + \epsilon^2\omega_2 + \epsilon^3\omega_3 + \ldots\right)^2 \\ &\cdot \left(\ddot{y}_0 + \epsilon\ddot{y}_1 + \epsilon^2\ddot{y}_2 + \epsilon^3\ddot{y}_3 + \ldots\right) + y_0 + \epsilon y_1 + \epsilon^2 y_2 \\ &+ \epsilon^3 y_3 + \ldots = \epsilon F(y_0,\dot{y}_0) + \epsilon^2\frac{\partial F(y_0,\dot{y}_0)}{\partial\epsilon} + \frac{\epsilon^3}{2} \\ &\cdot \frac{\partial^2 F(y_0,\dot{y}_0)}{\partial\epsilon^2} + \ldots. \end{aligned} \quad (9)$$

Then we put all different powers of ϵ to zero, and we obtain (10), (11), (12), (13), and (15), such that

$$\ddot{y}_0 + y_0 = 0, \tag{10}$$

$$\ddot{y}_1 + y_1 = -2\omega_1 \ddot{y}_0 + F(y_0, \dot{y}_0) =: G_1(y_0(\theta,0), \dot{y}_0(\theta,0)) = G_1(\theta), \tag{11}$$

$$\ddot{y}_2 + y_2 = -2\omega_1 \ddot{y}_1 - (\omega_1^2 + 2\omega_2)\ddot{y}_0 + F_y(y_0, \dot{y}_0) y_1 + F_{\dot{y}}(y_0, \dot{y}_0)(\omega_1 \dot{y}_0 + \dot{y}_1) =: G_2(y_0(\theta,0), y_1(\theta,0), \dot{y}_0(\theta,0), \dot{y}_1(\theta,0)) = G_2(\theta), \tag{12}$$

$$\ddot{y}_3 + y_3 =: G_3(y_0(\theta,0), y_1(\theta,0), y_2(\theta,0); \dot{y}_0(\theta,0), \dot{y}_1(\theta,0), \dot{y}_2(\theta,0)) = G_3(\theta), \tag{13}$$

$$\ldots\ldots\ldots \tag{14}$$

$$\ddot{y}_n + y_n = G_n(y_0(\theta,0), y_1(\theta,0), \ldots, y_{n-1}(\theta,0); \dot{y}_0(\theta,0), \dot{y}_1(\theta,0), \ldots, \dot{y}_{n-1}(\theta,0)) = G_n(\theta); \tag{15}$$

note here that $G_i, i = 1, .., n$, is also an analytical function of $y_0, y_1, \ldots, y_{i-1}; \dot{y}_0, \dot{y}_1, \ldots, \dot{y}_{i-1}$.

To calculate an approximate periodic solutions of (8), we must solve (11), (12), (13), and (15). The following proposition gives the general formula of periodic solutions. Although these results are in [17], they are not detailed.

Proposition 2. *We consider the following equation:*

$$\ddot{y} + y = G(\theta), \quad y(0) = 0, \ \dot{y}(0) = 0, \ \text{with } G(\theta) \neq 0, \tag{16}$$

and the solution of problem (16) is

$$y(\theta) = \int_0^\theta \sin(\theta - \tau) G(\tau)\, d\tau. \tag{17}$$

Moreover, problem (16) has a periodic solution $y_1(\theta, 0)$ if and only if

$$\int_0^{2\pi} F(A\cos\theta, -A\sin\theta)\sin\theta\, d\theta = 0,$$
$$2\pi\omega_1 A + \int_0^{2\pi} F(A\cos\theta, -A\sin\theta)\cos\theta\, d\theta = 0. \tag{18}$$

Proof. We know that the solution of (16) is $y(\theta) = C_1\cos\theta + C_2\sin\theta + y_p(\theta)$ such that $y_p(\theta) = C_1(\theta)\cos\theta + C_2(\theta)\sin\theta$. By variation of constants we find

$$C_1'(\theta)\cos\theta + C_2'(\theta)\sin\theta = 0,$$
$$-C_1'(\theta)\sin\theta + C_2'(\theta)\cos\theta = G(\theta)$$
$$\Downarrow$$
$$-C_1'(\theta) = -\sin\theta G(\theta) \Longrightarrow C_1(\theta) = -\int_0^\theta \sin\tau G(\tau)\, d\tau, \quad C_1(0) = 0,$$
$$-C_2'(\theta) = \cos\theta G(\theta) \Longrightarrow C_1(\theta) = \int_0^\theta \cos\tau G(\tau)\, d\tau, \quad C_2(0) = 0 \tag{19}$$

$\Longrightarrow y_p(\theta) = (-\int_0^\theta \sin\tau G(\tau)d\tau)\cos\theta + (\int_0^\theta \cos\tau G(\tau)d\tau)\sin\theta = \int_0^\theta(-\sin\tau\cos\theta + \cos\tau\sin\theta)G(\tau)d\tau$.

$\Longrightarrow y_p(\theta) = \int_0^\theta \sin(\theta-\tau)G(\tau)d\tau \Longrightarrow y(\theta) = C_1\cos\theta + C_2\sin\theta + \int_0^\theta \sin(\theta-\tau)G(\tau)d\tau$ with the initial values $y(0) = 0$, $\dot{y}(0) = 0$; we have $C_1 = C_2 = 0$, so we deduce that problem (16) admits (17) as a solution.

Moreover, (16) gives

$$\dot{y}_1 = y_2$$
$$\dot{y}_2 = -y_1 + G(\tau). \tag{20}$$

On the other hand, the condition of periodicity for the new variable θ can be expressed as $y(\theta) = y(\theta + 2\pi)$, so the corresponding conditions for $y_n(\theta)$ are $y_n(\theta) = y_n(\theta + 2\pi), n = 1, 2, \ldots$

$$\Longrightarrow \begin{cases} y_1(2\pi) = y_1(0) = 0 \\ y_2(2\pi) = y_2(0) = 0 \end{cases} \tag{21}$$

which yields to the periodicity condition $\int_\theta^{\theta+2\pi}\sin(\theta - \tau)G(\tau)d\tau = 0$,

$$\Longrightarrow \begin{cases} \int_0^{2\pi} \cos\theta G(\theta)\, d\theta = 0, \\ \int_0^{2\pi} \sin\theta G(\theta)\, d\theta = 0. \end{cases} \tag{22}$$

According to (11) we have $G(\theta) = -2\omega_1\ddot{y}_0 + F(y_0, \dot{y}_0)$, $y_0 = A\cos\theta$; we rewrite (22) as

$$\Longrightarrow \begin{cases} \displaystyle\int_0^{2\pi} \cos\theta \left[2\omega_1 A\cos\theta + F(A\cos\theta, -A\sin\theta)\right] d\theta = 0, \\ \displaystyle\int_0^{2\pi} \sin\theta \left[2\omega_1 A\cos\theta + F(A\cos\theta, -A\sin\theta)\right] d\theta = 0, \end{cases}$$
$$\Longrightarrow \begin{cases} \displaystyle 2\omega_1 \pi A + \int_0^{2\pi} \cos\theta F(A\cos\theta, -A\sin\theta)\, d\theta = 0, \\ \displaystyle\int_0^{2\pi} \sin\theta F(A\cos\theta, -A\sin\theta)\, d\theta = 0, \end{cases} \tag{23}$$

which is required.

2.3.1. Example. We apply the method of Lindstedt to (4) with the initial values $y(0,\epsilon) = A$, $\dot{y}(0,\epsilon) = 0$, and we calculate $y_i(\theta, 0)$, $i = 1, 2, 3$ according to Proposition 2. Thus, we find that the fourth approximation of the periodic solution of (4) is

$$\begin{aligned} y(\theta,\epsilon) &= A\cos\theta + \epsilon\left(\frac{A^2}{6}\right)(-3 + 4\cos\theta - \cos 2\theta) \\ &+ \epsilon^2\left(\frac{A^3}{3}\right)\left[-2 + \left(\frac{61}{24}\right)\cos\theta - \left(\frac{2}{3}\right)\cos 2\theta\right. \\ &\left. - \left(\frac{1}{8}\right)\cos 3\theta\right] + \epsilon^3\left[-\frac{23}{24}A^4 + \frac{659}{540}A^4\cos\theta\right. \\ &\left. - \frac{1}{3}A^4\cos 2\theta + \frac{A^4}{12}\cos 3\theta - \frac{13}{1080}A^4\cos 4\theta\right] \\ &+ O\left(\epsilon^4\right), \end{aligned} \tag{24}$$

with $\theta = \omega(\epsilon)t$ such that $\omega(\epsilon) = 1 - \epsilon^2(A^2/6) - \epsilon^3((2/9)A^3) + O(\epsilon^4)$.

Remark 3. Although the calculation of $y_3(\theta)$ is very long, usually in applications, the fourth approximation is among the high orders that are often useful. For this reason, we give its equation in the next proposition.

Remark 4. The Lindstedt method gives only periodic solutions.

3. Our Results

3.1. General Formula. Practically, for many considerations we are forced to use a small number of terms in the perturbation expansion. We note here that the second and third terms are determined by (11) and (12) in [17]. In the following proposition, (13) which determines the fourth term is explicitly stated.

Proposition 5. *The general formula of (13) is*

$$\begin{aligned} \ddot{y}_3 + y_3 &= G_3(\theta) \\ &= -2\omega_1\ddot{y}_2 - \left(\omega_1^2 + 2\omega_2\right)\ddot{y}_1 \\ &- (2\omega_3 + 2\omega_1\omega_2)\ddot{y}_0 + y_2 F_y(y_0, \dot{y}_0) \\ &+ \frac{{y_1}^2}{2} F_{yy}(y_0, \dot{y}_0) \\ &+ + y_1(\omega_1\dot{y}_0 + \dot{y}_1) F_{y\dot{y}}(y_0, \dot{y}_0) \\ &+ (\omega_2\dot{y}_0 + \omega_1\dot{y}_1 + \dot{y}_2) F_{\dot{y}}(y_0, \dot{y}_0) \\ &+ \frac{1}{2}(\omega_1\dot{y}_0 + \dot{y}_1)^2 F_{\dot{y}\dot{y}}(y_0, \dot{y}_0). \end{aligned} \tag{25}$$

Proof. First, (9) gives

$$\begin{aligned} &\epsilon^3\left(\ddot{y}_3 + y_3 + 2\omega_1\ddot{y}_2 + \left({\omega_1}^2 + 2\omega_2\right)\ddot{y}_1\right. \\ &\quad\left. + (2\omega_3 + 2\omega_1\omega_2)\ddot{y}_0\right) = \frac{\epsilon^3}{2}\frac{\partial^2 F(y_0, \dot{y}_0)}{\partial\epsilon^2} \Longrightarrow \\ &\ddot{y}_3 + y_3 = -2\omega_1\ddot{y}_2 - \left({\omega_1}^2 + 2\omega_2\right)\ddot{y}_1 - (2\omega_3 + 2\omega_1\omega_2) \\ &\quad\cdot\ddot{y}_0 + \frac{1}{2}\frac{\partial^2 F(y_0, \dot{y}_0)}{\partial\epsilon^2}, \end{aligned} \tag{26}$$

such that

$$\begin{aligned} \frac{\partial^2 F(y, \omega\dot{y})}{\partial\epsilon^2} &= \frac{\partial}{\partial\epsilon}\left(\frac{\partial F}{\partial y}\frac{\partial y}{\partial\epsilon} + \frac{\partial F}{\partial\dot{y}}\frac{\partial\omega\dot{y}}{\partial\epsilon}\right) \\ &= \frac{\partial y}{\partial\epsilon}\left(\frac{\partial}{\partial\epsilon}\frac{\partial F}{\partial y}\right) + \frac{\partial^2 y}{\partial\epsilon^2}\frac{\partial F}{\partial y} \\ &+ \frac{\partial\omega\dot{y}}{\partial\epsilon}\left(\frac{\partial}{\partial\epsilon}\frac{\partial F}{\partial\dot{y}}\right) + \frac{\partial^2\omega\dot{y}}{\partial\epsilon^2}\frac{\partial F}{\partial\dot{y}} \\ &= \frac{\partial y}{\partial\epsilon}\left(\frac{\partial^2 F}{\partial y^2}\frac{\partial y}{\partial\epsilon} + \frac{\partial^2 F}{\partial y\partial\dot{y}}\frac{\partial\omega\dot{y}}{\partial\epsilon}\right) + \frac{\partial^2 y}{\partial\epsilon^2}F_y \\ &+ \frac{\partial^2\omega\dot{y}}{\partial\epsilon^2}F_{\dot{y}} \\ &+ \frac{\partial^2\omega\dot{y}}{\partial\epsilon}\left(\frac{\partial^2 F}{\partial y\partial\dot{y}}\frac{\partial y}{\partial\epsilon} + \frac{\partial^2 F}{\partial y^2}\frac{\partial\omega\dot{y}}{\partial\epsilon}\right) \\ &= \left(\frac{\partial y}{\partial\epsilon}\right)^2 F_{yy} + 2\frac{\partial y}{\partial\epsilon}\frac{\partial\omega\dot{y}}{\partial\epsilon}F_{y\dot{y}} + \frac{\partial^2 y}{\partial\epsilon}F_y \\ &+ \frac{\partial^2\omega\dot{y}}{\partial\epsilon^2}F_{\dot{y}} + \left(\frac{\partial\omega\dot{y}}{\partial\epsilon}\right)^2 F_{\dot{y}\dot{y}}, \end{aligned} \tag{27}$$

with

$$F_{yy} \equiv \frac{\partial^2 F}{\partial y^2}, \quad F_{y\dot{y}} \equiv \frac{\partial^2 F}{\partial y \partial \dot{y}}, \quad F_{\dot{y}\dot{y}} \equiv \frac{\partial^2 F}{\partial \dot{y} \partial \dot{y}}. \tag{28}$$

On the other hand, in the third order we have

$$y = y_0 + \epsilon y_1 + \epsilon^2 y_2 + \epsilon^3 y_3 \Longrightarrow$$

$$\frac{\partial y}{\partial \epsilon} = y_1 + 2\epsilon y_2 + 3\epsilon^2 y_3 \Longrightarrow$$
$$\frac{\partial^2 y}{\partial \epsilon^2} = 2y_2 + 6\epsilon y_3, \tag{29}$$

and

$$\begin{aligned}\frac{\partial \omega \dot{y}}{\partial \epsilon} &= \frac{\partial \omega}{\partial \epsilon}\dot{y} + \frac{\partial \dot{y}}{\partial \epsilon}\omega \\ &= \left(\omega_1 + 2\epsilon\omega_2 + 3\epsilon^2\omega_3\right)\left(\dot{y}_0 + \epsilon\dot{y}_1 + \epsilon^2\dot{y}_2 + \epsilon^3\dot{y}_3\right) \\ &\quad + \left(\dot{y}_1 + 2\epsilon\dot{y}_2 + 3\epsilon^2\dot{y}_3\right)\left(1 + \epsilon\omega_1 + \epsilon^2\omega_2 + \epsilon^3\omega_3\right),\end{aligned} \tag{30}$$

and also

$$\begin{aligned}\frac{\partial^2 \omega \dot{y}}{\partial \epsilon^2} &= \left(2\omega_2 + 6\epsilon\omega_3\right)\left(\dot{y}_0 + \epsilon\dot{y}_1 + \epsilon^2\dot{y}_2 + \epsilon^3\dot{y}_3\right) + 2\left(\dot{y}_1 + 2\epsilon\dot{y}_2 + 3\epsilon^2\dot{y}_3\right)\left(\omega_1 + 2\epsilon\omega_2 + 3\epsilon^2\omega_3\right) \\ &\quad + \left(2\dot{y}_2 + 6\epsilon\dot{y}_3\right)\left(1 + \epsilon\omega_1 + \epsilon^2\omega_2 + \epsilon^3\omega_3\right) \Longrightarrow \\ \frac{1}{2}\frac{\partial^2 F\left(y_0, \dot{y}_0\right)}{\partial \epsilon^2} &= \frac{{y_1}^2}{2}F_{yy}\left(y_0, \dot{y}_0\right) + \left(\omega_1\dot{y}_0 + \dot{y}_1\right)y_1 F_{y\dot{y}}\left(y_0, \dot{y}_0\right) \\ &\quad + y_2 F_y\left(y_0, \dot{y}_0\right) + \left(\omega_2\dot{y}_0 + \omega_1\dot{y}_1 + \dot{y}_2\right)F_{\dot{y}}\left(y_0, \dot{y}_0\right) + \left(\omega_1\dot{y}_0 + \dot{y}_1\right)^2 F_{\dot{y}\dot{y}}\left(y_0, \dot{y}_0\right),\end{aligned} \tag{31}$$

and when we substitute (31) into (26), we get (25).

3.2. Main Result. In this essential part of our work, we deal with some nonclassical equations, more general than (1), and also different from the equation studied in [2]. We consider equations in the following form:

$$\frac{d^2\tilde{y}}{dt^2}(t,\epsilon) + \tilde{y}(t,\epsilon) = g(\epsilon)F\left(\tilde{y}(t,\epsilon), \frac{d\tilde{y}}{dt}(t,\epsilon)\right), \quad 0 < \epsilon << 1, \tag{32}$$

with $\tilde{y}(0,\epsilon) = A$, $(d\tilde{y}/dt)(0,\epsilon) = 0$, where ϵ is a small positive parameter and F is supposed to be an analytical function of $\tilde{y}(t,\epsilon)$ and $d\tilde{y}/dt(t,\epsilon)$.

To compute an uniformly approximate periodic solution, a new variable $\tilde{\theta} = \tilde{\omega}t$ is introduced, and both $\tilde{y}$ and $\tilde{\omega}$ are expanded in powers of ϵ as follows:

$$\tilde{y}\left(\tilde{\theta},\epsilon\right) = \tilde{y}_0\left(\tilde{\theta},0\right) + \epsilon\tilde{y}_1\left(\tilde{\theta},0\right) + \dots + \epsilon^n\tilde{y}_n\left(\tilde{\theta},0\right) + \dots, \tag{33}$$

with

$$\tilde{\omega}(\epsilon) = 1 + \epsilon\tilde{\omega}_1 + \dots + \epsilon^n\tilde{\omega}_n + \dots. \tag{34}$$

We note that, in this step, $\tilde{\omega}_j$ are unknowns, and we obtain them by elimination of the secular terms.

To use the uniformly approximate periodic solution (33), we give firstly the general formula of $\tilde{y}_0(\tilde{\theta},0)$, $\tilde{y}_1(\tilde{\theta},0)$, and $\tilde{y}_2(\tilde{\theta},0)$ in the following proposition.

Proposition 6. *The terms $\tilde{y}_0(\tilde{\theta},0)$, $\tilde{y}_1(\tilde{\theta},0)$ and $\tilde{y}_2(\tilde{\theta},0)$ are, respectively, solutions of (35), (36), and (37) such that*

$$\ddot{\tilde{y}}_0 + \tilde{y}_0 = 0, \tag{35}$$

$$\ddot{\tilde{y}}_1 + \tilde{y}_1 = \tilde{G}_1\left(\tilde{\theta}\right), \tag{36}$$

$$\ddot{\tilde{y}}_2 + \tilde{y}_2 = \tilde{G}_2\left(\tilde{\theta}\right), \tag{37}$$

with $\tilde{G}_1(\tilde{\theta}) = -2\tilde{\omega}_1\ddot{\tilde{y}}_0 + c_1 F(\tilde{y}_0(\tilde{\theta},0), \dot{\tilde{y}}_0(\tilde{\theta},0))$,

$$\begin{aligned}\tilde{G}_2\left(\tilde{\theta}\right) &= -2\tilde{\omega}_1\ddot{\tilde{y}}_0 + c_1 F\left(\tilde{y}_0\left(\tilde{\theta},0\right), \dot{\tilde{y}}_0\left(\tilde{\theta},0\right)\right) \\ &\quad - 2\tilde{\omega}_1\ddot{\tilde{y}}_1 - \left(\tilde{\omega}_1^2 + 2\tilde{\omega}_2\right)\ddot{\tilde{y}}_0 \\ &\quad + +c_1\left(F_{\tilde{y}}\left(\tilde{y}_0\left(\tilde{\theta},0\right), \dot{\tilde{y}}_0\left(\tilde{\theta},0\right)\right)\tilde{y}_1\right. \\ &\quad + F_{\dot{\tilde{y}}}\left(\tilde{y}_0\left(\tilde{\theta},0\right), \dot{\tilde{y}}_0\left(\tilde{\theta},0\right)\right)\left(\tilde{\omega}_1\dot{\tilde{y}}_0 + \dot{\tilde{y}}_1\right)\Big) \\ &\quad + c_2 F\left(\tilde{y}_0\left(\tilde{\theta},0\right), \dot{\tilde{y}}_0\left(\tilde{\theta},0\right)\right).\end{aligned} \tag{38}$$

Proof. Equation (32) will be written as

$$\frac{d^2\tilde{y}(\tilde{\theta},\epsilon)}{dt^2} + \tilde{y}(\tilde{\theta}) = \sum_{k\geq 1}\epsilon^k c_k F\left(\tilde{y}(\tilde{\theta},\epsilon), \frac{d\tilde{y}(\tilde{\theta},\epsilon)}{dt}\right)$$

$$= \epsilon\left(\sum_{k\geq 0}\epsilon^k c_{k+1} F\left(\tilde{y}(\tilde{\theta},\epsilon), \frac{d\tilde{y}(\tilde{\theta},\epsilon)}{dt}\right)\right) \tag{39}$$

$$= \epsilon K\left(\tilde{y}(\tilde{\theta},\epsilon), \frac{d\tilde{y}(\tilde{\theta},\epsilon)}{dt}\right),$$

where K is an analytical function of y and dy/dt.

Substituting (33) into (32), we have

$$\left(1+\epsilon\tilde{\omega}_1+\epsilon^2\tilde{\omega}_2+\epsilon^3\tilde{\omega}_3+\dots\right)^2\left(\ddot{\tilde{y}}_0+\epsilon\ddot{\tilde{y}}_1+\epsilon^2\ddot{\tilde{y}}_2\right.$$

$$\left.+\epsilon^3\ddot{\tilde{y}}_3+\dots\right)+\tilde{y}_0+\epsilon\tilde{y}_1+\epsilon^2\tilde{y}_2+\epsilon^3\tilde{y}_3+\dots$$

$$=\epsilon K\left(\tilde{y}_0,\dot{\tilde{y}}_0\right)+\epsilon^2\frac{\partial K\left(\tilde{y}_0,\dot{\tilde{y}}_0\right)}{\partial\epsilon}+\frac{\epsilon^3}{2}\frac{\partial^2 K\left(\tilde{y}_0,\dot{\tilde{y}}_0\right)}{\partial\epsilon^2}$$

$$+\dots,=\epsilon c_1 F\left(\tilde{y}_0,\dot{\tilde{y}}_0\right)+\epsilon^2\left[c_1\frac{\partial F\left(\tilde{y}_0,\dot{\tilde{y}}_0\right)}{\partial\epsilon}\right. \tag{40}$$

$$\left.+c_2 F\left(\tilde{y}_0,\dot{\tilde{y}}_0\right)\right]+\dots,=\epsilon c_1 F\left(\tilde{y}_0,\dot{\tilde{y}}_0\right)$$

$$+\epsilon^2\left[c_1\left(F_{\tilde{y}}\left(\tilde{y}_0,\dot{\tilde{y}}_0\right)\tilde{y}_1\right.\right.$$

$$\left.\left.+F_{\dot{\tilde{y}}}\left(\tilde{y}_0,\dot{\tilde{y}}_0\right)\left(\overline{\omega_1}\dot{\tilde{y}}_0+\dot{\tilde{y}}_1\right)\right)+c_2 F\left(\tilde{y}_0,\dot{\tilde{y}}_0\right)\right]+\cdots,$$

then we put all different powers of ϵ to zero, and we obtain (35), (36), (37), and so on.

The aim of this study is to construct a new approach to (2), which gives a closer approximate solution of (2) more than an approximate solution of (1). The relations between an approximate solution of (32) and that of (1) are determined by the following lemma.

Lemma 7. *If the function $g(\epsilon)$ is expanded in powers of ϵ with $g(0)=0$ i.e $g(\epsilon)=\sum_{k\geq 1}\epsilon^k c_k$, where c_k are real constants, we have*

(1) $\tilde{y}_0(\tilde{\theta},0)=y_0(\tilde{\theta},0)$.

(2) $\tilde{y}_1(\tilde{\theta},0)=c_1 y_1(\tilde{\theta},0)$.

(3) $\tilde{y}_2(\tilde{\theta},0)=c_1^2 y_2(\tilde{\theta},0)+c_2 y_1(\tilde{\theta},0)$.

Proof. (1) Equation (35) gives $\tilde{y}_0(\tilde{\theta},0) = A\cos\tilde{\theta} = A\cos(\tilde{\omega}(0)t) = A\cos t = A\cos(\omega(0)t) = A\cos\theta = y_0(\theta,0)$.

(2) When we apply the periodicity condition (22) to (36), we have

$$\int_0^{2\pi}\cos\tilde{\theta}\tilde{G}_1\left(\tilde{\theta}\right)d\tilde{\theta}=0,$$

$$\int_0^{2\pi}\sin\tilde{\theta}\tilde{G}_1(\theta)\,d\tilde{\theta}=0.$$

$$\Downarrow$$

$$\int_0^{2\pi}\cos\tilde{\theta}\left[-2\tilde{\omega}_1\ddot{\tilde{y}}_0+c_1F\left(\tilde{y}_0,\dot{\tilde{y}}_0\right)\right]d\tilde{\theta}=0,$$

$$\int_0^{2\pi}\sin\tilde{\theta}\left[-2\tilde{\omega}_1\ddot{\tilde{y}}_0+c_1F\left(\tilde{y}_0,\dot{\tilde{y}}_0\right)\right]d\tilde{\theta}=0. \tag{41}$$

$$\Downarrow$$

$$\tilde{\omega}_1=\frac{-c_1}{2\pi A}\int_0^{2\pi}\cos\tilde{\theta}F\left(\tilde{y}_0,\dot{\tilde{y}}_0\right)d\tilde{\theta}=c_1\omega_1,$$

$$\int_0^{2\pi}\sin\tilde{\theta}F\left(\tilde{y}_0,\dot{\tilde{y}}_0\right)d\tilde{\theta}=0.$$

On the other hand, according to (17) the solution $\tilde{y}_1(\tilde{\theta},0)$ of (36) is given by

$$\tilde{y}_1\left(\tilde{\theta},0\right)=\int_0^{\tilde{\theta}}\sin\left(\tilde{\theta}-\tilde{\tau}\right)\tilde{G}_1(\tilde{\tau})\,d\tilde{\tau}$$

$$=c_1\int_0^{\tilde{\theta}}\sin\left(\tilde{\theta}-\tilde{\tau}\right)G_1(\tilde{\tau})\,d\tilde{\tau}=c_1y_1\left(\tilde{\theta},0\right). \tag{42}$$

(3) When we apply the periodicity condition (22) to (12), we have

$$\int_0^{2\pi}\cos\tilde{\theta}\tilde{G}_2\left(\tilde{\theta}\right)d\tilde{\theta}=0,$$

$$\int_0^{2\pi}\sin\tilde{\theta}\tilde{G}_2(\theta)\,d\tilde{\theta}=0.$$

$$\Downarrow$$

$$\int_0^{2\pi}\cos\tilde{\theta}\left[c_1^2G_2\left(\tilde{\theta}\right)+c_2F\left(\tilde{y}_0,\dot{\tilde{y}}_0\right)\right]d\tilde{\theta}=0,$$

$$\int_0^{2\pi}\sin\tilde{\theta}\left[c_1^2G_2\left(\tilde{\theta}\right)+c_2F\left(\tilde{y}_0,\dot{\tilde{y}}_0\right)\right]d\tilde{\theta}=0. \tag{43}$$

$$\Downarrow$$

$$\tilde{\omega}_2=c_1^2\omega_2+c_2\omega_1,$$

$$\int_0^{2\pi}\sin\tilde{\theta}F\left(\tilde{y}_0,\dot{\tilde{y}}_0\right)d\tilde{\theta}=0.$$

On the other hand, according to (17) the solution $\widetilde{y}_2(\widetilde{\theta},0)$ of (11) is given by

$$\begin{aligned}\widetilde{y}_2(\widetilde{\theta},0) &= \int_0^{\widetilde{\theta}} \sin(\widetilde{\theta}-\widetilde{\tau})\widetilde{G}_2(\widetilde{\tau})\,d\widetilde{\tau} \\ &= c_1^2\int_0^{\widetilde{\theta}} \sin(\widetilde{\theta}-\widetilde{\tau})G_2(\widetilde{\tau})\,d\widetilde{\tau} \\ &+ c_2\int_0^{\widetilde{\theta}} \sin(\widetilde{\theta}-\widetilde{\tau})\left[-2\widetilde{\omega}_1\ddot{\widetilde{y}}_0 + c_1F(\widetilde{y}_0,\dot{\widetilde{y}}_0)\right]d\widetilde{\tau} \Longrightarrow \\ \widetilde{y}_2(\widetilde{\theta},0) &= c_1^2 y_2(\widetilde{\theta},0) + c_2 y_1(\widetilde{\theta},0).\end{aligned} \tag{44}$$

Theorem 8. *If the function $g(\epsilon)$ is expanded in powers of ϵ with $g(0)=0$ i.e $g(\epsilon)=\sum_{k\geq 1}\epsilon^k c_k$, where c_k are real constants, so one has the following:*

(1) If $c_1 \neq 0$, with $|c_1| < 1$, the approximate solutions of (32) are closer to the solutions of (2) more than the approximate solutions of (1).

(2) If $c_1 = 0$ and $|c_2| < 1$, with $c_2 \neq 0$, the approximate solutions of (32) are closer to the solutions of (2) more than the approximate solutions of (1).

Moreover, the approximate solutions of (32) are closer to the solutions of (2) more than the approximate solutions of (1) where $c_1 = 0, c_2 \neq 0$.

Proof. According to the results given by Lemma 7, we have the following:

(1) If $|c_1| < 1$ and $c_1 \neq 0$ we have

$$\begin{aligned}&\left|y_0(\theta,0)-\widetilde{y}(\widetilde{\theta},\epsilon)\right| \\ &= \left|y_0(\theta,0)-y_0(\widetilde{\theta},0)-\epsilon\widetilde{y}_1(\widetilde{\theta},0)-o(\epsilon)\right| \\ &= \left|\epsilon\widetilde{y}_1(\widetilde{\theta},0)+o(\epsilon)\right| = \left|c_1\epsilon y_1(\widetilde{\theta},0)+o(\epsilon)\right| \\ &= |c_1|\left|\epsilon y_1(\widetilde{\theta},0)+o(\epsilon)\right| < \left|\epsilon y_1(\widetilde{\theta},0)+o(\epsilon)\right| \\ &= \left|y_0(\theta,0)-y(\widetilde{\theta},\epsilon)\right|.\end{aligned} \tag{45}$$

So $\widetilde{y}(\widetilde{\theta},\epsilon)$ is an approximation to $y_0(\theta,0)$ closer than $y(\widetilde{\theta},\epsilon)$.

(2) If $c_1 = 0$ and $|c_2| < 1$ with $c_2 \neq 0$, we have

$$\begin{aligned}&\left|y_0(\theta,0)-\widetilde{y}(\widetilde{\theta},\epsilon)\right| \\ &= \left|y_0(\theta,0)-y_0(\widetilde{\theta},0)-\epsilon^2\widetilde{y}_2(\widetilde{\theta},0)-o(\epsilon^2)\right| \\ &= \left|\epsilon^2\widetilde{y}_2(\widetilde{\theta},0)+o(\epsilon^2)\right| = \left|\epsilon^2 c_2 y_1(\widetilde{\theta},0)+o(\epsilon^2)\right| \\ &= |c_2|\left|\epsilon^2 y_1(\widetilde{\theta},0)+o(\epsilon^2)\right| \\ &= |c_2|\left|\epsilon^2\left(y_0(\theta,0)-y(\widetilde{\theta},\epsilon)+\epsilon^2 y_2(\widetilde{\theta},0)\right)\right| \\ &< \left|\epsilon^2\left(y_0(\theta,0)-y(\widetilde{\theta},\epsilon)+\epsilon^2 y_2(\widetilde{\theta},0)\right)\right| \\ &< \epsilon^2\left|y_0(\theta,0)-y(\widetilde{\theta},\epsilon)\right|+\epsilon^4\left|y_2(\widetilde{\theta},0)\right|.\end{aligned} \tag{46}$$

Since $|y_0(\theta,0)-y(\widetilde{\theta},\epsilon)| = o(\epsilon)$, for ϵ small enough, there exists a positive real constant C such that $|y_0(\theta,0)-y(\widetilde{\theta},\epsilon)| \leq C\epsilon$.

Let h be a function defined by $h(\epsilon) := (\epsilon^2-1)C + \epsilon^3|y_2(\widetilde{\theta},0)|$. So, h is continuous with $h(0) = -C < 0$, then $\exists\,\epsilon_0 > 0$ such that $\forall\epsilon \in]0,\epsilon_0[$, we have $h(\epsilon) < 0$.

Therefore, for all $\epsilon \in]0,\epsilon_0[$ we get

$$\begin{aligned}&(\epsilon^2-1)\left|y_0(\theta,0)-y(\widetilde{\theta},\epsilon)\right|+\epsilon^4\left|y_2(\widetilde{\theta},0)\right| \\ &\leq \epsilon\left((\epsilon^2-1)C+\epsilon^3\left|y_2(\widetilde{\theta},0)\right|\right) = \epsilon h(\epsilon) < 0\end{aligned} \tag{47}$$

which implies from what precedes that $|y_0(\theta,0)-\widetilde{y}(\widetilde{\theta},\epsilon)| < |y_0(\theta,0)-y(\widetilde{\theta},\epsilon)|$. So $\widetilde{y}(\widetilde{\theta},\epsilon)$ is an approximation to $y_0(\theta,0)$ closer than $y(\widetilde{\theta},\epsilon)$.

Moreover, if $c_1 = 0$ and $c_2 \neq 0$, then $\exists\,\epsilon_0(< 1/|c_2|) > 0$ such that, $\forall\epsilon \in]0,\epsilon_0[$,

$$\begin{aligned}&\left|y_0(\theta,0)-\widetilde{y}(\widetilde{\theta},\epsilon)\right| \\ &= \left|y_0(\theta,0)-y_0(\widetilde{\theta},0)-\epsilon^2\widetilde{y}_2(\widetilde{\theta},0)-o(\epsilon^2)\right| \\ &= \left|\epsilon^2\widetilde{y}_2(\widetilde{\theta},0)+o(\epsilon^2)\right| = \left|\epsilon^2 c_2 y_1(\widetilde{\theta},0)+o(\epsilon^2)\right| \\ &= \epsilon|c_2|\left|\epsilon y_1(\widetilde{\theta},0)+o(\epsilon)\right| \\ &= \epsilon|c_2|\left|y_0(\theta,0)-y(\widetilde{\theta},\epsilon)\right| \\ &< \left|y_0(\theta,0)-y(\widetilde{\theta},\epsilon)\right|.\end{aligned} \tag{48}$$

So $\widetilde{y}(\widetilde{\theta},\epsilon)$ is an approximation to $y_0(\theta,0)$ closer than $y(\widetilde{\theta},\epsilon)$.

Remark 9. Let g be a real function such that $g(\epsilon) = \epsilon^m h(\epsilon), m \in \mathbb{N}^*$, where $h(\epsilon)$ can be expanded in powers of ϵ as $h(\epsilon)=\sum_{k\geq 1}\epsilon^k a_k$, with a_k real constants.

(1) If $m = 1$ or 2, according to the conditions of Theorem 8, we conclude that $\widetilde{y}(\widetilde{\theta},\epsilon)$ is an approximation to $y_0(\theta,0)$ closer than $y(\widetilde{\theta},\epsilon)$.

(2) If $m > 2$ (in the case where $m = 3$, the fourth term is given by Proposition 5), we can expect that $\widetilde{y}(\widetilde{\theta},\epsilon)$ is an approximation to $y_0(\theta,0)$ closer than $y(\widetilde{\theta},\epsilon)$.

Remark 10. We note here that, in the fractional case, the existence of a positive solution of (32) is studied in [18].

Remark 11. Although the Lindstedt-Poincaré method gives uniformly valid asymptotic expansions for periodic solutions of weakly nonlinear oscillations, i.e., $0 < \epsilon < C_1$, the technique does not work if the amplitude of the oscillation is a function of time (see [16, 17]).

Conflicts of Interest

The author declares that they have no conflicts of interest.

References

[1] B. Hong, D. Lu, and Y. Liu, "Homotopic Approximate Solutions for the General Perturbed Nonlinear Schrödinger Equation," *Mathematical & Computational Applications*, vol. 20, no. 1, pp. 50–61, 2015, http://www.mdpi.com/2297-8747/20/1/50.

[2] H. Hu and Z. G. Xiong, "Comparison of two Lindstedt-Poincaré-type perturbation methods," *Journal of Sound and Vibration*, vol. 278, no. 1-2, pp. 437–444, 2004.

[3] V. Zakharchenko and I. Kovalenko, "Best Approximation of the Fractional Semi-Derivative Operator by Exponential Series," *Mathematics*, vol. 6, no. 1, p. 12, 2018.

[4] E. A. Pimentel and E. V. Teixeira, "Sharp Hessian integrability estimates for nonlinear elliptic equations: an asymptotic approach," *Journal de Mathématiques Pures et Appliquées*, vol. 106, no. 4, pp. 744–767, 2016.

[5] G. Nemes, "On the coefficients of an asymptotic expansion related to Somos' quadratic recurrence constant," *Applicable Analysis and Discrete Mathematics*, vol. 5, no. 1, pp. 60–66, 2011.

[6] H. Belbachir and D. Berkane, "Asymptotic expansion for the sum of inverses of arithmetical functions involving iterated logarithms," *Applicable Analysis and Discrete Mathematics*, vol. 5, no. 1, pp. 80–86, 2011.

[7] L. Cădariu, L. Găvruţa, and P. Găvruţa, "Weighted space method for the stability of some nonlinear equations," *Applicable Analysis and Discrete Mathematics*, vol. 6, no. 1, pp. 126–139, 2012.

[8] M. S. Moslehian and T. M. Rassias, "Stability of functional equations in non-Archimedean spaces," *Applicable Analysis and Discrete Mathematics*, vol. 1, no. 2, pp. 325–334, 2007.

[9] P. Cai and J. Cai, "Asymptotic solutions and comparisons of a generalized Van del Pol oscillator with slowly varying parameter," *Mathematical & Computational Applications*, vol. 14, no. 2, pp. 159–168, 2009.

[10] J. Cai, S. Chen, and C. Yang, "Numerical verification and comparison of error of asymptotic expansion solution of the Duffing equation," *Mathematical & Computational Applications*, vol. 13, no. 1, pp. 23–29, 2008.

[11] M. Pakdemirli, M. M. Karahan, and H. Boyaci, "A New Perturbation Algorithm With Better Convergence Properties: Multiple Scales Lindstedt Poincare Method," *Mathematical & Computational Applications*, vol. 14, no. 1, pp. 31–44, 2009.

[12] M. Pakdemirli, M. M. Karahan, and H. Boyaci, "Forced vibrations of strongly nonlinear systems with multiple scales Lindstedt Poincare method," *Mathematical & Computational Applications*, vol. 16, no. 4, pp. 879–889, 2011.

[13] A. Berretti and G. Gentile, "Scaling properties for the radius of convergence of Lindstedt series: generalized standard maps," *Journal de Mathématiques Pures et Appliquées*, vol. 79, no. 7, pp. 691–713, 2000.

[14] C. Guo, D. O'Regan, and R. P. Agarwal, "Existence of multiple periodic solutions for a class of first-order neutral differential equations," *Applicable Analysis and Discrete Mathematics*, vol. 5, no. 1, pp. 147–158, 2011.

[15] M. Demiralp and S. Tuna, "Zero interval limit perturbation expansion for the spectral entities of Hilbert-Schmidt operators combined with most dominant spectral component extraction: formulation and certain technicalities," *Journal of Mathematical Chemistry*, vol. 55, no. 6, pp. 1253–1277, 2017.

[16] F. Verhulst, *Nonlinear Differential Equations and Dynamical Systems*, Universitext, Springer, Berlin, Heidelberg, 1996.

[17] R. E. Mickens, *Oscillations in Planar Dynamics Systems*, World Scientific, Singapore, 1996.

[18] C. S. Goodrich, "On positive solutions to nonlocal fractional and integer-order difference equations," *Applicable Analysis and Discrete Mathematics*, vol. 5, no. 1, pp. 122–132, 2011.

13

Univalent and Starlike Properties for Generalized Struve Function

Aaisha Farzana Habibullah,[1] Adolf Stephen Bhaskaran,[1] and Jeyaraman Muthusamy Palani[2]

[1]Department of Mathematics, Madras Christian College, Tambaram, Chennai, Tamil Nadu 600 059, India
[2]Department of Mathematics, L. N. Government College, Ponneri, Chennai, Tamil Nadu 601 204, India

Correspondence should be addressed to Aaisha Farzana Habibullah; h.aaisha@gmail.com

Academic Editor: Teodor Bulboaca

We derive conditions on the parameters p, b, and c so that the function $zw_{p,b,c}(z)$, where $w_{p,b,c}(z)$ is the normalized form of generalized Struve function, belongs to the class $S_1^*(\alpha)$. Also, some sufficient conditions for the function $zw_{p,b,c}(z)$, to be in the class $\mathscr{U}(\lambda)$, are obtained.

1. Introduction and Preliminaries

Let $\mathbb{U} := \{z : |z| < 1\}$ denote the unit disc in the complex plane $\mathbb{C}$ and let $\mathscr{A}$ denote the class of functions which are analytic and of the form

$$f(z) = z + \sum_{n=2}^{\infty} a_n z^n \tag{1}$$

and normalized by the conditions $f(0) = 0$ and $f'(0) = 1$. Let $\mathscr{S}$ denote the class of functions such that

$$\mathscr{S} = \{f : f \in \mathscr{A} \text{ and } f \text{ univalent in } \mathbb{U}\}. \tag{2}$$

Suppose that f and g are two analytic functions in $\mathbb{U}$, and g is univalent in $\mathbb{U}$. We say that f is subordinate to g, written $f(z) \prec g(z)$ or $f \prec g$, if and only if $f(0) = g(0)$ and $f(\mathbb{U}) \subset g(\mathbb{U})$.

A function $f \in \mathscr{A}$ belongs to the class of starlike functions of order α denoted by $S^*(\alpha)$ if

$$\operatorname{Re}\left(\frac{zf'(z)}{f(z)}\right) > \alpha, \quad z \in \mathbb{U}. \tag{3}$$

A subclass of the class of starlike functions denoted by $S_1^*(\alpha)$ for $0 \le \alpha < 1$ consists of functions for which

$$\left|\frac{zf'(z)}{f(z)} - 1\right| < 1 - \alpha, \quad z \in \mathbb{U}. \tag{4}$$

We also note that $S_1^*(0) \subset S^*$. In [1–3] the authors have discussed the coefficient bounds and other extremal properties of the class $S_1^*(\alpha)$.

For $0 < \lambda \le 1$, consider the class

$$\mathscr{U}(\lambda) = \left\{f \in \mathscr{A} : \left|\left(\frac{z}{f(z)}\right)^2 f'(z) - 1\right| < \lambda \text{ for } z \in \mathbb{U}\right\}. \tag{5}$$

From [4], we have the strict inclusion $\mathscr{U}(1) := \mathscr{U} \subset \mathscr{S}$. Very recently, Obradović et al. [5] have discussed the geometric behaviour of functions in $\mathscr{U}(\lambda)$. Also the class has been widely studied by many authors in [4, 6–8].

Let us consider the following second-order linear nonhomogenous differential equation (for more details see [9, 10]):

$$z^2u''(z) + bzu'(z) + \left[cz^2 - p^2 + (1-b)p\right]u(z) = \frac{4(z/2)^{p+1}}{\sqrt{\pi}\,\Gamma(p + b/2)}, \tag{6}$$

where $b, p, c \in \mathbb{C}$. The function $u_{p,b,c}(z)$ which is called the generalized Struve function of order p is defined as a particular solution of (6) and has the series representation as follows:

$$u_{p,b,c}(z) = \sum_{n\geq 0} \frac{(-1)^n (c)^n}{\Gamma(n+3/2)\Gamma(p+n+(b+2)/2)} \left(\frac{z}{2}\right)^{2n+p+1}, \quad z \in \mathbb{C}, \tag{7}$$

where Γ stands for the Euler gamma function.

Now, we consider the function $w_{p,b,c}(z)$ defined in terms of the generalized Struve function $u_{p,b,c}(z)$ by the transformation:

$$w_{p,b,c}(z) = 2^P \sqrt{\pi}\Gamma\left(p + \frac{b+2}{2}\right) z^{(-p-1)/2} u_{p,b,c}(\sqrt{z}). \tag{8}$$

By using the well-known Pochammer symbol (or the shifted factorial) $(\chi)_n$ defined for $\chi, n \in \mathbb{C}$ in terms of the Euler Γ-function, we have

$$(\chi)_n = \frac{\Gamma(\chi+n)}{\Gamma(\chi)} = \begin{cases} 1, & \text{if } n = 0,\ \chi \in \mathbb{C}^*, \\ \chi(\chi+1)\cdots(\chi+n-1), & \text{if } n \in \mathbb{N},\ \chi \in \mathbb{C}. \end{cases} \tag{9}$$

We obtain the following series representation for the function $w_{p,b,c}$ given by (8):

$$w_{p,b,c}(z) = \sum_{n\geq 0} \frac{(-c/4)^n}{(3/2)_n (k)_n} z^n, \tag{10}$$

where $k = p + (b+2)/2 \neq 0, -1, -2, \ldots$. Also, this function is analytic on $\mathbb{C}$ and satisfies the following second-order inhomogeneous differential equation:

$$4z^2 w''(z) + 2(2p+b+3) zw'(z) + (cz + 2p + b) w(z) = (2p+b), \tag{11}$$

where $p, b, c \in \mathbb{C}$.

Recently, the class $\mathcal{U}(\lambda)$ and its generalizations have been widely studied by many authors [4–8, 11]. By applying the admissible function method authors in [12] have obtained conditions on the triplet a, b, and c such that $zF_{a,b,c}(z)$ is in the class $\mathcal{U}$, where $F_{a,b,c}(z)$ is the confluent hypergeometric function. In [13], the authors have derived conditions on the parameters a, b, and c such that the function $z_2F_1(a,b,c)$ is in $S_1^*(0)$, where ${}_2F_1(a,b,c;z)$ is the Gaussian hypergeometric function. Moreover, in [9, 10] Yagmur and Orhan have obtained sufficient conditions for the generalized Struve function to be convex, starlike, and univalent. Most of these results were motivated by the research on geometric properties of Gaussian and confluent hypergeometric function.

Motivated by the above-mentioned works, in this paper we use the method of differential subordination to show that $zw_{p,b,c}(z)$ is in the class $S^*(\alpha)$ and also we provide sufficient conditions for the function $zw_{p,b,c}(z)$ to be in the class $\mathcal{U}$ and hence univalent.

To prove our main results, we will need the following lemmas.

Lemma 1 (see [14]). *Let (nonconstant) function $p(z)$ be analytic in $\mathbb{U}$ with $p(0) = 0$, $p(z) \neq 0$ ($z \in \mathbb{U}$).*

If $|p(z)|$ attains its maximum value on the circle $|z| = r < 1$ at a point $z_0 \in \mathbb{U}$, then

$$z_0 p'(z_0) = kp(z_0), \quad \Re\left(1 + \frac{z_0 p''(z_0)}{p'(z_0)}\right) \geq k, \tag{12}$$

where k is a real number and $k \geq 1$.

Lemma 2 (see [15]). *If an analytic function f has the form $f(z) = z + a_2 z^2 + \cdots$, $z \in \mathbb{U}$ and satisfies the condition*

$$\left|\frac{z^2 f'(z)}{f^2(z)} - 1\right| < 1 \tag{13}$$

then f is univalent in $\mathbb{U}$.

Lemma 3 (see [15]). *Let β be a complex number, $\Re(\beta) > 0$, and let α be a complex number, $|\alpha| \leq 1$, $\alpha \neq -1$, and $h(z) = z + a_2 z^2 + \cdots$ a regular function on $\mathbb{U}$. If*

$$\left|\alpha |z|^{2\beta} + \left(1 - |z|^{2\beta}\right) \frac{zh''(z)}{\beta h'(z)}\right| \leq 1 \tag{14}$$

for all $z \in \mathbb{U}$, then the function

$$F_\beta(z) = \left(\beta \int_0^z t^{\beta-1} h'(t)\, dt\right)^{1/\beta} = z + \frac{2a_2}{(1+\beta)} z^2 + \left(\frac{3a_3}{2+\beta} + \frac{2\beta(1-\beta)a_2^2}{(\beta+1)^3}\right) z^3 + \cdots \tag{15}$$

is regular and univalent in $\mathbb{U}$.

Lemma 4 (see [16]). *Let $\Omega \subset \mathbb{C}$ and let q be analytic and univalent on $\mathbb{U}$ except for those $\xi \in \partial\mathbb{U}$ for which $\lim_{z\in\xi} q(z) = \infty$. Suppose that $\psi : \mathbb{C}^3 \times \mathbb{U} \to \mathbb{C}$ satisfies the condition*

$$\psi\left(q(\zeta), m\zeta q'(\zeta), \zeta^2 q''(\zeta); z\right) \notin \Omega, \tag{16}$$

where $q(z)$ is finite, $m \geq n \geq 1$, and $|\xi| = 1$. If p and q analytic in $\mathbb{U}$, $p(z) = p(0) + p_n z^n + \cdots$, $p(0) = q(0)$, and further if

$$\psi\left(p(z), zp'(z), z^2 p''(z); z\right) \in \Omega \tag{17}$$

then $p(z) \prec q(z)$ in $\mathbb{U}$.

Suppose that $p(z)$ is analytic in $\mathbb{U}$ with $p(z) = p_n z^n + p_{n+1} z^{n+1} + \cdots$ and $q(z) = Mz$. Then the condition (16) reduces to a simple form:

$$\psi\left(Me^{i\theta}, Ke^{i\theta}, L; z\right) \notin \Omega \tag{18}$$

whenever $K \geq nM$, $\operatorname{Re}(Le^{-i\theta}) \geq (n-1)K$, $z \in \mathbb{U}$, and $\theta \in \mathbb{R}$.

2. Main Results

Theorem 5. *Let $\delta > 0$, $p, b \in \mathbb{R}$, and $c \in \mathbb{C}$ such that $|c|\delta < 3[(2p+b)+2] - |c|\delta$; then*

$$\left|w_{p,b,c}(\delta z) - 1\right| < \frac{|c|\,\delta}{3\left[(2p+b)+2\right] - |c|\,\delta}, \quad z \in \mathbb{U}. \tag{19}$$

Proof. Let $p(z) = w_{p,b,c}(\delta z) - 1$, $\delta > 0$; then $p(z)$ is analytic in $\mathbb{U}$ with $p(0) = 0$. Since the function $w_{p,b,c}(z)$ satisfies the differential equation (11) and $w_{p,b,c}(\delta z) = p(z) + 1$, $p(z)$ satisfies the following inhomogeneous differential equation:

$$\begin{aligned} &4z^2 p''(z) + 2(2p+b+3) zp'(z) \\ &\quad + (c\delta z + 2p + b)(p(z) + 1) = (2p+b). \end{aligned} \tag{20}$$

Using Lemma 4, we will show that $|p(z)| < M$, where $M = |c|\delta / 3[(2p+b)+2] - |c|\delta$. For this, if we let

$$\begin{aligned} \psi(r,s,t;z) &= 4t + 2(2p+b+3)s \\ &\quad + (c\delta z + 2p + b)(1+r) - (2p+b) \end{aligned} \tag{21}$$

and $\Omega = \{0\}$, then it is sufficient to prove that $\psi(Me^{i\theta}, Ke^{i\theta}, L; z) \notin \Omega$ whenever $K \geq M$, $\operatorname{Re}(Le^{-i\theta}) \geq 0$, $z \in \mathbb{U}$, and θ is real, we have that

$$\begin{aligned} &\left|\psi\left(Me^{i\theta}, Ke^{i\theta}, L; z\right)\right| = \left|4L + 2(2p+b+3)Ke^{i\theta}\right. \\ &\quad \left.+ (c\delta z + 2p + b)\left(1 + Me^{i\theta}\right) - (2p+b)\right| = \left|4Le^{-i\theta}\right. \\ &\quad + 2(2p+b+3)K + (c\delta z + 2p + b)\left(M + e^{-i\theta}\right) \\ &\quad \left.- (2p+b)e^{-i\theta}\right| \geq 4\operatorname{Re}\left(Le^{-i\theta}\right) + 2(2p+b+3)M \\ &\quad + \operatorname{Re}\left((c\delta z + 2p + b)M\right) \\ &\quad + \operatorname{Re}\left((c\delta z + 2p + b)e^{-i\theta}\right) - \operatorname{Re}\left((2p+b)e^{-i\theta}\right) \\ &\quad \geq \left[2(2p+b+3) + (2p+b)\right]M - \operatorname{Re}\left(c\delta z e^{-i\theta}\right) \\ &\quad - \operatorname{Re}(c\delta z M) \geq 3\left[(2p+b)+2\right]M - |c|\,\delta M - |c|\,\delta \\ &\quad = 0. \end{aligned} \tag{22}$$

In the last stage of the inequalities we have used the definition of M and shown that

$$\psi\left(Me^{i\theta}, Ke^{i\theta}, L; z\right) \notin \Omega = \{0\} \tag{23}$$

whenever $K \geq M$, θ is real, and $z \in \mathbb{U}$. Hence, $|p(z)| < M$.

Choosing $\delta = 1$ in the above theorem we have the following Corollary.

Corollary 6. *Let $p, b \in \mathbb{R}$, and $c \in \mathbb{C}$ be such that $3[(2p+b)+2] - 2|c| > 0$; then*

$$\left|w_{p,b,c}(z) - 1\right| < \frac{|c|}{3\left[(2p+b)+2\right] - |c|}, \quad z \in \mathbb{U}. \tag{24}$$

The next results give sufficient conditions for the function $g_{p,b,c}(z) = zw_{p,b,c}(z)$ to be starlike and univalent in the open unit disc.

Theorem 7. *If $p, b \in \mathbb{R}$, and $c \in \mathbb{C}$ such that $2|c| < 3[(2p+b)+2]$ and*

$$\begin{aligned} &\frac{(2p+b)|c|}{\left[3\left[(2p+b)+2\right] - 2|c|\right]} + |c| \\ &\quad \leq (1-\alpha)\left[2(2p+b+3) - 4(1-\alpha)\right], \end{aligned} \tag{25}$$

then the function $g_{p,b,c}(z)$ belongs to the class $S_1^*(\alpha)$.

Proof. For $g_{p,b,c}(z) = zw_{p,b,c}(z)$ to be in the class $S_1^*(\alpha)$ we need to prove that

$$\left|\frac{zg'_{p,b,c}(z)}{g_{p,b,c}} - 1\right| < (1-\alpha), \quad z \in \mathbb{U}. \tag{26}$$

Upon setting

$$\frac{\left[zw_{p,b,c}(z)\right]'}{w_{p,b,c}(z)} - 1 = \tilde{q}(z) \tag{27}$$

we observe that $\tilde{q}(z)$ is analytic in $\mathbb{U}$ and $\tilde{q}(0) = \tilde{q}'(0) - 1 = 0$. Also

$$\begin{aligned} zw'_{p,b,c}(z) &= \tilde{q}(z)\,w_{p,b,c}(z), \\ z^2 w''_{p,b,c}(z) &= \left[(\tilde{q}(z) - 1)\,\tilde{q}(z) + z\tilde{q}(z)\right] w_{p,b,c}(z). \end{aligned} \tag{28}$$

Now, since $w_{p,b,c}(z)$ satisfies the inhomogeneous differential equation (11), in terms of $\tilde{q}(z)$ we see that $\tilde{q}(z)$ satisfies the following equation:

$$\psi\left(\tilde{q}(z), z\tilde{q}'(z); z\right) = 0, \tag{29}$$

where

$$\begin{aligned} \psi(r,s;z) &= 4\left[r(r-1) + s\right] + 2(2p+b+3)r \\ &\quad + (cz + 2p + b) - (2p+b)\frac{1}{g_{p,b,c}(z)}. \end{aligned} \tag{30}$$

Now, we claim that $|\tilde{q}(z)| < 1 - \alpha$, $0 \leq \alpha < 1$. By using Lemma 4 with $\Omega = \{0\}$, $n = 1$, and $q(z) = (1-\alpha)z$, we need to show that

$$\psi\left((1-\alpha)e^{i\theta}, Ke^{i\theta}; z\right) \notin \Omega, \tag{31}$$

where θ is real, $K \geq (1-\alpha)$, and $z \in \mathbb{U}$:

$$\psi\left((1-\alpha)e^{i\theta}, Ke^{i\theta}; z\right) = e^{i\theta}C - D, \tag{32}$$

where

$$\begin{aligned} C &= 4\left[(1-\alpha)\left[(1-\alpha)e^{i\theta}-1\right]+K\right] \\ &\quad + 2(2p+b+3)(1-\alpha), \\ D &= (2p+b)\left[\frac{1}{g_{p,b,c}(z)}-1\right] - cz. \end{aligned} \tag{33}$$

Also, when $\delta = 1$ in Theorem 5 we have that if $2|c| < 3(2p+b)+6$ then

$$\left|w_p(z)-1\right| < \frac{|c|}{3(2p+b)+6} - |c| \leq 1. \tag{34}$$

Also,

$$\begin{aligned} |C| &\geq \operatorname{Re} C > 4(1-\alpha)\left[(1-\alpha)e^{i\theta}-1\right] + (1-\alpha)4 \\ &\quad + 2(2p+b+3)(1-\alpha) \geq (1-\alpha)\left[4(1-\alpha)\cos\theta\right. \\ &\quad - 4(1-\alpha) + 4(1-\alpha) + 2(2p+b+3)\left.\right] \geq (1-\alpha) \\ &\quad \cdot \left[2(2p+b+3) - 4(1-\alpha)\right] = \widetilde{\beta}, \text{ say}. \end{aligned} \tag{35}$$

Now, if we show that $|D| < \widetilde{\beta}$, then we have

$$\left|\psi(r,s;z)\right| \geq \left|e^{i\theta}C - D\right| \geq |C| - |D| = 0. \tag{36}$$

In order to establish that we need the following results which state that for $r < 1$ and $|w-1| < r$ if and only if

$$\left|\frac{1}{w} - \frac{1}{1-r^2}\right| < \frac{r}{1-r^2}. \tag{37}$$

In particular, $|w-1| < r$ implies that

$$\left|\frac{1}{w} - 1\right| < \frac{r}{1-r}. \tag{38}$$

When $r = |c|/(3[(2p+b)+2] - |c|)$, we have

$$\left|\frac{1}{g_{p,b,c}(z)} - 1\right| < \frac{r}{1-r} = \frac{|c|}{3(2p+b)+6-2|c|}. \tag{39}$$

Using the last inequality, we have

$$\begin{aligned} |D| &\leq (2p+b)\left|\frac{1}{g_{p,b,c}(z)} - 1\right| + |c|\,|z| \\ &< \frac{(2p+b)|c|}{3[(2p+b)+2] - 2|c|} + |c| \end{aligned} \tag{40}$$

and so we have $|D| < \widetilde{\beta}$ provided that

$$\frac{(2p+b)|c|}{3[(2p+b)+2]-2|c|} + |c| \leq \widetilde{\beta} \tag{41}$$

holds, where $\widetilde{\beta}$ is given by (35). And, we observe that the above condition is stated in the theorem. Thus, $|\widetilde{q}(z)| < (1-\alpha)$ in $\mathbb{U}$ and hence $g_{p,b,c}(z)$ belongs to $S_1^*(\alpha)$.

Theorem 8. *Let $p, b \in \mathbb{R}$, and $c \in \mathbb{C}$. If $g_{p,b,c}(z)$ satisfies any one of the following inequalities:*

$$\left|\frac{z^2 g'_{p,b,c}(z)}{\left[g_{p,b,c}(z)\right]^2}\left[\frac{\left(zg_{p,b,c}(z)\right)''}{g'_{p,b,c}(z)} - \frac{2zg'_{p,b,c}(z)}{g_{p,b,c}(z)}\right]\right| < 1 \quad (z \in \mathbb{U}), \tag{42}$$

$$\left|\frac{\left[g_{p,b,c}(z)\right]^2}{z^2 g'_{p,b,c}(z)}\left[\frac{\left[zg_{p,b,c}(z)\right]''}{g'_{p,b,c}(z)} - \frac{2zg'_{p,b,c}(z)}{g_{p,b,c}(z)}\right]\right| < \frac{1}{4} \quad (z \in \mathbb{U}), \tag{43}$$

$$\left|\frac{\left[zg_{p,b,c}(z)\right]''/g'_{p,b,c}(z) - 2zg'_{p,b,c}(z)/g_{p,b,c}(z)}{z^2 g'_{p,b,c}(z)/\left(g_{p,b,c}(z)\right)^2 - 1}\right| < \frac{1}{2} \quad (z \in \mathbb{U}), \tag{44}$$

$$\operatorname{Re}\left[\frac{z^2 g_{p,b,c}(z)}{\left(g_{p,b,c}(z)\right)^2}\left[\frac{\left[zg_{p,b,c}(z)\right]''/g'_{p,b,c}(z) - 2zg'_{p,b,c}(z)/g_{p,b,c}(z)}{z^2 g'_{p,b,c}(z)/\left[g_{p,b,c}(z)\right]^2}\right]\right] < 1 \quad (z \in \mathbb{U}), \tag{45}$$

then $g_{p,b,c}$ is in $\mathcal{U}$ and hence $g_{p,b,c}(z)$ is univalent in $\mathbb{U}$.

Proof. Define a function $p(z)$ by

$$\frac{z^2 g'_{p,b,c}(z)}{\left[g_{p,b,c}(z)\right]^2} = 1 + p(z) \quad (z \in \mathbb{U}), \tag{46}$$

and then $p(z)$ is analytic in $\mathbb{U}$ and $p(0) = 0$. Differentiating (46) gives

$$\frac{\left[zg_{p,b,c}(z)\right]''}{g'_{p,b,c}(z)} - \frac{2zg'_{p,b,c}(z)}{g_{p,b,c}(z)} = \frac{zp'(z)}{1+p(z)}. \tag{47}$$

Hence, from (46) and (47), we have

$$\begin{aligned}
E_1(z) &= \frac{z^2 g'_{p,b,c}(z)}{\left(g_{p,b,c}(z)\right)^2}\left(\frac{\left[zg_{p,b,c}(z)\right]''}{g'_{p,b,c}(z)} - \frac{2zg'_{p,b,c}(z)}{g_{p,b,c}(z)}\right) = zp'(z), \\
E_2(z) &= \frac{\left(g_{p,b,c}(z)\right)^2}{z^2 g'_{p,b,c}(z)}\left(\frac{\left[zg_{p,b,c}(z)\right]''}{g'_{p,b,c}(z)} - \frac{2zg'_{p,b,c}(z)}{g_{p,b,c}(z)}\right) = \frac{zp'(z)}{\left(1+p(z)\right)^2}, \\
E_3(z) &= \frac{\left[zg_{p,b,c}(z)\right]''/g'_{p,b,c}(z) - 2zg'_{p,b,c}(z)/g_{p,b,c}(z)}{z^2 g'_{p,b,c}(z)/\left(g_{p,b,c}(z)\right)^2 - 1} = \frac{zp'(z)}{p(z)}\frac{1}{\left(1+p(z)\right)}, \\
E_4(z) &= \frac{z^2 g'_{p,b,c}(z)}{\left(g_{p,b,c}(z)\right)^2}\left(\frac{\left[zg_{p,b,c}(z)\right]''/g'_{p,b,c}(z) - 2zg'_{p,b,c}(z)/g_{p,b,c}(z)}{z^2 g'_{p,b,c}(z)/\left(g_{p,b,c}(z)\right)^2 - 1}\right) = \frac{zp'(z)}{p(z)}.
\end{aligned} \tag{48}$$

Now suppose that there exists $z_0 \in \mathbb{U}$ such that

$$\max_{|z|<|z_0|}|p(z)| = |p(z_0)| = 1, \tag{49}$$

and then from Lemma 1, we have

$$z_0 p'(z_0) = kp(z_0). \tag{50}$$

Therefore, letting $p(z_0) = e^{i\theta}$ in each of (48), we obtain that

$$\begin{aligned}
|E_1(z_0)| &= |z_0 p'(z_0)| = |ke^{i\theta}| \geq 1, \\
|E_2(z_0)| &= \left|\frac{z_0 p'(z_0)}{\left(1+p(z_0)\right)^2}\right| = \frac{k}{|1+e^{i\theta}|^2} \geq \frac{1}{4}, \\
|E_3(z_0)| &= \left|\frac{z_0 p'(z_0)}{p(z_0)}\frac{1}{1+p(z_0)}\right| = \frac{k}{|1+e^{i\theta}|} \geq \frac{1}{2}, \\
\operatorname{Re}\{E_4(z_0)\} &= \operatorname{Re}\left\{\frac{z_0 p'(z_0)}{p(z_0)}\right\} = k \geq 1,
\end{aligned} \tag{51}$$

which contradicts our assumption (42)–(45), respectively. Therefore, $|w(z)| < 1$ for all $z \in \mathbb{U}$; then from (46) we have

$$\left|\frac{z^2 g'_{p,b,c}(z)}{\left(g_{p,b,c}(z)\right)^2} - 1\right| = |w(z)| < 1, \tag{52}$$

which implies $g_{p,b,c}$ is in the class $\mathcal{U}$ and hence univalent.

Theorem 9. *Let $c > 0$, $d \geq 0$, such that $c + 2d \leq 1$. If $g_{p,b,c}(z)$ satisfies the inequality*

$$\Re\left\{\frac{\left[zg_{p,b,c}(z)\right]''}{g'_{p,b,c}(z)} - \frac{2zg'_{p,b,c}(z)}{g_{p,b,c}(z)}\right\} < \frac{c+d}{(1+c)(1-d)}, \tag{53}$$

then $g_{p,b,c}(z)$ is univalent in $\mathbb{U}$.

Proof. Define a function $q(z)$ as follows:

$$\frac{z^2 g'_{p,b,c}(z)}{\left(g_{p,b,c}(z)\right)^2} = \frac{1+aq(z)}{1-bq(z)} \quad (z \in \mathbb{U}). \tag{54}$$

We see that $q(z)$ is analytic in $\mathbb{U}$ and $q(0) = 0$. Differentiation of (54) gives

$$\begin{aligned}
&\frac{\left[zg_{p,b,c}(z)\right]''}{g'_{p,b,c}(z)} - \frac{2zg'_{p,b,c}(z)}{g_{p,b,c}(z)} \\
&= \frac{(c+d)\,zq'(z)}{(1+cq(z))(1-dq(z))} = E_5(z), \text{ say.}
\end{aligned} \tag{55}$$

Now suppose there exists $z_0 \in \mathbb{U}$ such that

$$\max_{|z|<|z_0|}|q(z)| = |q(z_0)| = 1. \tag{56}$$

Then from Lemma 1, we have

$$z_0 q'(z_0) = kq(z_0). \tag{57}$$

Now letting $q(z_0) = e^{i\theta}$, $(\theta \in [0, 2\pi])$ in (55), we have

$$\begin{aligned}
\operatorname{Re}(E_5(z_0)) &= k(a+b) \\
&\cdot \operatorname{Re}\left(\frac{q(z_0)}{(1+cq(z_0))(1-dq(z_0))}\right) = k \\
&\cdot \operatorname{Re}\left(\frac{1}{(1-dq(z_0))} - \frac{1}{(1+cq(z_0))}\right) = k \\
&\cdot \operatorname{Re}\left(\frac{1-de^{i\theta}}{1+d^2-2d\cos\theta} - \frac{1+ce^{i\theta}}{1+c^2+2c\cos\theta}\right) \\
&= k\left(\frac{1}{2+(d^2-1)/(1-d\cos\theta)}\right. \\
&\quad \left. - \frac{1}{2+(c^2-1)/(1+c\cos\theta)}\right),
\end{aligned} \tag{58}$$

where $\theta \neq \cos^{-1}(-1/c)$ and $\theta \neq \cos^{-1}(1/d)$. Since $k \geq 1$ we have

$$\operatorname{Re}(E(z_0)) > \frac{c+d}{(1+c)(1-d)}. \tag{59}$$

This contradicts the hypothesis and therefore $|q(z)| < 1$ for all $z \in \mathbb{U}$. Thus,

$$\left|\frac{z^2 g'_{p,b,c}(z)}{(g_{p,b,c}(z))^2} - 1\right| = \left|\frac{(c+d)\,q(z)}{1-dq(z)}\right| < \frac{c+d}{1-d} \leq 1 \qquad (z \in \mathbb{U}). \tag{60}$$

In view of Lemma 2 it implies that $g_{p,b,c}$ is in $\mathcal{U}$ and hence univalent.

Theorem 10. *Let $M \geq 1$, ν be a real number such that $\nu \geq 2M + 1$ and let γ be a complex number which satisfies the inequality*

$$|\gamma| \leq 1 - \frac{1}{\nu}(2M+1). \tag{61}$$

If $g_{p,b,c}$ is univalent in $\mathbb{U}$ and $|g_{p,b,c}(z)| \leq M$ for all $z \in \mathbb{U}$, then the function

$$E(z) = \left\{\nu \int_0^z t^{\nu-1} \frac{g_{p,b,c}(t)}{t} dt\right\}^{1/\nu} \tag{62}$$

is univalent in $\mathbb{U}$, where the values of the complex powers are taken with their principal values.

Proof. Define a function

$$p(z) = \int_0^z \frac{g_{p,b,c}(t)}{t} dt. \tag{63}$$

Then we have $p(0) = p'(0) - 1 = 0$.

Also

$$p'(z) = \frac{g_{p,b,c}(z)}{z}, \tag{64}$$

$$\frac{zp''(z)}{p'(z)} = \frac{zg'_{p,b,c}(z)}{g_{p,b,c}(z)} - 1. \tag{65}$$

From (65), we have

$$\begin{aligned}
\left|\frac{zp''(z)}{p'(z)}\right| &\leq \left|\frac{zg'_{p,b,c}(z)}{g_{p,b,c}(z)}\right| + 1 \\
&= \left|\frac{z^2 g'_{p,b,c}(z)}{(g_{p,b,c}(z))^2}\right| \left|\frac{g_{p,b,c}(z)}{z}\right| + 1.
\end{aligned} \tag{66}$$

From the hypothesis, we have $|g_{p,b,c}(z)| \leq M$ $(z \in \mathbb{U})$; then, by the Schwarz Lemma (cf [17], we obtain that

$$|g_{p,b,c}(z)| \leq M|z| \quad (z \in \mathbb{U}). \tag{67}$$

Now, since $g_{p,b,c}(z)$ is univalent in $\mathbb{U}$

$$\begin{aligned}
\left|\frac{zp''(z)}{p'(z)}\right| &\leq \left|\frac{z^2 g'_{p,b,c}(z)}{(g_{p,b,c}(z))^2}\right| M + 1 \\
&\leq \left|\frac{z^2 g'_{p,b,c}(z)}{(g_{p,b,c}(z))^2} - 1\right| M + M + 1 \leq 2M + 1.
\end{aligned} \tag{68}$$

Using (68), we have

$$\begin{aligned}
\left|\gamma |z|^{2\nu} + (1 - |z|^{2\nu}) \frac{zp''(z)}{\nu p'(z)}\right| &\leq |\gamma| + \frac{1}{\nu}\left|\frac{zp''(z)}{\nu p'(z)}\right| \\
&\leq |\gamma| + \frac{1}{\beta}(2M+1).
\end{aligned} \tag{69}$$

So, from (61) we have

$$\left|\gamma |z|^{2\nu} + (1 - |z|^{2\nu}) \frac{zp''(z)}{\nu p'(z)}\right| \leq 1. \tag{70}$$

Applying Lemma 3, we obtain the function $E(z)$ defined by (62) which is univalent in $\mathbb{U}$.

Competing Interests

The authors declare that there is no conflict of interests regarding the publication of this paper.

Acknowledgments

The work of the first author was supported by the *Department of Science and Technology*, India, with reference to the Sanction Order no. SR/DST-WOS A/MS-10/2013(G). The work of the second author was supported by the grant given under *UGC Minor Research Project F. no. 5599/15 (MRP-SEM/UGC SERO)*. The work of the third author was supported by the grant given under *UGC Minor Research Project, 2014-15/MRP-5591/15(SERO/UGC)*.

References

[1] P. J. Eenigenburg, "A class of starlike mappings of the unit disk," *Compositio Mathematica*, vol. 24, pp. 235–238, 1972.

[2] H. Silverman, "Subclasses of starlike functions," *Revue Roumaine de Mathématique Pures et Appliquées*, vol. 23, no. 7, pp. 1093–1099, 1978.

[3] D. J. Wright, "On class of starlike functions," *Compositio Mathematica*, vol. 21, pp. 122–124, 1969.

[4] L. A. Aksentev, "Sufficient conditions for univalence of regular functions," *Izvestiya Vysshikh Uchebnykh Zavedenii Matematika*, no. 4, pp. 3–7, 1958 (Russian).

[5] M. Obradović, S. Ponnusamy, and K. J. Wirths, "Geometric Studies on the Class $U(\lambda)$," *Bulletin of the Malaysian Mathematical Sciences Society Series 2*, vol. 39, no. 3, pp. 1259–1284, 2016.

[6] M. Obradovic, "A class of univalent functions," *Hokkaido Mathematical Journal*, vol. 27, no. 2, pp. 329–335, 1998.

[7] M. Obradovic and S. Ponnusamy, "New criteria and distortion theorems for univalent functions," *Complex Variables, Theory and Application*, vol. 44, no. 3, pp. 173–191, 2001.

[8] M. Obradovic and S. Ponnusamy, "Univalence and starlikeness of certain transforms defined by convolution of analytic functions," *Journal of Mathematical Analysis and Applications*, vol. 336, no. 2, pp. 758–767, 2007.

[9] N. Yagmur and H. Orhan, "Starlikeness and convexity of generalized Struve functions," *Abstract and Applied Analysis*, vol. 2013, Article ID 954513, 6 pages, 2013.

[10] H. Orhan and N. Yagmur, "Geometric properties of generalized Struve functions," in *Proceedings of the International Congress in Honour of Professor Hari M. Srivastava*, Bursa, Turkey, August 2012.

[11] S. P. Goyal, R. Kumar, and T. Bubloaca, "Majorization problems and integral transforms for a class of univalent functions with missing coefficients," *Boletim da Sociedade Paranaense de Matemática*, vol. 33, no. 2, pp. 217–230, 2015.

[12] M. Obradovic and S. Ponnusamy, "Univalency and convolution results associated with confluent hypergeometric functions," *Houston Journal of Mathematics*, vol. 35, no. 4, pp. 1313–1328, 2009.

[13] P. Hästö, S. Ponnusamy, and M. Vuorinen, "Starlikeness of the Gaussian hypergeometric functions," *Complex Variables and Elliptic Equations*, vol. 55, no. 1–3, pp. 173–184, 2010.

[14] S. S. Miller and P. T. Mocanu, "Second order differential inequalities in the complex plane," *Journal of Mathematical Analysis and Applications*, vol. 65, no. 2, pp. 289–305, 1978.

[15] M. Nunokawa, "On some angular estimates of analytic functions," *Mathematica Japonica*, vol. 41, no. 2, pp. 447–452, 1995.

[16] S. S. Miller and P. T. Mocanu, "Differential subordinations and inequalities in the complex plane," *Journal of Differential Equations*, vol. 67, no. 2, pp. 199–211, 1987.

[17] Z. Nehari, *Conformal Mapping*, McGraw Hill Book Comp, New York, NY, USA, 1952, (Dover, 1975).

14

On Solving of Constrained Convex Minimize Problem Using Gradient Projection Method

Taksaporn Sirirut and Pattanapong Tianchai

Faculty of Science, Maejo University, Chiangmai 50290, Thailand

Correspondence should be addressed to Pattanapong Tianchai; pattana@mju.ac.th

Academic Editor: Jean Michel Rakotoson

Let C and Q be closed convex subsets of real Hilbert spaces H_1 and H_2, respectively, and let $g : C \longrightarrow \mathbb{R}$ be a strictly real-valued convex function such that the gradient ∇g is an $1/L$-ism with a constant $L > 0$. In this paper, we introduce an iterative scheme using the gradient projection method, based on Mann's type approximation scheme for solving the constrained convex minimization problem (CCMP), that is, to find a minimizer $q \in C$ of the function g over set C. As an application, it has been shown that the problem (CCMP) reduces to the split feasibility problem (SFP) which is to find $q \in C$ such that $Aq \in Q$ where $A : H_1 \longrightarrow H_2$ is a linear bounded operator. We suggest and analyze this iterative scheme under some appropriate conditions imposed on the parameters such that another strong convergence theorems for the CCMP and the SFP are obtained. The results presented in this paper improve and extend the main results of Tian and Zhang (2017) and many others. The data availability for the proposed SFP is shown and the example of this problem is also shown through numerical results.

1. Introduction

Throughout this paper, we always assume that C be a closed convex subset of a real Hilbert space H with inner product and norm are denoted by $\langle \cdot, \cdot \rangle$ and $\| \cdot \|$, respectively. Let $g : C \longrightarrow \mathbb{R}$ be a strictly real-valued convex function.

Consider the following constrained convex minimization problem (CCMP):

$$\min_{x \in C} \quad g(x). \tag{1}$$

Assume that (1) is consistent (that is, the CCMP has a solution) and we use $\mathcal{U}$ to denote its solution set. If g is Fréchet differentiable, then the gradient projection algorithm (GPA) is usually applied to solving the CCMP (1), which generates a sequence $\{x_n\}$ through the recursion:

$$x_{n+1} = P_C (I - \lambda \nabla g) x_n, \quad \forall n = 0, 1, 2, \ldots, \tag{2}$$

or more generally,

$$x_{n+1} = P_C (I - \lambda_n \nabla g) x_n, \quad \forall n = 0, 1, 2, \ldots, \tag{3}$$

where the initial guess $x_0 \in C$ is chosen arbitrarily, the parameters λ or λ_n are positive real number, and P_C is the metric projection from H onto C. It is well known that the convergence of algorithms (2) and (3) depends on the behavior of the gradient ∇g. It is known from Levitin and Polyak [1] that if ∇g is α-strongly monotone and L-Lipschitzian, that is, there exists the constants $\alpha > 0$ and $L > 0$ such that

$$\langle \nabla g(x) - \nabla g(y), x - y \rangle \geq \alpha \|x - y\|^2, \quad \forall x, y \in C, \tag{4}$$

and

$$\|\nabla g(x) - \nabla g(y)\| \leq L \|x - y\|, \quad \forall x, y \in C, \tag{5}$$

respectively, then, for $0 < \lambda < 2\alpha/L^2$, the operator

$$T = P_C (I - \lambda \nabla g) \tag{6}$$

is a contraction; hence, the sequence $\{x_n\}$ defined by the GPA (2) converges in norm to the unique minimizer of the CCMP (1). More generally, for $0 < \lambda_n < 2\alpha/L^2$ for all $n = 0, 1, 2, \ldots$, the operator

$$G_n = P_C (I - \lambda_n \nabla g) \tag{7}$$

is a contraction; if the sequence $\{\lambda_n\}$ is chosen satisfying the property

$$0 < \liminf_{n\to\infty} \lambda_n \le \limsup_{n\to\infty} \lambda_n < \frac{2\alpha}{L^2}, \quad (8)$$

then the sequence $\{x_n\}$ defined by the GPA (3) converges in norm to the unique minimizer of the CCMP (1). However, if the gradient ∇g fails to be α-strongly monotone (it means that the gradient ∇g only satisfies the L-Lipschitzian condition), then the operators T and G_n defined by (6) and (7), respectively, may fail to be contraction; consequently, the sequence $\{x_n\}$ generated by algorithms (2) and (3) may fail to converge strongly (see also Xu [2]) in the setting of infinite-dimensional real Hilbert space, but still converge weakly as the following statement.

Theorem 1 (see [1, 2]). *Assume that the CCMP (1) is the unique consistent. Let the gradient ∇g satisfy the L-Lipschitzian condition and the sequence of the parameter $\{\lambda_n\}$ satisfies the following condition:*

$$0 < a \le \lambda_n \le b < \frac{2}{L}, \quad (9)$$

for all $n = 0, 1, 2, \ldots$, where a and b are the constants. Then the sequence generated by the GPA (3) converges weakly to the minimizer of the CCMP (1). Indeed, the results of this theorem still hold on the gradient ∇g which satisfies an $1/L$-inverse strongly monotone with $L > 0$ (in brief, we denote $1/L$-ism), that is, $\langle \nabla g(x) - \nabla g(y), x - y\rangle \ge (1/L)\|\nabla g(x) - \nabla g(y)\|^2$ for all $x, y \in C$, because the class of L-Lipschitzian mapping contains the class of $1/L$-ism mapping.

We observe from Theorem 1 that if the parameter $\{\lambda_n\}$ converges to $\lambda \in (0, 2/L)$ such that $\{\lambda_n\}$ satisfies the condition (9) then $q \in C$ solves the CCMP (1) which is the unique consistent if and only if q solves the fixed-point equation

$$q = P_C (I - \lambda \nabla g)\, q. \quad (10)$$

It is well known that the gradient-projection algorithm is very useful in dealing with the CCMP (1) and has extensively been studied (see [1–9] and the references therein). It has recently been applied to solve the split feasibility problems (SFP) (see [10–15]) which find applications in image reconstructions and the intensity modulated radiation therapy (see [13–18]). We now consider the following regularized minimization (that is, the CCMP (1) has the unique minimizer solution) problem:

$$\min_{x\in C} \; g_{\beta_n}(x) := g(x) + \frac{\beta_n}{2}\|x\|^2, \quad (11)$$

where $\beta_n > 0$ for all $n = 0, 1, 2, \ldots$ and $g : C \longrightarrow \mathbb{R}$ is a continuous differentiable function, and we also consider the regularized gradient-projection algorithm which generates a sequence $\{x_n\}$ by the following recursive formula:

$$\begin{aligned} x_{n+1} &= P_C \left(I - \lambda \nabla g_{\beta_n}\right) x_n \\ &= P_C \left(I - \lambda (\nabla g + \beta_n I)\right) x_n, \quad \forall n = 0, 1, 2, \ldots. \end{aligned} \quad (12)$$

Many researchers studied the strong convergency theorems for solving the CCMP (1) using the sequence $\{x_n\}$ which is generated by algorithm (12) for their proposal on the gradient ∇g which is the class of nonexpansive mapping and the class of L-Lipschitzian mapping (see [19–25]) and in case the gradient ∇g is the class of $1/L$-ism mapping such that $L > 0$, Xu (2010) introduced the sequence $\{x_n\}$ which is generated by algorithm (12), and he proved that this sequence $\{x_n\}$ converges weakly to the minimizer of the CCMP (1) in the setting of infinite-dimensional real Hilbert space (see [15]) under some appropriate condition.

Recently, Tian and Zhang (2017) introduced the sequence $\{x_n\}$ generated by algorithm (12), and they proved that this sequence $\{x_n\}$ converges strongly to the minimizer of the CCMP (1) in the same setting of infinite-dimensional real Hilbert space (see [26]) under the control conditions:

(i) $0 < \lambda < 2/(L + 2)$.

(ii) $\{\beta_n\} \subset (0, 1)$, $\lim_{n\to\infty}\beta_n = 0$ and $\sum_{n=0}^{\infty} \beta_n = \infty$.

(iii) $\sum_{n=0}^{\infty} |\beta_{n+1} - \beta_n| < \infty$.

In this paper, under the motivated and the inspired by above results, we introduce new iterative scheme, based on Mann's type approximation scheme for solving the CCMP (1) in the case of the gradient ∇g being the class of $1/L$-ism mapping such that $L > 0$ as follows:

$$\begin{aligned} & x_0 \in C, \\ & x_{n+1} = \alpha_n x_n + (1 - \alpha_n)\, P_C (I - \lambda (\nabla g + \beta_n I))\, x_n, \quad (13) \\ & \qquad \forall n = 0, 1, 2, \ldots, \end{aligned}$$

under the mild some appropriate conditions of the parameters $\{\alpha_n\}$, $\{\beta_n\}$, and λ, we obtain a strong convergency theorem to solve the CCMP (1), in which condition (iii) $\sum_{n=0}^{\infty} |\beta_{n+1} - \beta_n| < \infty$ of Tian and Zhang to be removed. In Section 4 of the applications, it has been shown that the CCMP (1) reduces to the split feasibility problem (SFP) and the data availability for the proposed SFP is shown in Section 5, and the example of this problem is also shown in Section 6 through numerical results.

2. Preliminaries

Let C be a nonempty closed convex subset of a real Hilbert space H. If $g : C \longrightarrow \mathbb{R}$ is a differentiable function, then we denote ∇g the gradient of the function g. We will also use the notation: $\longrightarrow$ to denote the strong convergency, $\rightharpoonup$ to denote the weak convergency, and $\mathrm{Fix}(T) = \{x : x = Tx\}$ to denote the fixed point set of the mapping T.

Recall that the metric projection $P_C : H \longrightarrow C$ is defined as follows: for each $x \in H$, $P_C x$ is the unique point in C satisfying

$$\|x - P_C x\| = \inf \{\|x - y\| : y \in C\}. \quad (14)$$

Let $g : C \longrightarrow \mathbb{R}$ be a function. Recall that the function g is a strictly real-valued convex function if

$$\begin{aligned} & g(\lambda x + (1 - \lambda) y) < \lambda g(x) + (1 - \lambda) g(y), \\ & \qquad \forall \lambda \in [0, 1], \; \forall x, y \in C, \end{aligned} \quad (15)$$

such that $x \neq y$. We collect together some known lemmas and definitions which are our main tool in proving our results.

Lemma 2. *Let H be a real Hilbert space. Then, for all $x, y \in H$,*

$$\|x+y\|^2 = \|x\|^2 + 2\langle x, y\rangle + \|y\|^2. \quad (16)$$

Lemma 3 (see [27]). *Let C be a nonempty closed convex subset of a real Hilbert space H. Then,*

$$z = P_C x \Longleftrightarrow \langle x - z, z - y\rangle \geq 0, \quad \forall x \in H,\ y \in C. \quad (17)$$

Definition 4. Let H be a real Hilbert space. The operator $T : H \longrightarrow H$ is called

(i) L-Lipschitzian with $L > 0$ if

$$\|Tx - Ty\| \leq L\|x - y\|, \quad \forall x, y \in H, \quad (18)$$

(ii) k-contraction with a positive real number k such that $k \in (0,1)$ if

$$\|Tx - Ty\| \leq k\|x - y\|, \quad \forall x, y \in H, \quad (19)$$

(iii) nonexpansive if

$$\|Tx - Ty\| \leq \|x - y\|, \quad \forall x, y \in H, \quad (20)$$

(iv) monotone if

$$\langle Tx - Ty, x - y\rangle \geq 0, \quad \forall x, y \in H, \quad (21)$$

(v) α-strongly monotone if

$$\langle Tx - Ty, x - y\rangle \geq \alpha\|x - y\|^2, \quad \forall x, y \in H, \quad (22)$$

(vi) α-inverse strongly monotone (α-ism) if

$$\langle Tx - Ty, x - y\rangle \geq \alpha\|Tx - Ty\|^2, \quad \forall x, y \in H, \quad (23)$$

(vii) firmly nonexpansive if

$$\|Tx - Ty\|^2 \leq \|x - y\|^2 - \|(I - T)x - (I - T)y\|^2, \quad \forall x, y \in H. \quad (24)$$

Lemma 5 (see [27]). *Let C be a nonempty closed convex subset of a real Hilbert space H. Then, for $x, y \in C$ and $\alpha \in [0,1]$, we have*

$$\|\alpha x + (1-\alpha)y\|^2 = \alpha\|x\|^2 + (1-\alpha)\|y\|^2 - \alpha(1-\alpha)\|x - y\|^2. \quad (25)$$

Lemma 6 (see [28]). *Let H be a real Hilbert space and $T : H \longrightarrow H$ be an operator. The following statements are equivalent:*

(i) *T is firmly nonexpansive,*

(ii) *$\|Tx - Ty\|^2 \leq \langle x - y, Tx - Ty\rangle$, $\forall x, y \in H$,*

(iii) *$I - T$ is firmly nonexpansive.*

Lemma 7 (see [28]). *Let H and K be two real Hilbert spaces and let $T : K \longrightarrow K$ be a firmly nonexpansive mapping such that $\|(I-T)x\|$ is a convex function from K to $\overline{\mathbb{R}} = [-\infty, +\infty]$. Let $A : H \longrightarrow K$ be a bounded linear operator and $f(x) = (1/2)\|(I - T)Ax\|^2$ for all $x \in H$. Then,*

(i) *$\nabla f(x) = A^*(I-T)Ax$ for all $x \in H$ where A^* is adjoint operator of A,*

(ii) *∇f is $\|A\|^2$-Lipschitzian.*

Lemma 8 (see [29, 30]). *Let C be a nonempty closed convex subset of a real Hilbert space H. Let $\{T_n\}$ and φ be two classes of nonexpansive mappings from C into C such that*

$$\emptyset \neq Fix(\varphi) = \bigcap_{n=0}^{\infty} Fix(T_n). \quad (26)$$

Then, for any bounded sequence $\{z_n\} \subset C$, we have

(i) *if $\lim_{n\longrightarrow\infty}\|z_n - T_n z_n\| = 0$ then $\lim_{n\longrightarrow\infty}\|z_n - T z_n\| = 0$ for all $T \in \varphi$, which is called that the NST-condition(I),*

(ii) *if $\lim_{n\longrightarrow\infty}\|z_{n+1} - T_n z_n\| = 0$ then $\lim_{n\longrightarrow\infty}\|z_n - T_m z_n\| = 0$ for all $m \in \mathbb{N} \cup \{0\}$, which is called that the NST-condition (II).*

Lemma 9 (see [31] (demiclosedness principle)). *Let C be a nonempty closed convex subset of a real Hilbert space H and let $S : C \longrightarrow C$ be a nonexpansive mapping with $Fix(S) \neq \emptyset$. If the sequence $\{x_n\} \subset C$ converges weakly to x and the sequence $\{(I - S)x_n\}$ converges strongly to y. Then, $(I - S)x = y$, in particular, if $y = 0$ then $x \in Fix(S)$.*

Lemma 10 (see [32]). *Let $\{a_n\}$ be a sequence of nonnegative real number such that*

$$a_{n+1} \leq (1 - \gamma_n)a_n + \gamma_n\delta_n, \quad \forall n = 0, 1, 2, \ldots, \quad (27)$$

where $\{\gamma_n\}$ is a sequence in $(0,1)$ and $\{\delta_n\}$ is a sequence in $\mathbb{R}$ such that

(i) *$\sum_{n=0}^{\infty}\gamma_n = \infty$;*

(ii) *$\limsup_{n\longrightarrow\infty}\delta_n \leq 0$ or $\sum_{n=0}^{\infty}|\gamma_n\delta_n| < \infty$.*

Then, $\lim_{n\longrightarrow\infty} a_n = 0$.

3. Main Result

Throughout this paper, we let C be a nonempty closed convex subset of a real Hilbert space H. First, we will show that G_n which is defined by

$$G_n x = \alpha_n x + (1 - \alpha_n)P_C(I - \lambda(\nabla g + \beta_n I))x, \quad \forall x \in C,\ n = 0, 1, 2, \ldots \quad (28)$$

has the unique fixed point under the conditions $0 < \lambda < 2/(L+2)$, $0 \leq \alpha_n < 1$ and $0 < \beta_n < 1$ where $g : C \longrightarrow \mathbb{R}$ be a strictly real-valued convex function such that ∇g is $1/L$-ism with $L > 0$. Since, ∇g is $1/L$-ism and the nonexpansiveness of P_C. Then, for each $x, y \in C$, we have

$$\begin{aligned}
&\|P_C(I-\lambda(\nabla g+\beta_n I))x - P_C(I-\lambda(\nabla g+\beta_n I))y\|^2 \\
&\le \|(I-\lambda(\nabla g+\beta_n I))x-(I-\lambda(\nabla g+\beta_n I))y\|^2 \\
&= \|(1-\lambda\beta_n)(x-y)-\lambda(\nabla g(x)-\nabla g(y))\|^2 \\
&= (1-\lambda\beta_n)^2\|x-y\|^2+\lambda^2\|\nabla g(x)-\nabla g(y)\|^2 \\
&\quad -2\lambda(1-\lambda\beta_n)\langle x-y,\nabla g(x)-\nabla g(y)\rangle \\
&\le (1-\lambda\beta_n)^2\|x-y\|^2+\lambda^2\|\nabla g(x)-\nabla g(y)\|^2 \\
&\quad -\frac{2}{L}\lambda(1-\lambda\beta_n)\|\nabla g(x)-\nabla g(y)\|^2 \\
&= (1-\lambda\beta_n)^2\|x-y\|^2 \\
&\quad +\left(\lambda^2-\frac{2}{L}\lambda(1-\lambda\beta_n)\right)\|\nabla g(x)-\nabla g(y)\|^2 \\
&\le (1-\lambda\beta_n)^2\|x-y\|^2 \\
&\quad +\left(\lambda^2-\frac{2}{L}\lambda+\frac{2}{L}\lambda^2\right)\|\nabla g(x)-\nabla g(y)\|^2 \\
&= (1-\lambda\beta_n)^2\|x-y\|^2 \\
&\quad -\lambda\left(\frac{2}{L}(1-\lambda)-\lambda\right)\|\nabla g(x)-\nabla g(y)\|^2 \\
&\le (1-\lambda\beta_n)^2\|x-y\|^2.
\end{aligned} \tag{29}$$

Therefore,

$$\begin{aligned}
&\|P_C(I-\lambda(\nabla g+\beta_n I))x-P_C(I-\lambda(\nabla g+\beta_n I))y\| \\
&\le (1-\lambda\beta_n)\|x-y\|.
\end{aligned} \tag{30}$$

It follows that, for $x \neq y$, by (30) we have

$$\begin{aligned}
&\|G_n x - G_n y\| = \|(\alpha_n x \\
&+(1-\alpha_n)P_C(I-\lambda(\nabla g+\beta_n I))x)-(\alpha_n y \\
&+(1-\alpha_n)P_C(I-\lambda(\nabla g+\beta_n I))y)\| = \|\alpha_n(x \\
&-y)+(1-\alpha_n)(P_C(I-\lambda(\nabla g+\beta_n I))x \\
&-P_C(I-\lambda(\nabla g+\beta_n I))y)\| \le \alpha_n\|x-y\|+(1 \\
&-\alpha_n)\|P_C(I-\lambda(\nabla g+\beta_n I))x-P_C(I \\
&-\lambda(\nabla g+\beta_n I))y\| \le \alpha_n\|x-y\|+(1-\alpha_n)(1 \\
&-\lambda\beta_n)\|x-y\| < \alpha_n\|x-y\|+(1-\alpha_n)\|x-y\| \\
&= \|x-y\|.
\end{aligned} \tag{31}$$

That is,

$$\|G_n x - G_n y\| < \|x-y\|, \quad \forall x \neq y. \tag{32}$$

So, G_n is a contraction, therefore, by Banach's contraction principle, G_n has the unique fixed point. Therefore, G_n is well-defined.

Let $\mathcal{U}$ be the solution set of the CCMP (1). It is clear that $\mathcal{U}$ is a closed and convex sets. We now ready to present my main results as follows.

Theorem 11. *Let C be a nonempty closed convex subset of a real Hilbert space H, $g : C \longrightarrow \mathbb{R}$ is a strictly real-valued convex function such that the gradient ∇g is $1/L$-ism with $L > 0$. Assume that $\mathcal{U} \neq \emptyset$ and let $\{x_n\} \subset C$ be a sequence generated by*

$$\begin{aligned}
&x_0 \in C, \\
&x_{n+1} = \alpha_n x_n + (1-\alpha_n) P_C(I-\lambda(\nabla g+\beta_n I))x_n, \\
&\qquad \forall n = 0,1,2,\ldots,
\end{aligned} \tag{33}$$

where $\lambda \in (0, 2/(2+L))$ and $\{\alpha_n\} \subset [0,1)$, $\{\beta_n\} \subset (0,1)$ satisfy the following conditions:

(i) *$\alpha_n \le \beta_n^m$ such that $m > 1$ for all $n = 0,1,2,\ldots$,*
(ii) *$\lim_{n\to\infty}\beta_n = 0$ and $\sum_{n=0}^{\infty}\beta_n = \infty$,*

then the sequence $\{x_n\}$ converges strongly to $q \in \mathcal{U}$, which is the unique minimizer of the CCMP (1).

Proof. We divide the proof into 4 steps.

Step 1. We will show that $\{x_n\}$ is bounded. Let $p \in \mathcal{U}$. By the strictly convexity of g, we have that $\mathcal{U}$ is a singleton set. Noticing from $1/L$-ism of ∇g that ∇g is L-Lipschitzian. So, by (10), we have $p = P_C(I-\lambda\nabla g)p$. Therefore, by (30), we have

$$\begin{aligned}
&\|x_{n+1}-p\| = \|(\alpha_n x_n \\
&+(1-\alpha_n)P_C(I-\lambda(\nabla g+\beta_n I))x_n)-p\| \\
&= \|(\alpha_n x_n+(1-\alpha_n)P_C(I-\lambda(\nabla g+\beta_n I))x_n) \\
&-(\alpha_n+(1-\alpha_n))p\| = \|\alpha_n(x_n-p)+(1-\alpha_n) \\
&\cdot(P_C(I-\lambda(\nabla g+\beta_n I))x_n-p)\| = \|\alpha_n(x_n-p) \\
&+(1-\alpha_n)(P_C(I-\lambda(\nabla g+\beta_n I))x_n \\
&-P_C(I-\lambda\nabla g)p)\| \le \alpha_n\|x_n-p\|+(1-\alpha_n) \\
&\cdot\|P_C(I-\lambda(\nabla g+\beta_n I))x_n-P_C(I-\lambda\nabla g)p\| \\
&\le \alpha_n\|x_n-p\|+(1-\alpha_n) \\
&\cdot(\|P_C(I-\lambda(\nabla g+\beta_n I))x_n \\
&-P_C(I-\lambda(\nabla g+\beta_n I))p\| \\
&+\|P_C(I-\lambda(\nabla g+\beta_n I))p-P_C(I-\lambda\nabla g)p\|) \\
&\le \alpha_n\|x_n-p\|+(1-\alpha_n)((1-\lambda\beta_n)\|x_n-p\| \\
&+\|(I-\lambda(\nabla g+\beta_n I))p-(I-\lambda\nabla g)p\|) = \alpha_n\|x_n \\
&-p\|+(1-\alpha_n)((1-\lambda\beta_n)\|x_n-p\|+\lambda\beta_n\|p\|)
\end{aligned}$$

$$= \alpha_n \|x_n - p\| + (1-\alpha_n)(1-\lambda\beta_n)\|x_n - p\| + (1 - \alpha_n)\lambda\beta_n\|p\| = (\alpha_n + (1-\alpha_n)(1-\lambda\beta_n))\|x_n - p\| + (1-\alpha_n)\lambda\beta_n\|p\| = (1-(1-\alpha_n)\lambda\beta_n)\|x_n - p\| + (1-\alpha_n)\lambda\beta_n\|p\| \le \max\{\|x_n - p\|, \|p\|\}$$
$$\vdots$$
$$\le \max\{\|x_0 - p\|, \|p\|\}. \quad (34)$$

It follows that $\{x_n\}$ is bounded, and so are $\{P_C(I-\lambda(\nabla g + \beta_n I))x_n\}$ and $\{P_C(I-\lambda\nabla g)x_n\}$.

Step 2. We will show that $\lim_{n\to\infty}\|x_n - P_C(I-\lambda\nabla g)x_n\| = 0$. Since,

$$\begin{aligned}\|x_{n+1} - P_C(I-\lambda\nabla g)x_n\| &= \|\alpha_n x_n + (1-\alpha_n) \cdot P_C(I-\lambda(\nabla g+\beta_n I))x_n - P_C(I-\lambda\nabla g)x_n\| \\ &= \|\alpha_n x_n + (1-\alpha_n)P_C(I-\lambda(\nabla g+\beta_n I))x_n - (\alpha_n + (1-\alpha_n))P_C(I-\lambda\nabla g)x_n\| \\ &= \|\alpha_n(x_n - P_C(I-\lambda\nabla g)x_n) + (1-\alpha_n) \cdot (P_C(I-\lambda(\nabla g+\beta_n I))x_n - P_C(I-\lambda\nabla g)x_n)\| \\ &\le \alpha_n\|x_n - P_C(I-\lambda\nabla g)x_n\| + (1-\alpha_n) \cdot \|P_C(I-\lambda(\nabla g+\beta_n I))x_n - P_C(I-\lambda\nabla g)x_n\| \le \alpha_n\|x_n - P_C(I-\lambda\nabla g)x_n\| \\ &\quad + (1-\alpha_n)\|(I-\lambda(\nabla g+\beta_n I))x_n - (I-\lambda\nabla g)x_n\| \\ &= \alpha_n\|x_n - P_C(I-\lambda\nabla g)x_n\| + (1-\alpha_n)\lambda\beta_n\|x_n\|.\end{aligned} \quad (35)$$

Therefore, by conditions (i) and (ii), we have

$$\lim_{n\to\infty}\|x_{n+1} - P_C(I-\lambda\nabla g)x_n\| = 0. \quad (36)$$

Since, ∇g is $1/L$-ism, then we have

$$\begin{aligned}\|P_C(I-\lambda\nabla g)x - P_C(I-\lambda\nabla g)y\|^2 &\le \|(I-\lambda\nabla g)x - (I-\lambda\nabla g)y\|^2 \\ &= \|(x-y) - \lambda(\nabla g(x) - \nabla g(y))\|^2 \\ &= \|x-y\|^2 - 2\lambda\langle x-y, \nabla g(x) - \nabla g(y)\rangle + \lambda^2\|\nabla g(x) - \nabla g(y)\|^2 \\ &\le \|x-y\|^2 - \frac{2\lambda}{L}\|\nabla g(x) - \nabla g(y)\|^2 + \lambda^2\|\nabla g(x) - \nabla g(y)\|^2 \\ &= \|x-y\|^2 - \lambda\left(\frac{2}{L} - \lambda\right)\|\nabla g(x) - \nabla g(y)\|^2 \\ &\le \|x-y\|^2.\end{aligned} \quad (37)$$

Hence,

$$\|P_C(I-\lambda\nabla g)x - P_C(I-\lambda\nabla g)y\| \le \|x-y\|, \quad (38)$$

that is, $P_C(I-\lambda\nabla g)$ is a nonexpansive. Therefore, by (36) and NST-condition (II) in Lemma 8, we have

$$\lim_{n\to\infty}\|(I-P_C(I-\lambda\nabla g))x_n\| = \lim_{n\to\infty}\|x_n - P_C(I-\lambda\nabla g)x_n\| = 0. \quad (39)$$

Step 3. Let $q \in \mathcal{U}$. Since, $\mathcal{U}$ is a singleton set, we have

$$\langle -q, q - y\rangle \ge 0, \quad \forall y \in \mathcal{U}. \quad (40)$$

Therefore, by Lemma 3, we have $q = P_{\mathcal{U}}(0)$. We will show that $\limsup_{n\to\infty}\langle -q, x_{n+1} - q\rangle \le 0$. From (10) we have

$$q = P_C(I-\lambda\nabla g)q. \quad (41)$$

Since $\{x_n\}$ is bounded, we consider a subsequence $\{x_{n_{i_j}}\}$ of $\{x_{n_i}\}$; there exists a subsequence $\{x_{n_{i_j}}\}$ of $\{x_{n_i}\}$ which converges weakly to z. It follows by the demiclosedness to the zero in Lemma 9 and (39) that $z = P_C(I-\lambda\nabla g)z$. So, by (10) we have $z \in \mathcal{U}$ (indeed, $z = q$). Therefore, by (40), we have

$$\begin{aligned}\limsup_{n\to\infty}\langle -q, x_{n+1} - q\rangle &= \limsup_{j\to\infty}\left\langle -q, x_{n_{i_j}+1} - q\right\rangle \\ &= \max\left\{\langle -q, z - q\rangle : \text{ for all } x_{n_{i_j}} \rightharpoonup z \text{ as } j \longrightarrow \infty\right\} \\ &\le 0.\end{aligned} \quad (42)$$

Step 4. We will show that $\{x_n\}$ converges strongly to q. By (30), Lemma 5, condition (i), and the linearity orthogonal projection of P_C, we have

$$\begin{aligned}\|x_{n+1} - q\|^2 &= \|\alpha_n x_n + (1-\alpha_n)P_C(I-\lambda(\nabla g+\beta_n I)) \cdot x_n - q\|^2 = \|\alpha_n x_n + (1-\alpha_n)P_C(I - \lambda(\nabla g+\beta_n I))x_n - (\alpha_n + (1-\alpha_n))q\|^2 \\ &= \|\alpha_n(x_n - q) + (1-\alpha_n) \cdot (P_C(I-\lambda(\nabla g+\beta_n I))x_n - q)\|^2 \le \alpha_n\|x_n - q\|^2 \\ &\quad + (1-\alpha_n)\|P_C(I-\lambda(\nabla g+\beta_n I))x_n - q\|^2 \\ &= \alpha_n\|x_n - q\|^2 + (1-\alpha_n)\langle P_C(I-\lambda(\nabla g+\beta_n I)) \cdot x_n - q, P_C(I-\lambda(\nabla g+\beta_n I))x_n - q\rangle = \alpha_n\|x_n\end{aligned}$$

$$- q\|^2 + (1-\alpha_n)\Big\langle P_C(I-\lambda(\nabla g+\beta_n I))x_n - q, \frac{x_{n+1}-\alpha_n x_n}{1-\alpha_n} - q\Big\rangle = \alpha_n\|x_n-q\|^2 + (1-\alpha_n) \cdot \Big(\frac{1}{1-\alpha_n}\langle P_C(I-\lambda(\nabla g+\beta_n I))x_n - q, x_{n+1} - q\rangle + \frac{\alpha_n}{1-\alpha_n}\langle P_C(I-\lambda(\nabla g+\beta_n I))x_n - q, q - x_n\rangle\Big) = \alpha_n\|x_n-q\|^2 + \langle P_C(I-\lambda(\nabla g+\beta_n I)) \cdot x_n - P_C(I-\lambda\nabla g)q, x_{n+1}-q\rangle + \alpha_n\langle P_C(I - \lambda(\nabla g+\beta_n I))x_n - q, q - x_n\rangle \le \alpha_n\|x_n-q\|^2 + \langle P_C(I-\lambda(\nabla g+\beta_n I))x_n - P_C(I - \lambda(\nabla g+\beta_n I))q, x_{n+1}-q\rangle + \langle P_C(I - \lambda(\nabla g+\beta_n I))q - P_C(I-\lambda\nabla g)q, x_{n+1}-q\rangle + \alpha_n\|P_C(I-\lambda(\nabla g+\beta_n I))x_n - q\|\cdot\|q-x_n\| \le \alpha_n\|x_n-q\|^2 + \|P_C(I-\lambda(\nabla g+\beta_n I))x_n - P_C(I-\lambda(\nabla g+\beta_n I))q\|\cdot\|x_{n+1}-q\| + \lambda\beta_n\langle -q, x_{n+1}-q\rangle + \alpha_n\|P_C(I-\lambda(\nabla g+\beta_n I))x_n - q\|\cdot\|q - x_n\| \le \beta_n^m\|x_n-q\|^2 + (1-\lambda\beta_n)\|x_n-q\|\cdot\|x_{n+1} - q\| + \lambda\beta_n\langle -q, x_{n+1}-q\rangle + \beta_n^m\|P_C(I - \lambda(\nabla g+\beta_n I))x_n - q\|\cdot\|q-x_n\| = \beta_n^m\|x_n-q\|^2 + \frac{1-\lambda\beta_n}{2}\big(\|x_n-q\|^2 + \|x_{n+1}-q\|^2 - (\|x_n-q\| - \|x_{n+1}-q\|)^2\big) + \lambda\beta_n\langle -q, x_{n+1}-q\rangle + \beta_n^m\|P_C(I - \lambda(\nabla g+\beta_n I))x_n - q\|\cdot\|q-x_n\| \le \beta_n^m\|x_n-q\|^2 + \frac{1-\lambda\beta_n}{2}\|x_n-q\|^2 + \frac{1}{2}\|x_{n+1}-q\|^2 + \lambda\beta_n\langle -q, x_{n+1}-q\rangle + \beta_n^m\|P_C(I-\lambda(\nabla g+\beta_n I))x_n - q\|\cdot\|q - x_n\|, \tag{43}$$

and, therefore,

$$\|x_{n+1}-q\|^2 \le 2\beta_n^m\|x_n-q\|^2 + (1-\lambda\beta_n)\|x_n-q\|^2 + 2\lambda\beta_n\langle -q, x_{n+1}-q\rangle + 2\beta_n^m\|P_C(I-\lambda(\nabla g+\beta_n I))x_n - q\|\cdot\|q-x_n\| = (1-\lambda\beta_n)\|x_n-q\|^2 + \lambda\beta_n\Big(\frac{2\beta_n^{m-1}}{\lambda}\|x_n-q\|^2 + 2\langle -q, x_{n+1}-q\rangle + \frac{2\beta_n^{m-1}}{\lambda}\|P_C(I-\lambda(\nabla g+\beta_n I))x_n - q\| \cdot\|q-x_n\|\Big) = (1-\gamma_n)\|x_n-q\|^2 + \gamma_n\delta_n, \tag{44}$$

where $\gamma_n = \lambda\beta_n$ and

$$\delta_n = \frac{2\beta_n^{m-1}}{\lambda}\|x_n-q\|^2 + 2\langle -q, x_{n+1}-q\rangle + \frac{2\beta_n^{m-1}}{\lambda}\|P_C(I-\lambda(\nabla g+\beta_n I))x_n - q\| \cdot\|q-x_n\|. \tag{45}$$

It is easy to see that $\sum_{n=0}^{\infty}\gamma_n = \infty$ and $\limsup_{n\to\infty}\delta_n \le 0$. Therefore, by Lemma 10, we obtain $\{x_n\}$ converges strongly to q. This completes the proof.

Notice that when $\alpha_n = 0$ for all $n = 0,1,2,\ldots$ then the result of Theorem 11 can be reduced into the result of Tian and Zhang [26] without the control condition $\sum_{n=0}^{\infty}|\beta_{n+1}-\beta_n| < \infty$ as follows.

Corollary 12 (see [26]). *Let C be a nonempty closed convex subset of a real Hilbert space H and $g : C \longrightarrow \mathbb{R}$ be a strictly real-valued convex function such that the gradient ∇g is $1/L$-ism with $L > 0$. Assume that $\mathcal{U} \ne \emptyset$ and let $\{x_n\} \subset C$ be a sequence generated by*

$$x_0 \in C, \quad x_{n+1} = P_C(I-\lambda(\nabla g+\beta_n I))x_n, \quad \forall n = 0,1,2,\ldots, \tag{46}$$

where $\lambda \in (0, 2/(2+L))$ and $\{\beta_n\} \subset (0,1)$ satisfies the following conditions: $\lim_{n\to\infty}\beta_n = 0$ and $\sum_{n=0}^{\infty}\beta_n = \infty$, then the sequence $\{x_n\}$ converges strongly to $q \in \mathcal{U}$, which is the unique minimizer of the CCMP (1).

4. Applications

Let C and Q be closed convex subsets of real Hilbert spaces H_1 and H_2, respectively, and $A : H_1 \longrightarrow H_2$ be a bounded linear operator. We now consider the split feasibility problem (SFP) which introduced in 1994 by Censor and Elfving [13], where this problem is to find an element $q \in C$ such that $Aq \in Q$. Define the convex function $g : C \longrightarrow \mathbb{R}$ as follows:

$$g(x) = \frac{1}{2}\|Ax - P_Q Ax\|^2. \tag{47}$$

It follows by Lemma 7 that the gradient of g as $\nabla g = A^*(I - P_Q)A$ where A^* is the adjoint operator of A, and ∇g is $1/\|A\|^2$-ism. We have the consequence results as follows.

Theorem 13. *Let C and Q be closed convex subsets of real Hilbert spaces H_1 and H_2, respectively, and let $A : H_1 \longrightarrow$*

H_2 *be a bounded linear operator. Suppose that the SFP has a nonempty solution. Let* $\{x_n\} \subset C$ *be a sequence generated by*

$$x_0 \in C,$$

$$x_{n+1} = \alpha_n x_n + (1-\alpha_n) P_C (I - \lambda (A^* (I - P_Q) A + \beta_n I)) x_n, \quad \forall n = 0, 1, 2, \ldots, \tag{48}$$

where $\lambda \in (0, 2/(2 + \|A\|^2))$ *and* $\{\alpha_n\} \subset [0, 1), \{\beta_n\} \subset (0, 1)$ *satisfy the following conditions:*

(i) $\alpha_n \leq \beta_n^m$ *such that* $m > 1$ *for all* $n = 0, 1, 2, \ldots,$

(ii) $\lim_{n \to \infty} \beta_n = 0$ *and* $\sum_{n=0}^{\infty} \beta_n = \infty$,

then the sequence $\{x_n\}$ *converges strongly to* q*, which is the unique minimizer of the minimum-norm solution of the SFP.*

Corollary 14. *Let* C *and* Q *be closed convex subsets of real Hilbert spaces* H_1 *and* H_2*, respectively, and let* $A : H_1 \longrightarrow H_2$ *be a bounded linear operator. Suppose that the SFP has a nonempty solution. Let* $\{x_n\} \subset C$ *be a sequence generated by*

$$x_0 \in C,$$

$$x_{n+1} = P_C (I - \lambda (A^* (I - P_Q) A + \beta_n I)) x_n, \quad \forall n = 0, 1, 2, \ldots, \tag{49}$$

where $\lambda \in (0, 2/(2 + \|A\|^2))$ *and* $\{\beta_n\} \subset (0, 1)$ *satisfies the following conditions:* $\lim_{n \to \infty} \beta_n = 0$ *and* $\sum_{n=0}^{\infty} \beta_n = \infty$*, then the sequence* $\{x_n\}$ *converges strongly to* q*, which is the unique minimizer of the minimum-norm solution of the SFP.*

5. Data Availability

In order of the feasible solution, all algorithms of the iterations have to compute many inner iterations to find the appropriate result, and stack overflow often occurs in which a computer program makes too many subroutine calls and its call stack runs out of space when the parameters of iterations have using many stack arrays to compute the feasible solution.

To avoid the stack overflow, we introduce how to do the mathematical programming without using the stack arrays of its parameters for solving the SFP of the algorithm in Corollary 14. Indeed, the situation of the stack overflow may have occurred from calculating the floating point numbers or the significant decimal digits; to avoid it we ought to be careful of that by always using digit precision command such as the command *N[expr_]* in Mathematica, and the command *digits(num_)* in Matlab, and also define all matrix in the regular command type without using the matrix palette to avoid it.

Some mathematical software has a command to give the total number of seconds of CPU time used and the total number of seconds since the beginning of computation in the session such as the commands *TimeUsed*[] and *AbsoluteTime*[] in Mathematica, respectively, and the commands *cputime* and *tic/toc* in Matlab, respectively.

We now give the formulation of orthogonal projection P_C where C is a simply closed convex sets as follows, and in the case that C is not a simply closed convex sets, for instance, C is a halfspace, we can found more the formulation in [33].

Proposition 15. *For* $x \in \mathbb{R}^N$ *we have*

(i) *if* $C = \mathbb{R}^N$ *then* $P_C x = x$,

(ii) *if* $C = \{b\}$ *such that* $b \in \mathbb{R}^M$ *then* $P_C x = b$,

(iii) *if* $C = \{x \in \mathbb{R}^N : \|x\|_2 \leq \rho, \rho > 0\}$ *then*

$$P_C x = \begin{cases} \dfrac{\rho x}{\|x\|_2}, & x \notin C, \\ x, & x \in C. \end{cases} \tag{50}$$

Proof. Obviously, the results (i) and (ii) hold by the definition of orthogonal projection of P_C, and the result (iii) also holds by the normal vector of the boundary points set of C.

We are ready to introduce how to do the mathematical programming without using the stack arrays of its parameters for solving the SFP of the algorithm in Corollary 14 as follows. Suppose that the SFP has the unique consistent. Taking $H_1 = (\mathbb{R}^N, \|\cdot\|_2)$ and $H_2 = (\mathbb{R}^M, \|\cdot\|_2)$ into Corollary 14. Let the sets C and Q, the operator A, the sequence $\{x_n\}$, and the parameters $\{\beta_n\}, \lambda$ satisfy the conditions in Corollary 14. We have that $\{x_n\}$ is a convergent sequence, and so it is a Cauchy sequence. Hence, we can choose the stopping criteria $\epsilon > 0$ which satisfies $\|x_{n+1} - x_n\|_2 < \epsilon$ for stopping the program, and also the approximate solution refers to the last iteration. Steps of the mathematical programming of the algorithm in Corollary 14 are shown as follows:

Mathematical programming for the split feasibility problem

Finding the solution of an augmented matrix equation $A_{M \times N} X_{N \times 1} = B_{M \times 1}$.

Step 1. Declare of all parameters $A_{M \times N}, B_{M \times 1}$, the starting point $x\text{Start}_{N \times 1}$ and ϵ.

We set $M = 3, N = 3$ and $b = B_{M \times 1}$.

The example of the commands in Mathematica is shown as follows.

$A = \{\{1, 2, -1\}, \{1, 1, -1\}, \{1, -2, 1\}\}; b = \{\{1\}, \{0\}, \{3\}\};$
$xStart = \{\{0\}, \{0\}, \{0\}\}; \epsilon = 10^{-6};$

Step 2. Define the formulations of the orthogonal projections of P_C and P_Q where

$$C = \left\{x \in \mathbb{R}^N : \|x\|_2 \leq \rho, \rho > 0\right\},$$
$$Q = \left\{x \in \mathbb{R}^M : \|x\|_2 \leq \sigma, \sigma > 0\right\}. \tag{51}$$

If we choose $C = \mathbb{R}^N$ and $Q = \{b\}$ such that $b \in \mathbb{R}^M$ then the orthogonal projections of P_C and P_Q are easy to calculated, and, hence, we do not need to define its formulations in this step, and we can put directly its formulations to process.

The example of the commands in Mathematica is shown as follows.

```
Pc[u_] := If[Norm[u] > ρ, ρu/(Norm[u]//N), u];
Pq[u_] := If[Norm[u] > σ, σu/(Norm[u]//N), u];
```

Step 3. Set the starting index $n = 0$ and fix parameter $\lambda \in (0, 2/(2 + \|A\|^2))$. If the parameter λ is not a fix number such that it is a sequence, then we must lie it in the while loop of step 4.

We set $\lambda = 2/((2 + \|A\|^2) + 10^2)$.

The example of the commands in Mathematica is shown as follows.

```
n = 0; λ = 2/((2 + (Norm[A]//N)^2) + 10^2);
```

Step 4. Start to calculate the iterations of the sequence $\{x_n\}$ such that

$$x_{n+1} = P_C (I - \lambda (A^* (I - P_Q) A + \beta_n I)) x_n \tag{52}$$

using the while loop. Set the parameter $\{\beta_n\} \subset (0, 1)$ for all $n = 0, 1, 2, \ldots$ into the while loop such that it satisfies the following conditions: $\lim_{n\to\infty}\beta_n = 0$ and $\sum_{n=0}^{\infty} \beta_n = \infty$. If $\|x_{n+1} - x_n\|_2 < \epsilon$ then we break the while loop for approximate feasible solution, which is referred to in the last iteration.

It well known that, in the case of finite-dimensional real space, $A^* = A^T$ where A^T stands for matrix transposition of A, and, hence, the algorithm in Corollary 14 can be reduced to

$$x_{n+1} = \begin{cases} P_C\left(x_n - \lambda A^T A x_n + \lambda A^T P_Q A x_n - \lambda\beta_n x_n\right) & \text{for the closed balls } C \text{ and } Q, \\ P_C\left(x_n - \lambda A^T A x_n + \lambda A^T b - \lambda\beta_n x_n\right) & \text{for the closed ball } C \text{ and } Q = \{b\}, \\ x_n - \lambda A^T A x_n + \lambda A^T b - \lambda\beta_n x_n & \text{for } C = \mathbb{R}^N \text{ and } Q = \{b\} \end{cases} \tag{53}$$

where $b \in \mathbb{R}^M$. We set $\beta_n = 1/(n+2)$ for all $n = 0, 1, 2, \ldots$ and instead of x_n and x_{n+1} with the variables *xOld* and *xNew*, respectively, and also instead of β_n with β in the while loop for avoidance using stack arrays of the parameters.

The example of the commands in Mathematica is shown as follows.

```
xOld = xStart;
While[True, β = 1/(n + 2);
    (* We set C = ℝ^N and Q = {b} where b ∈ ℝ^M.*)
    xNew = xOld − λ((Transpose[A].A).xOld) + λ(Transpose[A].b) − λ β xOld;
    Print["x[", n + 1, "] =", Transpose[xNew]//N//MatrixForm, ", Error =", ScientificForm[Norm[xNew − xOld]]];
    If[Norm[xNew − xOld] < ε,
        Print["− − −Conclusion − − −"];
        Print["Starting point x[0] =", Transpose[xStart]//MatrixForm];
        Print["Approximate Solution x[", n + 1, "] =", Transpose[xNew]//N//MatrixForm];
        Print["TimeUsed =", TimeUsed[], " second"];
        Print["− − −End − − −"];
        Break[],
        (* Goto Next Loop *)
        xOld = xNew; n + +;
    ];
];
```

Step 5. Clear memory of the system.

The example of the command in Mathematica is shown as follows.

```
Quit[];
```

6. Numerical Results

In this section, we give some insight into the behavior of the algorithm presented in Corollary 14. We implemented them in Mathematica to solve and run on a computer Intel(R) Core (TM) i3 processor 2.00 GHz. We use $\|x_{n+1} - x_n\|_2 < \epsilon$ as the stopping criteria.

Throughout the computational experiments, the parameters used in those algorithms were sets as $\epsilon = 10^{-6}$, $\beta_n = 1/(n + 2)$ and $\lambda = 2/((2 + \|A\|^2) + 10^2)$ for all $n = 0, 1, 2, \ldots$, where A is a bounded linear operator. In the results report below, all CPU times reported are in seconds. The approximate solution is referred to the last iteration.

Example 1. Find the solution of linear equation system as follows:

$$\begin{aligned} x + 2y - z &= 1, \\ x + y - z &= 0, \\ x - 2y + z &= 3, \end{aligned} \tag{54}$$

where $x, y, z \in \mathbb{R}$.

Let $H_1 = H_2 = (\mathbb{R}^3, \|\cdot\|_2)$. Take $A = \begin{pmatrix} 1 & 2 & -1 \\ 1 & 1 & -1 \\ 1 & -2 & 1 \end{pmatrix}$, $C = \mathbb{R}^3$ and $Q = \{b : b = (1, 0, 3)^T\}$ into Corollary 14. We have chosen $x_0 \in C$ arbitrarily and

Table 1: Results for Example 1 using algorithm in Corollary 14.

Starting points x_0	*Number of iterations*	*CPU(s)*	*Approximate solution* q
$(0,0,0)^T$	6015	2.393	$(1.999190, 0.997529, 2.995430)^T$
$(1,1,1)^T$	5986	2.293	$(1.999180, 0.997518, 2.995410)^T$
$(2,2,2)^T$	5954	2.333	$(1.999180, 0.997506, 2.995380)^T$
$(2,3,1)^T$	5963	2.543	$(1.999180, 0.997509, 2.995390)^T$
$(3,2,1)^T$	5970	2.533	$(1.999180, 0.997512, 2.995400)^T$

$$\begin{aligned} x_{n+1} &= P_C\left(I - \lambda\left(A^T(I - P_Q)A + \beta_n I\right)\right)x_n \\ &= x_n - \lambda A^T A x_n + \lambda A^T P_Q A x_n - \lambda \beta_n x_n \\ &= x_n - \lambda A^T A x_n + \lambda A^T b - \lambda \beta_n x_n \end{aligned} \quad (55)$$

where $b = (1,0,3)^T$ for all $n = 0,1,2,\ldots$. As $n \longrightarrow \infty$, we have $x_n \longrightarrow q$ such that q is the our solution. The numerical results are listed in Table 1.

7. Conclusion

In this paper, we obtain an iterative scheme using the gradient projection method based on Mann's approximation method for solving the constrained convex minimization problem (CCMP) and also solving the split feasibility problem (SFP) such that another strong convergence theorems for the CCMP and the SFP are obtained.

Conflicts of Interest

The authors declare that there are no conflicts of interest regarding the publication of this paper.

Acknowledgments

The authors would like to acknowledge the Science Achievement Scholarship of Thailand (SAST) and the Faculty of Science, Maejo University, for financial support.

References

[1] E. S. Levitin and B. T. Polyak, "Constrained minimization methods," *Zhurnal Vychislitel'noi Matematiki i Matematicheskoi Fiziki*, vol. 6, no. 5, pp. 787–823, 1966.

[2] H.-K. Xu, "Averaged mappings and the gradient-projection algorithm," *Journal of Optimization Theory and Applications*, vol. 150, no. 2, pp. 360–378, 2011.

[3] P. Kumam, "A new hybrid iterative method for solution of equilibrium problems and fixed point problems for an inverse strongly monotone operator and a nonexpansive mapping," *Journal of Applied Mathematics and Computing*, vol. 29, no. 1-2, pp. 263–280, 2009.

[4] Y. Yao and H.-K. Xu, "Iterative methods for finding minimum-norm fixed points of nonexpansive mappings with applications," *Journal of Mathematical Programming and Operations Research*, vol. 60, no. 6, pp. 645–658, 2011.

[5] S. He and W. Sun, "New hybrid steepest descent algorithms for variational inequalities over the common fixed points set of infinite nonexpansive mappings," *WSEAS Transactions on Mathematics*, vol. 11, no. 2, pp. 83–92, 2012.

[6] M. Su and H.-K. Xu, "Remarks on the gradient-projection algorithm," *Journal of Nonlinear Analysis and Optimization*, vol. 1, no. 1, pp. 35–43, 2010.

[7] R.-X. Ni, "Strong convergence of a hybrid projection algorithm for approximation of a common element of three sets in Banach spaces," *WSEAS Transactions on Mathematics*, vol. 12, no. 3, pp. 296–306, 2013.

[8] P. Kumam, "A hybrid approximation method for equilibrium and fixed point problems for a monotone mapping and a nonexpansive mapping," *Nonlinear Analysis: Hybrid Systems*, vol. 2, no. 4, pp. 1245–1255, 2008.

[9] P. H. Calamai and J. J. Moré, "Projected gradient methods for linearly constrained problems," *Mathematical Programming*, vol. 39, no. 1, pp. 93–116, 1987.

[10] F. Wang and H.-K. Xu, "Approximating curve and strong convergence of the CQ algorithm for the split feasibility problem," *Journal of Inequalities and Applications*, vol. 2010, Article ID 102085, 13 pages, 2010.

[11] J. S. Jung, "Strong convergence of composite iterative methods for equilibrium problems and fixed point problems," *Applied Mathematics and Computation*, vol. 213, no. 2, pp. 498–505, 2009.

[12] J. S. Jung, "Strong convergence of iterative methods for k-strictly pseudo-contractive mappings in Hilbert spaces," *Applied Mathematics and Computation*, vol. 215, no. 10, pp. 3746–3753, 2010.

[13] Y. Censor and T. Elfving, "A multiprojection algorithm using Bregman projections in a product space," *Numerical Algorithms*, vol. 8, no. 2–4, pp. 221–239, 1994.

[14] H.-K. Xu, "A variable Krasnosel'skii–Mann algorithm and the multiple-set split feasibility problem," *Inverse Problems*, vol. 22, no. 6, pp. 2021–2034, 2006.

[15] H.-K. Xu, "Iterative methods for the split feasibility problem in infinite-dimensional Hilbert spaces," *Inverse Problems*, vol. 26, no. 10, Article ID 105018, 2010.

[16] C. Byrne, "A unified treatment of some iterative algorithms in signal processing and image reconstruction," *Inverse Problems*, vol. 20, no. 1, pp. 103–120, 2004.

[17] H. Songnian and G. Jun, "Algorithms for finding the minimum norm solution of hierarchical fixed point problems," *WSEAS Transactions on Mathematics*, vol. 12, no. 3, pp. 317–328, 2013.

[18] G. Lopez, V. Martin, and H.-K. Xu, "Perturbation techniques for nonexpansive mappings with applications," *Nonlinear Analysis: Real World Applications*, vol. 10, no. 4, pp. 2369–2383, 2009.

[19] L.-C. Ceng, Q. H. Ansari, and J.-C. Yao, "Extragradient-projection method for solving constrained convex minimiza-

tion problems," *Numerical Algebra, Control and Optimization*, vol. 1, no. 3, pp. 341–359, 2011.

[20] Y. Yao, S. M. Kang, W. Jigang, and P.-X. Yang, "A Regularized Gradient Projection Method for the Minimization Problem," *Journal of Applied Mathematics*, vol. 2012, Article ID 259813, 9 pages, 2012.

[21] Y. Yao, Y.-C. Liou, and C.-F. Wen, "Variant gradient projection methods for the minimization problems," *Abstract and Applied Analysis*, vol. 2012, Article ID 792078, 16 pages, 2012.

[22] M. Tian and M. Li, "A general iterative method for constrained convex minimization problems in hilbert spaces," *WSEAS Transactions on Mathematics*, vol. 13, pp. 271–281, 2014.

[23] G. Cai and Y. Shehu, "An iterative algorithm for fixed point problem and convex minimization problem with applications," *Fixed Point Theory and Applications*, vol. 2015, no. 7, 17 pages, 2015.

[24] C. D. Enyi and M. E. Soh, "Modified gradient-projection algorithm for solving convex minimization problem in Hilbert spaces," *IAENG International Journal of Applied Mathematics*, vol. 44, no. 3, 7 pages, 2014.

[25] Y. Wu and L. Shi, "Projection and contraction methods for constrained convex minimization problem and the zero points of maximal monotone operator," *Journal of Nonlinear Sciences and Applications. JNSA*, vol. 10, no. 2, pp. 637–646, 2017.

[26] M. Tian and H.-F. Zhang, "Regularized gradient-projection methods for finding the minimum-norm solution of the constrained convex minimization problem," *Journal of Inequalities and Applications*, vol. 2017, no. 13, 12 pages, 2017.

[27] W. Takahashi, *Introduction to Nonlinear and Convex Analysis*, Yokohama Publishers, Yokohama, Japan, 2009.

[28] J. F. Tang, S. S. Chang, and F. Yuan, "A strong convergence theorem for equilibrium problems and split feasibility problems in Hilbert spaces," *Fixed Point Theory and Applications*, vol. 2014, no. 36, 16 pages, 2014.

[29] K. Nakajo, K. Shimoji, and W. Takahashi, "Strong convergence to common fixed points of families of nonexpansive mappings in Banach spaces," *Journal of Nonlinear and Convex Analysis. An International Journal*, vol. 8, no. 1, pp. 11–34, 2007.

[30] W. Takahashi, Y. Takeuchi, and R. Kubota, "Strong convergence theorems by hybrid methods for families of nonexpansive mappings in Hilbert spaces," *Journal of Mathematical Analysis and Applications*, vol. 341, no. 1, pp. 276–286, 2008.

[31] H. S. Hundal, "An alternating projection that does not converge in norm," *Nonlinear Analysis: Theory, Methods & Applications*, vol. 57, no. 1, pp. 35–61, 2004.

[32] H. K. Xu, "Viscosity approximation methods for nonexpansive mappings," *Journal of Mathematical Analysis and Applications*, vol. 298, no. 1, pp. 279–291, 2004.

[33] P. Tianchai, "Gradient projection method with a new step size for the split feasibility problem," *Journal of Inequalities and Applications*, vol. 2018, no. 120, 22 pages, 2018.

15

Oscillatory and Asymptotic Behavior of a First-Order Neutral Equation of Discrete Type with Variable Several Delay under Δ Sign

Radhanath Rath [1,2] **and Chittaranjan Behera** [3]

[1] VSSUT, Burla, Sambalpur, 768018 Odisha, India
[2] Former Principal Khallikote Autonomous College, Berhampur, 760001 Odisha, India
[3] Department of Mathematics, Silicon Institute of Technology, Bhubaneswar, Odisha, India

Correspondence should be addressed to Radhanath Rath; radhanathmath@yahoo.co.in

Academic Editor: Hans Engler

We obtain necessary and sufficient conditions so that every solution of neutral delay difference equation $\Delta(y_n - \sum_{j=1}^{k} p_n^j y_{n-m_j}) + q_n G(y_{\sigma(n)}) = f_n$ oscillates or tends to zero as $n \longrightarrow \infty$, where $\{q_n\}$ and $\{f_n\}$ are real sequences and $G \in C(\mathbb{R}, \mathbb{R})$, $xG(x) > 0$, and $m_1, m_2, \ldots, m_k$ are positive integers. Here Δ is the forward difference operator given by $\Delta x_n = x_{n+1} - x_n$, and $\{\sigma_n\}$ is an increasing unbounded sequences with $\sigma_n \le n$. This paper complements, improves, and generalizes some past and recent results.

1. Introduction

Consider the neutral delay difference equation of first order

$$\Delta\left(y_n - \sum_{j=1}^{k} p_n^j y_{n-m_j}\right) + q_n G\left(y_{\sigma(n)}\right) = f_n \qquad (1)$$

where Δ is the forward difference operator given by $\Delta x_n = x_{n+1} - x_n$, q_n and f_n are members of infinite real sequences, and m_j are positive integers. Further, assume $\{p_n^j\}$ are real sequences for each $j \in 1, 2, \ldots k$ and that $G \in C(\mathbb{R}, \mathbb{R})$ and $\sigma(n) \le n$ are monotonic increasing sequences which are unbounded.

We study the oscillatory behavior of solutions of neutral difference equation (1) under the following assumptions.

(H1) $xG(x) > 0$ for $x \neq 0$.

(H2) There exists a bounded sequence $\{F_n\}$ such that $\Delta F_n = f_n$.

(H3) The sequence $\{F_n\}$ in (H2) satisfies $\lim_{n \longrightarrow \infty} F_n = 0$.

(H4) $q_n > 0$, $\sum_{n=n_0}^{\infty} q_n = \infty$.

In addition to the above we assume some new conditions on p_n^j (see (12), (22), (26), and (30) in next section). It is important to note that our results hold good for the solutions of the neutral equation

$$\Delta\left(y_n - \sum_{j=1}^{k} p_n^j y_{n-m_j}\right) + \sum_{j=1}^{l} q_n^j G\left(y_{\sigma_j(n)}\right) = f_n \qquad (2)$$

under the assumption

$$\sum_{n=n_0}^{\infty}\left(\sum_{j=1}^{l} q_n^j\right) = \infty, \quad q_n^j > 0 \qquad (3)$$

instead of (H4). The following neutral difference equations/delay difference equations are obtained as particular case of (2).

$$\Delta\left(y_n - p_n y_{n-m}\right) + \sum_{j=1}^{l} q_n^j G\left(y_{\sigma_j(n)}\right) = f_n, \qquad (4)$$

$$\Delta\left(y_n - p_n y_{n-m}\right) + q_n G\left(y_{\sigma(n)}\right) = f_n, \qquad (5)$$

$$\Delta(y_n) + \sum_{j=1}^{l} q_n^j G\left(y_{\sigma_j(n)}\right) = 0, \tag{6}$$

and

$$\Delta(y_n) + q_n G\left(y_{\sigma(n)}\right) = 0. \tag{7}$$

The neutral difference equations (5) are seen as the discrete analogue of the neutral differential equations

$$(y(t) - p(t)\,y(t-\tau))' + q(t)\,G(y(t-\sigma)) = f(t). \tag{8}$$

The oscillatory and asymptotic behavior of delay difference equations and neutral difference equations have been intensively studied in recent years due to its various application in different field of science and technology [1]. It is observed that several articles (see [2–4]) exist in literature for the study of neutral difference equations/delay difference equations with several delay, i.e., for (4) or (6), respectively. However study of neutral equations with several delay term under Δ symbol, i.e., (1) or (2), seems to be relatively scarce in literature. Use of lemmas from [1, Lemma 1.5.1 and 1.5.2] or its discrete analogue (see [5]) plays an important role in studying (4) [6], (5) [7], and (8) [8]. In this context, one may note these lemmas cannot be applied to the study of (1) or (2). Hence study of (1) and (2) needs a different approach.

The work in this paper complements and generalizes the work in [3, 9]. This can be verified that the results in [3, 9] which are concerned with the study of (6) and (7) cannot be applied to the delay difference equation

$$\Delta(y_n) + \left(e^{-2} - e^{-3}\right) y_{n-2} = 0. \tag{9}$$

which has a solution $y_n = e^{-n}$ tending to zero. It is because the primary assumption,

$$\liminf_{k\longrightarrow\infty} \sum_{i=\sigma(k)}^{k-1} q_i > \frac{1}{e}, \tag{10}$$

is not satisfied. However, note that (10) implies (H4) and (H4) is satisfied in (9) and hence the results of this paper give an answer to the behavior of solutions of neutral equations like (9). While working on nonlinear neutral equations most of the authors [7, 8, 10–12] assume the condition that G is nondecreasing unlike this paper.

Let n_0 be a fixed nonnegative integer. Let $\max\{m_1, m_2, \ldots, m_k\} = m$ and $\rho = \min\{\sigma(n_0), n_0 - m\}$. By a solution of (1) we mean a real sequence $\{y_n\}$ which is defined for all positive integers $n \geq \rho$ and satisfies (1) for $n \geq n_0$. Clearly if the initial condition

$$y_n = a_n \quad \text{for } \rho \leq n \leq n_0, \tag{11}$$

is given then (1) has a unique solution satisfying the given initial condition (11). A solution $\{y_n\}$ of (1) is said to be oscillatory if, for every positive integer $n_0 > 0$, there exists $n \geq n_0$ such that $y_n y_{n+1} \leq 0$; otherwise $\{y_n\}$ is said to be nonoscillatory. In the sequel, unless otherwise specified, when we write a functional inequality, it will be assumed to hold for all n sufficiently large. Here we assume the existence of solution of (1) and study its oscillatory and asymptotic behavior.

2. Sufficient Condition

In this section we present some results which prove that (H4) is sufficient for any solution of (2) to be oscillatory or tending to zero as $n \longrightarrow \infty$. Moreover we give some examples to illustrate and signify our results. Our first result and the subsequent ones are as follows.

Theorem 1. *Suppose that (H1)-(H4) hold. Assume that there exists a positive constant p such that the sequences $\{p_n^j\}$ for $j = 1, 2, \ldots, k$ satisfy the condition*

$$p_n^j \geq 0, \quad \textit{for every } j = 1, 2, \ldots, k$$
$$\textit{and } \sum_{j=1}^{k} \limsup_{n\longrightarrow\infty} p_n^j < p < 1. \tag{12}$$

Then every solution of (1) oscillates or tends to zero as $n \longrightarrow \infty$.

Proof. Let $y = \{y_n\}$ be any solution of (1) for $n \geq n_0$, where n_0 is a fixed positive integer. If it oscillates then there is nothing to prove; otherwise, it leads to two distinct possibilities, either $y_n > 0$ or $y_n < 0$. for $n \geq n_1 > n_0$. Consider the first one, i.e., $y_n > 0$ eventually. There exits positive integer $n_2 \geq n_1$ such that $y_n > 0$, $y_{n-m_j} > 0$ for each j, and $y_{\sigma(n)} > 0$ for $n \geq n_2$. For $n \geq n_2$, let

$$z_n = y_n - \sum_{j=1}^{k} p_n^j y_{n-m_j}, \tag{13}$$

and

$$w_n = z_n - F_n. \tag{14}$$

From (1), (13), and (14), it follows due to (H1) that

$$\Delta w_n = -q_n G\left(y_{\sigma(n)}\right) \leq 0. \tag{15}$$

Then there exists $n_3 \geq n_2$ such that w_n is monotonic and is of constant sign for $n \geq n_3$. For the sake of a contradiction assume that y_n is not bounded. Then there exists a subsequence $\{y_{n_r}\}$ such that

$$n_r \longrightarrow \infty, \quad y_{n_r} \longrightarrow \infty \quad \text{as } r \longrightarrow \infty, \tag{16}$$

and

$$y(n_r) = \max\{y_n : n_3 \le n \le n_r\}. \tag{17}$$

Since $\sigma(n) \longrightarrow \infty$ as $n \longrightarrow \infty$, we may choose r large enough so that $\sigma(n_r) \ge n_3$. For $0 < \epsilon$, because of (H3), we can find a positive integer n_4 such that $n \ge n_4 \ge n_3$ implies $|F_n| < \epsilon$. As (12) holds, then using (13), (14), and (17) we obtain

$$\begin{aligned} w_{n_r} &= y_{n_r} - \sum_{j=1}^{k} p_n^j y_{n_r-m_j} - F_{n_r} \ge \left(1 - \sum_{j=1}^{k} p_n^j\right) y_{n_r} - \epsilon \\ &> (1-p)\, y_{n_r} - \epsilon. \end{aligned} \tag{18}$$

Taking $r \longrightarrow \infty$, we find $\lim_{n\longrightarrow\infty} w_n = \infty$, a contradiction as w_n is monotonic decreasing. Hence y_n is bounded which implies w_n and z_n are bounded and $\lim_{n\longrightarrow\infty} w_n$ exists. Further it follows that $\liminf_{n\longrightarrow\infty} y_n$ and $\limsup_{n\longrightarrow\infty} y_n$ exist. We claim $\liminf_{n\longrightarrow\infty} y_n = 0$. Otherwise, let $y_n \ge \alpha > 0$. Next boundedness of y_n yields $y_n \le \beta$. Hence we have $0 \le \alpha < y_n \le \beta$, which will be used for bounding the G term in (1) from below.

From the continuity of G and assumption (H1) it follows that there exists a positive lower bound m for G on $[\alpha, \beta]$. Hence there exists n_5 such that $G(y_{\sigma(n)}) > m > 0$ for $n > n_5$. Then summing (15) from $n = n_5$ to $s - 1$ we obtain

$$w_{n_5} - w_s = \sum_{j=n_5}^{s-1} q_j G\left(y_{\sigma(j)}\right) \ge m \sum_{j=n_5}^{\infty} q_j. \tag{19}$$

Since the left hand side is the member of a bounded sequence, while the right hand side approaches $+\infty$, we have a contradiction. This yields $\liminf_{n\longrightarrow\infty} y_n = 0$. From (H3), monotonic nature of w_n and (14), it follows that $\lim_{n\longrightarrow\infty} z_n$ exists finitely. Let $\lim_{n\longrightarrow\infty} z_n = \delta$. If $\delta > 0$, then

$$\begin{aligned} 0 < \delta &= \liminf_{n\longrightarrow\infty} z_n = \liminf_{n\longrightarrow\infty} \left(y_n - \sum_{j=1}^{k} p_n^j y_{n-m_j} \right) \\ &\le \liminf_{n\longrightarrow\infty} y_n + \limsup_{n\longrightarrow\infty} \left(-\sum_{j=1}^{k} p_n^j y_{n-m_j} \right) \\ &= -\liminf_{n\longrightarrow\infty} \left(\sum_{j=1}^{k} p_n^j y_{n-m_j} \right) \le -\sum_{j=1}^{k} \liminf_{n\longrightarrow\infty} \left(p_n^j y_{n-m_j} \right) \\ &\le -\sum_{j=1}^{k} \left(\liminf_{n\longrightarrow\infty} p_n^j \right) \left(\liminf_{n\longrightarrow\infty} y_{n-m_j} \right) \le 0, \end{aligned} \tag{20}$$

a contradiction. If $\delta \le 0$ then

$$\begin{aligned} 0 \ge \delta &= \limsup_{n\longrightarrow\infty} z_n = \limsup_{n\longrightarrow\infty} \left(y_n - \sum_{j=1}^{k} p_n^j y_{n-m_j} \right) \\ &\ge \limsup_{n\longrightarrow\infty} y_n + \liminf_{n\longrightarrow\infty} \left(-\sum_{j=1}^{k} p_n^j y_{n-m_j} \right) \\ &\ge \limsup_{n\longrightarrow\infty} y_n - \limsup_{n\longrightarrow\infty} \left(\sum_{j=1}^{k} p_n^j y_{n-m_j} \right) \\ &\ge \limsup_{n\longrightarrow\infty} y_n - \sum_{j=1}^{k} \limsup_{n\longrightarrow\infty} \left(p_n^j y_{n-m_j} \right) \\ &\ge \limsup_{n\longrightarrow\infty} y_n - \sum_{j-1}^{k} \limsup_{n\longrightarrow\infty} p_n^j \limsup_{n\longrightarrow\infty} y_{n-m_j} \\ &\ge \limsup_{n\longrightarrow\infty} y_n \left(1 - \sum_{j=1}^{k} \limsup_{n\longrightarrow\infty} p_n^j \right) \\ &\ge \limsup_{n\longrightarrow\infty} y_n (1-p). \end{aligned} \tag{21}$$

Hence $\limsup_{n\longrightarrow\infty} y_n \le 0$, by (12), which implies the desired result $\lim_{n\longrightarrow\infty} y_n = 0$. If $y_n < 0$ for $n > n_1$ then proceeding as above we can arrive at $\lim_{n\longrightarrow\infty} y_n = 0$. Thus the theorem is proved. □

Theorem 2. *Suppose that (H1)–(H4) hold. Assume that there exists a positive constant p such that the sequences $\{p_n^j\}$ for $j = 1, 2, \ldots, k$ satisfy the condition*

$$\begin{aligned} &p_n^j \le 0, \quad \text{for every } j = 1, 2, \ldots, k \\ &\text{and } \sum_{j=1}^{k} \liminf_{n\longrightarrow\infty} p_n^j > -p > -1. \end{aligned} \tag{22}$$

Then every solution of (1) oscillates or tends to zero as $n \longrightarrow \infty$.

Proof. Proceeding as in the proof of Theorem 1 and setting z_n, w_n as in (13) and (14), respectively, we obtain (15) and further prove y_n is bounded with $\liminf_{n\longrightarrow\infty} y_n = 0$. From (H3) and the fact that w_n is monotonic it follows that $\lim_{n\longrightarrow\infty} w_n =$

$\lim_{n\longrightarrow\infty} z_n = \delta \in \mathbb{R}$. As $z_n \geq 0$, so $\delta \geq 0$. We claim $\delta = 0$; if not then $\delta > 0$, and this implies

$$\begin{aligned}
\delta &= \liminf_{n\longrightarrow\infty} z_n = \liminf_{n\longrightarrow\infty}\left(y_n - \sum_{j=1}^{k} p_n^j y_{n-m_j}\right) \\
&\leq \liminf_{n\longrightarrow\infty} y_n + \limsup_{n\longrightarrow\infty}\left(-\sum_{j=1}^{k} p_n^j y_{n-m_j}\right) \\
&\leq \sum_{j=1}^{k} \limsup_{n\longrightarrow\infty}\left(-p_n^j\right) \limsup_{n\longrightarrow\infty}\left(y_{n-m_j}\right) \\
&= \sum_{j=1}^{k} - \liminf_{n\longrightarrow\infty}\left(p_n^j\right) \limsup_{n\longrightarrow\infty}\left(y_{n-m_j}\right) \\
&\leq p\limsup_{n\longrightarrow\infty}\left(y_n\right) \leq p\alpha.
\end{aligned} \tag{23}$$

Hence we get

$$\alpha \geq \frac{\delta}{p} > \delta. \tag{24}$$

Again

$$\begin{aligned}
\delta &= \limsup_{n\longrightarrow\infty} z_n = \limsup_{n\longrightarrow\infty}\left(y_n - \sum_{j=1}^{k} p_n^j y_{n-m_j}\right) \\
&\geq \limsup_{n\longrightarrow\infty} y_n + \liminf_{n\longrightarrow\infty}\left(-\sum_{j=1}^{k} p_n^j y_{n-m_j}\right) \\
&= \limsup_{n\longrightarrow\infty} y_n + \liminf_{n\longrightarrow\infty}\left(\sum_{j=1}^{k}\left(-p_n^j\right) y_{n-m_j}\right) \\
&\geq \limsup_{n\longrightarrow\infty} y_n + \sum_{j=1}^{k}\liminf_{n\longrightarrow\infty}\left(\left(-p_n^j\right) y_{n-m_j}\right) \\
&\geq \limsup_{n\longrightarrow\infty} y_n + \sum_{j=1}^{k}\liminf_{n\longrightarrow\infty}\left(-p_n^j\right)\liminf_{n\longrightarrow\infty} y_{n-m_j} \\
&= \limsup_{n\longrightarrow\infty} y_n = \alpha,
\end{aligned} \tag{25}$$

a contradiction, due to inequality (24). Hence we conclude $\delta - 0$ and from $z_n > y_n$, it follows that $\lim_{n\longrightarrow\infty} y_n \leq 0$. Hence $\lim_{n\longrightarrow\infty} y_n = 0$.

The proof for the case $y_n < 0$ for large n is similar. Hence the theorem is proved.

Remark 3. Theorems 1 and 2 hold good for $k = 0$ and $k = 1$. Hence these results could be compared with results concerned with the difference equations (4), (5), (6), and (7).

Theorem 4. *Suppose that (H1)-(H4) hold. Assume that there exists a positive constant p such that the sequences $\{p_n^j\}$ for $j = 1, 2, \ldots, k$ satisfy the condition*

$$p_n^j < 0 \quad \text{for every } j = 1, 2, \ldots, k \text{ and there exists, } i \in \{1, 2, 3, \ldots, k\} \text{ such that } \limsup_{n\longrightarrow\infty} p_n^i - \sum_{j\neq i}\liminf_{n\longrightarrow\infty} p_n^j < -1. \tag{26}$$

Then every solution of (1) oscillates or tends to zero as $n \longrightarrow \infty$.

Proof. Proceeding as in the proof of Theorem 1 and setting z_n, w_n as in (13) and (14), respectively, we obtain (15) and further prove y_n is bounded with $\liminf_{n\longrightarrow\infty} y_n = 0$. From (H3) and that w_n is monotonic it follows that $\lim_{n\longrightarrow\infty} w_n = \lim_{n\longrightarrow\infty} z_n = \delta \in \mathbb{R}$. As $z_n \geq 0$, so $\delta \geq 0$. We claim $\delta = 0$. If not, then $\delta > 0$, and this implies

$$\begin{aligned}
\delta &= \liminf_{n\longrightarrow\infty} z_n = \liminf_{n\longrightarrow\infty}\left(y_n - \sum_{j=1}^{k} p_n^j y_{n-m_j}\right) \\
&\leq \limsup_{n\longrightarrow\infty}\left(y_n + \sum_{j\neq i} - p_n^j y_{n-m_j}\right) \\
&\quad + \liminf_{n\longrightarrow\infty}\left(-p_n^i y_{n-m_i}\right) \\
&\leq \limsup_{n\longrightarrow\infty} y_n + \limsup_{n\longrightarrow\infty}\sum_{j\neq i} - p_n^j y_{n-m_j} \\
&\quad + \limsup_{n\longrightarrow\infty}\left(-p_n^j\right)\liminf_{n\longrightarrow\infty}\left(y_{n-m_j}\right) \\
&\leq \limsup_{n\longrightarrow\infty} y_n + \sum_{j\neq i1}\limsup_{n\longrightarrow\infty}\left(-p_n^j\right)\limsup_{n\longrightarrow\infty}\left(y_{n-m_j}\right) \\
&\leq \limsup_{n\longrightarrow\infty}\left(y_n\right)\left[1 - \sum_{j\neq i}\liminf_{n\longrightarrow\infty} p_n^j\right].
\end{aligned} \tag{27}$$

Again we have

$$\begin{aligned}
\delta &= \limsup_{n\longrightarrow\infty} z_n = \limsup_{n\longrightarrow\infty}\left(y_n - \sum_{j=1}^{k} p_n^j y_{n-m_j}\right) \\
&\geq \liminf_{n\longrightarrow\infty} y_n + \limsup_{n\longrightarrow\infty}\left(-\sum_{j=1}^{k} p_n^j y_{n-m_j}\right) \\
&= \limsup_{n\longrightarrow\infty}\left(-p_n^i y_{n-mi}\right) + \liminf_{n\longrightarrow\infty}\sum_{j\neq i}\left(-p_n^j y_{n-m_j}\right) \\
&\geq \limsup_{n\longrightarrow\infty} y_{n-m_j}\liminf_{n\longrightarrow\infty}\left(-p_n^i\right) \\
&\quad + \sum_{j\neq i}\liminf_{n\longrightarrow\infty}\left(\left(-p_n^j\right) y_{n-m_j}\right)
\end{aligned}$$

$$\geq \limsup_{n\to\infty} y_n \left(-\limsup_{n\to\infty} p_n^i\right) + \sum_{j\neq i} \liminf_{n\to\infty} \left(-p_n^j\right) \liminf_{n\to\infty} y_{n-m_j} \geq \limsup_{n\to\infty} y_n \left(-\limsup_{n\to\infty} p_n^i\right). \quad (28)$$

From (27) and (28), it follows that

$$\limsup_{n\to\infty} y_n \left(\left(\sum_{j\neq i} \liminf_{n\to\infty} p_n^j\right) - 1 - \limsup_{n\to\infty} p_n^i\right) \leq 0 \quad (29)$$

Using (26), we obtain $\lim_{n\to\infty} y_n = 0$. Thus the theorem is proved.

Next, we intend to present a result where p_n^j, $j = 1, 2, 3, \ldots, k$, satisfy the following condition:

$$p_n^j > 0 \quad \text{for every } j = 1, 2, \ldots, k \text{ and there exists, } i \in \{1, 2, 3, \ldots, k\} \text{ such that } \liminf_{n\to\infty} p_n^i - \sum_{j\neq i} \limsup_{n\to\infty} p_n^j > 1. \quad (30)$$

For that purpose we give an example which would lead us to our next result.

Example 5. Consider the first-order neutral delay difference equation with several delays and variable coefficients

$$\Delta\left[y_n - \left(1 + 2^{-n}\right) y_{n-1} - \left(4 + 2^{-n}\right) y_{n-2}\right] + 2^{(2n+1)/3} y_{n-4} = 0. \quad (31)$$

Note that p_n^j satisfy (30) for the above neutral delay difference equation (31). This neutral delay difference equation has an unbounded solution $y_n = 2^n$ tending to ∞ as $n \longrightarrow \infty$ unlike other results presented so far.

The above example is the motivating point to the statement of our next result. Since the proof is almost similar to that of Theorem 4, it is omitted.

Theorem 6. *Suppose that (H1)–(H4) hold. Assume that there exists a positive constant p such that the sequences* $\{p_n^j\}$ *for* $j = 1, 2, \ldots, k$ *satisfy the condition (30). Then every bounded solution of (1) oscillates or tends to zero as* $n \longrightarrow \infty$.

Remark 7. The above Theorems 4 and 6 hold for $k = 1$ but not for $k = 0$. Hence these results can be compared with results concerned with neutral delay difference equations (4) and (5).

Few examples are noted below to illustrate our results and establish its significance.

Example 8. Consider the first-order neutral delay difference equation

$$\Delta\left(y_n - p_n^1 y_{n-1} - p_n^2 y_{n-4}\right) + \frac{97}{60 y_{n-2}^3} = 0, \quad n \geq 5, \quad (32)$$

where

$$p_n^1 = \begin{cases} \dfrac{1}{4}, & \text{if } n \text{ is odd,} \\ \dfrac{1}{5}, & \text{if } n \text{ is even,} \end{cases} \quad (33)$$

and

$$p_n^2 = \begin{cases} \dfrac{1}{3}, & \text{if } n \text{ is odd,} \\ \dfrac{1}{2}, & \text{if } n \text{ is even.} \end{cases} \quad (34)$$

The neutral delay difference equation (32) satisfies all the conditions of Theorem 1. As such, it has an oscillatory solution $y_n = (-1)^n$.

Example 9. Consider the first-order inhomogeneous neutral delay difference equation

$$\Delta\left(y_n + p_n^1 y_{n-4} + p_n^2 y_{n-5}\right) + \frac{3}{4} 2^{n-4} y_{n-3}^3 = -\frac{3}{2^{n+1}} - \frac{3}{2^{2n-2}}, \quad n \geq 5, \quad (35)$$

where $p_n^1 = 2^{-n} + 1/16$ and $p_n^2 = 2^{-n} + 1/32$. This neutral delay difference equation satisfies all the conditions of Theorem 2. As such, it has a bounded positive solution $y_n = 2^{-n}$ tending to zero as $n \longrightarrow \infty$. Note that, no result in the papers cited under reference can be applied to the neutral delay difference equations (32) and (35).

Remark 10. Results of [3, 9] cannot be applied to the delay difference equation (9), because the condition (10) is not satisfied. However, due to Remark 3, Theorem 1 can be applied to the delay equation (9) as all the conditions are satisfied and as such the delay equation has a positive bounded solution e^{-n} tending to zero as $n \longrightarrow \infty$. Thus our work complements the work in [3, 9]. Further, since we do not assume G is nondecreasing, our Theorems 1, 2, 4, and 6 improve and generalize the results in [7].

3. Necessary Conditions

In this section we show that (H4) is necessary for every solution of (1) to be oscillatory or tending to zero as $n \longrightarrow \infty$. For this, we need the following lemma.

Lemma 11 (Krasnoselskii's fixed point theorem [13]). *Let X be a Banach space and S be a bounded closed convex subset of*

X. Let A, B be operators from S to X such that $Ax + By \in S$ for every pair of $x, y \in S$. If A is a contraction and B is completely continuous then the equation

$$Ax + Bx = x \tag{36}$$

has a solution in S.

Theorem 12. *Assume that (H2) holds. Further, assume that one of the conditions of (12) and (22) hold. Then (H4) is a necessary condition for all solution of (1) to be oscillatory or tending to zero as $n \longrightarrow \infty$.*

Proof. Suppose the condition (12) holds. The proof for the case when (22) holds would follow on similar lines. Assume for the sake of contradiction that (H4) does not hold. Hence

$$\sum_{n=n_0}^{\infty} q_n < \infty. \tag{37}$$

Thus, all we need to show is the existence of a bounded solution y_n of (1) with $\liminf_{n\longrightarrow\infty} y_n > 0$. From (H2), we find a positive constant c and a positive integer $n_1 > n_0 > 0$ such that

$$|F_n| < c \quad for\ n \ge n_1. \tag{38}$$

Choose a positive constant L such that $L \ge 5c/1 - p$. Since $G \in C(\mathbb{R}, \mathbb{R})$, let

$$\mu = \max\{|G(x)| : c \le x \le L\}. \tag{39}$$

Let

$$\eta = \max\{m_1, m_2, \dots, m_k\}. \tag{40}$$

Then using (37) one can fix $n_2 > n_1$ such that for $n \ge n_2$ it follows that

$$\mu \sum_{i=n}^{\infty} q_i < c. \tag{41}$$

Choose $N_1 > n_2$ such that

$$N_0 = \min\{\sigma(N_1), N_1 - \eta\}. \tag{42}$$

Let $X = \ell_{\infty}^{N_0}$, Banach space of real bounded sequences $x = \{x_n\}$ with $x_1 = x_2 = \cdots = x_{N_0}$ and supremum norm

$$\|x\| = \sup(|x_n| : n \ge N_0). \tag{43}$$

Define

$$S = \{y \in X : c \le y_n \le L, n \ge N_0\}. \tag{44}$$

Clearly S is a bounded closed and convex subset of X. Now we define two operators A and $B : S \longrightarrow X$ as follows. For $y \in S$, define

$$(Ay)_n = \begin{cases} (Ay)_{N_1}, & N_0 \le n \le N_1 \\ \sum_{j=1}^{k} p_n^j y_{n-m_j} + F_n + 3c, & n \ge N_1. \end{cases}$$

$$(By)_n = \begin{cases} (By)_{N_1}, & N_0 \le n \le N_1 \\ \sum_{i=n}^{\infty} q_i G(y_{\sigma(i)}), & n \ge N_1. \end{cases} \tag{45}$$

First we show that if $x, y \in S$ then $Ax + By \in S$. Hence, for $x = \{x_n\}$ and $y = \{y_n\} \in S$ and for $n \ge N_1$ we obtain

$$\begin{aligned}(Ax)_n + (By)_n &\le \sum_{j=1}^{k} p_n^j x_{n-m_j} + 3c + \sum_{i=n}^{\infty} q_i |G(y_{\sigma(i)})| \\ &+ |F_n| \le pL + 5c \le L.\end{aligned} \tag{46}$$

On the other hand

$$\begin{aligned}(Ax)_n + (By)_n &\ge 3c - \sum_{i=n}^{\infty} q_i |G(y_{\sigma(i)})| - |F_n| \\ &\ge 3c - c - c \ge c.\end{aligned} \tag{47}$$

Hence

$$c \le (Ax)_n + (By)_n \le L \quad \text{for } n \ge N_1. \tag{48}$$

Thus, we proved that $Ax + By \in S$ for any $x, y \in S$. Next we show that A is a contraction on S. In fact for $x, y \in S$ and $n \ge N_1$ we have

$$\begin{aligned}\|(Ax)_n - (Ay)_n\| &\le \sum_{j=1}^{k} |p_n^j| |x_{n-m_j} - y_{n-m_j}| \\ &\le p\|x - y\|\end{aligned} \tag{49}$$

This implies A is a contraction because $0 < p < 1$. Next we show that B is completely continuous. For this as a first step we show that B is continuous. Suppose the sequence $x^l \equiv \{x_n^l\} \longrightarrow x \equiv \{x_n\}$ in S as $l \longrightarrow \infty$ (with l taken from the index set). Since S is closed then $x \in S$. For $n \ge N_1$ we have

$$|(Bx^l)_n - (Bx)_n| \le \sum_{i=n}^{\infty} q_i |G(x_{\sigma(i)}^l) - G(x_{\sigma(i)})| \tag{50}$$

Since G is continuous, therefore $|G(x_{\sigma(i)}^l) - G(x_{\sigma(i)})| \longrightarrow 0$ as $l \longrightarrow \infty$. Hence B is continuous. Next what remained to show is BS is relatively compact. Using the result [14, Theorem 3.3], we need only show that BS is uniformly cauchy. Let $x \equiv \{x_n\}$ be a sequence in S. From (H2) and (37), it follows that, for $\epsilon > 0$, there exists $N^* \ge N_1$ such that, for $n \ge N^*$,

$$\sum_{i=n}^{\infty} q_i |G(x_{\sigma(i)})| < \frac{\epsilon}{2}. \tag{51}$$

Then for $n_3 > n_4 \ge N^*$ we have

$$|(Bx)_{n_3} - (Bx)_{n_4}| < \epsilon. \tag{52}$$

Thus BS is uniformly cauchy. Hence it is relatively compact. Then by Lemma 11, we can find x^0 in S such that $Ax^0 + Bx^0 = x^0$. Clearly, $(x^0)_n$ is a bounded, positive solution of (1) with limit infimum greater than or equal to $c > 0$. Thus the theorem is proved.

Theorem 13. *Assume that (H2) holds. Further assume that one of the conditions of (26) and (30) holds. Then (H4) is a necessary condition for all solution of (1) to be oscillatory or tending to zero as $n \longrightarrow \infty$.*

Proof. Suppose that p_n satisfies (30). The proof for the case when (26) holds is similar. Assume for the sake of contradiction that (H4) does not hold. Hence (37) holds. Thus, all we need to show is the existence of a bounded solution y_n of (1) with $\liminf_{n\longrightarrow\infty} y_n > 0$. From (H2), we find a positive constant L and a positive integer $n_1 > n_0 > 0$ such that

$$|F_n| < L \quad \text{for } n \geq n_1. \tag{53}$$

By (30), we can find a small positive real ϵ, a lower bound c for p_n^i, and upper bounds d_j for p_n^j ($j \neq i$ and $1 \leq j \leq k$) such that $c - \sum d_j - 1 = \epsilon$. Let $\sum d_j = d$. Hence $c = d + 1 + \epsilon$. Next choose an upper bound b for p_n^i such that $b < (c^2 - c)/d$. The nonexistence of such an upper bound for p_n^i would lead to the fact that, for all $\delta > 0, b = c + \delta$ and $b \geq (c^2 - c)/d$. Taking $\delta = \epsilon$, we have $\epsilon^2 + \epsilon \leq 0$, a contradiction. Choose a real λ as follows:

$$0 < \lambda = \frac{(L+\epsilon)\, d + (c-1)(c+L+\epsilon)}{c^2 - (bd + c)}. \tag{54}$$

Let

$$H = \frac{b\lambda + L + \epsilon}{c - 1}. \tag{55}$$

From (54) and (55) it follows that

$$\lambda - \frac{Hd + L + \epsilon}{c} = 1. \tag{56}$$

Since $G \in C(\mathbb{R}, \mathbb{R})$, let

$$\mu = \max\{|G(x)| : 1 \leq x \leq H\}. \tag{57}$$

Let $\eta = \max\{m_1, m_2, \ldots, m_k\}$. Then using (37), one can fix $n_2 > n_1$ such that for $n \geq n_2$ it follows that

$$\mu \sum_{i=n}^{\infty} q_i < \epsilon. \tag{58}$$

Choose $N_1 > n_2$ such that

$$N_0 = \min\{\sigma(N_1), N_1 - \eta\}. \tag{59}$$

Let $X = \ell_\infty^{N_0}$, Banach space of real bounded sequences $x = \{x_n\}$ with $x_1 = x_2 = \cdots = x_{N_0}$ and supremum norm

$$\|x\| = \sup(|x_n| : n \geq N_0). \tag{60}$$

Define

$$S = \{y \in X : 1 \leq y_n \leq H, n \geq N_0\}. \tag{61}$$

Clearly S is a bounded closed and convex subset of X. Now we define two operators A and $B : S \longrightarrow X$ as follows. For $y \in S$, define

$$(Ay)_n = \begin{cases} (Ay)_{N_1}, & N_0 \leq n \leq N_1 \\ \dfrac{y_{n+m_i}}{p_{n+m_i}^i} - \dfrac{\sum_{j\neq i} p_{n+m_i}^j y_{n-m_j+m_i}}{p_{n+m_i}^i} + \dfrac{b\lambda}{p_{n+m_i}^i} - \dfrac{F_{n+m_i}}{p_{n+m_i}^i}, & n \geq N_1. \end{cases}$$

$$(By)_n = \begin{cases} (By)_{N_1}, & N_0 \leq n \leq N_1 \\ \dfrac{-\sum_{j=n+m_i}^{\infty} q_j G(y_{\sigma(j)})}{p_{n+m_i}^i}, & n \geq N_1. \end{cases} \tag{62}$$

Proceeding as in the proof of above theorem we show that (i) if $x, y \in S$ then $Ax + By > 1$ by (56) and $Ax + By < H$ by (55), so that $Ax + By \in S$, (ii) $\|Ax_n - Ay_n\| < [(d+1)/c]\|x_n - y_n\|$, hence A is a contraction on S, and (iii) B is completely continuous. This completes the proof of the theorem.

Remark 14. For the results in this section, we assume none of conditions (H3), G is nondecreasing, and $xG(x) > 0$, whereas the authors [7, 8] assumed these three conditions in their corresponding results. Hence the results of this article generalize and improve the corresponding results of these papers.

Combining all the above results, i.e., Theorems 1, 2, 4, 6, 12, and 13, we obtain the following theorem.

Theorem 15. *Suppose that (H1)-(H3) hold. Assume p_n^j in (1) to satisfy one of the four conditions (12), (22), (26), and (30). Then (H4) is both necessary and sufficient condition for every solution of (1) to be oscillatory or tending to zero as $n \longrightarrow \infty$.*

Remark 16. The results of this work hold for $G(x) = x$ and $f(x) = 0$, i.e., for the linear homogeneous equation associated with (1).

Disclosure

This work is done for the Ph.D. thesis work of the second author.

Conflicts of Interest

The authors declare that they have no conflicts of interest.

Acknowledgments

The authors are thankful to Professor Prayag Prasad Mishra for his valuable guidance during the completion of this paper.

References

[1] I. Gyori and G. Ladas, *Oscillation Theory of Delay-Differential Equations with Applications*, Clarendon Press, Oxford, UK, 1991, https://global.oup.com/academic/product/oscillation-theory-of-delay-differential-equations-9780198535829?lang=en&cc=es.

[2] G. E. Chatzarakis and G. N. Miliaras, "Asymptotic behavior in neutral difference equations with variable coefficients and more than one delay arguments," *Journal of Mathematical and Computational Science*, vol. 1, no. 1, pp. 32–52, 2011, http://scik.org/index.php/jmcs/article/view/16.

[3] R. Koplatadze and S. Pinelas, "Oscillation Criteria for the First-Order Linear Difference Equations with Several Delay Arguments," *Journal of Mathematical Sciences*, vol. 208, no. 5, pp. 571–592, 2015.

[4] R. Koplatadze, "On asymptotic behavior of solutions of first order difference equations with several delay," *Bulletin. Tbilisi International Centre of Mathematics and Informatics*, vol. 21, no. 2, pp. 117–123, 2017, https://www.emis.de/journals/TICMI/vol21_2/p117-121.pdf.

[5] R. P. Agarwal, *Difference Equations and Inequalities*, Marcel Dekker, New York, NY. USA, 2nd edition, 2000, https://doc.lagout.org/science/0_Computer%20Science/3_Theory/Mathematics/Difference%20Equations%20and%20Inequalities%20-%20Theory.pdf.

[6] R. N. Rath, N. Misra, and S. K. Rath, "Sufficient conditions for oscillatory behaviour of a first order neutral difference equation with oscillating coefficients," *Acta Mathematica Academiae Paedagogicae Nyiregyhaziensis*, vol. 25, no. 1, pp. 55–63, 2009, https://www.emis.de/journals/AMAPN/vol25_1/7.html.

[7] N. Parhi and A. K. Tripathy, "Oscillation of forced nonlinear neutral delay difference equations of first order," *Czechoslovak Mathematical Journal*, vol. 53, no. 1, pp. 83–101, 2003.

[8] N. Parhi and R. N. Rath, "Oscillation criteria for forced first order neutral differential equations with variable coefficients," *Journal of Mathematical Analysis and Applications*, vol. 256, no. 2, pp. 525–541, 2001.

[9] G. E. Chatzarakis, R. Koplatadze, and I. P. Stavroulakis, "Optimal oscillation criteria for first order difference equations with delay argument," *Pacific Journal of Mathematics*, vol. 235, no. 1, pp. 15–33, 2008, https://msp.org/pjm/2008/235-1/pjm-v235-n1-p02-p.pdf.

[10] R. N. Rath, S. Padhi, and B. L. Barik, "Oscillatory and asymptotic behaviour of a homogeneous neutral delay difference equation of second order," *Bulletin of the Institute of Mathematics, Academia Sinica*, vol. 3, no. 3, pp. 453–467, 2008, https://web.math.sinica.edu.tw/bulletin_ns/20083/2008310.pdf.

[11] E. Thandapani, R. Arul, and P. S. Raja, "Oscillatory and asymptotic behavior of solutions of non-homogeneous neutral difference equations," *Studies of the University of Zilina Mathematical Series*, vol. 15, no. 1, pp. 67–82, 2002, https://frcatel.fri.uniza.sk/studies/.

[12] E. Thandapani, R. Arul, and P. . Raja, "Oscillation of first order neutral delay difference equations," *Applied Mathematics E-Notes*, vol. 3, pp. 88–94, 2003, https://www.emis.de/journals/AMEN/.

[13] L. H. Erbe, Q. K. Kong, and B. G. Zhang, *Oscillation Theory for Functional Differential Equations*, Marcel Dekker, New York, NY, USA, 1995, https://www.hindawi.com/journals/aaa/2011/591254/ref/.

[14] S. S. Cheng and W. T. Patula, "An existence theorem for a nonlinear difference equation," *Nonlinear Analysis. Theory, Methods & Applications*, vol. 20, no. 3, pp. 193–203, 1993.

Optimal Form for Compliance of Membrane Boundary Shift in Nonlinear Case

Y. Yousfi,[1] **I. Hadi**,[2] **and A. Benbrik**[1]

[1]*Department of Mathematics, Faculty of Science, University Mohammed 1st, Oujda, Morocco*
[2]*Moroccan Ministry of National Education, Higher Education and Research, Morocco*

Correspondence should be addressed to Y. Yousfi; yousfi.98@gmail.com

Academic Editor: Vladimir V. Mityushev

In this work, we search the existence shifting compliance optimal form of some boundary membrane, which is not elastic and not isotropic, generating nonlinear PDE. An optimal form of the elastic membrane described by the p-Laplacian is investigated. The boundary perturbation method due to Hadamard is applied in Sobolev spaces.

1. Introduction and Preliminaries

In this work we will study the geometric shape optimization of forms, where the main idea is to vary the edge position of a form, without changing its topology which remains the same. We use a membrane model as shown in Figure 1. At rest the membrane occupies a reference domain Ω whose edge is divided into three disjoint parts:

$$\partial\Omega = \Gamma \cup \Gamma_N \cup \Gamma_D \tag{1}$$

where Γ is the free variable part, Γ_D is the fixed part of the boundary (Dirichlet boundary conditions), and Γ_N is also the free part of the boundary on which we apply the efforts $g \in L^p(\Gamma_N)^N$ (Neumann boundary condition). The three of parts of the boundary are supposed to be nonzero surface measurements, as we suppose that the free boundary variable Γ responds to homogenous Neumann condition. So the vertical displacement u is the solution of the following membrane model:

$$\begin{aligned} -\Delta u &= 0 \quad in\ \Omega \\ u &= 0 \quad on\ \Gamma_D \\ \frac{\partial u}{\partial n} &= g \quad on\ \Gamma_N \\ \frac{\partial u}{\partial n} &= 0 \quad on\ \Gamma \end{aligned} \tag{2}$$

We want to minimize the compliance defined by $J(\Omega) = \int_{\Gamma_N} gudx$ whenever $x \in \Omega$.

The shape optimization problem is $\inf_{\Omega\in U_{ad}} J(\Omega)$ where it remains to define the set of admissible forms.

1.1. Existence under a Condition of Regularity. The main idea of this section is to apply a regularity constraint on all the admissible forms U_{ad}, to demonstrate a result of existence of optimal forms. The results and demonstrations are mainly due to F. Murat and J. Simons [1, 2]. It rests on a very significant restriction of U_{ad}; in other words, Ω is obtained by applying a regular diffeomorphism T to the reference domain Ω_0. We first define a diffeomorphism set:

$$\begin{aligned} \tau = \Big\{T \text{ such that } &(T - Id) \\ &\in W^{1,\infty}\left(\mathbb{R}^N, \mathbb{R}^N\right); \left(T^{-1} - Id\right) \in W^{1,\infty}\left(\mathbb{R}^N, \mathbb{R}^N\right) \end{aligned} \tag{3}$$

Then we define a set of the admissible forms obtained by deformation of Ω:

$$C(\Omega_0) = \{\Omega \text{ such that } \exists T \in \tau; \Omega = T(\Omega_0)\} \tag{4}$$

Finally we introduce a pseudo-distance on $C(\Omega_0)$:

$$\begin{aligned} d(\Omega_1, \Omega_2) = \inf_{T\in\tau/T(\Omega_1)=\Omega_0} \Big(&\|(T - Id)\|_{W^{1,\infty}(\mathbb{R}^N,\mathbb{R}^N)} \\ &+ \left\|\left(T^{-1} - Id\right)\right\|_{W^{1,\infty}(\mathbb{R}^N,\mathbb{R}^N)}\Big) \end{aligned} \tag{5}$$

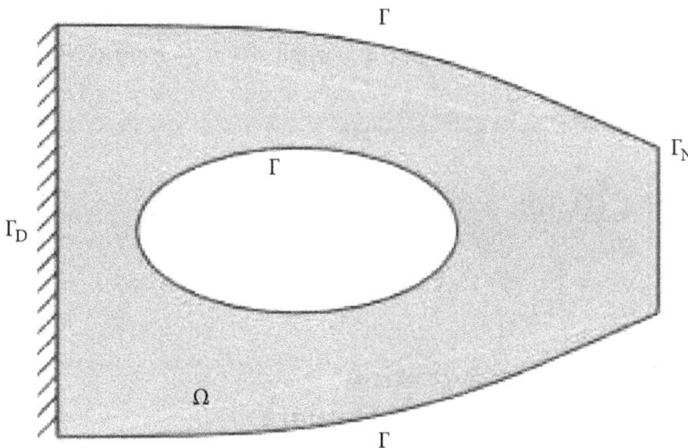

FIGURE 1: Membrane.

Thus, we introduce a condition of uniform regularity of the permissible forms, that is to say, Ω open sets close to Ω_0 in the sense of pseudo-distance; for each $R > 0$ we pose $U_{ad} = \{\Omega \in C(\Omega_0)$ such that $d(\Omega, \Omega_0) \leq R; \Gamma_D \cup \Gamma_N \subset \partial\Omega$ and $\int_\Omega dx = V_0\}$ where V_0 is an imposed volume. The result is the following theorem.

Theorem 1. *For all objective functions, the shape optimization problem* $\inf_{\Omega \in U_{ad}} J(\Omega)$ *admits at least a minimum point.*

1.2. Derivation from the Domain. The boundary variation method that we study is a classical idea well known and used before by Hadamard [3] in 1907 and many others as [4–12]. We will adopt the same representation as F. Murat and J. Simons [1]. In fact, let Ω_0 be an open regular bounded referential domain of $\mathbb{R}^N$ and the admissible form class $C(\Omega_0)$ composed of the open sets such as $\Omega = T(\Omega_0)$ where T is a Lipschitz diffeomorphism and

$$T = Id + \theta$$
$$\text{with } \theta \in W^{1,\infty}\left(\mathbb{R}^N, \mathbb{R}^N\right) \tag{6}$$

where Id is the identity application, and we note $\Omega = (Id + \theta)(\Omega_0)$ defined by

$$\Omega = \{x + \theta(x) \ \ such\ that\ \ x \in \Omega_0\} \tag{7}$$

Lemma 2 (See [13]). *For all* $\theta \in W^{1,\infty}(\mathbb{R}^N, \mathbb{R}^N)$ *satisfying* $\|\theta\|_{W^{1,\infty}(\mathbb{R}^N,\mathbb{R}^N)} < 1$, *the application* $T = Id + \theta$ *is a bijection of* $\mathbb{R}^N$ *and* $T \in \tau$.

Definition 3. Let J be application from $C(\Omega_0)$ to $\mathbb{R}$. One says that it is differentiable with respect to the domain at Ω_0 if the function

$$\Theta \longrightarrow J\left((Id + \theta)(\Omega_0)\right) \tag{8}$$

is Frechet differentiable at 0 in the Banach space $W^{1,\infty}(\mathbb{R}^N, \mathbb{R}^N)$. i.e., $\exists L$, a linear continuous form on $W^{1,\infty}(\mathbb{R}^N, \mathbb{R}^N)$, such that

$$J\left((Id + \theta)(\Omega_0)\right) = J(\Omega_0) + J'(\Omega_0) + \circ(\theta),$$
$$\text{with } \lim_{\theta \longrightarrow O} \frac{\circ(\theta)}{\|\theta\|_{W^{1,\infty}}} = 0. \tag{9}$$

The linear form $J'(\Omega_0)$ depends only on the normal component of θ on the boundary of Ω_0.

Proposition 4. *Let* Ω_0 *be a regular bounded open set of* $\mathbb{R}^N$. *Let* J *be a differentiable application on* Ω_0. *If* $\theta_1, \theta 2 \in W^{1,\infty}(\mathbb{R}^N, \mathbb{R}^N)$ *are such that* $\theta_1 - \theta_2 \in C^1(\mathbb{R}^N, \mathbb{R}^N)$ *such that* $\theta_1.n = \theta_2.n$ *on* $\partial\Omega_0$, *then the derivative* $J'(\Omega_0)$ *is verifying:*

$$J'(\Omega_0)(\theta_1) = J'(\Omega_0)(\theta_2) \tag{10}$$

1.3. Derivation of Integrals. Since the compliance J is defined by surface or volume integrals then its differentiation devotes the following tools.

Lemma 5 (see [1, 7]). *Let* Ω_0 *be an open set of* $\mathbb{R}^N$. *Let* $T \in \tau$ *and* $1 \leq p \leq +\infty$. *Then* $f \in L^p(T(\Omega_0))$ *iff* $foT \in L^p(\Omega_0)$ *and one has*

$$\int_{T(\Omega_0)} f\,dx = \int_{\Omega_0} foT\,|\det \nabla T|\,dx$$
$$\text{and } \int_{T(\Omega_0)} f\left|\det(\nabla T)^{-1}\right| dx = \int_{\Omega_0} foTdx \tag{11}$$

On the other hand $f \in W^{1,p}(T(\Omega_0))$ *iff* $foT \in W^{1,p}(T(\Omega_0))$ *and one has*

$$(\nabla f)\,oT = \left((\nabla T)^{-1}\right)^t (\nabla foT) \tag{12}$$

Proposition 6 (See [13]). *Let* Ω_0 *be a regular bounded open set of* $\mathbb{R}^N$. *Let* $f \in W^{1,1}(\mathbb{R}^N)$ *and* J *be an application from* $C(\Omega_0)$ *to* $\mathbb{R}$ *defined by* $J(\Omega) = \int_\Omega f(x)dx$. *Then* J *is differentiable in* Ω_0 *and*

$$J'(\Omega_0)(\theta) = \int_{\Omega_0} \operatorname{div}(\theta(x) f(x))\,dx$$
$$= \int_{\partial\Omega_0} \theta(x)\,n(x)\,f(x)\,dS \tag{13}$$
$$\text{with } \theta \in W^{1,\infty}\left(\mathbb{R}^N, \mathbb{R}^N\right)$$

Now we move to a lemma on the change of variables in surfaces integrals.

Lemma 7 (See [1, 7]). *Let* Ω_0 *be an open set of* $\mathbb{R}^N$. *Let* $T \in \tau$ *be a class* $\mathscr{C}^1$ *diffeomorphism of* $\mathbb{R}^N$, *and* $1 \leq p \leq +\infty$. *Then* $f \in L^p(T(\Omega_0))$ *iff* $foT \in L^p(\Omega_0)$ *and one has*

$$\int_{\partial T\Omega_0} fdS = \int_{\partial\Omega_0} f \circ T\left|\det(\nabla T)^{-1}\right| \left|\left((\nabla T)^{-1}\right)^t n\right|_{\mathbb{R}^N} dS \tag{14}$$

where n *is the external normal to* $\partial\Omega_0$.

The surface integral derivative of a function with respect to the domain is given by the following proposition.

Proposition 8 (See [13]). *Let Ω_0 be a regular bounded open set of $\mathbb{R}^N$. Let $f \in W^{2,1}(\mathbb{R}^N)$ and J be an application from $C(\Omega_0)$ to $\mathbb{R}$ defined by $J(\Omega) = \int_\Omega f(x)dx$. Then J is differentiable in Ω_0 and*

$$J'(\Omega_0)(\theta) = \int_{\partial\Omega_0} (\nabla f.\theta + f(\operatorname{div}(\theta) - \nabla\theta n.n)) = \int_{\partial\Omega_0} \theta.n\left(\frac{\partial f}{\partial n} + Hf\right)dS \quad (15)$$

with $\theta \in \mathscr{C}^1(\mathbb{R}^N, \mathbb{R}^N)$

where H is the average curvature of $\partial\Omega_0$ defined by $H = divn$.

1.4. Derivation of a Domain Dependent Function. In this section we try to derive a function depending on the domain; for this we use the Eulerian U or Lagrangian L derivative. The second is a more reliable concept than the first. Let $u(x, \Omega)$ be a function defined for all $x \in \Omega$ and depending on Ω. It represents a solution of an PDE posed in Ω. In a point x belonging to both Ω_0 and $\Omega = (Id + \theta)(\Omega_0)$, we can calculate the differential $u(\Omega, x)$:

$$u((Id+\theta)(\Omega_0), x) = u(\Omega_0, x) + U(\theta, x) + o(\theta)$$
$$\text{such that } \lim_{\theta \mapsto 0} \frac{\|o(\theta)\|}{\|\theta\|} = 0 \quad (16)$$

U is a linear continuous form in θ; it represents a directional derivative in the direction θ. This definition makes sense in the case where $x \in \Omega$, but it poses a problem if $x \in \Omega_0$. Then in this case we use the Lagrangian derivative; for this we build the transported $\overline{u}(\theta)$ on Ω_0.

By changing variables we obtain $\overline{u}(\theta, x) = u((Id + \theta)(\Omega_0)o(Id+\theta)) = u((Id+\theta)(\Omega_0), x + \theta(x))$.

To arrive at the derivative Lagrangian by drifting $\overline{u}(\theta, x)$ with respect to θ

$$\overline{u}(\theta, x) = u(\theta, x) + L(\theta, x) + o(\theta)$$
$$\text{such that } \lim_{\theta \longmapsto 0} \frac{\|o(\theta)\|}{\|\theta\|} = 0. \quad (17)$$

L is a linear continuous form in θ; it represents a directional derivative in the direction θ.

There is a relation between these two derivatives $L(\theta, x) = U(\theta, x) + \theta.\nabla u(\Omega_0, x)$.

Proposition 9. *Let Ω_0 be a regular bounded open set of R^N. Let $f(\Omega)$ be an application from $C(\Omega_0)$ to $L^1(\mathbb{R})$; one defines its transpose from $W^{1,1}(\mathbb{R}^N)$ to $L^1(\mathbb{R})\overline{f}(\theta, x) = f((Id + \theta)(\Omega_0)o(Id+\theta))$ which we suppose to be derivable in 0 and L is considered as its derivative. So the application J_1 from $C(\Omega_0)$ to $\mathbb{R}$ defined by $J_1(\Omega) = \int_\Omega f(\Omega)dx$ is differentiable in Ω_0 and for all $\theta \in W^{1,\infty}(\mathbb{R}^N, \mathbb{R}^N)$ one has*

$$J_1'(\Omega_0)(\theta) = \int_{\partial\Omega} (f(\Omega_0)\operatorname{div}(\theta) + L(\theta))dS \quad (18)$$

In the same way, if $\overline{f}(\theta)$ is derivable as an application from $\mathscr{C}^1(\mathbb{R}^N, \mathbb{R}^N)$ to $L^1(\partial\Omega_0)$, so the application J_2 from $C(\Omega_0)$ to $\mathbb{R}$ defined by $J_2(\Omega) = \int_{\partial\Omega} f(\Omega)dx$ is differentiable in Ω_0 and for all $\theta \in \mathscr{C}^1(\mathbb{R}^N, \mathbb{R}^N)$ one has

$$J_2'(\Omega_0)(\theta) = \int_{\partial\Omega} (f(\Omega_0)(\operatorname{div}(\theta) - \nabla\theta n.n + Y(\theta))dS \quad (19)$$

2. Deriving an Equation with respect to the Domain

2.1. Dirichlet Conditions. We consider the following equation with Dirichlet boundary conditions:

$$-\Delta_p(u) = f \quad \text{in } \Omega$$
$$u = 0 \quad \text{on } \partial\Omega \quad (20)$$

With Ω_0 a regular bounded open set in $\mathbb{R}^N$, $f \in W^{1,p}(\mathbb{R}^N)$ and $\Delta_p(u) = \operatorname{div}(|\nabla u|^{p-2}\nabla u)$ with $1 < p < \infty$.

Equation (20) admits a unique solution in $W_0^{1,p}(\Omega)$.

Remark. For $p = 2$ we obtain the linear operator "*Laplacian*".

The variational formulation of problem (20) is as follows: $\forall v \in W_0^{1,p}(\Omega)$ we have $\int_\Omega -\Delta_p u.v = \int_\Omega f.v$ or $\operatorname{div}(f.F) = \nabla f.F + f\operatorname{div}(F)$; it implies that

$$f\operatorname{div}(F) = \operatorname{div}(f.F) - \nabla f.F$$

$$\text{So} \int_\Omega -\Delta_p u.v = \int_\Omega -\operatorname{div}\left(|\nabla u|^{p-2}\nabla u\right).v = \int_\Omega -\operatorname{div}\left(v.|\nabla u|^{p-2}\nabla u\right) + \int_\Omega \nabla v \nabla u^{p-2}\nabla u \quad (21)$$

Using the Green formula we obtain

$$= \int_{\partial\Omega} v.|\nabla u|^{p-2}\nabla u.ndS + \int_\Omega |\nabla u|^{p-2}\nabla u \nabla v \quad (22)$$

but $v \in W_0^{1,p}(\Omega)$; it implies that $v = 0$ on $\partial\Omega$. So the first term equals zero. Then

$$= \int_\Omega |\nabla u|^{p-2}\nabla u\nabla v = \int_\Omega f.v \quad \forall v \in W_0^{1,p}(\Omega) \quad (23)$$

Proposition 10. *Let $\Omega = (Id+\theta)(\Omega_0)$; $u(\Omega)$ is the solution of the problem (20). We define its transported on Ω_0 by*

$$\overline{u}(\theta) = u((Id+\theta)(\Omega_0) \circ (Id+\theta) \quad \in W_0^{1,p}(\Omega_0) \quad (24)$$

Then the application $\theta \longrightarrow \overline{u}(\theta)$ from $W^{1,\infty}(\mathbb{R}^N, \mathbb{R}^N)$ to $W^{1,p}(\Omega_0)$ is differentiable in 0 and its directional derivative *called Lagrangian derivative* $L =< \overline{u'}(0), \theta >$ *is the unique solution of*

$$-\Delta_p L - \operatorname{div}\left[\alpha_x \nabla L(\theta,x)\right] - \operatorname{div}\left(|\nabla L(\theta,x)|^{p-2}\nabla\overline{u}(0,x)\right) = g' + \operatorname{div}(f.\theta) \quad in\ \Omega \tag{25}$$

$$L = 0 \quad on\ \partial\Omega$$

with $g' = -\operatorname{div}[(\lambda_{(x,\theta)} + \alpha_x \rho_\theta)\nabla\overline{u}(0,x)]$.

Proof. We consider a test function $w = v \circ (Id + \theta) \in W_0^{1,p}(\Omega_0) \Longrightarrow v = w \circ (Id+\theta)^{-1} \in W_0^{1,p}(\Omega)$. Let $U \in W_0^{1,p}(\Omega_0)$ such that u is a solution of problem (20) satisfying $u = U \circ (Id+\theta)^{-1}$. We remark that w and U are independent of θ. By a change of variable $x = T(y)$ and the Lemma 5, (23) becomes

$$(23) \Longleftrightarrow$$

$$\int_{T(\Omega_0)} |\nabla u|^{p-2}\nabla u \nabla v dx = \int_{T(\Omega_0)} f v dx \Longleftrightarrow$$

$$\int_{\Omega_0} |\nabla u \circ T|^{p-2}(\nabla u)\circ T(\nabla v)\circ T\,|\det \nabla T|\,dy = \int_{\Omega_0}|f\circ T|.(v\circ T)\,|\det\nabla T|\,dy \Longleftrightarrow$$

$$\int_{\Omega_0}\left|\left((\nabla T)^{-1}\right)^t \circ \nabla(u\circ T)\right|^{p-2}\left((\nabla T)^{-1}\right)^t\nabla(u\circ T) \cdot \left((\nabla T)^{-1}\right)^t\nabla(v\circ T)\,|\det(\nabla T)|\,dy = \int_{\Omega_0}|f\circ T|.(v\circ T)\,|\det\nabla T|\,dy$$

$$(23) \Longleftrightarrow \tag{26}$$

$$\int_{\Omega_0}\left|\left((\nabla T)^{-1}\right)^t\nabla(\overline{u})\right|^{p-2}(\nabla T)^{-1} \cdot \left((\nabla T)^{-1}\right)^t\nabla(\overline{u})\nabla w\,|\det(\nabla T)|\,dy = \int_{\Omega_0}|f\circ T|.w\,|\det\nabla T|\,dy \Longleftrightarrow$$

$$\int_{\Omega_0}|\det(\nabla T)|\left|\left((\nabla T)^{-1}\right)^t\nabla(\overline{u})\right|^{p-2}(\nabla T)^{-1} \cdot \left((\nabla T)^{-1}\right)^t\nabla(\overline{u})\nabla w dy = \int_{\Omega_0}|f\circ T|.w\,|\det\nabla T|\,dy$$

We pose $A(\theta,\overline{u}) = |\det(\nabla T)|\left|\left((\nabla T)^{-1}\right)^t\nabla(\overline{u})\right|^{p-2} \cdot (\nabla T)^{-1}\left((\nabla T)^{-1}\right)^t\nabla(\overline{u})$

and since $T = Id+\theta$ we have $\nabla T = I + \nabla\theta$; it implies that

$$A(\theta,\overline{u}) = |\det(I+\nabla\theta)|\left|\left((I+\nabla\theta)^{-1}\right)^t\nabla(\overline{u})\right|^{p-2} \cdot (I+\nabla\theta)^{-1}\left((I+\nabla\theta)^{-1}\right)^t\nabla(\overline{u}) \tag{27}$$

Then (23)$\Longleftrightarrow \int_{\Omega_0} A(\theta,\overline{u})\nabla w dy = \int_{\Omega_0}|f\circ T|.w|\det\nabla T|dy$; then we drift with respect to θ in 0.

On the other hand the application $\theta \longrightarrow A(\theta,\overline{u})$ from $W^{1,\infty}(\mathbb{R}^N,\mathbb{R}^N)$ to $L^\infty(\mathbb{R}^N,\mathbb{R}^N)$ is differentiable in 0.

In fact $\det(I+\nabla\theta) = 1 + \operatorname{div}(\theta) + \circ(\theta)$ with $\lim_{\theta\longrightarrow 0}(\|\circ(\theta)\|_{L^\infty(\mathbb{R}^N)}/\|\theta\|_{W^{1,\infty}(\mathbb{R}^N,\mathbb{R}^N)}) = 0$. Therefore $|\det(I+\nabla\theta)| = 1 + \operatorname{div}(\theta) + \circ(\theta)$ because θ is small enough.

$$\begin{aligned} &\text{We have } (I+\nabla\theta)^{-1} = I - \nabla\theta + \circ(\theta)\\ &\text{and } \left((I+\nabla\theta)^{-1}\right)^t = I - (\nabla\theta)^t + \circ(\theta)\\ &\text{And } \overline{u}(\theta,x) = \overline{u}(0,x) + L(\theta,x) + \circ(\theta) \Longrightarrow\\ &\nabla\overline{u}(\theta,x) = \nabla\overline{u}(0,x) + \nabla L(\theta,x) + \circ(\theta)\end{aligned} \tag{28}$$

By using [14] we find

$$\begin{aligned}\left|\left((1-\nabla\theta)^{-1}\right)^t\nabla\overline{u}\right|^{p-2} &= \left|\left[I - (\nabla\theta)^t + \circ(\theta)\right] \cdot \left[\nabla\overline{u}(0,x) + \nabla L(\theta,x) + \circ(\theta)\right]\right|^{p-2} = \left|\nabla\overline{u}(0,x) - (\nabla\theta)^t\nabla\overline{u}(0,x) + \nabla L(\theta,x) + \circ(\theta)\right|^{p-2} = B(\theta,\overline{u})\\ &= |\nabla\overline{u}(0,x)|^{p-2} - \left|(\nabla\theta)^t\nabla\overline{u}(0,x)\right|^{p-2} + |\nabla L(\theta,x)|^{p-2} + \circ(\theta)\end{aligned} \tag{29}$$

On the other hand $|\det(I+\nabla\theta)|(I+\nabla\theta)^{-1}((I+\nabla\theta)^{-1})^t = (1+\operatorname{div}(\theta))I - \nabla\theta - (\nabla\theta)^t + \circ(\theta)$.

Thus

$$\begin{aligned} A(\theta,\overline{u}) &= B(\theta,\overline{u})\left[(1+\operatorname{div}(\theta))I - \nabla\theta - (\nabla\theta)^t + \circ(\theta)\right]\left[\nabla\overline{u}(0,x) + \nabla L(\theta,x) + \circ(\theta)\right] = B(\theta,\overline{u}) \cdot \left[\nabla\overline{u}(0,x) + \nabla L(\theta,x) + \left(\operatorname{div}\theta I - \nabla\theta - (\nabla\theta)^t\right)(\nabla\overline{u}(0,x) + \nabla L(\theta,x)) + \circ(\theta)\right]\\ &= B(\theta,\overline{u})\left[\nabla\overline{u}(0,x) + \nabla L(\theta,x) + \left(\operatorname{div}\theta I - \nabla\theta - (\nabla\theta)^t\right)\nabla\overline{u}(0,x) + \circ(\theta)\right]\\ &= |\nabla\overline{u}(0,x)|^{p-2}\nabla\overline{u}(0,x) + |\nabla\overline{u}(0,x)|^{p-2}\nabla L(\theta,x) + \left(\operatorname{div}(\theta)I - \nabla\theta - (\nabla\theta)^t\right)|\nabla\overline{u}(0,x)|^{p-2}\nabla\overline{u}(0,x)\\ &\quad - \left|(\nabla\theta)^t\nabla\overline{u}(0,x)\right|^{p-2}\nabla\overline{u}(0,x) - \left|(\nabla\theta)^t\nabla\overline{u}(0,x)\right|^{p-2}\nabla L(\theta,x) - \left|(\nabla\theta)^t\nabla\overline{u}(0,x)\right|^{p-2}\left(\operatorname{div}(\theta)I - \nabla\theta - (\nabla\theta)^t\right)\end{aligned}$$

$$\begin{aligned}
&\cdot \nabla \overline{u}(0,x) + |\nabla L(\theta,x)|^{p-2} \nabla \overline{u}(0,x) \\
&+ |\nabla L(\theta,x)|^{p-2} \nabla L(\theta,x) + |\nabla L(\theta,x)|^{p-2} \big(\operatorname{div}(\theta) I \\
&- \nabla\theta - (\nabla\theta)^t\big) \nabla \overline{u}(0,x) + \circ(\theta) = |\nabla \overline{u}(0,x)|^{p-2} \\
&\cdot \nabla \overline{u}(0,x) + |\nabla L(\theta,x)|^{p-2} \nabla L(\theta,x) \\
&- \left|(\nabla\theta)^t \nabla \overline{u}(0,x)\right|^{p-2} \nabla \overline{u}(0,x) + |\nabla L(\theta,x)|^{p-2} \\
&\cdot \nabla \overline{u}(0,x) - \left|(\nabla\theta)^t \nabla \overline{u}(0,x)\right|^{p-2} \big(\operatorname{div}(\theta) I - \nabla\theta \\
&- (\nabla\theta)^t\big) \nabla \overline{u}(0,x) + |\nabla L(\theta,x)|^{p-2} \big(\operatorname{div}(\theta) I - \nabla\theta \\
&- (\nabla\theta)^t\big) \nabla \overline{u}(0,x) + |\nabla \overline{u}(0,x)|^{p-2} \big[\nabla L(\theta,x) \\
&+ \big(\operatorname{div}(\theta) I - \nabla\theta - (\nabla\theta)^t\big) \nabla \overline{u}(0,x)\big] \\
&- \left|(\nabla\theta)^t \nabla \overline{u}(0,x)\right|^{p-2} \nabla L(\theta,x) + \circ(\theta)
\end{aligned} \tag{30}$$

Therefore, we have

$$\begin{aligned}
&\int_{\Omega_0} A(\theta,\overline{u}) \nabla w - \int_{\Omega_0} |\nabla \overline{u}(0,x)|^{p-2} \nabla \overline{u}(0,x) .\nabla w \\
&= \int_{\Omega_0} |\nabla L(\theta,x)|^{p-2} \nabla L(\theta,x) \nabla w + \Big(|\nabla \overline{u}(0,x)|^{p-2} \\
&- \left|(\nabla\theta)^t \nabla \overline{u}(0,x)\right|^{p-2}\Big) \nabla L(\theta,x) \\
&+ \Big[|\nabla L(\theta,x)|^{p-2} \big(I + \operatorname{div}(\theta) I - \nabla\theta - (\nabla\theta)^t\big) \\
&- \left|(\nabla\theta)^t \nabla \overline{u}(0,x)\right|^{p-2} \big(I + \operatorname{div}(\theta) I - \nabla\theta - (\nabla\theta)^t\big) \\
&+ |\overline{u}(0,x)|^{p-2} \big(\operatorname{div}(\theta) I - \nabla\theta - (\nabla\theta)^t\big)\Big] \nabla \overline{u}(0,x) \\
&= \int_{\Omega_0} \operatorname{div}(f.\theta) .w
\end{aligned} \tag{31}$$

Then afterwards we put

$$\rho_\theta = \operatorname{div}(\theta) - \nabla\theta - (\nabla\theta)^t$$

$$\lambda_{(x,\theta)} = \left|(\nabla\theta)^t \nabla \overline{u}(0,x)\right|^{p-2}$$

$$\alpha_x = |\nabla \overline{u}(0,x)|^{p-2}$$

And $g = -\operatorname{div}\left[\left(\lambda_{(x,\theta)}(I + \rho_\theta) + \alpha_x \rho_\theta\right) \nabla \overline{u}(0,x)\right]$

$$\begin{aligned}
&\text{Thus } \int_{\Omega_0} A(\theta,\overline{u}) \nabla w \\
&- \int_{\Omega_0} |\nabla \overline{u}(0,x)|^{p-2} \nabla \overline{u}(0,x) .\nabla w \\
&= \int_{\Omega_0} |\nabla L(\theta,x)|^{p-2} \nabla L(\theta,x) \nabla w \\
&+ \int_{\Omega_0} \left(\alpha_x - \lambda_{(x,\theta)}\right) \nabla L(\theta,x) \nabla w \\
&+ \int_{\Omega_0} \left[|\nabla L(\theta,x)|^{p-2} (I + \rho_\theta)\right] \nabla \overline{u}(0,x) \nabla w \\
&- \int_{\Omega_0} \left(\lambda_{(x,\theta)}(I + \rho_\theta) + \alpha_x \rho_\theta\right) \nabla \overline{u}(0,x) \nabla w + \circ(\theta) \\
&= \int_{\Omega_0} \operatorname{div}(f.\theta) w + \circ(\theta) \Longleftrightarrow \\
&\int_{\Omega_0} -\Delta_p L.w + \int_{\Omega_0} -\operatorname{div}\left[\left(\alpha_x - \lambda_{(x,\theta)}\right) \nabla L(\theta,x)\right] w \\
&+ \int_{\Omega_0} -\operatorname{div}\left(|\nabla L(\theta,x)|^{p-2} (I + \rho_\theta) \nabla \overline{u}(0,x)\right) w \\
&+ \int_{\Omega_0} g.w = \int_{\Omega_0} \operatorname{div}(f.\theta) w
\end{aligned} \tag{32}$$

Thus $\forall w \in W_0^{1,p}(\Omega)$ L the Lagrangian of $\overline{u}$ is a solution of the following differential equation:

$$\begin{aligned}
&-\Delta_p L - \operatorname{div}\left[\left(\alpha_x - \lambda_{(x,\theta)}\right) \nabla L(\theta,x)\right] \\
&\quad - \operatorname{div}\left(|\nabla L(\theta,x)|^{p-2} (I + \rho_\theta) \nabla \overline{u}(0,x)\right) = g \\
&\quad + \operatorname{div}(f.\theta) \quad in\ \Omega \\
&L = 0 \quad on\ \partial\Omega
\end{aligned} \tag{33}$$

Remark 11. When $p > 2$ we will have:

(i) $\lambda_{(x,\theta)}.\nabla L(\theta,x) = \circ(\theta)$

(ii) $|L(\theta,x)|^{p-2} \rho_\theta = \circ(\theta)$

(iii) $\lambda_{(x,\theta)}.\rho_\theta = \circ(\theta)$

Then L, the Lagrangian of $\overline{u}$, will be solution of the following reduced differential equation:

$$\begin{aligned}
&-\Delta_p L - \operatorname{div}\left[\alpha_x \nabla L(\theta,x)\right] \\
&\quad - \operatorname{div}\left(|\nabla L(\theta,x)|^{p-2} \nabla \overline{u}(0,x)\right) = g' + \operatorname{div}(f.\theta) \\
&\qquad in\ \Omega
\end{aligned} \tag{34}$$

$$L = 0 \quad on\ \partial\Omega$$

with $g' = -\operatorname{div}[(\lambda_{(x,\theta)} + \alpha_x \rho_\theta) \nabla \overline{u}(0,x)]$

2.2. Neumann Conditions. We consider the following equation with Neumann boundary conditions (see [15]):

$$\begin{aligned}
&-\Delta_p(u) + |u|^{p-2} u = f \quad in\ \Omega \\
&|\nabla u|^{p-2} \frac{\partial u}{\partial n} = g \quad on\ \partial\Omega
\end{aligned} \tag{35}$$

where Ω_0 a regular bounded open set in $\mathbb{R}^N$, $f \in W^{1,p}(\mathbb{R}^N)$, $g \in W^{2,p}(\mathbb{R}^N)$, and $\Delta_p(u) = \operatorname{div}(|\nabla u|^{p-2}\nabla u)$ with $1 < p < \infty$.

The variational formulation of problem (35) is to find $u \in W^{1,p}(\Omega)$ such that $\forall v \in W^{1,p}(\Omega)$ (v represents a test function) $\int_\Omega -\Delta_p(u).vdx + \int_\Omega |u|^{p-2}u.vdx = \int_\Omega f.vdx$ or $\operatorname{div}(f.F) = \nabla f.F + f\operatorname{div}(F)$.

It implies that $f\operatorname{div}(F) = \operatorname{div}(f.F) - \nabla f.F$:

$$\begin{aligned}
&\text{Then } \int_\Omega -\operatorname{div}\left(|\nabla u|^{p-2}\nabla u\right).vdx + \int_\Omega |u|^{p-2}u.vdx \\
&\quad = \int_\Omega f.vdx \Longleftrightarrow \\
&\int_\Omega |\nabla u|^{p-2}\nabla u.\nabla v - \int_\Omega \operatorname{div}\left(v|\nabla u|^{p-2}\nabla u\right) \\
&\quad + \int_\Omega |u|^{p-2}u.vdx = \int_\Omega f.vdx.
\end{aligned} \tag{36}$$

By using the Green formula we find

$$\begin{aligned}
&\int_\Omega |\nabla u|^{p-2}\nabla u.\nabla vdx - \int_{\partial\Omega} v|\nabla u|^{p-2}\nabla u.ndS \\
&\quad + \int_\Omega |u|^{p-2}u.vdx = \int_\Omega f.vdx \Longleftrightarrow \\
&\int_\Omega |\nabla u|^{p-2}\nabla u.\nabla vdx + \int_\Omega |u|^{p-2}u.vdx = \int_\Omega f.vdx \\
&\quad + \int_{\partial\Omega} g.vdS \quad \text{and } \forall v \in W^{1,p}(\Omega)
\end{aligned} \tag{37}$$

Proposition 12. *Let $\Omega = (Id+\theta)(\Omega_0)$; $u(\Omega)$ is the solution of problem (35). We define its transport on Ω_0 by $\overline{u}(\theta) = u((Id+\theta)(\Omega_0) \circ (Id+\theta) \in W_0^{1,p}(\Omega_0)$.*

Then the application $\theta \longrightarrow \overline{u}(\theta)$ from $\mathscr{C}^1(\mathbb{R}^N,\mathbb{R}^N)$ to $W^{1,\infty}(\Omega_0)$ is differentiable in 0 on the direction θ and its directional derivative called Lagrangian derivative of u is $L =< \overline{u'}(0), \theta >$ where $L \in W^{1,p}(\Omega_0)$ is the unique solution of

$$\begin{aligned}
&-\Delta_p L + |\nabla L(\theta,x)|^{p-2}\nabla L(\theta,x) = F \\
&\quad + \operatorname{div}\left((\alpha_x - \lambda_{(x,\theta)})\nabla L(\theta,x)\right) \\
&\quad + \operatorname{div}\left[|\nabla L(\theta,x)|^{p-2}(I+\rho_\theta)\nabla\overline{u}(0,x)\right] \\
&\quad - |L(\theta,x)|^{p-2}\overline{u}(0,x) - |\overline{u}(0,x)|^{p-2}L(\theta,x) \\
&\quad - \left(|\overline{u}(0,x)|^{p-2} + |\nabla L(\theta,x)|^{p-2}\right) \\
&\quad \cdot \left(\overline{u}(0,x)|^{p-2} + L(\theta,x)\right)\operatorname{div}\theta \quad \text{in } \Omega \\
&G_\theta - (\alpha_x - \lambda_{(x,\theta)})\frac{\partial L(\theta,x)}{\partial n} - |\nabla L(\theta,x)|^{p-2} \\
&\quad \cdot (I+\rho_\theta)\frac{\partial\overline{u}(0,x)}{\partial n} - |L(\theta,x)|^{p-2}\frac{\partial L(\theta,x)}{\partial n} = 0 \\
&\qquad \text{on } \partial\Omega
\end{aligned} \tag{38}$$

where $F = -\operatorname{div}((\lambda_{(x,\theta)}(I+\rho_\theta) + \alpha_x\rho_\theta)\nabla\overline{u}(0,x) - f.\theta)$.

Proof. We make a change of variable $x = y + \theta(y)$ where $y \in \Omega_0$ and $x \in \Omega$ in the variational formulation (37). We pose $w(y) = v(x) \Longleftrightarrow w(y) = v(Id+\theta)(y) \Longleftrightarrow v = w \circ (Id+\theta)^{-1}$ noticing that w does not depend on θ. Thus by drifting the variational formulation, we obtain by using Lemmas 5 and 7

$$\begin{aligned}
&(37) \Longleftrightarrow \\
&\int_{\Omega_0} A(\theta,\overline{u})\nabla wdy \\
&\quad + \int_{\Omega_0} |\overline{u}(\theta,x)|^{p-2}\overline{u}(\theta,x).w|\det\nabla T|dy \\
&\quad = \int_{\Omega_0} (f\circ T).w|\det\nabla T|dy \\
&\quad + \int_{\partial\Omega} g\circ Tw|\det\nabla T|\left|\left((\nabla T)^{-1}\right)^t.n\right|_{\mathbb{R}^N} dS
\end{aligned} \tag{39}$$

Or

$$\begin{aligned}
&\int_{\Omega_0} A(\theta,\overline{u})\nabla wdy \\
&\quad = \int_{\Omega_0} |\nabla\overline{u}(0,x)|^{p-2}\nabla\overline{u}(0,x).\nabla w \\
&\quad + \int_{\Omega_0} |\nabla L(\theta,x)|^{p-2}\nabla L(\theta,x).\nabla w \\
&\quad + \int_{\Omega_0} (\alpha_x - \lambda_{(x,\theta)})\nabla L(\theta,x)\nabla w \\
&\quad + \int_{\Omega_0} \left[|\nabla L(\theta,x)|^{p-2}(I+\rho_\theta)\right]\nabla\overline{u}(0,x)\nabla w \\
&\quad - \int_{\Omega_0} (\lambda_{(x,\theta)}(I+\rho_\theta) + \alpha_x\rho_\theta)\nabla\overline{u}(0,x)\nabla w \\
&\quad + \circ(\theta)
\end{aligned} \tag{40}$$

We have by using [14]

$$\begin{aligned}
&\int_{\Omega_0} |\overline{u}(\theta,x)|^{p-2}\overline{u}(\theta,x).w|\det\nabla T|dy \\
&\quad = \int_{\Omega_0} |\overline{u}(0,x) + L(\theta,x)|^{p-2} \\
&\quad \cdot (\overline{u}(0,x) + L(\theta,x)).w.(1+\operatorname{div}(\theta))dy + \circ(\theta) \\
&\quad = \int_{\Omega_0} \left(|\overline{u}(0,x)|^{p-2} + |L(\theta,x)|^{p-2}\right) \\
&\quad \cdot (\overline{u}(0,x) + L(\theta,x)).w.(1+\operatorname{div}(\theta))wdy + \circ(\theta) \\
&\quad = \int_{\Omega_0} |\overline{u}(0,x)|^{p-2}\overline{u}(0,x)wdy + \int_{\Omega_0} |L(\theta,x)|^{p-2} \\
&\quad \cdot L(\theta,x)wdy + \int_{\Omega_0} |L(\theta,x)|^{p-2}\overline{u}(0,x)wdy \\
&\quad + \int_{\Omega_0} |\overline{u}(0,x)|^{p-2}L(\theta,x)wdy
\end{aligned}$$

$$+\int_{\Omega_0}\left(|\overline{u}(0,x)|^{p-2}+|L(\theta,x)|^{p-2}\right)\cdot(\overline{u}(0,x)+L(\theta,x)).w.\mathrm{div}(\theta)\,wdy+\circ(\theta) \quad (41)$$

And also $\int_{\Omega_0}(f\circ T).w\,|\det\nabla T|\,dy-\int_{\Omega_0}fwdy$

$$=\int_{\Omega_0}\mathrm{div}(f.\theta)\,wdy+\circ(\theta) \quad (42)$$

Then $\int_{\partial\Omega}g\circ Tw\,|\det\nabla T|\left|\left((\nabla T)^{-1}\right)^t.n\right|_{\mathbb{R}^N}dS$

$$-\int_{\partial\Omega_0}gwdS=\int_{\partial\Omega_0}(\nabla g\theta+g(\mathrm{div}\theta-\nabla\theta n.n))\cdot wdS+\circ(\theta)$$

And

$$\int_{\Omega_0}A(\theta,\overline{u})\nabla wdy+\int_{\Omega_0}|\overline{u}(\theta,x)|^{p-2}\cdot\overline{u}(\theta,x).w\,|\det\nabla T|\,dy-\int_{\Omega_0}|\overline{u}(\theta,x)|^{p-2}\nabla\overline{u}(\theta,x).\nabla w-\int_{\Omega_0}|\overline{u}(\theta,x)|^{p-2}\cdot\overline{u}(\theta,x).wdy=\int_{\Omega_0}|\nabla L(\theta,x)|^{p-2}\nabla L(\theta,x).\nabla w$$
$$-\int_{\Omega_0}|L(\theta,x)|^{p-2}L(\theta,x).wdy+\int_{\Omega_0}(\alpha_x-\lambda_{(x,\theta)})\nabla L(\theta,x)\nabla w+\int_{\Omega_0}\left[|\nabla L(\theta,x)|^{p-2}(I+\rho_\theta)\right]\nabla\overline{u}(0,x)\nabla w \quad (43)$$
$$-\int_{\Omega_0}(\lambda_{(x,\theta)}(I+\rho_\theta)+\alpha_x\rho_\theta)\nabla\overline{u}(0,x)\nabla w+\int_{\Omega_0}|L(\theta,x)|^{p-2}\overline{u}(0,x)\,wdy+\int_{\Omega_0}|\overline{u}(\theta,x)|^{p-2}\cdot L(\theta,x)\,wdy+\int_{\Omega_0}\left(|\overline{u}(0,x)|^{p-2}+|L(\theta,x)|^{p-2}\right)$$
$$\cdot(\overline{u}(0,x)+L(\theta,x)).w.\mathrm{div}(\theta)\,wdy=\int_{\Omega_0}\mathrm{div}(f.\theta)\cdot wdy+\int_{\partial\Omega_0}(\nabla g\theta+g(\mathrm{div}\theta-\nabla\theta n.n))\,wdS+\circ(\theta)$$

For

$$\int_{\Omega_0}-\Delta_pL.wdy+\int_{\Omega_0}|L(\theta,x)|^{p-2}L(\theta,x)\,wdy-\int_{\Omega_0}\mathrm{div}((\alpha_x-\lambda_{(x,\theta)})\nabla L(\theta,x))\,wdy$$
$$-\int_{\Omega_0}\mathrm{div}\left(|\nabla L(\theta,x)|^{p-2}(I+\rho_\theta)\nabla\overline{u}(0,x)\right)wdy+\int_{\Omega_0}\mathrm{div}((\lambda_{(x,\theta)}(I+\rho_\theta)+\alpha_x\rho_\theta)\nabla\overline{u}(0,x))\,wdy$$
$$+\int_{\Omega_0}|L(\theta,x)|^{p-2}\overline{u}(0,x)\,wdy+\int_{\Omega_0}|\overline{u}(\theta,x)|^{p-2}\cdot L(\theta,x)\,wdy+\int_{\Omega_0}\left(|\overline{u}(0,x)|^{p-2}+|L(\theta,x)|^{p-2}\right)$$
$$\cdot(\overline{u}(0,x)+L(\theta,x)).w.\mathrm{div}(\theta)\,wdy-\int_{\Omega_0}\mathrm{div}(f.\theta)\cdot wdy=\int_{\partial\Omega_0}(\nabla g\theta+g(\mathrm{div}\theta-\nabla\theta n.n))\,wdS$$
$$-\int_{\partial\Omega_0}|L(\theta,x)|^{p-2}\frac{\partial L(\theta,x)}{\partial n}wdS-\int_{\partial\Omega_0}(\alpha_x-\lambda_{(x,\theta)})\frac{\partial L(\theta,x)}{\partial n}wdS-\int_{\partial\Omega_0}|\nabla L(\theta,x)|^{p-2}(I+\rho_\theta)\frac{\partial\overline{u}(0,x)}{\partial n}wdS$$
$$+\int_{\partial\Omega_0}|\nabla L(\theta,x)|^{p-2}(I+\rho_\theta)\frac{\partial\overline{u}(0,x)}{\partial n}wdS+\int_{\partial\Omega_0}(\lambda_{(x,\theta)}(I+\rho_\theta)+\alpha_x\rho_\theta)\frac{\partial\overline{u}(0,x)}{\partial n}wdS$$
$$=\left[G_\theta-(\alpha_x-\lambda_{(x,\theta)})\frac{\partial L(\theta,x)}{\partial n}-|\nabla L(\theta,x)|^{p-2}(I+\rho_\theta)\frac{\partial\overline{u}(0,x)}{\partial n}-|L(\theta,x)|^{p-2}\frac{\partial L(\theta,x)}{\partial n}\right]\cdot wdS \quad (44)$$

We note
$$G_\theta=\nabla g\theta+g(\mathrm{div}\theta-\nabla\theta n.n)+(\lambda_{(x,\theta)}(I+\rho_\theta)+\alpha_x\rho_\theta)\frac{\partial\overline{u}(0,x)}{\partial n} \quad (45)$$

So the transported derivative of $\overline{u}(\theta,x)$ at 0 in the direction θ is the Lagrangian $L=<\overline{u'}(0),\theta>$ which is the solution of the following equation:

$$-\Delta_pL+|\nabla L(\theta,x)|^{p-2}\nabla L(\theta,x)=F+\mathrm{div}((\alpha_x-\lambda_{(x,\theta)})\nabla L(\theta,x))+\mathrm{div}\left[|\nabla L(\theta,x)|^{p-2}(I+\rho_\theta)\nabla\overline{u}(0,x)\right]$$
$$-|L(\theta,x)|^{p-2}\overline{u}(0,x)-|\overline{u}(0,x)|^{p-2}L(\theta,x)-\left(|\overline{u}(0,x)|^{p-2}+|\nabla L(\theta,x)|^{p-2}\right)\cdot\left(\overline{u}(0,x)|^{p-2}+L(\theta,x)\right)\mathrm{div}\theta \quad in\ \Omega$$

$$G_\theta - (\alpha_x - \lambda_{(x,\theta)}) \frac{\partial L(\theta, x)}{\partial n} - |\nabla L(\theta, x)|^{p-2} \cdot (I + \rho_\theta) \frac{\partial \overline{u}(0, x)}{\partial n} - |L(\theta, x)|^{p-2} \frac{\partial L(\theta, x)}{\partial n} = 0 \quad \text{on } \partial\Omega \tag{46}$$

where $F = -\text{div}((\lambda_{(x,\theta)}(I + \rho_\theta) + \alpha_x \rho_\theta)\nabla\overline{u}(0, x) - f.\theta)$.

3. Optimality Condition

To calculate the optimality conditions of the following problem $\inf_{\Omega \in U_{ad}} J(\Omega)$ with $U_{ad} = \{\Omega \in C(\Omega_0) \text{ and } \int_x dx = V_0\}$, where $C(\Omega_0)$ is the set of admissible forms obtained by diffeomorphism, the cost function $J(\Omega)$ is the compliance defined by

$$J(\Omega) = \int_\Omega |u(\Omega) - u(\Omega_0)|^p \, dx \tag{47}$$

to reach a target displacement $u(\Omega_0) \in L^p(\mathbb{R}^N)$ where the function $u(\Omega)$ is solution of the boundary problem posed (resp., Dirichlet or Neumann boundary conditions).

We consider the following boundary value problems.

Dirichlet Boundary Condition

$$\begin{aligned} -\Delta_p(u) &= f \quad \text{in } \Omega \\ u &= 0 \quad \text{on } \partial\Omega \end{aligned} \tag{48}$$

where $f \in W^{1,p}(\mathbb{R}^N)$.

Neumann Boundary Condition

$$\begin{aligned} -\Delta_p(u) + |u|^{p-2} u &= f \quad \text{in } \Omega \\ |\nabla u|^{p-2} \frac{\partial u}{\partial n} &= g \quad \text{on } \partial\Omega \end{aligned} \tag{49}$$

where $f \in W^{1,p}(\mathbb{R}^N)$ and $g \in W^{2,p}(\mathbb{R}^N)$.

The problems admit a unique solution $u(\Omega)$.

Theorem 13. *Let Ω_0 be a regular bounded open set. The cost function $J(\Omega)$ is differentiable and its derivative is*

$$J'(\Omega_0)(\theta) = \int_{\Omega_0} |u(\Omega_0) - u_0|^{p-2} \operatorname{div}(\theta) + p|u(\Omega_0) - u_0|^{p-2} (u(\Omega_0) - u_0) \cdot (L - \theta.\nabla u_0) \, dx \tag{50}$$

L is the Lagrangian derivative and also solution of

$$-\Delta_p L - \operatorname{div}[\alpha_x \nabla L(\theta, x)] - \operatorname{div}(|\nabla L(\theta, x)|^{p-2} \nabla\overline{u}(0, x)) = g' + \operatorname{div}(f.\theta) \quad \text{in } \Omega \tag{51}$$

$$L = 0 \quad \text{on } \partial\Omega$$

with $g' = -\text{div}[(\lambda_{(x,\theta)} + \alpha_x \rho_\theta)\nabla\overline{u}(0, x)]$.

Proof. By applying Proposition 4 for the compliance we obtain

$$\begin{aligned} J'(\Omega_0)(\theta) &= \int_{\Omega_0} |u(\Omega_0) - u_0|^{p-2} \operatorname{div}(\theta) + \int_{\Omega_0} |u(\Omega_0) - u_0|^p \theta \\ J'(\Omega_0)(\theta) &= \int_{\Omega_0} |u(\Omega_0) - u_0|^{p-2} \operatorname{div}(\theta) + p\int_{\Omega_0} |u(\Omega_0) - u_0|^{p-2} (u(\Omega_0) - u_0) \cdot (\nabla u - \nabla u_0) \, dx \\ J'(\Omega_0)(\theta) &= \int_{\Omega_0} |u(\Omega_0) - u_0|^{p-2} \operatorname{div}(\theta) + p|u(\Omega_0) - u_0|^{p-2} (u(\Omega_0) - u_0) \cdot (L - \theta.\nabla u_0) \, dx \end{aligned} \tag{52}$$

where L is the Lagrangian derivative, solution of the PDE.

Remark 14. From extensive literature which deals with optimum condition calculus for problems as $\inf_{\Omega \in \mathcal{U}_a d} J(\Omega)$ we can cite

(i) $J_1(\Omega) = \int_{\partial\Omega_0} |u(\Omega_0) - u_0|^p dS$

(ii) $J_2(\Omega) = \int_{\partial\Omega_0} |\nabla u(\Omega_0) - \nabla u_0|^p dS + \int_{\partial\Omega_0} |u(\Omega_0) - u_0|^p dS$

So, to calculate the gradient of each compliance we use the same argument by the propositions [13].

Conflicts of Interest

The authors declare that they have no conflicts of interest.

References

[1] F. Murat and J. Simon, "Etudes de problèmes d'optimal design," in *Lecture Notes in Computer Science*, vol. 41, pp. 54–62, Springer, Berlin, Germany, 1976.

[2] F. Murat and L. Tartar, Sur le côntrole par un domaine géométrique, Internal Report No 76 015, Laboratoire d'Analyse Numérique de l'Université Paris, vol. 6, 1976.

[3] J. Hadamard, "Mémoire sur le problème d'analyse relatif à l'èquilibre des plaques élastiques encastrées," *Bulletin de la Sociètè Mathèmatique de France*, 1907.

[4] D. Bucur and G. Buttazzo, *Variational Methods in Shape Optimization Problems, Progress in Nonlinear Differential Equations and Their Applications*, vol. 65, Birkhauser, Basel, Switzerland, 2005.

[5] D. Daners, "Domain perturbation for linear and nonlinear parabolic equations," *Journal of Differential Equations*, vol. 129, no. 2, pp. 358–402, 1996.

[6] A. Dervieux and B. Palmerio, "Une formule de hadamard dans des problemes d'optimal design," in *Optimization Techniques, Lecture Notes in Computer Science*, J. Cea, Ed., vol. 41, pp. 63–76, Springer, Berlin, Germany, 1976.

[7] M. Henrot and P. Michel, *Variation et Optimisation de Formes, Collection Mathématique & Applications*, vol. 48, Springer, Paris, France, 2005.

[8] O. Pironneau, *Optimal Shape Design for Elliptic Systems*, Springer, New York, NY, USA, 1984.

[9] B. Rousselet, "Shape design sensitivity of a membrane," *Journal of Optimization Theory and Applications*, vol. 40, no. 4, pp. 595–623, 1983.

[10] J. Sokolowski and J.-P. Zolesio, *Introduction to Shape Optimization, Shape Sensivity Analysis*, Springer Series in Computational Mathematics, Berlin, Germany, 1992.

[11] V. Sverak, "On optimal shape design," *Journal de Mathématiques Pures et Appliquées*, vol. 72, no. 6, pp. 537–551, 1993.

[12] J.-P. Zolésio, "The material derivative (or speed) method fos shape optimization," in *Optimization of Distributed Parameter Structures, vol. II*, NATO Adv. Study Inst. Ser. E: Appl. Sci. 50, pp. 1089–1151, 1981.

[13] G. Allaire, *Conception Optimal de Structures, Collection Mathématique & Applications*, vol. 58, Springer, Paris, France, 2007.

[14] H. Brézis and E. Lieb, "A relation between pointwise convergence of functions and convergence of functionals," *Proceedings of the American Mathematical Society*, vol. 88, no. 3, pp. 486–490, 1983.

[15] A. Mercaldo, J. D. Rossi, S. Segura de Leon, and C. Trombetti, "Behaviour of p-Laplacian problems with Neumann boundary conditions when p goes to 1," *Communications on Pure and Applied Analysis*, vol. 12, no. 1, pp. 253–267, 2013.

17

An Extension Theorem for a Sequence of Krein Space Contractions

Gerald Wanjala

Department of Mathematics and Statistics, Sultan Qaboos University, P.O. Box 36, Al-Khod, 123 Muscat, Oman

Correspondence should be addressed to Gerald Wanjala; wanjalag@yahoo.com

Academic Editor: Gelu Popescu

Consider Krein spaces $\mathcal{U}$ and $\mathcal{Y}$ and let $\mathcal{H}_k$ and $\mathcal{K}_k$ be regular subspaces of $\mathcal{U}$ and $\mathcal{Y}$, respectively, such that $\mathcal{H}_k \subset \mathcal{H}_{k+1}$ and $\mathcal{K}_k \subset \mathcal{K}_{k+1}$ $(k \in \mathbb{N})$. For each $k \in \mathbb{N}$, let $A_k : \mathcal{H}_k \to \mathcal{K}_k$ be a contraction. We derive necessary and sufficient conditions for the existence of a contraction $B : \mathcal{U} \to \mathcal{Y}$ such that $B|_{\mathcal{H}_k} = A_k$. Some interesting results are proved along the way.

1. Introduction

A number of extensions and generalizations of classical function theoretic interpolation problems are driven by factors such as applications in systems and control theory. Many such extensions and generalizations make use of the commutant lifting theorem in one way or another. In particular, the commutant lifting theorem in the Hilbert space case which was obtained by Sz.-Nagy and Foias has been used to solve extension problems like the ones of Nevanlinna-Pick, Nudelman, Nehari, and many others. Extensions of this theorem to an indefinite setting are given in [1–5]. In [6] (see also [7, 8]), a time-variant version of the commutant lifting theorem is developed. This time-variant version is called the *three-chain completion theorem* and is used to solve a number of nonstationary norm constrained interpolation problems on Hilbert spaces. Recall (see [6]) that the given data for the three-chain completion problem are bounded linear operators

$$A_k : \mathcal{H}_k \longrightarrow \mathcal{K}_k \ominus \mathcal{M}_k \quad (k \in \mathbb{Z}), \tag{1}$$

where $\mathcal{H}_k \subset \mathcal{U}$ and $\mathcal{M}_k \subset \mathcal{K}_k \subset \mathcal{Y}$ for $k \in \mathbb{Z}$ are Hilbert spaces satisfying the inclusion relations

$$\begin{gathered} \mathcal{H}_{k-1} \subset \mathcal{H}_k, \\ \mathcal{M}_{k-1} \subset \mathcal{M}_k, \\ \mathcal{K}_{k-1} \subset \mathcal{K}_k \\ (k \in \mathbb{Z}). \end{gathered} \tag{2}$$

Given operators (1) and tolerance γ the problem is to find an operator $B : \mathcal{U} \to \mathcal{Y}$ such that $\|B\| \le \gamma$ and

$$\begin{gathered} B\mathcal{H}_k \subset \mathcal{K}_k, \\ \left(I - P_{\mathcal{M}_k}\right) B\big|_{\mathcal{H}_k} = A_k \\ (k \in \mathbb{Z}). \end{gathered} \tag{3}$$

As far as we know, no extension of this theorem to an indefinite setting has been developed so far.

The three-chain theorem mentioned above is the motivation for the extension problem considered in this paper. For $k \in \mathbb{N}$, consider a sequence of Krein space contractions $A_k : \mathcal{H}_k \to \mathcal{K}_k$, where $\mathcal{H}_k$ and $\mathcal{K}_k$ are nested regular subspaces of some Krein spaces $\mathcal{U}$ and $\mathcal{Y}$, respectively. The problem is to find a contraction $B : \mathcal{U} \to \mathcal{Y}$ such that

$$Bh = A_k h \tag{4}$$

for all $h \in \mathcal{H}_k$.

In order to keep this paper as self-contained as possible, we briefly outline some definitions and some elementary facts about Krein spaces and bounded linear operators defined on them. This is done in Section 2. Some important results are stated and proved in Section 3. An extension theorem is considered in Section 4 while Section 5 contains some simple application of the extension theorem discussed in Section 4.

2. Preliminaries

In this section we recall some definitions and basic notions of indefinite metric space theory. More details can be obtained from [2, 9–12].

Let $\mathscr{K}$ be a linear space and let $\langle\cdot,\cdot\rangle$ be a sesquilinear form on $\mathscr{K}$. This sesquilinear form is called an indefinite metric on $\mathscr{K}$. If $\mathscr{K}$ admits a direct orthogonal sum decomposition

$$\mathscr{K} = \mathscr{K}^{+} \oplus \mathscr{K}^{-} \tag{5}$$

in which $(\mathscr{K}^{\pm}, \pm\langle\cdot,\cdot\rangle)$ are Hilbert spaces, then $\mathscr{K}$ (or $(\mathscr{K}, \langle\cdot,\cdot\rangle)$) is called a *Krein space*. Decomposition (5), which is not unique in general, is called a *fundamental decomposition* and gives rise to orthogonal projections from $\mathscr{K}$ onto $\mathscr{K}^{\pm}$, which we denote by $P^{\pm}$, respectively. The self-adjoint and unitary operator J on $\mathscr{K}$ defined by $J = P^{+} - P^{-}$, is called a *fundamental symmetry* on $\mathscr{K}$.

The space $\mathscr{K}$ with the inner product

$$[x, y] = \langle Jx, y\rangle = \langle x_{+}, y_{+}\rangle - \langle x_{-}, y_{-}\rangle, \quad x_{\pm}, y_{\pm} \in \mathscr{K}^{\pm} \tag{6}$$

is a Hilbert space. This inner product is used to define the norm of an element x of the Krein space $\mathscr{K}$ by

$$\|x\|^{2} = [x, x]. \tag{7}$$

Topological notions such as convergence and continuity are understood to be with respect to this norm topology. The inner product $[\cdot,\cdot]$ in (6) depends on decomposition (5), as does the norm $\|\cdot\|$ in (7), but the norms generated by different decompositions of $\mathscr{K}$ are equivalent. We will denote the norm $\|\cdot\|$ in (7) by $\|\cdot\|_{J}$ where clarity is needed.

An *orthogonal projection* in a Krein space is a bounded self-adjoint operator R in $\mathscr{K}$ such that $R^{2} = R$. Note that the norm of an orthogonal projection in a Krein space need not be less than 1. The range $\mathscr{M}$ of an orthogonal projection R in a Krein space $\mathscr{K}$ is a closed subspace of $\mathscr{K}$ and the space $\mathscr{K}$ can be decomposed as

$$\mathscr{K} = \mathscr{M} \oplus \mathscr{M}^{\perp}, \tag{8}$$

where $\mathscr{M}^{\perp} = \{l \in \mathscr{K} : \langle l, m\rangle = 0\ \forall m \in \mathscr{M}\}$. On the other hand, given a closed subspace $\mathscr{M}$ of $\mathscr{K}$ such that decomposition (8) holds, then $\mathscr{M}$ is the range of an orthogonal projection in $\mathscr{K}$. In this case, $(\mathscr{M}, \langle\cdot,\cdot\rangle)$ is again a Krein space. Unlike in the Hilbert space case, the decomposition (8) need not hold for a given closed subspace. Closed subspaces for which this decomposition holds are referred to as *regular subspaces*.

Let $\mathscr{H}$ and $\mathscr{K}$ be Krein spaces and let $T : \mathscr{H} \to \mathscr{K}$ be a bounded linear operator. We say that T is a *contraction* if for all $x \in \mathscr{H}$, $\langle Tx, Tx\rangle_{\mathscr{K}} \leq \langle x, x\rangle_{\mathscr{H}}$. If both T and its Krein space adjoint T^{*} are contractions, then T is called a *bicontraction*.

Let $\mathscr{H}$ and $\mathscr{K}$ be Krein spaces and let $T \in B(\mathscr{H}, \mathscr{K})$, the space of bounded linear operators from $\mathscr{H}$ into $\mathscr{K}$. A column extension of T is an operator of the form

$$C = \begin{pmatrix} T \\ E \end{pmatrix} : \mathscr{H} \longrightarrow \begin{pmatrix} \mathscr{K} \\ \mathscr{E} \end{pmatrix}, \tag{9}$$

where $\mathscr{E}$ is a Krein space and $E \in B(\mathscr{H}, \mathscr{E})$. By a row extension of T we shall mean an operator of the form

$$L = \begin{pmatrix} T & E \end{pmatrix} : \begin{pmatrix} \mathscr{H} \\ \mathscr{E} \end{pmatrix} \longrightarrow \mathscr{K}, \tag{10}$$

where $\mathscr{E}$ is a Krein space and $E \in B(\mathscr{E}, \mathscr{K})$. It is shown in [2] (see also [3]) that if T is a contraction, then there exist contractive row and column extensions of T where the extension space is a Hilbert space. Contractive 2×2 matrix extensions of a contractive operator $T : \mathscr{H} \to \mathscr{K}$ where the extending space $\mathscr{E}$ is a Hilbert space are thoroughly discussed in [2] where Lemma 1, Theorem 2, and Lemma 3 can be found. Results more general than those provided by Lemma 1 and Theorem 2 can be found in [1] while Lemma 3 can also be found in [4].

Lemma 1. *Let $\mathscr{H}$ and $\mathscr{K}$ be Krein spaces and let $T \in B(\mathscr{H}, \mathscr{K})$ be a contraction. Let*

$$C = \begin{pmatrix} T \\ \widetilde{E}^{*} \end{pmatrix} \in B\left(\mathscr{H}, \mathscr{K} \oplus \widetilde{\mathscr{E}}\right) \tag{11}$$

be a contractive column extension of T with $\widetilde{\mathscr{E}}$ a Hilbert space. If norms are computed relative to some fixed fundamental decompositions of $\mathscr{H}$ and $\mathscr{K}$ and the induced fundamental decomposition of $\mathscr{K} \oplus \widetilde{\mathscr{E}}$, then

$$\|C\|^{2} \leq 1 + 2\|T\|^{2}. \tag{12}$$

The above norm estimate enables one to fix a bound for the norm of the operator C. The following lemma is helpful in finding 2×2 matrix extensions of a contractive operator T. See [13] for a similar result in a Hilbert space setting.

Theorem 2. *Let $\mathscr{H}$, $\mathscr{K}$, and $\mathscr{E}$ be Krein spaces and let $\widetilde{\mathscr{E}}$ be a Hilbert space. Assume that*

$$\begin{gathered} \begin{pmatrix} C_{11} & C_{12} \end{pmatrix} : \begin{pmatrix} \mathscr{H} \\ \mathscr{E} \end{pmatrix} \longrightarrow \mathscr{K}, \\ \begin{pmatrix} C_{11} \\ C_{21} \end{pmatrix} : \mathscr{H} \longrightarrow \begin{pmatrix} \mathscr{K} \\ \widetilde{\mathscr{E}} \end{pmatrix} \end{gathered} \tag{13}$$

are contractions. Then there exists an operator $C_{22} : \mathscr{E} \to \widetilde{\mathscr{E}}$ such that

$$C = \begin{pmatrix} C_{11} & C_{12} \\ C_{21} & C_{22} \end{pmatrix} : \begin{pmatrix} \mathscr{H} \\ \mathscr{E} \end{pmatrix} \longrightarrow \begin{pmatrix} \mathscr{K} \\ \widetilde{\mathscr{E}} \end{pmatrix} \tag{14}$$

is a contraction. If

$$\begin{pmatrix} C_{11} & C_{12} \end{pmatrix} \tag{15}$$

is a bicontraction, C_{22} may be chosen such that C is a bicontraction.

We conclude this section by stating the following lemma.

Lemma 3. *Let $\mathcal{N}_1, \mathcal{N}_2, \ldots$ be regular subspaces of a Krein space $\mathcal{H}$ such that $\mathcal{N}_1 \supset \mathcal{N}_2 \supset \cdots$. Suppose that the projections $P_1, P_2, \ldots$ of $\mathcal{H}$ onto the subspaces $\mathcal{N}_1, \mathcal{N}_2, \ldots$ are uniformly bounded. Then*

$$\mathcal{N} = \bigcap_{n=1}^{\infty} \mathcal{N}_n, \qquad \mathcal{M} = \overline{\operatorname{span}}\, \mathcal{N}_n^{\perp}, \qquad n = 1, 2, \ldots \tag{16}$$

are regular subspaces of $\mathcal{H}$ with $\mathcal{N} = \mathcal{M}^{\perp}$. If P is the projection of $\mathcal{H}$ onto $\mathcal{N}$, then $P = \lim_{n\to\infty} P_n$ with convergence in the strong operator topology.

3. Some Preliminary Results

In this section we prove some useful results regarding sequences of regular subspaces. See [4, Corollary 3.3 and Lemma 3.4] for closely related results.

Theorem 4. *For $n \geq 1$, let $\mathcal{L}_n$ be a sequence of regular subspaces of a Krein space $\mathcal{H}$ such that $\mathcal{L}_n \subset \mathcal{L}_{n+1}$, $n \in \mathbb{N}$. For each $n \in \mathbb{N}$, let P_n be the projection of $\mathcal{H}$ onto $\mathcal{L}_n$. If the projections P_n are uniformly bounded then $\mathcal{L} = \overline{\operatorname{span}}\, \mathcal{L}_n$ is a regular subspace of $\mathcal{H}$.*

Proof. Set $\mathcal{N}_1 = \mathcal{L}_1^{\perp}, \mathcal{N}_2 = \mathcal{L}_2^{\perp}, \ldots$. Then $\mathcal{N}_1, \mathcal{N}_2, \ldots$ is a sequence of regular subspaces of $\mathcal{H}$ such that $\mathcal{N}_1 \supset \mathcal{N}_2 \supset \cdots$. Let $Q_1, Q_2, \ldots$ be the projections of $\mathcal{H}$ onto the subspaces $\mathcal{N}_1, \mathcal{N}_2, \ldots$. Then $Q_n = (1 - P_n)$ and so $\|Q_n\| \leq 1 + \|P_n\|$. Since the projections P_n are uniformly bounded we see that the projections Q_n are also uniformly bounded. Hence $\mathcal{L}$ is a regular subspace of $\mathcal{H}$ by Lemma 3.

Theorem 5. *For $n \geq 1$, let $\mathcal{L}_n$ be regular subspaces of a Krein space $\mathcal{K}$ such that $\mathcal{L}_n \subset \mathcal{L}_{n+1}$. Let P_n be the projection of $\mathcal{K}$ onto $\mathcal{L}_n$. If $\mathcal{L}_n = \mathcal{L}_n^{+} \oplus \mathcal{L}_n^{-}$ is a fundamental decomposition of $\mathcal{L}_n$, we denote by $Q_n^{\pm}$ the projection of $\mathcal{L}_n$ onto $\mathcal{L}_n^{\pm}$. Let*

$$\mathcal{L} = \overline{\operatorname{span}}\, \mathcal{L}_n. \tag{17}$$

Then the following are equivalent:

(1) *$\mathcal{L}$ is regular and there exists a fundamental symmetry J on $\mathcal{L}$ such that $J \mid \mathcal{L}_n$ is a fundamental symmetry on $\mathcal{L}_n$.*

(2) *There exist fundamental decompositions $\mathcal{L}_n = \mathcal{L}_n^{+} \oplus \mathcal{L}_n^{-}$ such that*

  (a) $\mathcal{L}_n^{\pm} \subset \mathcal{L}_{n+1}^{\pm}$,

  (b) $\sup \|Q_n^{\pm} P_n\|_{\widetilde{J}} < \infty$,

where $\widetilde{J}$ is any fundamental symmetry on $\mathcal{K}$.

Proof. Suppose that (1) holds and let $\mathcal{L} = \mathcal{L}_+ \oplus \mathcal{L}_-$ be the fundamental decomposition of $\mathcal{L}$ which gives rise to the fundamental symmetry $J = Q_+ - Q_-$ with the stated property, where $Q_{\pm}\mathcal{L} = \mathcal{L}_{\pm}$. Then $J \mid \mathcal{L}_n = Q_n^{+} - Q_n^{-}$ where $Q_n^{\pm}\mathcal{L}_n = \mathcal{L}_n^{\pm}$. To show that (a) holds, we let $x \in \mathcal{L}_n^{+}$. Then

$$x = Q_n^{+}x - Q_n^{-}x = Jx = Q_{n+1}^{+}x - Q_{n+1}^{-}x. \tag{18}$$

Hence $x = Q_{n+1}^{+}x - Q_{n+1}^{-}x$ and so $(I - Q_{n+1}^{+})x = -Q_{n+1}^{-}x$. This means that $Q_{n+1}^{-}x = -Q_{n+1}^{-}x$ and that $Q_{n+1}^{-}x = 0$. Therefore $x = Q_{n+1}^{+}x \in \mathcal{L}_{n+1}^{+}$. Similarly, we have that $\mathcal{L}_n^{-} \subset \mathcal{L}_{n+1}^{-}$. Hence (a) holds. To show that (b) also holds, we let

$$y \in \mathcal{K} = \mathcal{L}_n \oplus (\mathcal{L} \ominus \mathcal{L}_n) \oplus \mathcal{L}^{\perp}. \tag{19}$$

Let $J_n = J \mid \mathcal{L}_n$, $\widehat{J}_n = J \mid (\mathcal{L} \ominus \mathcal{L}_n)$ and let $J_{\perp}$ be a fundamental symmetry on $\mathcal{L}^{\perp}$. Then $\widetilde{J} := J + J_{\perp}$ is a fundamental symmetry on $\mathcal{K}$. Now,

$$\begin{aligned} \|y\|_{\mathcal{K},\widetilde{J}}^2 &= \langle \widetilde{J}y, y \rangle \\ &= \langle \widetilde{J}(x_n + y_n + z_n), (x_n + y_n + z_n) \rangle \\ &= \langle J_n x_n, x_n \rangle + \langle \widehat{J}_n y_n, y_n \rangle + \langle J_{\perp} z_n, z_n \rangle, \end{aligned} \tag{20}$$

where $y = x_n + y_n + z_n$ with $x_n \in \mathcal{L}_n$, $y_n \in (\mathcal{L} \ominus \mathcal{L}_n)$ and $z_n \in \mathcal{L}^{\perp}$. Therefore

$$\begin{aligned} \|Q_n^{+}P_n y\|_{\mathcal{L}_n,J_n}^2 &= \|Q_n^{+}x_n\|_{\mathcal{L}_n,J_n}^2 = \langle J_n Q_n^{+}x_n, Q_n^{+}x_n \rangle \\ &= \langle Q_n^{+}x_n, Q_n^{+}x_n \rangle = \langle Q_n^{+}x_n, x_n \rangle \\ &= \langle (Q_n^{+} - Q_n^{-}) x_n, x_n \rangle + \langle Q_n^{-}x_n, x_n \rangle \\ &\leq \langle (Q_n^{+} - Q_n^{-}) x_n, x_n \rangle = \langle J_n x_n, x_n \rangle \\ &\leq \langle J_n x_n, x_n \rangle + \langle \widehat{J}_n y_n, y_n \rangle + \langle J_{\perp} z_n, z_n \rangle \\ &= \|y\|_{\mathcal{K}}^2. \end{aligned} \tag{21}$$

Hence $\|Q_n^{+}P_n\| \leq 1$. Similarly $\|Q_n^{-}P_n\| \leq 1$.

Conversely, assume that (2) holds. We start by showing that $\mathcal{L}$ is regular. To do this we note that $P_n = (Q_n^{+} + Q_n^{-})P_n$. So $\|P_n\| \leq \|Q_n^{+}P_n\| + \|Q_n^{-}P_n\| \leq M_1$ for some constant $M_1 > 0$. Hence $\mathcal{L} = \overline{\operatorname{span}}\, \mathcal{L}_n$ is regular by Theorem 4.

Next set $\mathcal{N}_n = (\mathcal{L}_n^{+})^{\perp}$ and let R_n be the projection of $\mathcal{K}$ onto $\mathcal{N}_n = (\mathcal{L}_n^{+})^{\perp}$. Then $R_n = I - Q_n^{+}P_n$. Hence $\|R_n\| \leq 1 + \|Q_n^{+}P_n\| \leq M_2$ for some constant $M_2 > 0$ since (2) holds. Hence $\mathcal{N}^{+} := \bigcap \mathcal{N}_n$ and $\mathcal{M}^{+} := \overline{\operatorname{span}}\, \mathcal{L}_n^{+}$ are regular subspaces by Lemma 3. Let $x \in \mathcal{M}^{+}$. Then there exists a sequence $x_n \in \operatorname{span} \mathcal{L}_n^{+}$ such that $x_n \to x$. Hence $\langle x_n, x_n \rangle \to \langle x, x \rangle$. This means that $\langle x, x \rangle \geq 0$ since $\langle x_n, x_n \rangle \geq 0$. Hence $\mathcal{M}^{+}$ is regular and nonnegative and so it is a uniformly positive subspace of $\mathcal{L}$. Similarly $\mathcal{M}^{-} := \overline{\operatorname{span}}\, \mathcal{L}_n^{-}$ is a uniformly negative subspace of $\mathcal{L}$. We now show that

$$\mathcal{L} = \mathcal{M}^{+} \oplus \mathcal{M}^{-}. \tag{22}$$

Let $x \in \mathcal{M}^+$. Then there exists a sequence $x_n \in \operatorname{span} \mathcal{L}_n^+$ such that $x_n \to x$. If $y \in \mathcal{M}^-$ then there exists a sequence $y_n \in \operatorname{span} \mathcal{L}_n^-$ such that $y_n \to y$. Hence $\langle x, y\rangle = \lim\langle x_n, y_n\rangle = 0$. This implies that $\mathcal{M}^+ \perp \mathcal{M}^-$. Next, assume that $u \in \mathcal{L} \ominus (\mathcal{M}^+ \oplus \mathcal{M}^-)$. Then $u \perp \mathcal{M}^+$ and $u \perp \mathcal{M}^-$. This in turn means that $u \perp \mathcal{L}_n^+$ and $u \perp \mathcal{L}_n^-$ and so $u \perp \mathcal{L}_n$. Let $v \in \mathcal{L}$. Then there exists a sequence $v_n \in \operatorname{span} \mathcal{L}_n$ such that $v_n \to v$. Hence $\langle u, v\rangle = \lim\langle u, v_n\rangle = 0$ for all $v \in \mathcal{L}$ and so $v = 0$. Hence $\mathcal{L} = \mathcal{M}^+ \oplus \mathcal{M}^-$. This implies that $\mathcal{M}^+$ is maximal positive and $\mathcal{M}^-$ is maximal negative. Hence $\mathcal{M}^+$ is a maximal uniformly positive space and $\mathcal{M}^-$ is a maximal uniformly negative space. Since $\mathcal{M}^+ \perp \mathcal{M}^-$, $\mathcal{M}^+$ is uniformly positive and $\mathcal{M}^-$ is uniformly negative, there exists a fundamental decomposition $\mathcal{L} = \mathcal{L}_+ \oplus \mathcal{L}_-$ such that $\mathcal{M}^\pm \subset \mathcal{L}_\pm$. But $\mathcal{M}^\pm$ maximal implies $\mathcal{M}^\pm = \mathcal{L}_\pm$. Hence $\mathcal{L} = \mathcal{M}^+ \oplus \mathcal{M}^-$ is a fundamental decomposition of $\mathcal{L}$. This decomposition gives rise to a fundamental symmetry $J = Q_+ - Q_-$ where $Q_\pm\mathcal{L} = \mathcal{M}^\pm$.

We show that $J \mid \mathcal{L}_n = Q_n^+ - Q_n^-$. Let $x \in \mathcal{L}_n \subset \mathcal{L}$. Then $x = x_+ + x_-$, $x_\pm \in \mathcal{L}_n^\pm \subset \overline{\operatorname{span}}\, \mathcal{L}_n^\pm = \mathcal{M}^\pm$. Hence $Q_+x = Q_+(x_+ + x_-) = x_+ = Q_n^+x$. Similarly, $Q_-x = Q_n^-x$ and so $J \mid \mathcal{L}_n = Q_n^+ - Q_n^-$.

Corollary 6. *For $n \geq 1$, let $\mathcal{K}$, $\mathcal{L}_n$, $\mathcal{L}$, and P_n be as in Theorem 5. Then there exists a fundamental symmetry J on $\mathcal{K}$ which commutes with the projections P_n onto $\mathcal{L}_n$ if and only if $\mathcal{L}$ is regular and there exists a fundamental symmetry J' on $\mathcal{L}$ such that $J' \mid \mathcal{L}_n$ is a fundamental symmetry on $\mathcal{L}_n$.*

Proof. Suppose that $\mathcal{L}$ is regular and J' exists which has the stated property. Let $\mathcal{L} = \mathcal{L}_+ \oplus \mathcal{L}_-$ be the fundamental decomposition which gives rise to J'. Since $\mathcal{L}$ is regular there exists a fundamental decomposition $\mathcal{K} = \mathcal{K}_+ \oplus \mathcal{K}_-$ such that $\mathcal{L}_\pm \subset \mathcal{K}_\pm$. This gives rise to J on $\mathcal{K}$ such that $J \mid \mathcal{L}_n$ is a fundamental symmetry on $\mathcal{L}_n$. Let $f \in \mathcal{K}$. Then $f = f_n + f_n^\perp$, where $f_n \in \mathcal{L}_n$ and $f_n^\perp \in \mathcal{L}_n^\perp$. Now, $JP_nf = Jf_n = f_n^+ - f_n^-$ (where $f_n^\pm \in \mathcal{L}_n^\pm$, $\mathcal{L}_n = \mathcal{L}_n^+ \oplus \mathcal{L}_n^-$). On the other hand,

$$\begin{aligned} P_nJf &= P_nJ\left(f_n + f_n^\perp\right) \\ &= P_nJ\left[\left(f_n^+ + f_n^-\right) + \left(f_n^{\perp +} + f_n^{\perp -}\right)\right] \\ &= P_n\left[J\left(f_n^+ + f_n^-\right) + J\left(f_n^{\perp +} + f_n^{\perp -}\right)\right] \\ &= P_n\left(f_n^+ - f_n^-\right) + P_n\left(f_n^{\perp +} - f_n^{\perp -}\right) = f_n^+ - f_n^-. \end{aligned} \tag{23}$$

Hence $JP_n = P_nJ$.

Conversely, assume that there exists a fundamental symmetry J on $\mathcal{K}$ that commutes with the projections P_n. To show that P_n's are uniformly bounded we consider the fundamental decomposition $\mathcal{K} = \mathcal{K}_+ \oplus \mathcal{K}_-$ which gives rise to the fundamental symmetry J. Let

$$P_n = \begin{pmatrix} A_{11}^n & A_{12}^n \\ A_{21}^n & A_{22}^n \end{pmatrix} : \begin{pmatrix} \mathcal{K}_+ \\ \mathcal{K}_- \end{pmatrix} \longrightarrow \begin{pmatrix} \mathcal{L}_n^+ \\ \mathcal{L}_n^- \end{pmatrix} \tag{24}$$

be the matrix representation of P_n with respect to these decompositions. Then the commutativity condition

$$\begin{pmatrix} 1 & 0 \\ 0 & -1 \end{pmatrix}\begin{pmatrix} A_{11}^n & A_{12}^n \\ A_{21}^n & A_{22}^n \end{pmatrix} = \begin{pmatrix} A_{11}^n & A_{12}^n \\ A_{21}^n & A_{22}^n \end{pmatrix}\begin{pmatrix} 1 & 0 \\ 0 & -1 \end{pmatrix} \tag{25}$$

implies that $A_{21}^n = A_{12}^n = 0$ and so the matrix representation of P_n is diagonal; that is to say,

$$P_n = \begin{pmatrix} A_{11}^n & 0 \\ 0 & A_{22}^n \end{pmatrix} : \begin{pmatrix} \mathcal{K}_+ \\ \mathcal{K}_- \end{pmatrix} \longrightarrow \begin{pmatrix} \mathcal{L}_n^+ \\ \mathcal{L}_n^- \end{pmatrix}. \tag{26}$$

Since $\|A_{11}^n\| \leq 1$ and $\|A_{22}^n\| \leq 1$ for all $n \in \mathbb{N}$, we see that P_n's are uniformly bounded. The uniform boundedness of P_n's implies that $\mathcal{L}$ is regular. From the matrix representation of P_n above we see that $A_{11}^n = Q_n^+P_n$ and $A_{22}^n = Q_n^-P_n$ and so $\sup\|Q_n^\pm P_n\| < \infty$. Hence condition (b) in Theorem 5 holds. Since, for each $n \in \mathbb{N}$, $\mathcal{L}_n$ is a regular subspace and $\mathcal{L}_n \subset \mathcal{L}_{n+1}$, we have that, for any fundamental decomposition $\mathcal{L}_n = \mathcal{L}_n^+ \oplus \mathcal{L}_n^-$, there exist a fundamental decomposition

$$\mathcal{L}_{n+1} = \mathcal{L}_{n+1}^+ \oplus \mathcal{L}_{n+1}^- \tag{27}$$

such that $\mathcal{L}_n^\pm \subset \mathcal{L}_{n+1}^\pm$. Hence condition (a) in Theorem 5 also holds. This shows that part (1) of Theorem 5 holds and this completes the proof.

4. An Extension Theorem for a Sequence of Contractions

In this section, we formulate and give a proof of an extension theorem for a sequence of contractions defined on a nested sequence of regular subspaces. Please refer to [14, Lemma 3.1] and [5] for closely related results.

Theorem 7. *Let $\mathcal{U}$ and $\mathcal{Y}$ be Krein spaces and let $\mathcal{H}_k \subset \mathcal{U}$ and $\mathcal{K}_k \subset \mathcal{Y}$, $k \in \mathbb{N}$, be sequences of regular subspaces satisfying*

$$\begin{aligned} \mathcal{H}_k &\subset \mathcal{H}_{k+1}, \\ \mathcal{K}_k &\subset \mathcal{K}_{k+1}. \end{aligned} \tag{28}$$

Let $P_{\mathcal{H}_k}$ and $P_{\mathcal{K}_k}$ be the orthogonal projections of $\mathcal{U}$ onto $\mathcal{H}_k$ and $\mathcal{Y}$ onto $\mathcal{K}_k$, respectively, and let $\widetilde{\mathcal{H}} = \overline{\operatorname{span}}\, \mathcal{H}_k$ and $\widetilde{\mathcal{K}} = \overline{\operatorname{span}}\, \mathcal{K}_k$. Assume that there exist fundamental symmetries J on $\mathcal{U}$ and J' on $\mathcal{Y}$ such that J commutes with $P_{\mathcal{H}_k}$ and J' commutes with $P_{\mathcal{K}_k}$ and that $\mathcal{U} \ominus \widetilde{\mathcal{H}}$, $\mathcal{Y} \ominus \widetilde{\mathcal{K}}$, $\mathcal{H}_{k+1} \ominus \mathcal{H}_1$, and $\mathcal{K}_{k+1} \ominus \mathcal{K}_1$ are all Hilbert space.

For each $k \in \mathbb{N}$, let

$$A_k : \mathcal{H}_k \longrightarrow \mathcal{K}_k \tag{29}$$

be a contraction. Then there exists a contraction $B : \mathcal{U} \to \mathcal{Y}$ such that

$$B|_{\mathcal{H}_k} = A_k, \quad k \in \mathbb{N}, \tag{30}$$

if and only if

$$\begin{aligned} A_{k+1}|_{\mathcal{H}_k} &= A_k, \\ \sup_{k\in\mathbb{N}} \|A_k\| &< \infty. \end{aligned} \tag{31}$$

Proof. First, let us assume that such an extension $B \in B(\mathcal{U}, \mathcal{Y})$ exists and take $h_k \in \mathcal{H}_k$. Since $\mathcal{H}_k \subset \mathcal{H}_{k+1}$ and (30) holds, we have that

$$A_{k+1}h_k = Bh_k = A_k h_k \tag{32}$$

and therefore $A_{k+1}|_{\mathcal{H}_k} = A_k$. Hence the first condition in (31) holds. To show that the second condition also holds, we first note that since J and J' commute with $P_{\mathcal{H}_k}$ and $P_{\mathcal{K}_k}$, respectively, Corollary 6 ensures that the subspaces $\widetilde{\mathcal{H}}$ and $\widetilde{\mathcal{K}}$ are regular and that there exist fundamental symmetries $J_{\widetilde{\mathcal{H}}}$ on $\widetilde{\mathcal{H}}$ and $J'_{\widetilde{\mathcal{K}}}$ on $\widetilde{\mathcal{K}}$ such that $J_{\widetilde{\mathcal{H}}} \mid \mathcal{H}_k$ is a fundamental symmetry on $\mathcal{H}_k$ and $J'_{\widetilde{\mathcal{K}}} \mid \mathcal{K}_k$ is a fundamental symmetry on $\mathcal{K}_k$. Let $h \in \mathcal{H}_k$. Then we have that $A_k h Bh$. Hence,

$$\begin{aligned} \|A_k h\|_{\widetilde{\mathcal{K}}} &= \|A_k h\|_{\mathcal{K}_k} = \left\langle J'_{\widetilde{\mathcal{K}}} A_k h, A_k h\right\rangle \\ &= \left\langle J'_{\widetilde{\mathcal{K}}} Bh, Bh\right\rangle = \|Bh\|_{\widetilde{\mathcal{K}}} = \|B\|\,\|h\|_{\widetilde{\mathcal{H}}}. \end{aligned} \tag{33}$$

Hence, $\|A_k\| \le \|B\|$ for all $k \in \mathbb{N}$ and so the norms of A_k's are uniformly bounded.

Conversely, let $A_k : \mathcal{H}_k \to \mathcal{K}_k$, $k \in \mathbb{N}$, be contractions satisfying both conditions in (31). Since $A_{k+1}|_{\mathcal{H}_k} = A_k$, we can decompose the operator

$$A_{k+1}|_{\mathcal{H}_k} : \mathcal{H}_k \longrightarrow \begin{pmatrix} \mathcal{K}_k \\ \mathcal{K}_{k+1} \ominus \mathcal{K}_k \end{pmatrix} \tag{34}$$

as $A_{k+1}|_{\mathcal{H}_k} = \begin{pmatrix} A_k \\ 0 \end{pmatrix}$. The fact that A_k is a contraction and $\mathcal{H}_{k+1} \ominus \mathcal{H}_k$ is a Hilbert space implies that the operator

$$\begin{pmatrix} A_k & 0 \end{pmatrix} : \begin{pmatrix} \mathcal{H}_k \\ \mathcal{H}_{k+1} \ominus \mathcal{H}_k \end{pmatrix} \longrightarrow \mathcal{K}_k \tag{35}$$

is a contraction. Since $A_{k+1}|_{\mathcal{H}_k} = \begin{pmatrix} A_k \\ 0 \end{pmatrix}$ is a contraction, Theorem 2 implies that there exists a bounded operator $c_{22} : \mathcal{H}_{k+1} \ominus \mathcal{H}_k \to \mathcal{K}_{k+1} \ominus \mathcal{K}_k$ such that

$$C_{k+1} = \begin{pmatrix} A_k & 0 \\ 0 & c_{22} \end{pmatrix} : \begin{pmatrix} \mathcal{H}_k \\ \mathcal{H}_{k+1} \ominus \mathcal{H}_k \end{pmatrix} \longrightarrow \begin{pmatrix} \mathcal{K}_k \\ \mathcal{K}_{k+1} \ominus \mathcal{K}_k \end{pmatrix} \tag{36}$$

is a contraction. For each $k = 0, 1, 2, \ldots$, the operator C_{k+1} is clearly a contractive extension of A_k.

We now show that, for $k = 0, 1, 2, \ldots$, the contractive liftings C_{k+1}'s of A_k's are uniformly bounded. Let M be the bound for the norm of A_k's. Then $\|\begin{pmatrix} A_k & 0 \end{pmatrix}\| = \|A_k\| \le M$ and so Lemma 1 implies that $\|C_{k+1}\|^2 \le 1 + 2\|A_k\|^2 \le 1 + 2M^2$. Hence C_{k+1}'s are uniformly bounded.

Define an operator $C'_\infty : \operatorname{span} \mathcal{H}_k \to \operatorname{span} \mathcal{K}_k$ by $C'_\infty x = C_{k+1}x$ for $x \in \mathcal{H}_k$. The operator C'_∞ is well defined. To see this, assume that $x \in \mathcal{H}_k$ and $x \in \mathcal{H}_l$. Then $C'_\infty x = C_{k+1}x$ and $C'_\infty x = C_{l+1}x$. For $k < l$, $C_{l+1}|_{\mathcal{H}_k} = A_l|_{\mathcal{H}_k} = A_k = C_{k+1}|_{\mathcal{H}_k}$. The operator C'_∞ is bounded since the contractions C_{k+1} are uniformly bounded. To see that this is the case, consider $x \in \mathcal{H}_k$. Then

$$\begin{aligned} [x, x]_{\widetilde{\mathcal{H}}} &= \langle J_{\widetilde{\mathcal{H}}} x, x\rangle_{\widetilde{\mathcal{H}}} = \langle J_{\widetilde{\mathcal{H}}} x, x\rangle_{\mathcal{H}_k} = \left\langle J_{\mathcal{H}_k} x, x\right\rangle_{\mathcal{H}_k} \\ &= [x, x]_{\mathcal{H}_k}. \end{aligned} \tag{37}$$

Hence for $x \in \operatorname{span} \mathcal{H}_k$, we see that

$$\begin{aligned} \left\|C'_\infty x\right\|_{\widetilde{\mathcal{K}}} &= \left\|C_{k+1}x\right\|_{\widetilde{\mathcal{K}}} = \left\|C_{k+1}x\right\|_{\mathcal{K}_k} \le \left\|C_{k+1}\right\| \|x\|_{\mathcal{H}_k} \\ &= \left\|C_{k+1}\right\| \|x\|_{\widetilde{\mathcal{H}}} \end{aligned} \tag{38}$$

and so C'_∞ is a bounded operator. Since it is defined on a dense set we can extend it by continuity to a bounded operator $C_\infty : \widetilde{\mathcal{H}} \to \widetilde{\mathcal{K}}$. The operator C_∞ is clearly an extension of A_k for each $k \ge 0$ since C_{k+1} is an extension of A_k for each $k \ge 0$. To show that C_∞ is a contraction, let $x \in \widetilde{\mathcal{H}}$. Then there exists a sequence $x_n \in \operatorname{span} \mathcal{H}_k$ such that $x_n \to x$. Since the inequality

$$\left\langle C'_\infty x_n, C'_\infty x_n\right\rangle \le \langle x_n, x_n\rangle \tag{39}$$

holds for each n we have

$$\begin{aligned} \langle C_\infty x, C_\infty x\rangle &= \lim_{n\to\infty} \left\langle C'_\infty x_n, C'_\infty x_n\right\rangle \\ &\le \lim_{n\to\infty} \langle x_n, x_n\rangle = \langle x, x\rangle. \end{aligned} \tag{40}$$

Hence $C_\infty : \widetilde{\mathcal{H}} \to \widetilde{\mathcal{K}}$ is a contraction.

Define a new operator $B : \mathcal{U} \to \mathcal{Y}$ by the matrix

$$B := \begin{pmatrix} C_\infty & E \\ 0 & X \end{pmatrix} : \begin{pmatrix} \widetilde{\mathcal{H}} \\ \mathcal{U} \ominus \widetilde{\mathcal{H}} \end{pmatrix} \longrightarrow \begin{pmatrix} \widetilde{\mathcal{K}} \\ \mathcal{Y} \ominus \widetilde{\mathcal{K}} \end{pmatrix}, \tag{41}$$

where

$$\begin{pmatrix} C_\infty & E \end{pmatrix} \tag{42}$$

is any contractive row extension of C_∞ and $X \in B((\mathcal{U} \ominus \widetilde{\mathcal{H}}), (\mathcal{Y} \ominus \widetilde{\mathcal{K}}))$. Since $\mathcal{Y} \ominus \widetilde{\mathcal{K}}$ is a Hilbert space, Theorem 2 guarantees the existence of X such that B is a contraction. Note that since $\mathcal{U} \ominus \widetilde{\mathcal{H}}$ is also a Hilbert space we may set $E = 0$, the zero operator, in the matrix representation of B and the above result still holds.

Clearly, $B \mid \mathcal{H}_k = A_k$ and so B is the required extension.

5. Applications

In this section we use Theorem 7 to solve a nonstationary extension problem in a Krein space setting. Let $\mathcal{U}_k$ and $\mathcal{Y}_j$ $(j, k \in \mathbb{N})$ be Krein spaces with fixed fundamental decompositions

$$\begin{aligned} \mathcal{U}_k &= \mathcal{U}_k^+ \oplus \mathcal{U}_k^-, \\ \mathcal{Y}_j &= \mathcal{Y}_j^+ \oplus \mathcal{Y}_j^- \end{aligned} \tag{43}$$

and let

$$f_{j,k} : \mathcal{U}_k \longrightarrow \mathcal{Y}_j \tag{44}$$

be bounded operators with matrix representations

$$f_{j,k} = \begin{pmatrix} f_{j,k}^{11} & f_{j,k}^{21} \\ f_{j,k}^{21} & f_{j,k}^{22} \end{pmatrix} : \begin{pmatrix} \mathcal{U}_k^+ \\ \mathcal{U}_k^- \end{pmatrix} \longrightarrow \begin{pmatrix} \mathcal{Y}_j^+ \\ \mathcal{Y}_j^- \end{pmatrix} \tag{45}$$

with $f^{11}_{j,k} = 0$ for $j > k$ and $f^{21}_{j,k} = 0$ for $j > 0$. Define the Krein spaces $\overrightarrow{\mathscr{U}}$ and $\overrightarrow{\mathscr{Y}}$ by

$$\begin{aligned} \overrightarrow{\mathscr{U}} &:= \mathscr{U}_0 \oplus \mathscr{U}_1 \oplus \mathscr{U}_2 \oplus \cdots, \\ \overrightarrow{\mathscr{Y}} &:= \mathscr{Y}_0 \oplus \mathscr{Y}_1 \oplus \mathscr{Y}_2 \oplus \cdots \end{aligned} \tag{46}$$

and let $G : \overrightarrow{\mathscr{U}}$ to $\overrightarrow{\mathscr{Y}}$ be a bounded operator such that

$$P_{\mathscr{Y}_j} G \Big|_{\mathscr{U}_k} = f_{j,k}. \tag{47}$$

We use Theorem 7 to establish conditions under which such an operator G is a contraction. Equality (47) implies that the operator G is of the form

$$G = \begin{pmatrix} \cdots & f_{4,0} & f_{3,0} & f_{2,0} & f_{1,0} & f_{0,0} \\ \cdots & f_{4,1} & f_{3,1} & f_{2,1} & f_{1,1} & f_{0,1} \\ \cdots & f_{4,2} & f_{3,2} & f_{2,2} & f_{1,2} & f_{0,2} \\ \cdots & f_{4,3} & f_{3,3} & f_{2,3} & f_{1,3} & f_{0,3} \\ ⋰ & \vdots & \vdots & \vdots & \vdots & \vdots \end{pmatrix} : \begin{pmatrix} \vdots \\ \mathscr{U}_4 \\ \mathscr{U}_3 \\ \mathscr{U}_2 \\ \mathscr{U}_1 \\ \mathscr{U}_0 \end{pmatrix} \longrightarrow \begin{pmatrix} \mathscr{Y}_0 \\ \mathscr{Y}_1 \\ \mathscr{Y}_2 \\ \mathscr{Y}_3 \\ \mathscr{Y}_4 \\ \vdots \end{pmatrix}. \tag{48}$$

Consider the decompositions

$$\begin{aligned} \overrightarrow{\mathscr{U}} &= \cdots \oplus \mathscr{U}_1^- \oplus \mathscr{U}_0^- \oplus \mathscr{U}_0^+ \oplus \mathscr{U}_1^+ \oplus \cdots, \\ \overrightarrow{\mathscr{Y}} &= \cdots \oplus \mathscr{Y}_1^- \oplus \mathscr{Y}_0^- \oplus \mathscr{Y}_0^+ \oplus \mathscr{Y}_1^+ \oplus \cdots. \end{aligned} \tag{49}$$

With the above decompositions, the operator G takes the form

$$G = \left(\begin{array}{ccccc|ccccc} \ddots & \vdots & \vdots & \vdots & \vdots & \vdots & \vdots & \vdots & \vdots & ⋰ \\ \cdots & f^{11}_{2,3} & f^{11}_{2,2} & 0 & 0 & 0 & 0 & 0 & 0 & \cdots \\ \cdots & f^{11}_{1,3} & f^{11}_{1,2} & f^{11}_{1,1} & 0 & 0 & 0 & 0 & 0 & \cdots \\ \cdots & f^{11}_{0,3} & f^{11}_{0,2} & f^{11}_{0,1} & f^{11}_{0,0} & f^{12}_{0,0} & f^{12}_{0,1} & f^{12}_{0,2} & f^{12}_{0,3} & \cdots \\ \hline \cdots & f^{21}_{0,3} & f^{21}_{0,2} & f^{21}_{0,1} & f^{21}_{0,0} & f^{22}_{0,0} & f^{22}_{0,1} & f^{22}_{0,2} & f^{22}_{0,3} & \cdots \\ \cdots & f^{11}_{1,3} & f^{11}_{1,2} & f^{11}_{1,1} & f^{11}_{1,0} & f^{22}_{1,0} & f^{22}_{1,1} & f^{22}_{1,2} & f^{22}_{1,3} & \cdots \\ ⋰ & \vdots & \vdots & \vdots & \vdots & \vdots & \vdots & \vdots & \vdots & \vdots \end{array}\right) : \left(\begin{array}{c} \vdots \\ \mathscr{U}_2^+ \\ \mathscr{U}_1^+ \\ \mathscr{U}_0^+ \\ \hline \mathscr{U}_0^- \\ \mathscr{U}_1^- \\ \mathscr{U}_2^- \\ \vdots \end{array}\right) \longrightarrow \left(\begin{array}{c} \vdots \\ \mathscr{Y}_2^+ \\ \mathscr{Y}_1^+ \\ \mathscr{Y}_0^+ \\ \hline \mathscr{Y}_0^- \\ \mathscr{Y}_1^- \\ \mathscr{Y}_2^- \\ \vdots \end{array}\right). \tag{50}$$

For each $k \in \mathbb{N}$, let A_k denote the operator

$$A_k = \left(\begin{array}{ccccc|ccccc} f^{11}_{k,k} & 0 & \cdots & 0 & 0 & 0 & 0 & 0 & 0 & \cdots \\ \vdots & \ddots & \vdots & \vdots & \vdots & \vdots & \vdots & \vdots & \vdots & \vdots \\ f^{11}_{2,k} & \cdots & f^{11}_{2,2} & 0 & 0 & 0 & 0 & 0 & 0 & \cdots \\ f^{11}_{1,k} & \cdots & f^{11}_{1,2} & f^{11}_{1,1} & 0 & 0 & 0 & 0 & 0 & \cdots \\ f^{11}_{0,k} & \cdots & f^{11}_{0,2} & f^{11}_{0,1} & f^{11}_{0,0} & f^{12}_{0,0} & f^{12}_{0,1} & f^{12}_{0,2} & f^{12}_{0,3} & \cdots \\ \hline f^{21}_{0,k} & \cdots & f^{21}_{0,2} & f^{21}_{0,1} & f^{21}_{0,0} & f^{22}_{0,0} & f^{22}_{0,1} & f^{22}_{0,2} & f^{22}_{0,3} & \cdots \\ f^{21}_{1,k} & \cdots & f^{21}_{1,2} & f^{21}_{1,1} & f^{21}_{1,0} & f^{22}_{1,0} & f^{22}_{1,1} & f^{22}_{1,2} & f^{22}_{1,3} & \cdots \\ \vdots & \vdots & \vdots & \vdots & \vdots & \vdots & \vdots & \vdots & \vdots & \vdots \end{array}\right) : \left(\begin{array}{c} \mathscr{U}_k \\ \vdots \\ \mathscr{U}_2^+ \\ \mathscr{U}_1^+ \\ \mathscr{U}_0^+ \\ \hline \mathscr{U}_0^- \\ \mathscr{U}_1^- \\ \mathscr{U}_2^- \\ \vdots \end{array}\right) \longrightarrow \left(\begin{array}{c} \mathscr{Y}_k \\ \vdots \\ \mathscr{Y}_2^+ \\ \mathscr{Y}_1^+ \\ \mathscr{Y}_0^+ \\ \hline \mathscr{Y}_0^- \\ \mathscr{Y}_1^- \\ \mathscr{Y}_2^- \\ \vdots \end{array}\right). \tag{51}$$

With these notations, we are in a position to state the following theorem.

Theorem 8. *Let $\mathscr{U}_k$ and $\mathscr{Y}_j$ be Krein spaces with decompositions (43) with operators $f_{j,k} : \mathscr{U}_k \to \mathscr{Y}_j$ in (44). Then there exists a contraction $G : \overrightarrow{\mathscr{U}} \to \overrightarrow{\mathscr{Y}}$ satisfying (47) if and only if, for each $k \in \mathbb{N}$, the operator A_k in (51) defines a contraction whose norm has an upper bound independent of k.*

Proof. Assume that a contraction $G : \overrightarrow{\mathscr{U}} \to \overrightarrow{\mathscr{Y}}$ satisfying (47) exists and consider the operator matrices (50) and (51). We see that, for $k \in \mathbb{N}$,

$$A_k = P_{\mathscr{K}_k} G\big|_{\mathscr{H}_k}, \tag{52}$$

where

$$\mathscr{H}_k = \begin{pmatrix} \mathscr{U}_k \\ \vdots \\ \mathscr{U}_1^+ \\ \mathscr{U}_0^+ \\ \hline \mathscr{U}_0^- \\ \mathscr{U}_1^- \\ \vdots \end{pmatrix}, \qquad \mathscr{K}_k = \begin{pmatrix} \mathscr{Y}_k \\ \vdots \\ \mathscr{Y}_1^+ \\ \mathscr{Y}_0^+ \\ \hline \mathscr{Y}_0^- \\ \mathscr{Y}_1^- \\ \vdots \end{pmatrix}. \tag{53}$$

Since $\overrightarrow{\mathscr{Y}} \ominus \mathscr{K}_k$ is a Hilbert space, we see that $\|P_{\mathscr{K}_k}\| \le 1$ and so

$$\|A_k\| \le \left\|P_{\mathscr{K}_k}\right\| \|G\| \le \|G\|. \tag{54}$$

This shows that the operators A_k are uniformly bounded. Since G is a contraction and $\overrightarrow{\mathscr{Y}} \ominus \mathscr{K}_k$ is a Hilbert space, (52) implies that A_k is a contraction for each $k \in \mathbb{N}$.

To prove the reverse implication, we assume that for each $k \in \mathbb{N}$ the operator $A_k : \mathscr{H}_k \to \mathscr{K}_k$ is a contraction of norm at most γ where $0 < \gamma < \infty$. Since $\operatorname{span} \mathscr{H}_k = \overrightarrow{\mathscr{U}}$ and $\operatorname{span} \mathscr{K}_k = \overrightarrow{\mathscr{Y}}$, it follows that $\overline{\operatorname{span}}\, \mathscr{H}_k$ and $\overline{\operatorname{span}}\, \mathscr{K}_k$ are regular subspaces. By construction, the subspaces $\mathscr{H}_k$ and $\mathscr{K}_k$ satisfy the conditions of Corollary 6 and hence do satisfy all the conditions specified in Theorem 7. Since, for each $k \in \mathbb{N}$, the operators

$$A_k : \mathscr{H}_k \longrightarrow \mathscr{K}_k \tag{55}$$

are defined by (51) and are assumed to be uniformly bounded, condition (31) in Theorem 7 is also fulfilled (it can be easily seen that the first condition in (31) is satisfied by the operators A_k). Since (31) is satisfied for all $k \in \mathbb{N}$, it follows that there exists a contraction B satisfying (30) and therefore must be of the form (50). This concludes the proof.

Conflicts of Interest

The author declares that there are no conflicts of interest regarding the publication of this paper.

References

[1] G. Arsene, T. Constantinescu, and A. Gheondea, "Lifting of operators and prescribed numbers of negative squares," *Michigan Mathematical Journal*, vol. 34, no. 2, pp. 201–216, 1987.

[2] M. A. Dritschel and J. Rovnyak, "Extension theorems for contraction operators on Krein spaces," *Operator Theory: Advances and Applications*, vol. 47, pp. 221–305, 1990.

[3] M. A. Dritschel, "A lifting theorem for bicontractions on Kreĭn spaces," *Journal of Functional Analysis*, vol. 89, no. 1, pp. 61–89, 1990.

[4] A. Gheondea, "Canonical forms of unbouded unitary operators in Krein operators," *Publications of the Research Institute for Mathematical Sciences*, vol. 24, no. 2, pp. 205–224, 1988.

[5] A. Gheondea, "On the structure of linear contractions with respect to fundamental decompositions," *Operator Theory: Advances and Applications*, vol. 48, pp. 275–295, 1990.

[6] C. Foias, A. E. Frazho, I. Gohberg, and M. . Kaashoek, "Metric constrained interpolation, commutant lifting and systems," *Operator Theory: Advances and Applications*, vol. 100, 1998.

[7] C. Foias, A. E. Frazho, I. Gohberg, and M. . Kaashoek, "Parameterization of all solutions of the three chains completion problem," *Integral Equations and Operator Theory*, vol. 29, no. 4, pp. 455–490, 1997.

[8] C. Foias, A. E. Frazho, I. Gohberg, and M. . Kaashoek, "A time-variant version of the commutant lifting theorem and nonstationary interpolation problems," *Integral Equations and Operator Theory*, vol. 28, no. 2, pp. 158–190, 1997.

[9] T. Ando, *Linear Operators on Krein Spaces*, Hokkaido University, Research Institute of Applied Electricity, Division of Applied Mathematics, Sapporo, Japan, 1979.

[10] T. J. Azizov and I. S. Iohvidov, "Linear operators in spaces with an indefinite metric," *Itogi Nauki i Techniki*, pp. 113–205, 272, 1979.

[11] J. Bognar, *Indefinite Inner Product Spaces*, Springer-Verlag, New York, NY, USA, 1974.

[12] I. S. Iohvidov, M. G. Krein, and H. Langer, *Introduction to The Spectral Theory of Operators in Spaces with An Indefinite Metric*, vol. 9 of *Mathematical Research*, Akademie-Verlag, Berlin, Germany, 1982.

[13] G. Arsene and A. Gheondea, "Completing matrix contractions," *The Journal of Operator Theory*, vol. 7, no. 1, pp. 179–189, 1982.

[14] T. Constantinescu and A. Gheondea, "Minimal signature in lifting of operators I," *The Journal of Operator Theory*, vol. 22, pp. 345–367, 1989.

18

Best Proximity Point Theorems for a Berinde MT-Cyclic Contraction on a Semisharp Proximal Pair

Chalongchai Klanarong[1] and Tanadon Chaobankoh[2]

[1] *Department of Mathematics, Faculty of Science, Mahasarakham University, Mahasarakham 44150, Thailand*
[2] *Center of Excellence in Mathematics and Applied Mathematics, Department of Mathematics, Faculty of Science, Chiang Mai University, Chiang Mai 50200, Thailand*

Correspondence should be addressed to Tanadon Chaobankoh; tanadon.c@cmu.ac.th

Academic Editor: Vladimir V. Mityushev

In this paper, a new type of non-self-mapping, called Berinde MT-cyclic contractions, is introduced and studied. Best proximity point theorems for this type of mappings in a metric space are presented. Some examples illustrating our main results are also given. Our results generalize and improve some known results in the literature.

1. Introduction and Preliminaries

Several problems in a real world can be modeled in the form of operator equations. An equation $Sx = x$, which is called the fixed point equation, is one of the important means for solving some problems among them. Fixed point theory is then considered as an important tool for solving such problems. The well-known fixed point theorem for contraction mappings was given by Banach [1]. It is known as the Banach contraction principle. This principle guarantees that each contraction mapping from a complete metric space X into itself always has a unique fixed point.

In 2004, Berinde [2] introduced and studied the concept of weak contraction mappings in the context of a complete metric space. Let (X, d) be a metric space. A mapping $S : X \longrightarrow X$ is called a *weak contraction* if there exist $a \in (0, 1)$ and $L \geq 0$ such that

$$d(Sx, Sy) \leq ad(x, y) + Ld(x, Sy), \quad \text{for all } x, y \in X. \tag{1}$$

A fixed point theorem of this type of mapping was proved in [2]. It extended and generalized that of the Banach contraction principle and others; see [2] and references therein.

On the other hand, if the fixed point equation $Sx = x$ does not have a solution, then $d(x, Sx) > 0$ for all $x \in X$. In this situation, it is natural to ask whether we can find an approximate solution x such that the error is

$$d(x, Sx) = \min_{y \in X} d(y, Sy). \tag{2}$$

In order to have a concrete lower bound, let us consider two nonempty subsets W and V of a metric space X and a mapping $S : W \longrightarrow V$. It is observed that $d(x, Sx) \geq d(W, V)$ for all $x \in W$, where $d(W, V) = \inf\{d(x, y) : x \in W \text{ and } y \in V\}$. So we are interested to find a point $x_0 \in W$ such that

$$d(x_0, Sx_0) = d(W, V). \tag{3}$$

Such point x_0 is called a *best proximity point* of the mapping S, and $d(x_0, Sx_0)$ is called the *global minimum value* of $d(W, V)$.

The best proximity point theorem was first studied by Fan [3], in 1969. He proved that if W is a nonempty compact convex subset in a normed space X and $S : W \longrightarrow X$ is a continuous mapping, then there exists $u \in W$ such that $\|u - Su\| = d(Su, W)$ where $d(Su, W) := \min\{\|Su - a\| : a \in W\}$. Especially, if $S(W) \subseteq W$, then we get that u is a fixed point of S.

Several years later, the above result has been studied and generalized by many researchers, such as Reich [4], Sehgal and Singh [5], Vetrivel et al. [6], Anuradha and Veeramani [7], Basha [8, 9], Kirk et al. [10], Raj [11], Gabeleh [12], Abkar and Gabeleh [13], Eldred and Veeramani [14], and Du and Lakzian [15] and references therein. Some recent research papers worth mentioning are [16–19].

Throughout this paper, we denote by W and V nonempty subsets of a metric space (X, d). We also require the following notions:

$$d(W,V) := \inf\{d(x,y) : x \in W \text{ and } y \in V\},$$
$$W_0 := \{x \in W : d(x,y) = d(W,V) \text{ for some } y \in V\}, \quad (4)$$
$$V_0 := \{y \in V : d(x,y) = d(W,V) \text{ for some } x \in W\}.$$

A mapping $S : W \cup V \longrightarrow W \cup V$ is called a *cyclic mapping* if $S(W) \subseteq V$ and $S(V) \subseteq W$. And a point $x \in W \cup V$ is said to be a *best proximity point* of S if $d(x, Sx) = d(W, V)$. In 2006, Eldred and Veeramani [14] introduced the concept of cyclic contraction and proved the existence of a best proximity point for this type of mapping on a complete metric space.

Definition 1 (see [14]). A mapping $S : W \cup V \longrightarrow W \cup V$ is called a *cyclic contraction* if the following conditions hold:

(i) S is a cyclic mapping;

(ii) there exists $k \in (0,1)$ such that $d(Sx, Sy) \leq kd(x,y) + (1-k)d(W,V)$ for all $x \in W$, $y \in V$.

The concept of *MT*-function was used by Reich [20] and Mizoguchi and Takahashi [21] to define a class of multivalued mappings which is more general than that of contraction mappings. After that Du [22, 23] studied the class of multivalued mappings generated by Mizoguchi and Takahashi functions (or MT-functions) and gave characterizations of MT-functions.

Definition 2 (see [22]). A function $\varphi : [0,\infty) \longrightarrow [0,1)$ is said to be an *MT-function* (or *R-function*) if $\limsup_{s \longrightarrow t^+} \varphi(s) < 1$ for all $t \in [0,\infty)$.

Theorem 3 (see [23]). *Let $\varphi : [0,\infty) \longrightarrow [0,1)$ be a function. Then the following statements are equivalent.*

(a) *φ is an MT-function; i.e., $\limsup_{s \longrightarrow t^+} \varphi(s) < 1$ for all $t \in [0,\infty)$.*

(b) *For each $t \in [0,\infty)$, there exist $r_t \in [0,1)$ and $\varepsilon_t > 0$ such that $\varphi(s) \leq r_t$ for all $s \in (t, t+\varepsilon_t)$.*

(c) *For any nonincreasing sequence $\{x_n\}_{n\in\mathbb{N}}$ in $[0,\infty)$, $0 \leq \sup_{n\in\mathbb{N}} \varphi(x_n) < 1$.*

It is clear that if φ is a nondecreasing function or a nonincreasing function, then φ is an MT-function. For more examples and details, see [15, 22, 23].

Consequently, Du and Lakzian [15] introduced MT-cyclic contractions with respect to φ and proved the existence and convergence theorems for this type of non-self-mapping in metric spaces.

Definition 4 (see [15]). If a map $S : W \cup V \longrightarrow W \cup V$ satisfies the following: then S is called an *MT-cyclic contraction with respect to φ* on $W \cup V$.

(i) S is a cyclic mapping;

(ii) there exists an MT-function $\varphi : [0,\infty) \longrightarrow [0,1)$ such that $d(Sx, Sy) \leq \varphi(d(x,y))d(x,y) + (1 - \varphi(d(x,y)))d(W,V)$ for all $x \in W$ and $y \in V$.

It is obvious that if $\varphi(t) = k$ with $k \in [0,1)$, then S is a cyclic contraction, and, hence, an MT-cyclic contraction with respect to φ which is more general than that of cyclic contraction. For example of an MT-cyclic contraction with respect to φ, but is not a cyclic contraction, see [15].

In 2009, Suzuki et al. [24] introduced the concept of the property UC of two nonempty subsets of a metric space as follows.

Definition 5 (see [24]). Let (W, V) be a pair of nonempty subsets of a metric space (X, d). The pair (W, V) is said to satisfy the *property UC* if $\{x_n\}$ and $\{z_n\}$ are sequences in W and $\{y_n\}$ is a sequence in V such that

$$\lim_{n\longrightarrow\infty} d(x_n, y_n) = d(W,V) = \lim_{n\longrightarrow\infty} d(z_n, y_n), \quad (5)$$

and then $\lim_{n\longrightarrow\infty} d(x_n, z_n) = 0$.

Later, in 2011, Kosuru and Veeramani [25] introduced the concept of semisharp proximal pair of two nonempty subsets of a metric space. This concept is again more general than that of the property UC.

Definition 6 (see [25]). Let (W, V) be a pair of nonempty subsets of a metric space (X, d). The pair (W, V) is said to be a *semisharp proximal pair* if for each $w \in W$ and $u \in V$ there exist at most one $x \in V$ and $y \in W$ such that $d(w, x) = d(W, V) = d(y, u)$.

Example 7. Consider the space X of all real valued continuous functions on $[0,1]$ with the supremum norm; i.e., $X = (C[0,1], \|\cdot\|_\infty)$. Set

$$W = \{f_t : t \in [0,1]\}$$
$$\text{and } V = \{g_t : t \in [0,1]\}, \quad (6)$$

where

$$f_t(x) = \begin{cases} t + x, & \text{if } x \in \left[0, \frac{1}{2}\right]; \\ t - x + 1, & \text{if } x \in \left[\frac{1}{2}, 1\right], \end{cases} \quad (7)$$

and

$$g_t(x) = \begin{cases} t + x + 1, & \text{if } x \in \left[0, \frac{1}{2}\right]; \\ t - x + 2, & \text{if } x \in \left[\frac{1}{2}, 1\right]. \end{cases} \quad (8)$$

It is easy to show that $\|f_t - g_t\| = 1$ and $\|f_t - g_k\| > 1$ for all $t \neq k \in [0,1]$. Hence $d(W, V) = 1$ and (W, V) is a semisharp proximal pair.

Note that the property UC implies semisharp proximality. In 2015, R. Espinola, et al. [26] introduced the concept of a proximally complete pair (W, V) of subsets of a metric space. They proved existences and convergence theorems of best proximity points for cyclic contraction mappings. They obtained a useful theorem presented as follows.

Theorem 8 (Espinola et al. [26, Theorem 3.3]). *Let (W, V) be a pair of complete subsets of a metric space satisfying the UC property. If $\{x_n\}$ is a sequence in $W \cup V$ with $x_{2n} \in W$ and $x_{2n+1} \in V$, for all $n \in \mathbb{N}$, then the sequences $\{x_{2n}\}$ and $\{x_{2n+1}\}$ have convergent subsequences in W and V, respectively.*

By those works mentioned above, we aim to introduce a new type of single-valued, non-self-mapping which is more general than that of contractions, cyclic contractions, and MT-cyclic contractions. The best proximity point theorems for this type of mappings in metric spaces will be investigated. Our main results extend and generalize those of Du and Lakzian [15], Eldred and Veeramani [14], and others.

2. Main Results

By using ideas of cyclic contractions, MT-functions, and weak contractions, we shall first introduce Berinde MT-cyclic contractions with respect to φ and prove the existence and convergence theorems for this type of non-self-mapping in metric spaces.

Definition 9. Let $S : W \cup V \longrightarrow W \cup V$ be a cyclic mapping. The mapping S is said to be a *Berinde MT-cyclic contraction with respect to φ* if there exists an MT-function $\varphi : [0, \infty) \longrightarrow [0, 1)$ and $L \geq 0$ such that

$$\begin{aligned} d(Sx, Sy) \leq{} & \varphi(d(x, y))\, d(x, y) \\ & + (1 - \varphi(d(x, y)))\, d(W, V) \\ & + L \min\{d(y, Sx), d(x, Sy)\}, \end{aligned} \tag{9}$$

for all $x \in W$ and $y \in V$.

It is easy to see that a Berinde MT-cyclic contraction with respect to φ can be reduced to an MT-cyclic contraction with respect to φ.

Remark 10. If $S : W \cup V \longrightarrow W \cup V$ is a Berinde MT-cyclic contraction with respect to φ, then S satisfies the following condition:

$$d(Sx, Sy) \leq d(x, y) + L \min\{d(y, Sx), d(x, Sy)\}, \tag{10}$$

for all $x \in W$ and $y \in V$. To see this, we can write (9) in the form $d(Sx, Sy) - d(W, V) \leq \varphi(d(x, y))[d(x, y) - d(W, V)] + L\min\{d(y, Sx), d(x, Sy)\}$, for all $x \in W$ and $y \in V$. Because of $\varphi(d(x, y)) < 1$, it follows that

$$\begin{aligned} & d(Sx, Sy) - d(W, V) \\ & \quad \leq d(x, y) - d(W, V) \\ & \quad\quad + L \min\{d(y, Sx), d(x, Sy)\}, \end{aligned} \tag{11}$$

and, for all $x \in W$ and $y \in V$, hence (10) is satisfied.

Example 11. Let l^∞ be the metric space consisting of all bounded real sequences with supremum metric d_∞ and let $\{e_n\}$ be the canonical basis of c_0, where c_0 is the space of all null sequences. Let $\{\tau_n\}$ be a sequence of positive real numbers satisfying $\tau_1 = \tau_2$ and $0 < a \leq \tau_{n+1} < \tau_n$ for $n \geq 2$ and for some positive real number a. Thus $\{\tau_n\}$ is convergent. Put $x_n = \tau_n e_n$ for $n \in \mathbb{N}$ and let $X = \{x_1, x_2, x_3, \ldots\}$. Then X is a bounded and complete subset of l^∞, and hence (X, d_∞) is a complete metric space with $d_\infty(x_n, x_m) = \tau_n$ if $m > n$.

Let $W = \{x_1, x_3, x_5, \ldots\}$, $V = \{x_1, x_2, x_4, \ldots\}$ and let $S : W \cup V \longrightarrow W \cup V$ be defined by

$$Sx_n = \begin{cases} x_1 & \text{if } n = 1, 2; \\ x_{n+3} & \text{if } n \geq 3, \end{cases} \tag{12}$$

and define $\varphi : [0, \infty) \longrightarrow [0, 1)$ by

$$\varphi(t) = \begin{cases} \dfrac{\tau_{n+3}}{\tau_n} & \text{if } t = \tau_n \text{ for some } n \in \mathbb{N} \text{ with } n \geq 3; \\ 0 & \text{otherwise.} \end{cases} \tag{13}$$

Then $d(W, V) = 0$ and $S(W) \subseteq V$, $S(V) \subseteq W$. Since $\limsup_{s \to t^+} \varphi(s) = 0 < 1$ for all $t \in [0, \infty)$, we have that φ is an MT-function. Next, we show that S is a Berinde MT-cyclic contraction with respect to φ. Obviously, $d(Sx_1, Sx_i)$ for $i = 1, 2, 3, \ldots$ satisfy (9) with $L \geq 1$. We will consider three cases as follows.

Case 1. For $i \geq 3$ and $x_i \in W$, we have

$$\begin{aligned} d(Sx_2, Sx_i) &= d(x_1, x_{i+3}) = \tau_1 \leq L\tau_2 \\ &= \varphi(d(x_2, x_i))\, d(x_2, x_i) \\ &\quad + (1 - \varphi(d(x_2, x_i)))\, d(W, V) \\ &\quad + L \min\{d(x_2, x_{i+3}), d(x_i, x_1)\}, \\ &\qquad \text{for } L \geq 1. \end{aligned} \tag{14}$$

Case 2. For $i \geq 4$ and $x_i \in V$, we get

$$\begin{aligned} d(Sx_3, Sx_i) &= d(x_6, x_{i+3}) = \tau_6 \\ &\leq \tau_6 + L \min\{d(x_3, x_{i+3}), d(x_i, x_6)\} \\ &= \varphi(d(x_3, x_i))\, d(x_3, x_i) \\ &\quad + (1 - \varphi(d(x_3, x_i)))\, d(W, V) \\ &\quad + L \min\{d(x_3, x_{i+3}), d(x_i, x_6)\}, \\ &\qquad \text{for } L \geq 0. \end{aligned} \tag{15}$$

Case 3. For $j > i \geq 4$ and $x_i, x_j \in W \cup V$, we have

$$\begin{aligned} d\left(Sx_i, Sx_j\right) &= d\left(x_{i+3}, x_{j+3}\right) = \tau_{i+3} \\ &\leq \tau_{i+3} \\ &\quad + L\min\left\{d\left(x_i, x_{j+3}\right), d\left(x_j, x_{i+3}\right)\right\} \\ &= \varphi\left(d\left(x_i, x_j\right)\right) d\left(x_i, x_j\right) \\ &\quad + \left(1 - \varphi\left(d\left(x_i, x_j\right)\right)\right) d\left(W, V\right) \\ &\quad + L\min\left\{d\left(x_i, x_{j+3}\right), d\left(x_j, x_{i+3}\right)\right\}, \\ &\qquad \text{for } L \geq 0. \end{aligned} \tag{16}$$

From all of the above cases, we can now conclude that S is a Berinde MT-cyclic contraction with respect to φ and $L \geq 1$. We note from Case 1 that S is not an MT-cyclic contraction with respect to φ.

Proposition 12. *Let W and V be nonempty subsets of a metric space (X, d) and $S : W \cup V \longrightarrow W \cup V$ be a Berinde MT-cyclic contraction with respect to φ. Then starting with any given $x_0 \in W \cup V$, define a sequence $\{x_n\}$ in $W \cup V$ by $x_{n+1} = Sx_n$ for all $n \geq 0$; we have $d(x_n, x_{n+1}) \longrightarrow d(W, V)$ as $n \longrightarrow \infty$.*

Proof. Let $x_0 \in W \cup V$ be given. Define a sequence $\{x_n\}$ in $W \cup V$ by

$$x_{n+1} = Sx_n, \quad \text{for each } n \geq 0. \tag{17}$$

Suppose that $x_0 \in W$ (when $x_0 \in V$ is similar); then $x_1 = Sx_0 \in V$, and so, $x_2 = Sx_1 \in W$. Since S is a Berinde MT-cyclic contraction with respect to φ, we have

$$\begin{aligned} d\left(x_1, x_2\right) &= d\left(Sx_0, Sx_1\right) \\ &\leq \varphi\left(d\left(x_0, x_1\right)\right) d\left(x_0, x_1\right) \\ &\quad + \left(1 - \varphi\left(d\left(x_0, x_1\right)\right)\right) d\left(W, V\right) \\ &\quad + L\min\left\{d\left(x_1, Sx_0\right), d\left(x_0, Sx_1\right)\right\} \\ &= \varphi\left(d\left(x_0, x_1\right)\right) d\left(x_0, x_1\right) \\ &\quad + \left(1 - \varphi\left(d\left(x_0, x_1\right)\right)\right) d\left(W, V\right) \\ &\quad + L\min\left\{d\left(x_1, x_1\right), d\left(x_0, x_2\right)\right\} \\ &= \varphi\left(d\left(x_0, x_1\right)\right) d\left(x_0, x_1\right) \\ &\quad + \left(1 - \varphi\left(d\left(x_0, x_1\right)\right)\right) d\left(W, V\right). \end{aligned} \tag{18}$$

Again, since $x_2 \in W$ and S is a cyclic mapping, we get $x_3 = Sx_2 \in V$. By the Berinde MT-cyclic contraction with respect to φ of S, we have

$$\begin{aligned} d\left(x_2, x_3\right) &= d\left(Sx_1, Sx_2\right) \\ &\leq \varphi\left(d\left(x_1, x_2\right)\right) d\left(x_1, x_2\right) \\ &\quad + \left(1 - \varphi\left(d\left(x_1, x_2\right)\right)\right) d\left(W, V\right) \\ &\quad + L\min\left\{d\left(x_1, Sx_2\right), d\left(x_2, Sx_1\right)\right\} \\ &= \varphi\left(d\left(x_1, x_2\right)\right) d\left(x_1, x_2\right) \\ &\quad + \left(1 - \varphi\left(d\left(x_1, x_2\right)\right)\right) d\left(W, V\right) \\ &\quad + L\min\left\{d\left(x_1, x_3\right), d\left(x_2, x_2\right)\right\} \\ &= \varphi\left(d\left(x_1, x_2\right)\right) d\left(x_1, x_2\right) \\ &\quad + \left(1 - \varphi\left(d\left(x_1, x_2\right)\right)\right) d\left(W, V\right). \end{aligned} \tag{19}$$

By induction, we can show that, for each $n \geq 1$,

$$\begin{aligned} d\left(x_n, x_{n+1}\right) &\leq \varphi\left(d\left(x_{n-1}, x_n\right)\right) d\left(x_{n-1}, x_n\right) \\ &\quad + \left(1 - \varphi\left(d\left(x_{n-1}, x_n\right)\right)\right) d\left(W, V\right). \end{aligned} \tag{20}$$

By Remark 10, we have

$$\begin{aligned} d\left(x_n, x_{n+1}\right) &= d\left(Sx_{n-1}, Sx_n\right) \\ &\leq d\left(x_{n-1}, x_n\right) \\ &\quad + L\min\left\{d\left(x_{n-1}, Sx_n\right), d\left(x_n, Sx_{n-1}\right)\right\} \\ &= d\left(x_{n-1}, x_n\right), \end{aligned} \tag{21}$$

for all $n \geq 1$. It means that $\{d(x_n, x_{n+1})\}$ is a nonincreasing sequence. By Theorem 3, we get $0 \leq \sup_{n\in\mathbb{N}}\varphi(d(x_n, x_{n+1})) < 1$.

Put $k := \sup_{n\in\mathbb{N}}\varphi(d(x_n, x_{n+1}))$. Thus $0 \leq \varphi(d(x_n, x_{n+1})) \leq k < 1$, for all $n \geq 1$. It follows from (20) that

$$d\left(x_n, x_{n+1}\right) \leq kd\left(x_{n-1}, x_n\right) + (1-k)\, d\left(W, V\right), \quad \text{for all } n \geq 1. \tag{22}$$

Hence for each $n \geq 1$, we have

$$\begin{aligned} d\left(x_n, x_{n+1}\right) &\leq kd\left(x_{n-1}, x_n\right) + (1-k)\, d\left(W, V\right) \\ &\leq k\left[k\left(d\left(x_{n-2}, x_{n-1}\right) + (1-k)\, d\left(W, V\right)\right)\right] \\ &\quad + (1-k)\, d\left(W, V\right) \\ &= k^2\left(d\left(x_{n-2}, x_{n-1}\right) + \left(1-k^2\right) d\left(W, V\right)\right) \\ &\;\;\vdots \\ &\leq k^n\left(d\left(x_0, x_1\right) + \left(1-k^n\right) d\left(W, V\right)\right). \end{aligned} \tag{23}$$

Since $d(W, V) \leq d(x_n, x_{n+1})$ and $k^n \longrightarrow 0$ as $n \longrightarrow \infty$, by taking $n \longrightarrow \infty$ in the above inequality, we obtain that

$$d\left(x_n, x_{n+1}\right) \longrightarrow d\left(W, V\right), \quad \text{as } n \longrightarrow \infty. \tag{24}$$

The proof is now completed.

Example 13. Let $\tau_n = 1/3 + 1/n$ for all $n \in \mathbb{N}$. Then $\lim_{n\longrightarrow\infty}\tau_n = 1/3$. Put $x_n = \tau_n e_n$ for $n \in \mathbb{N}$ and let $X = \{x_1, x_2, x_3, \ldots\}$ be a bounded and complete subset of l^∞. Then

(X, d_∞) be a complete metric space with $d_\infty(x_n, x_m) = \tau_n$ if $m > n$. Set $W = \{x_1, x_3, x_5, \ldots\}$ and $V = \{x_2, x_4, x_6, \ldots\}$. So $\lim_{n \to \infty} d(x_n, x_{n+1}) = \lim_{n\to\infty} \tau_n = 1/3 = d(W, V)$. Let $S : W \cap V \longrightarrow W \cap V$ be defined by

$$Sx_n = x_{n+1} \quad \text{for all } n \in \mathbb{N}. \tag{25}$$

It is easy to see that $S(W) = V$ and $S(V) \subset W$ and so S is a cyclic mapping. Define $\varphi : [0, \infty) \longrightarrow [0, 1)$ as

$$\varphi(t) = \begin{cases} \dfrac{\tau_{n+1}}{\tau_n} & \text{if } t = \tau_n \text{ for some } n \in \mathbb{N}; \\ 0 & \text{otherwise.} \end{cases} \tag{26}$$

Then φ is an MT-function. Now, we will show that S is a Berinde MT-cyclic contraction with respect to φ. For $m, n \in \mathbb{N}$ with $m > n$,

$$\begin{aligned} d(Sx_n, Sx_m) &= \tau_{n+1} \\ &< \tau_{n+1} + \frac{1}{3}\left(1 - \frac{\tau_{n+1}}{\tau_n}\right) \\ &\quad + L \min\{d(x_n, Sx_m), d(x_m, Sx_n)\} \\ &= \varphi(d(x_n, x_m)) d(x_n, x_m) \\ &\quad + (1 - \varphi(d(x_n, x_m))) d(W, V) \\ &\quad + L \min\{d(x_n, Sx_m), d(x_m, Sx_n)\}. \end{aligned} \tag{27}$$

Hence S is a Berinde MT-cyclic contraction with respect to φ. Therefore, all the assumptions of Proposition 12 hold.

The following result is obtained immediately from Proposition 12 because every nondecreasing function or nonincreasing function is an MT-function.

Corollary 14. *Let W and V be nonempty subsets of a metric space (X, d) and $S : W \cup V \longrightarrow W \cup V$ be a cyclic mapping. Let $x_0 \in W$ be given and define a sequence $\{x_n\}$ in $W \cup V$ by $x_{n+1} = Sx_n$, for all $n \geq 0$. Suppose that there exists a nondecreasing (or nonincreasing) function $\eta : [0, \infty) \longrightarrow [0, 1)$ such that for all $x \in W$ and $y \in V$*

$$\begin{aligned} d(Sx, Sy) &\leq \eta(d(x, y)) d(x, y) \\ &\quad + (1 - \eta(d(x, y))) d(W, V) \\ &\quad + L \min\{d(y, Sx), d(x, Sy)\}. \end{aligned} \tag{28}$$

Then $\lim_{n\to\infty} d(x_n, x_{n+1}) = d(W, V)$.

By Proposition 12, if S is a cyclic contraction or an MT-cyclic contraction with respect to φ, then we obtain directly the following results which were proved by Eldred and Veeramani [14] and Du and Lakzian [15], respectively.

Corollary 15 (Eldred and Veeramani [14, Proposition 3.1]). *Let W and V be nonempty subsets of a metric space (X, d) and $S : W \cup V \longrightarrow W \cup V$ be a cyclic contraction. Then starting with any $x_0 \in W$ and defining a sequence $\{x_n\}$ in $W \cup V$ by $x_{n+1} = Sx_n$, for all $n \geq 0$, then $\lim_{n\to\infty} d(x_n, x_{n+1}) = d(W, V)$.*

Corollary 16 (Du and Lakzian [15, Theorem 2.1]). *Let W and V be nonempty subsets of a metric space (X, d) and $S : W \cup V \longrightarrow W \cup V$ be an MT-cyclic contraction with respect to φ. Then starting with any $x_0 \in W \cup V$, define a sequence $\{x_n\}$ in $W \cup V$ by $x_{n+1} = Sx_n$ for all $n \geq 0$; we have $d(x_n, x_{n+1}) \longrightarrow d(W, V)$ as $n \longrightarrow \infty$.*

Observe that if W and V are nonempty subsets of a metric space (X, d) and $S : W \cup V \longrightarrow W \cup V$ is a cyclic mapping with $x_0 \in W$, define a sequence $\{x_n\}$ in $W \cup V$ by $x_{n+1} = Sx_n$, for all $n \geq 0$; then $\{x_{2n}\}$ and $\{x_{2n+1}\}$ are subsequences of $\{x_n\}$ in W and V, respectively. Similarly, if $x_0 \in V$, then $\{x_{2n}\}$ and $\{x_{2n+1}\}$ are subsequences of $\{x_n\}$ in V and W, respectively. Moreover, $d(W, V) \leq d(x_n, x_{n+1})$, for all $n \geq 0$.

Theorem 17. *Let W and V be nonempty subsets of a metric space (X, d) and $S : W \cup V \longrightarrow W \cup V$ be a cyclic mapping. Let $x_0 \in W$ be given and define a sequence $\{x_n\}$ in $W \cup V$ by $x_{n+1} = Sx_n$, for all $n \geq 0$. Suppose that the following conditions hold:*

(i) *$d(Sx, Sy) \leq d(x, y) + L \min\{d(x, Sy), d(y, Sx)\}$ for all $x \in W$ and $y \in V$ with $L \geq 0$;*

(ii) *$\{x_{2n}\}$ has a convergent subsequence in W;*

(iii) *$\lim_{n\to\infty} d(x_n, x_{n+1}) = d(W, V)$.*

Then there exists $x \in W$ such that $d(x, Sx) = d(W, V)$.

Proof. Let $x_0 \in W$ and $\{x_{2n_k}\}$ be a subsequence of $\{x_{2n}\}$ such that

$$\lim_{k\to\infty} x_{2n_k} = x, \quad \text{for some } x \in W. \tag{29}$$

In fact, for each $k \in \mathbb{N}$,

$$\begin{aligned} d(W, V) &\leq d(x, x_{2n_k - 1}) \\ &\leq d(x, x_{2n_k}) + d(x_{2n_k}, x_{2n_k - 1}). \end{aligned} \tag{30}$$

It follows by (iii) that

$$\lim_{k\to\infty} d(x, x_{2n_k - 1}) = d(W, V), \quad \text{as } k \longrightarrow \infty. \tag{31}$$

From (i), for each $k \in \mathbb{N}$, we have

$$\begin{aligned} d(W, V) &\leq d(Sx, x_{2n_k}) = d(Sx, Sx_{2n_k - 1}) \\ &\leq d(x, x_{2n_k - 1}) \\ &\quad + L \min\{d(x, x_{2n_k}), d(x_{2n_k - 1}, Sx)\} \\ &\leq d(x, x_{2n_k - 1}) + Ld(x, x_{2n_k}). \end{aligned} \tag{32}$$

Taking $k \longrightarrow \infty$ in the above inequality, we obtain that

$$d(x, Sx) = \lim_{k\to\infty} d(x_{2n_k}, Sx) = d(W, V). \tag{33}$$

The proof is completed.

Using the same proof as Theorem 17, we obtain a similar result.

Theorem 18. *Let W and V be nonempty subsets of a metric space (X, d) and $S : W \cup V \longrightarrow W \cup V$ be a cyclic mapping. Let $x_0 \in W$ be given and define a sequence $\{x_n\}$ in $W \cup V$ by $x_{n+1} = Sx_n$, for all $n \geq 0$. Suppose that the following conditions hold:*

(i) *$d(Sx, Sy) \leq d(x, y) + L \min\{d(x, Sy), d(y, Sx)\}$ for all $x \in W$ and $y \in V$ with $L \geq 0$;*

(ii) *$\{x_{2n+1}\}$ has a convergent subsequence in V;*

(iii) *$\lim_{n\longrightarrow\infty} d(x_n, x_{n+1}) = d(W, V)$.*

Then there exists $y \in V$ such that $d(y, Sy) = d(W, V)$.

Applying Proposition 12 and Theorems 17 and 18, we establish the following new best proximity point theorems for a Berinde MT-cyclic contraction with respect to φ.

Theorem 19. *Let W and V be nonempty subsets of a metric space (X, d) and $S : W \cup V \longrightarrow W \cup V$ be a Berinde MT-cyclic contraction with respect to φ. Let $x_0 \in W$ be given and define a sequence $\{x_n\}$ in $W \cup V$ by $x_{n+1} = Sx_n$, for all $n \geq 0$. If $\{x_{2n}\}$ has a convergent subsequence in W, then there exists $x \in W$ such that $d(x, Sx) = d(W, V)$.*

Theorem 20. *Let W and V be nonempty subsets of a metric space (X, d) and $S : W \cup V \longrightarrow W \cup V$ be a Berinde MT-cyclic contraction with respect to φ. Let $x_0 \in W$ be given and define a sequence $\{x_n\}$ in $W \cup V$ by $x_{n+1} = Sx_n$, for all $n \geq 0$. If $\{x_{2n+1}\}$ has a convergent subsequence in V, then there exists $y \in V$ such that $d(y, Sy) = d(W, V)$.*

By Theorems 19 and 20, we obtain the next result.

Corollary 21. *Let W and V be nonempty subsets of a metric space (X, d) and $S : W \cup V \longrightarrow W \cup V$ be a Berinde MT-cyclic contraction with respect to φ. Let $x_0 \in W$ be given and define a sequence $\{x_n\}$ in $W \cup V$ by $x_{n+1} = Sx_n$, for all $n \geq 0$. Suppose that W or V is compact. Then there exists $x \in W \cup V$ such that $d(x, Sx) = d(W, V)$.*

Theorem 22. *Let W and V be nonempty subsets of a metric space (X, d) such that (W, V) is a semisharp proximal pair and $S : W \cup V \longrightarrow W \cup V$ is a Berinde MT-cyclic contraction with respect to φ. Let $x_0 \in W$ be given and define a sequence $\{x_n\}$ in $W \cup V$ by $x_{n+1} = Sx_n$, for all $n \geq 0$. If $\{x_{2n}\}$ has a convergent subsequence in W, then the following hold:*

(i) *there exists $x \in W$ such that $d(x, Sx) = d(W, V)$;*

(ii) *x and Sx are fixed point of S^2 in W and V, respectively.*

Proof. By Theorem 19, there exists $x \in W$ such that $d(x, Sx) = d(W, V)$ and it follows that

$$\begin{aligned} d(W, V) &\leq d\left(Sx, S^2x\right) \\ &\leq \varphi(d(x, Sx))\, d(x, Sx) \\ &\quad + (1 - \varphi(d(x, Sx)))\, d(W, V) \\ &\quad + L \min\left\{d\left(x, S^2x\right), d(Sx, Sx)\right\} \\ &= \varphi(d(x, Sx))\, d(x, Sx) \\ &\quad + (1 - \varphi(d(x, Sx)))\, d(W, V) \\ &= d(W, V). \end{aligned} \tag{34}$$

Hence $d(Sx, S^2x) = d(W, V)$. In the semisharp proximality of (W, V), we have $S^2x = x$. Consider

$$\begin{aligned} d\left(x, S^2(Sx)\right) &= d\left(x, S\left(S^2x\right)\right) = d(x, Sx) \\ &= d(W, V), \end{aligned} \tag{35}$$

which implies that $S^2(Sx) = Sx$. Therefore, x and Sx are fixed points of S^2 in W and V, respectively.

Using the proof of Theorem 22, we obtain the following result.

Theorem 23. *Let W and V be nonempty subsets of a metric space (X, d) such that (W, V) is a semisharp proximal pair, and $S : W \cup V \longrightarrow W \cup V$ is a Berinde MT-cyclic contraction with respect to φ. Let $x_0 \in W$ be given and define a sequence $\{x_n\}$ in $W \cup V$ by $x_{n+1} = Sx_n$, for all $n \geq 0$. If $\{x_{2n+1}\}$ has a convergent subsequence in V, then the following hold:*

(i) *there exists $x \in V$ such that $d(x, Sx) = d(W, V)$;*

(ii) *x and Sx are fixed point of S^2 in V and W, respectively.*

By using Theorem 8, we have the following corollary.

Corollary 24. *Let W and V be nonempty complete subsets of a metric space (X, d) such that (W, V) satisfies the property UC, and $S : W \cup V \longrightarrow W \cup V$ is a Berinde MT-cyclic contraction with respect to φ. Let $x_0 \in W$ be given and define a sequence $\{x_n\}$ in $W \cup V$ by $x_{n+1} = Sx_n$, for all $n \geq 0$. Then the following hold:*

(i) *there exists $x \in W$ such that $d(x, Sx) = d(W, V)$;*

(ii) *x and Sx are fixed points of S^2 in W and V, respectively.*

We have discussed that, under some specific conditions, a Berinde MT-cyclic contraction with respect to φ can be reduced to a cyclic contraction or an MT-cyclic contraction with respect to φ. Thus, Theorems 17 and 19 are generalizations of the results proved by Eldred and Veeramani [14] and Du and Lakzian [15], respectively. Hence, the three following corollaries are obtained directly from those theorems.

Corollary 25 (Eldred and Veeramani [14, Proposition 3.2]). *Let W and V be nonempty closed subsets of a complete metric space X. Let $S : W \cup V \longrightarrow W \cup V$ be a cyclic contraction, $x_0 \in W$, and define $x_{n+1} = Sx_n$ for all $n \geq 0$. Suppose that $\{x_{2n}\}$ has a convergent subsequence in W. Then there exists $x \in W$ such that $d(x, Sx) = d(A, B)$.*

Corollary 26 (Du and Lakzian [15, Theorem 2.3]). *Let W and V be nonempty subsets of a metric space (X, d) and $S : W \cup$

$V \longrightarrow W \cup V$ be a cyclic mapping. Let $x_0 \in W$ be given and define a sequence $\{x_n\}$ in $W \cup V$ by $x_{n+1} = Sx_n$, for all $n \geq 0$. Suppose that the following conditions hold:

(i) *$d(Sx, Sy) \leq d(x, y)$ for all $x \in W$ and $y \in V$;*

(ii) *$\{x_{2n}\}$ has a convergent subsequence in W;*

(iii) *$\lim_{n\longrightarrow\infty} d(x_n, x_{n+1}) = d(W, V)$.*

Then there exists $x \in W$ such that $d(x, Sx) = d(W, V)$.

Corollary 27 (Du and Lakzian [15, Theorem 2.4]). *Let W and V be nonempty subsets of a metric space (X, d) and $S : W \cup V \longrightarrow W \cup V$ be a MT-cyclic contraction with respect to φ. Let $x_0 \in W$ be given and define a sequence $\{x_n\}$ in $W \cup V$ by $x_{n+1} = Sx_n$, for all $n \geq 0$. Suppose that $\{x_{2n}\}$ has a convergent subsequence in W. Then there exists $x \in W$ such that $d(x, Sx) = d(W, V)$.*

Conflicts of Interest

The authors declare that there are no conflicts of interest regarding the publication of this paper.

Acknowledgments

The authors would like to thank Professor Dr. Suthep Suantai for his helpful advice. This research is supported by Center of Excellence in Mathematics and Applied Mathematics, Chiang Mai University, and Faculty of Science, Chiang Mai University, Chiang Mai, Thailand.

References

[1] S. Banach, "Sur les opérations dans les ensembles abstraits et leur application aux équations intégrales," *Fundamenta Mathematicae*, vol. 3, pp. 133–181, 1922.

[2] V. Berinde, "Approximating fixed points of weak contractions using the Picard iteration," *Nonlinear Analysis Forum* , vol. 9, no. 1, pp. 43–53, 2004.

[3] K. Fan, "Extensions of two fixed point theorems of F. E. Browder," *Mathematische Zeitschrift*, vol. 112, pp. 234–240, 1969.

[4] S. Reich, "Approximate selections, best approximations, fixed points, and invariant sets," *Journal of Mathematical Analysis and Applications*, vol. 62, no. 1, pp. 104–113, 1978.

[5] V. M. Sehgal and S. P. Singh, "A theorem on best approximations," *Numerical Functional Analysis and Optimization*, vol. 10, no. 1-2, pp. 181–184, 1989.

[6] V. Vetrivel, P. Veeramani, and P. Bhattacharyya, "Some extensions of Fan's best approximation theorem," *Numerical Functional Analysis and Optimization*, vol. 13, no. 3-4, pp. 397–402, 1992.

[7] J. Anuradha and P. Veeramani, "Proximal pointwise contraction," *Topology and its Applications*, vol. 156, no. 18, pp. 2942–2948, 2009.

[8] S. Sadiq Basha, "Best proximity points: optimal solutions," *Journal of Optimization Theory and Applications*, vol. 151, no. 1, pp. 210–216, 2011.

[9] S. Sadiq Basha, "Best proximity point theorems," *Journal of Approximation Theory*, vol. 163, no. 11, pp. 1772–1781, 2011.

[10] W. A. Kirk, S. Reich, and P. Veeramani, "Proximinal retracts and best proximity pair theorems," *Numerical Functional Analysis and Optimization*, vol. 24, no. 7-8, pp. 851–862, 2003.

[11] V. S. Raj, "A best proximity point theorem for weakly contractive non-self-mappings," *Nonlinear Analysis: Theory, Methods & Applications*, vol. 74, no. 14, pp. 4804–4808, 2011.

[12] M. Gabeleh, "Global optimal solutions of non-self mappings," *"Politehnica" University of Bucharest. Scientific Bulletin. Series A. Applied Mathematics and Physics*, vol. 75, no. 3, pp. 67–74, 2013.

[13] A. Abkar and M. Gabeleh, "Best proximity points of non-self mappings," *TOP*, vol. 21, no. 2, pp. 287–295, 2013.

[14] A. A. Eldred and P. Veeramani, "Existence and convergence of best proximity points," *Journal of Mathematical Analysis and Applications*, vol. 323, no. 2, pp. 1001–1006, 2006.

[15] W.-S. Du and H. Lakzian, "Nonlinear conditions for the existence of best proximity points," *Journal of Inequalities and Applications*, vol. 2012, article 206, 2012.

[16] H. Aydi, A. Felhi, and E. Karapinar, "On common best proximity points for generalized α-ψ-proximal contractions," *Journal of Nonlinear Sciences and Applications. JNSA*, vol. 9, no. 5, pp. 2658–2670, 2016.

[17] A. Felhi, "A note on "Convergence and best proximity points for Berinde's cyclic contraction with proximally complete property"," *Mathematical Methods in the Applied Sciences*, vol. 41, no. 1, pp. 140–143, 2018.

[18] A. Felhi, "New best proximity results on partial ordered metric spaces," *Journal of Advanced Mathematical Studies*, vol. 10, no. 2, pp. 216–230, 2017.

[19] A. Felhi and H. Aydi, "Best proximity points and stability results for controlled proximal contractive set valued mappings," *Fixed Point Theory and Applications*, Paper No. 22, 23 pages, 2016.

[20] S. Reich, "Fixed points of contractive functions," *Bollettino della Unione Matematica Italiana*, vol. 5, pp. 26–42, 1972.

[21] N. Mizoguchi and W. Takahashi, "Fixed point theorems for multivalued mappings on complete metric spaces," *Journal of Mathematical Analysis and Applications*, vol. 141, no. 1, pp. 177–188, 1989.

[22] W. Du, "Some new results and generalizations in metric fixed point theory," *Nonlinear Analysis: Theory, Methods & Applications*, vol. 73, no. 5, pp. 1439–1446, 2010.

[23] W. Du, "On coincidence point and fixed point theorems for nonlinear multivalued maps," *Topology and its Applications*, vol. 159, no. 1, pp. 49–56, 2012.

[24] T. Suzuki, M. Kikkawa, and C. Vetro, "The existence of best proximity points in metric spaces with the property UC," *Nonlinear Analysis: Theory, Methods & Applications*, vol. 71, no. 7-8, pp. 2918–2926, 2009.

[25] G. S. Kosuru and P. Veeramani, "A note on existence and convergence of best proximity points for pointwise cyclic contractions," *Numerical Functional Analysis and Optimization*, vol. 32, no. 7, pp. 821–830, 2011.

[26] R. Espinola, G. S. Kosuru, and P. Veeramani, "Pythagorean property and best-proximity point theorems," *Journal of Optimization Theory and Applications*, vol. 164, no. 2, pp. 534–550, 2015.

19

New Extension of Beta Function and Its Applications

Mehar Chand,[1] **Hanaa Hachimi**,[2] **and Rekha Rani**[3]

[1]*Department of Mathematics, Baba Farid College, Bathinda 151001, India*
[2]*BOSS Team, GS laboratory, ENSA, Ibn Tofail University, Kenitra 14000, Morocco*
[3]*Department of Applied Sciences, Gurukashi University, Bathinda 151302, India*

Correspondence should be addressed to Mehar Chand; mehar.jallandhra@gmail.com

Academic Editor: Niansheng Tang

In the present paper, new type of extension of classical beta function is introduced and its convergence is proved. Further it is used to introduce the extension of Gauss hypergeometric function and confluent hypergeometric functions. Then we study their properties, integral representation, certain fractional derivatives, and fractional integral formulas and application of these functions.

1. Introduction and Preliminaries

No doubt the classical beta function $B(\alpha, \beta)$ is one of the most fundamental special functions, because of its precious role in several field of sciences such as mathematical, physical, and statistical sciences and engineering. In many areas of applied mathematics, different types of special functions have become necessary tool for the scientists and engineers. During the recent decades or so, numerous interesting and useful extensions of the different special functions (the Gamma and beta functions, the Gauss hypergeometric function, and so on) have been introduced by different authors [1–6].

In 1997 Choudhary et al. [1] introduced the following extension of classical beta function defined as

$$B_p(\alpha,\beta) = B(\alpha,\beta;p) = \int_0^1 t^{\alpha-1}(1-t)^{\beta-1}\exp\left(-\frac{p}{t(1-t)}\right)dt, \quad (\Re(p) \geq 0). \tag{1}$$

Further Chaudhry et al. [7, p. 591, Eqs. (2.1) and (2.2)] made use of the extended beta function $B_p(\alpha, \beta)$ in (1) to extend the Gauss hypergeometric function and confluent hypergeometric function as follows:

$$F_p(a,b;c;z) = \sum_{n=0}^{\infty}(a)_n\frac{B_p(b+n,c-b)}{B(b,c-b)}\frac{z^n}{n!}, \quad (|z|<1;\ \Re(c)>\Re(b)>0;\ \Re(p)\geq 0) \tag{2}$$

and

$$\Phi_p(b;c;z) = \sum_{n=0}^{\infty}\frac{B_p(b+n,c-b)}{B(b,c-b)}\frac{z^n}{n!}, \quad (|z|<1;\ \Re(c)>\Re(b)>0;\ \Re(p)\geq 0). \tag{3}$$

and present their Euler type integrals as follows:

$$F_p(a,b;c;z) = \frac{1}{B(b,c-b)} \cdot \int_0^1 t^{b-1}(1-t)^{c-b-1}(1-zt)^{-a}\exp\left[-\frac{p}{t(1-t)}\right]dt, \quad (p>0;\ p=0,\ |\arg(1-z)|<\pi;\ \Re(c)>\Re(b)>0) \tag{4}$$

and

$$\Phi_p(b;c;z) = \frac{1}{B(b,c-b)} \cdot \int_0^1 t^{b-1}(1-t)^{c-b-1}\exp\left[zt-\frac{p}{t(1-t)}\right]dt, \quad (p>0;\ p=0;\ \Re(c)>\Re(b)>0). \tag{5}$$

If we choose $p = 0$, the above definitions given in (1), (2), (3), (4), and (5) reduce to the following form, respectively:

$$B(\alpha,\beta) = \int_0^1 t^{\alpha-1}(1-t)^{\beta-1}\,dt, \quad (\Re(\alpha) > 0,\ \Re(\beta) > 0). \tag{6}$$

$$F(a,b;c;z) = \sum_{n=0}^{\infty}(a)_n \frac{B(b+n,c-b)}{B(b,c-b)}\frac{z^n}{n!}, \quad (|z| < 1;\ \Re(c) > \Re(b) > 0). \tag{7}$$

$$\Phi(b;c;z) = \sum_{n=0}^{\infty}\frac{B(b+n,c-b)}{B(b,c-b)}\frac{z^n}{n!}, \quad (|z| < 1;\ \Re(c) > \Re(b) > 0). \tag{8}$$

$$F(a,b;c;z) = \frac{1}{B(b,c-b)}\int_0^1 t^{b-1}(1-t)^{c-b-1}(1-zt)^{-a}\,dt, \quad (|\arg(1-z)| < \pi;\ \Re(c) > \Re(b) > 0). \tag{9}$$

$$\Phi(b;c;z) = \frac{1}{B(b,c-b)}\int_0^1 t^{b-1}(1-t)^{c-b-1}\exp(zt)\,dt, \quad (\Re(c) > \Re(b) > 0). \tag{10}$$

Gauss hypergeometric function and confluent hypergeometric function are special cases of the generalized hypergeometric series ${}_pF_q(p,q \in \mathbb{N})$ defined as (see [8, p.73]) and [9, pp. 71-75]:

$$ {}_pF_q\left[\begin{matrix}\alpha_1,\dots,\alpha_p; \\ \beta_1,\dots,\beta_q;\end{matrix}\ z\right] = \sum_{n=0}^{\infty}\frac{(\alpha_1)_n\cdots(\alpha_p)_n}{(\beta_1)_n\cdots(\beta_q)_n}\frac{z^n}{n!} = {}_pF_q(\alpha_1,\dots,\alpha_p;\beta_1,\dots,\beta_q;z), \tag{11}$$

where $(\xi)_n$ is the Pochhammer symbol defined (for $\xi \in \mathbb{C}$) by (see[9, p.2 and p.5])

$$(\xi)_n := \begin{cases} 1 & (n = 0) \\ \xi(\xi+1)\dots(\xi+n-1) & (n \in \mathbb{N}) \end{cases} \tag{12}$$

$$= \frac{\Gamma(\xi+n)}{\Gamma(\xi)} \quad (\xi \in \mathbb{C}\setminus\mathbb{Z}_0^-), \tag{13}$$

and $\mathbb{Z}_0^-$ denotes the set of nonpositive integers and $\Gamma(\xi)$ is familiar Gamma function.

The Fox-Wright function ${}_p\Psi_q$ is defined as (see, for details, Srivastava and Karisson [10])

$$ {}_p\Psi_q[z] = {}_p\Psi_q\left[\begin{matrix}(a_1,\alpha_1),\dots,(a_p,\alpha_p); \\ (b_1,\beta_1),\dots,(b_q,\beta_q);\end{matrix}\ z\right] = {}_p\Psi_q\left[\begin{matrix}(a_i,\alpha_i)_{1,p}; \\ (b_j,\beta_j)_{1,q};\end{matrix}\ z\right] = \sum_{n=0}^{\infty}\frac{\prod_{i=1}^{p}\Gamma(a_i+\alpha_i n)}{\prod_{j=1}^{q}\Gamma(b_j+\beta_j n)}\frac{z^n}{n!}, \tag{14}$$

where the coefficients $\alpha_1,\dots,\alpha_p,\ \beta_1,\dots,\beta_q \in \mathbb{R}^+$ such that

$$1 + \sum_{j=1}^{q}\beta_j - \sum_{i=1}^{p}\alpha_i \geq 0. \tag{15}$$

Motivated from the above literature, we introduce new extension of classical beta function in (16) and its convergence is studied in Theorem 1 in Section 2. Using MATLAB(R2015a), the numerical results and graphs are presented in Section 3 and also radius of convergence of new extension of classical beta function is discussed on the basis of numerical results established by using MATLAB software. We establish the integral representations and study the properties of new extension of classical beta function.

Using the new extended beta function, extension of the beta distribution is also introduced; Gauss hypergeometric function and confluent hypergeometric function are extended by employing the new extension of classical beta function. Then we have studied the generating relations, extension of Riemann-Lioville fractional derivative operator. Fractional integrals of extended hypergeometric functions and their image formulas in the form of beta transform, Laplace transform, and Whittaker transform have been also established. The solutions of fractional kinetic equations involving extended Gauss hypergeometric function and extended confluent hypergeometric function are established. The numerical results and graphical interpretation have made it easier to study the nature of these fractional kinetic equations.

2. Extension of Beta Function

In this section, we introduce new extension of classical beta function. Its convergence is proved mathematically; then numerical results are established for different values of parameters involved.

We introduce new extension of classical beta function as follows:

$$ {}^{MC}B_m(\alpha,\beta) = \int_0^1 t^{\alpha-1}(1-t)^{\beta-1}e^{mt(1-t)}\,dt, \tag{16}$$

where $\Re(\alpha) > 0, \Re(\beta) > 0, m \in \mathbb{C}; |m| < M$ (where M is positive number).

Theorem 1. *If $\Re(\alpha) > 0, \Re(\beta) > 0, m \in \mathbb{C}; |m| < M$ (where M is positive number), then the new extension of the classical beta function in equation (16) is convergent.*

Proof. We can write (16) as follows:

$$\begin{aligned} {}^{MC}B_m(\alpha,\beta) &= \int_0^1 t^{\alpha-1}(1-t)^{\beta-1}\sum_{n=0}^{\infty}\frac{[mt(1-t)]^n}{n!}dt \\ &= \sum_{n=0}^{\infty}\frac{m^n}{n!}\int_0^1 t^{\alpha+n-1}(1-t)^{\beta+n-1}dt, \end{aligned} \tag{17}$$

and further, using the definition of classical beta function (6), (17) reduces to

$${}^{MC}B_m(\alpha,\beta) = \sum_{n=0}^{\infty}\frac{m^n}{n!}B(\alpha+n,\beta+n). \tag{18}$$

In the above equation, ${}^{MC}B_m(\alpha,\beta)$ is in series form involving $B(\alpha+n,\beta+n)$ (where $n = 0,1,2,\ldots$) and in each term of the series, $B(\alpha+n,\beta+n)$ is convergent, since $\Re(\alpha+n) > 0$ and $\Re(\beta+n) > 0$ for $\Re(\alpha) > 0$ and $\Re(\beta) > 0$, which implies that each term of the series (18) exists.

Now we shall prove that ${}^{MC}B_m(\alpha,\beta)$ is convergent. m may be greater than or less than 0, so there are two cases as follows.

Case 1. If $m > 0$, then we need to prove that ${}^{MC}B_m(\alpha,\beta)$ is convergent.

Equation (18) can be written as

$$\begin{aligned} {}^{MC}B_m(\alpha,\beta) &= \sum_{n=0}^{\infty}a_n, \\ \text{where } a_n &= \frac{m^n}{n!}B(\alpha+n,\beta+n). \end{aligned} \tag{19}$$

Further,

$$\lim_{n\to\infty}\frac{a_n}{a_{n+1}} = \infty > 1. \tag{20}$$

By ratio test for positive series, ${}^{MC}B_m(\alpha,\beta)$ is convergent for $m > 0$.

Case 2. If $m < 0$, then we need to prove that the extension of classical beta function $B_m(\alpha,\beta)$ is convergent.

To prove this case, let $m = -p$ (where $p > 0$); then (18) becomes

$${}^{MC}B_m(\alpha,\beta) = \sum_{n=0}^{\infty}\frac{(-p)^n}{n!}B(\alpha+n,\beta+n). \tag{21}$$

Equation (21) can be written as

$$\begin{aligned} {}^{MC}B_m(\alpha,\beta) &= -\sum_{n=0}^{\infty}(-1)^{n-1}b_n, \\ \text{where } b_n &= \frac{p^n}{n!}B(\alpha+n,\beta+n). \end{aligned} \tag{22}$$

The series (22) is an alternating series; therefore

(1) $b_n > 0, \forall p > 0, \Re(\alpha,\beta) > 0$

(2) $b_n - b_{n+1} = (p^n/n!)B(\alpha+n,\beta+n)[1-(\alpha+n)(\beta+n)/(\alpha+\beta+2n)(\alpha+\beta+2n+1)] > 0 \Longrightarrow b_n$ is decreasing

(3) $\lim_{n\to\infty}b_n = 0$ if $p \le 2$ ($p^n/n! \longrightarrow 0$ as $n \longrightarrow \infty$ only if $p \le 2$ and $B(\alpha+n,\beta+n) \longrightarrow 0$ as $n \longrightarrow \infty$)

All the conditions of Leibniz's test for alternating series have been satisfied; therefore $B_m(\alpha,\beta)$ is convergent for $0 < p \le 2$ *i.e.* $-2 \le m < 0$.

From Cases 1 and 2 it is implied that the power series in (18) is convergent.

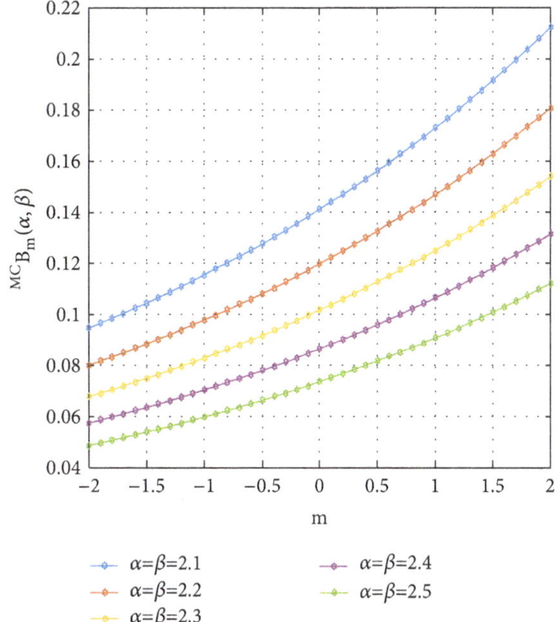

FIGURE 1: Graph of new extension of classical beta function for fixed value of α, β and $m = -2 : .5 : 2$.

3. Numerical Results and Graphs of New Extension of the Classical Beta Function

The numerical results of new extension of classical beta function have been calculated in this section. For this purpose we choose the values of variables α,β and parameter m as $\alpha,\beta \in [0,10]$ and $m \in [-2.0335, 2.0335]$. All the numerical values of new extension of the classical beta function are presented in Tables 1 and 2, from which we can easily observe that ${}^{MC}B_m(\alpha,\beta)$ does not exist at $\alpha = \beta = 0$ and it is also investigated that ${}^{MC}B_m(\alpha,\beta)$ does not exist for $m < -2.0335$ and $m > 2.0335$; ${}^{MC}B_m(\alpha,\beta) \longrightarrow \infty$ as $\alpha,\beta \longrightarrow 0$ and ${}^{MC}B_m(\alpha,\beta) \longrightarrow 0$ as $\alpha,\beta \longrightarrow \infty$, which implies that the behaviour of new extension of classical beta function is the same as that of classical beta function.

We also check the effect of m on the new extension of classical beta function. For this purpose, we fix the values of α and β as shown in Figure 1, then we plot the graph which depicts that ${}^{MC}B_m(\alpha,\beta)$ is an increasing function as the values of m increase. It is very clear from Figure 1 that for the graph of classical beta function, new extension of

TABLE 1: Numerical values of new extension of classical beta function ${}^{MC}B_m(\alpha, \beta)$.

α	β	$m = -2.0335$	$m = -2.0$	$m = -1.0$	$m = 0.0$
0.00	0.00	NaN	NaN	NaN	NaN
0.01	0.01	198.27177610	198.29548624	199.06411204	199.96757732
0.02	0.02	98.27460320	98.29781890	99.05065967	99.93608768
0.03	0.03	64.94418565	64.96691911	65.70435720	66.57217010
0.04	0.04	48.28052079	48.30278382	49.02519017	49.87579792
0.05	0.05	38.28360599	38.30541007	39.01314459	39.84694542
0.06	0.06	31.62010537	31.64146162	32.33487366	33.15225447
0.07	0.07	26.86144500	26.88236422	27.56179291	28.36312969
0.08	0.08	23.29333673	23.31382939	23.97960405	24.76526226
0.09	0.09	20.51895277	20.53902903	21.19146950	21.96180446
0.10	0.10	18.30019555	18.31986529	18.95928227	19.71463949
1.0	1.0	0.72102562	0.72477846	0.84887277	1.00000000
2.0	2.0	0.11166177	0.11238923	0.13665458	0.16666667
3.0	3.0	0.02163870	0.02179192	0.02692731	0.03333333
4.0	4.0	0.00455690	0.00459058	0.00572283	0.00714286
5.0	5.0	0.00100158	0.00100918	0.00126512	0.00158730
6.0	6.0	0.00022592	0.00022766	0.00028649	0.00036075
7.0	7.0	0.00005185	0.00005225	0.00006594	0.00008325
8.0	8.0	0.00001205	0.00001214	0.00001536	0.00001943
9.0	9.0	0.00000283	0.00000285	0.00000361	0.00000457
10.0	10.0	0.00000067	0.00000067	0.00000085	0.00000108

TABLE 2: Numerical values of new extension of classical beta function ${}^{MC}B_m(\alpha, \beta)$.

α	β	$m = 0.0$	$m = 1.0$	$m = 2.0$	$m = 2.0335$
0.00	0.00	NaN	NaN	NaN	NaN
0.01	0.01	199.96757732	201.03554301	202.30476477	202.35127432
0.02	0.02	99.93608768	100.98330172	102.22848838	102.27412810
0.03	0.03	66.57217010	67.59910183	68.82077895	68.86556756
0.04	0.04	49.87579792	50.88290361	52.08158267	52.12553836
0.05	0.05	39.84694542	40.83466875	42.01084757	42.05398809
0.06	0.06	33.15225447	34.12102693	35.27519021	35.31753282
0.07	0.07	28.36312969	29.31337104	30.44599075	30.48755229
0.08	0.08	24.76526226	25.69738092	26.80891675	26.84971362
0.09	0.09	21.96180446	22.87619790	23.96709764	24.00714582
0.10	0.10	19.71463949	20.61169459	21.68239456	21.72170964
1.0	1.0	1.00000000	1.18459307	1.41068613	1.41909001
2.0	2.0	0.16666667	0.20385173	0.25000000	0.25172603
3.0	3.0	0.03333333	0.04133360	0.05133577	0.05171118
4.0	4.0	0.00714286	0.00892526	0.01116423	0.01124846
5.0	5.0	0.00158730	0.00199311	0.00250456	0.00252383
6.0	6.0	0.00036075	0.00045452	0.00057299	0.00057746
7.0	7.0	0.00008325	0.00010515	0.00013287	0.00013392
8.0	8.0	0.00001943	0.00002458	0.00003112	0.00003137
9.0	9.0	0.00000457	0.00000579	0.00000734	0.00000740
10.0	10.0	0.00000108	0.00000137	0.00000174	0.00000176

classical beta function remains concave upward (or convex downward) for different values of α, β, and m. The value of m does not affect the nature of classical beta function; the main effect of the value of m is that it just pushes the curve up or drags down the curve from the curve of the classical beta function. In Figure 2, Mesh-Plot is established of new extension of classical beta function, which can be easily interpreted.

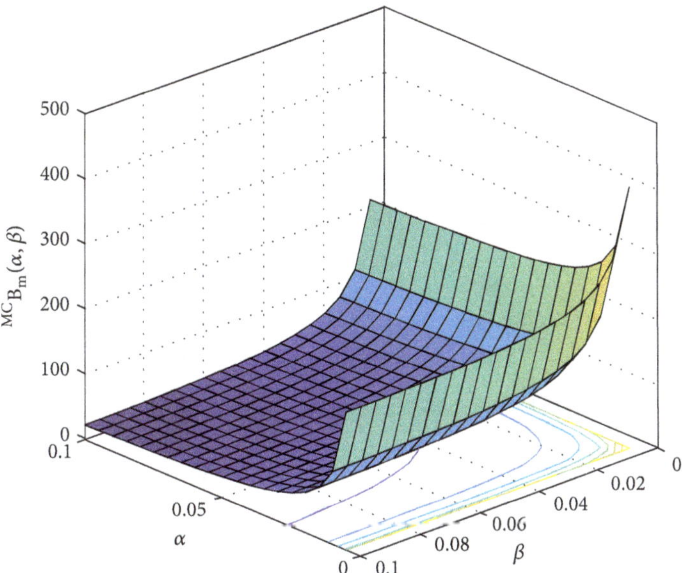

FIGURE 2: Mesh-Plot of the new extension of classical beta function.

From the above proof of radius of convergence of series and further numerical investigation of the power series in Tables 1 and 2, we find that the interval of convergence of the series is $[-2.0335, 2.0335]$, which implies that $^{MC}B_m(\alpha, \beta)$ is convergent for $|m| < M$, where M is positive number not greater than 2.0335.

Note 2. From the above discussion, it is easy to conclude that the value of $\Re(m)$ lies in the interval $[-2.0335, 2.0335]$; i.e., $-2.0335 \le \Re(m) \le 2.0335$.

Note 3. In the sequel of this paper, $|m| < M$ represents the circle of convergence and M is the radius of convergence of (16), where M is not greater than 2.0335.

Remark 4. For $\Re(\alpha) > 0$, $\Re(\beta) > 0$, $m \in \mathbb{C}$; $|m| < M$ (where M is positive number not greater than 2.0335), the new extension of classical beta function can be presented in the relation Fox-Wright function (see (14)) as follows:

$$\begin{aligned} ^{MC}B_m(\alpha, \beta) &= \sum_{n=0}^{\infty} \frac{m^n}{n!} \frac{\Gamma(\alpha+n)\Gamma(\beta+n)}{\Gamma(\alpha+\beta+2n)} \\ &= {}_2\Psi_1\left[\begin{matrix}(\alpha,1),(\beta,1)\\(\alpha+\beta,2)\end{matrix}; m\right]. \end{aligned} \tag{23}$$

The above result is obtained from (18).

4. Integral Representation of the New Extension of Classical Beta Function

The integral representation of the new extended beta function is important both to check whether the extension is natural and simple and for later use. It is also important to investigate the relationship between the classical beta function and the new extension of the classical beta function. In this connection, we first provide a relationship between them. The following integral formula is useful for further investigation [11]:

$$\int_0^{\infty} x^m \exp(-\beta x^n)\, dx = \frac{\Gamma(\gamma)}{n\beta^{\gamma}}, \quad \left(\text{where } \gamma = \frac{\mathrm{m}+1}{\mathrm{n}}\right) \text{ and } \Re(\beta) > 0. \tag{24}$$

Theorem 5 (relation between new extension of the classical beta function and the classical beta function). *If $\Re(\alpha+s) > 0$, $\Re(\beta+s) > 0$, $m \in \mathbb{C}$; $|m| < M$ (where M is positive number not greater than 2.0335), then we have the following relation:*

$$\int_0^{\infty} m^{s-1}\, {}^{MC}B_m(\alpha, \beta)\, dm = (-1)^s \Gamma(s) B(\alpha+s, \beta+s). \tag{25}$$

Proof. Multiplying both sides of (16) by m^{s-1}, then integrating with respect to m from $m = 0$ to $m = \infty$, we have

$$\begin{aligned} &\int_0^{\infty} m^{s-1}\, {}^{MC}B_m(\alpha, \beta)\, dm \\ &= \int_0^{\infty} m^{s-1} \left[\int_0^1 t^{\alpha-1}(1-t)^{\beta-1} \cdot \exp(mt(1-t))\, dt\right] dm, \end{aligned} \tag{26}$$

and interchanging the order of integration, (26) reduces to

$$\int_0^{\infty} m^{s-1}\, {}^{MC}B_m(\alpha, \beta)\, dm = \int_0^1 t^{\alpha-1}(1-t)^{\beta-1} \cdot \left[\int_0^{\infty} m^{s-1} \exp(mt(1-t))\, dm\right] dt, \tag{27}$$

and further using the formula given in (24), after simplification, (27) reduces to

$$\int_0^{\infty} m^{s-1}\,{}^{MC}B_m(\alpha,\beta)\,dm = (-1)^s\,\Gamma(s)\int_0^1 t^{\alpha+s-1}(1-t)^{\beta+s-1}\,dt, \tag{28}$$

and using the definition of classical beta function, we have the required result.

Remark 6. By setting $s = 1$, the result in (25) reduces to

$$\int_0^{\infty} {}^{MC}B_m(\alpha,\beta)\,dm = -B(\alpha+1,\beta+1), \qquad \Re(\alpha) > -1,\ \Re(\beta) > -1 \tag{29}$$

which gives the interesting relation between classical beta function and new extended beta function.

Remark 7. All the derivatives of the new extension of classical beta function with respect to the parameter m can be expressed in terms of the function as

$$\frac{\partial^n}{\partial^n m}\,{}^{MC}B_m(\alpha,\beta) = B(\alpha+n,\beta+n), \qquad \Re(\alpha+n) > 0,\ \Re(\beta+n) > 0. \tag{30}$$

Theorem 8 (integral representations of the new extension of the classical beta function). *If $\Re(\alpha) > 0$, $\Re(\beta) > 0$, $m \in \mathbb{C}$; $|m| < M$ (where M is positive number not greater than 2.0335), then we have the following relation:*

$${}^{MC}B_m(\alpha,\beta) = 2\int_0^{\pi/2} \cos^{2\alpha-1}\theta\,\sin^{2\beta-1}\theta\,\exp\left(m\cos^2\theta\,\sin^2\theta\right)d\theta, \tag{31}$$

$${}^{MC}B_m(\alpha,\beta) = \int_0^{\infty} \frac{u^{\alpha-1}}{(1+u)^{\alpha+\beta}}\exp\left(\frac{mu}{(1+u)^2}\right)du, \tag{32}$$

$${}^{MC}B_m(\alpha,\beta) = \frac{1}{2}\int_0^{\infty} \frac{u^{\alpha-1}+u^{\beta-1}}{(1+u)^{\alpha+\beta}}\exp\left(\frac{mu}{(1+u)^2}\right)du, \tag{33}$$

$${}^{MC}B_m(\alpha,\beta) = (c-a)^{1-\alpha-\beta}\cdot\int_a^c (u-a)^{\alpha-1}(c-u)^{\beta-1}\exp\left(\frac{m(u-a)(c-u)}{(c-a)^2}\right)du, \tag{34}$$

$${}^{MC}B_m(\alpha,\beta) = 2^{1-\alpha-\beta}\int_{-1}^{1} (1+t)^{\alpha-1}(1-t)^{\beta-1}\exp\left(\frac{m\left(1-t^2\right)}{4}\right)du, \tag{35}$$

$${}^{MC}B_m(\alpha,\beta) = 2^{1-\alpha-\beta}\int_{-\infty}^{\infty} \exp\left[(\alpha-\beta)x + \frac{m}{4\cosh^2 x}\right]\frac{dx}{(\cosh x)^{\alpha+\beta}}, \tag{36}$$

$${}^{MC}B_m(\alpha,\beta) = 2^{2-\alpha-\beta}\int_0^{\infty} \cosh((\alpha-\beta)x)\exp\left[\frac{m}{4\cosh^2 x}\right]\frac{dx}{(\cosh x)^{\alpha+\beta}}, \tag{37}$$

$${}^{MC}B_m(\alpha,\beta) = 2^{1-\alpha-\beta}\int_{-\infty}^{\infty} \exp\left[\frac{1}{2}(\alpha-\beta)x + \frac{m}{2\cosh x}\right]\frac{dx}{(\cosh x/2)^{\alpha+\beta}}, \tag{38}$$

$${}^{MC}B_m(\alpha,\beta) = 2^{2-\alpha-\beta}\int_0^{\infty} \frac{\cosh((\alpha-\beta)x/2)}{(\cosh x/2)^{\alpha+\beta}}\exp\left[\frac{m}{2\cosh x}\right]dx. \tag{39}$$

Proof. The result (31) can be easily obtained by setting $t = \cos^2\theta$ in (16); to prove (32) choose $t = u/(1+u)$; (33) can be easily obtained by applying the symmetric property in (32) then adding new one and (32); the result in (34) is obtained by taking $t = (u-a)/(c-a)$, and setting $a = -1, c = 1$ in (34) gives the result in (35) and to prove the result in (36) put $u = \tanh x$ in (35). The results in (37), (38), and (39) can be easily obtained from the result (36).

Remark 9 (useful inequalities). If $\Re(\alpha) > 0$, $\Re(\beta) > 0$, then we have the following inequality

$$\left|{}^{MC}B_m(\alpha,\beta)\right| \le 1.6626B(\alpha,\beta) \tag{40}$$

follows from the integral representation (32), since the function $\exp(mu/(1+u)^2)$ attains its maximum value 1.6626 at $u = 1$ and $m = 2.0335$.

5. Properties of the New Extension of the Classical Beta Function

Theorem 10 (functional relation). *If $\Re(\alpha) > 0$, $\Re(\beta) > 0$, $m \in \mathbb{C}$; $|m| < M$ (where M is positive number), then we have the following relation:*

$${}^{MC}B_m(\alpha,\beta+1) + {}^{MC}B_m(\alpha+1,\beta) = {}^{MC}B_m(\alpha,\beta). \tag{41}$$

Proof. Using the definition of new extension of beta function, LHS of (41) is equal to

$$\int_0^1 \left\{t^{\alpha-1}(1-t)^{\beta} + t^{\alpha}(1-t)^{\beta-1}\right\}e^{mt(1-t)}dt, \tag{42}$$

and after simplification (42) reduced to

$$\int_0^1 \left\{t^{\alpha-1}(1-t)^{\beta-1}\right\}e^{mt(1-t)}dt = {}^{MC}B_m(\alpha,\beta). \tag{43}$$

If we choose $m = 0$, we get the usual relation for the beta function from (41).

Theorem 11 (symmetry). *If $\Re(\alpha) > 0$, $\Re(\beta) > 0$, $m \in \mathbb{C}$; $|m| < M$ (where M is positive number), then we have the following relation:*

$${}^{MC}B_m(\alpha,\beta) = {}^{MC}B_m(\beta,\alpha). \tag{44}$$

Proof. From (18), we have

$$^{MC}B_m(\alpha,\beta)=\sum_{n=0}^{\infty}\frac{(m)^n}{n!}B(\alpha+n,\beta+n), \quad (45)$$

and since usual beta function is symmetric, i.e., $B(\alpha,\beta)=B(\beta,\alpha)$, using this property in the right-hand side of (45), then we have

$$\begin{aligned} ^{MC}B_m(\alpha,\beta)&=\sum_{n=0}^{\infty}\frac{m^n}{n!}B(\beta+n,\alpha+n)\\ &={}^{MC}B_m(\beta,\alpha). \end{aligned} \quad (46)$$

Theorem 12 (first summation relation). *If $\Re(\alpha)>0$, $\Re(1-\beta)>0$, $m\in\mathbb{C}$; $|m|<M$ (where M is positive real number), then we have the following relation:*

$$^{MC}B_m(\alpha,1-\beta)=\sum_{n=0}^{\infty}\frac{(\beta)_n}{n!}{}^{MC}B_m(\alpha+n,1). \quad (47)$$

Proof. The LHS of (47) can be written as

$$\begin{aligned} &^{MC}B_m(\alpha,1-\beta)\\ &=\int_0^1 t^{\alpha-1}(1-t)^{-\beta}\exp(mt(1-t))\,dt, \end{aligned} \quad (48)$$

and using the binomial series expansion $(1-t)^{-\beta}=\sum_{n=0}^{\infty}((\beta)_n/n!)t^n$ in (48) and then interchanging the order of summation and integration, the above result (48) reduced to the following form:

$$\begin{aligned} &^{MC}B_m(\alpha,1-\beta)\\ &=\sum_{n=0}^{\infty}\frac{(\beta)_n}{n!}\int_0^1 t^{\alpha+n-1}\exp(mt(1-t))\,dt\Longrightarrow \end{aligned} \quad (49)$$

$$^{MC}B_m(\alpha,1-\beta)=\sum_{n=0}^{\infty}\frac{(\beta)_n}{n!}{}^{MC}B_m(\alpha+n,1). \quad (50)$$

Theorem 13 (second summation relation). *If $\Re(\alpha)>0,\Re(\beta)>0,m\in\mathbb{C};|m|<M$ (where M is positive number), then we have the following relation:*

$$^{MC}B_m(\alpha,\beta)=\sum_{n=0}^{\infty}{}^{MC}B_m(\alpha+n,\beta+1). \quad (51)$$

Proof. The LHS of (47) can be written as

$$^{MC}B_m(\alpha,\beta)=\int_0^1 t^{\alpha-1}(1-t)^{\beta-1}\exp(mt(1-t))\,dt, \quad (52)$$

and using the binomial series expansion $(1-t)^{\beta-1}=(1-t)^{\alpha}\sum_{n=0}^{\infty}t^n(|t|<1)$ and interchanging the order of summation and integration, (52) reduces to

$$\begin{aligned} &^{MC}B_m(\alpha,\beta)\\ &=\sum_{n=0}^{\infty}\int_0^1 t^{\alpha+n-1}(1-t)^{\beta}\exp(mt(1-t))\,dt\Longrightarrow \end{aligned} \quad (53)$$

$$^{MC}B_m(\alpha,\beta)=\sum_{n=0}^{\infty}{}^{MC}B_m(\alpha+n,\beta+1). \quad (54)$$

Theorem 14 (separation). *If $\Re(\alpha)>0$, $\Re(\beta)>0$, $m\in\mathbb{C}$; $|m|<M$ (where M is positive number), then $^{MC}B_m(\alpha,\beta)$ can be separated into real and imaginary parts of m as follows:*

$$\begin{aligned} &^{MC}B_{r\cos\theta}(\alpha,\beta)\\ &=\int_0^1 t^{\alpha-1}(1-t)^{\beta-1}\exp(rt(1-t)\cos\theta)\,dt, \end{aligned} \quad (55)$$

$$\begin{aligned} &^{MC}B_{r\sin\theta}(\alpha,\beta)\\ &=\int_0^1 t^{\alpha-1}(1-t)^{\beta-1}\exp(rt(1-t)\sin\theta)\,dt, \end{aligned} \quad (56)$$

where $r=\sqrt{x^2+y^2}=|m|<M$ and $\theta=\tan^{-1}y/x$.

Proof. Since $m\in\mathbb{C}$, so let $m=x+iy$, where $x,y\in\mathbb{R}$ and also let $x+iy=r\cos\theta+ir\sin\theta\Longrightarrow r=\sqrt{x^2+y^2}$ and $\theta=\tan^{-1}y/x$; then from (16), we have

$$\begin{aligned} &^{MC}B_{(x+iy)}(\alpha,\beta)\\ &=\int_0^1 t^{\alpha-1}(1-t)^{\beta-1}\exp\left(re^{i\theta}t(1-t)\right)dt, \end{aligned} \quad (57)$$

and after simplification (57) reduces to

$$\begin{aligned} &^{MC}B_{(r\cos\theta+ir\sin\theta)}(\alpha,\beta)\\ &=\int_0^1 t^{\alpha-1}(1-t)^{\beta-1}\exp(rt(1-t)\cos\theta)\,dt\\ &\quad+i\int_0^1 t^{\alpha-1}(1-t)^{\beta-1}\exp(rt(1-t)\sin\theta)\,dt. \end{aligned} \quad (58)$$

Equating the real and imaginary parts of m only, we have the required results.

6. Applications of New Extension of the Classical Beta Function

It is expected that there will be many applications of the new extension of the classical beta function, e.g., new extension of the beta distribution, new extensions of Gauss hypergeometric functions and confluent hypergeometric function, generating relations, and extension of Riemann-Liouville derivatives. All these have been introduced in the following subsections.

6.1. The New Extension of the Beta Distribution. One application that springs to mind is to statistics. For example, the conventional beta distribution can be extended, by using our new extension of the classical beta function, to variables p and q with an infinite range. It appears that such an extension may be desirable for the project evaluation and review technique used in some special cases.

We define the extension of the beta distribution by

$$f(t) = \begin{cases} \dfrac{t^{p-1}(1-t)^{q-1}\exp(mt(1-t))}{{}^{MC}B_m(p,q)}, & 0<t<1 \\ 0, & \text{otherwise.} \end{cases} \tag{59}$$

A random variable X with probability density function (pdf) given in (59) will be said to have the extended beta distribution with parameters p and q, $-\infty < p,q < \infty$, and $|m| < M$ where M is positive number. If ν is any real number [12], then

$$E(X^\nu) = \frac{{}^{MC}B_m(p+\nu,q)}{{}^{MC}B_m(p,q)}. \tag{60}$$

In particular, for $\nu = 1$,

$$\mu = E(X) = \frac{{}^{MC}B_m(p+1,q)}{{}^{MC}B_m(p,q)} \tag{61}$$

represents the mean of the distribution and

$$\sigma^2 = E(X^2) - (E(X))^2 = \frac{{}^{MC}B_m(p,q)\,{}^{MC}B_m(p+2,q) - {}^{MC}B_m^2(p+1,q)}{{}^{MC}B_m^2(p,q)} \tag{62}$$

is a variance of the distribution.

The moment of generating function of the distribution is

$$M(t) = \sum_{n=0}^{\infty}\frac{t^n E(X^n)}{n!} = \frac{1}{{}^{MC}B_m(p,q)}\sum_{n=0}^{\infty}\frac{t^n}{n!}\,{}^{MC}B_m(p+n,q). \tag{63}$$

The commutative distribution of (59) can be written as

$$F(x) = \frac{{}^{MC}B_{m,x}(p,q)}{{}^{MC}B_m(p,q)}, \tag{64}$$

where

$${}^{MC}B_{m,x}(p,q) = \int_0^x t^{p-1}(1-t)^{q-1}\exp(mt(1-t))\,dt, \tag{65}$$

$$|m| < M, \quad -\infty < p,q < \infty$$

is the new extended incomplete beta function. For $m = 0$, we must have $p,q > 0$ in (65) for convergence, and ${}^{MC}B_{0,x}(p,q) = B_x(p,q)$, where $B_x(p,q)$ is the incomplete beta function [11] defined as

$$B_x(p,q) = {}^{MC}B_{0,x}(p,q) = \frac{x^p}{p}\,{}_2F_1(p,1-q;p=1;x). \tag{66}$$

It is to be noted that the problem of expressing ${}^{MC}B_{m,x}(p,q)$ in terms of other special functions remains open. Presumably, this distribution should be useful in extending the statistical results for strictly positive variables to deal with variables that can take arbitrarily large negative values as well.

6.2. Extensions of Gauss and Confluent Hypergeometric Function Using the New Extension of Beta Function. In this section, we extended the Gauss hypergeometric function and confluent hypergeometric function via new extension of classical beta function, which is defined as follows:

$${}^{MC}F_m(\alpha,\beta;\gamma;z) := \sum_{n=0}^{\infty}(\alpha)_n\frac{{}^{MC}B_m(\beta+n,\gamma-\beta)}{B(\beta,\gamma-\beta)}\frac{z^n}{n!}, \tag{67}$$

$$|z|<1;\ \Re(\gamma) > \Re(\beta) > 0;\ |m| < M \ (\text{where } M \text{ is a positive real number}).$$

$${}^{MC}\Phi_m(\beta;\gamma;z) := \sum_{n=0}^{\infty}\frac{{}^{MC}B_m(\beta+n,\gamma-\beta)}{B(\beta,\gamma-\beta)}\frac{z^n}{n!}, \tag{68}$$

$$|z|<1;\ \Re(\gamma) > \Re(\beta) > 0;\ |m| < M \ (\text{where } M \text{ is a positive real number}).$$

We call ${}^{MC}F_m(a,b;c;z)$ new extension of Gauss hypergeometric function and ${}^{MC}\Phi_m(b;c;z)$ new extension of confluent hypergeometric function.

Note 15. If we choose $m = 0$, the above two new extensions in (67) and (68) reduce to Gauss hypergeometric function and confluent hypergeometric function given in (7) and (8), respectively.

6.3. Numerical Results of New Extension of Gauss Hypergeometric Function and New Extension of Confluent Hypergeometric Function. intoHere, we present the numerical values

TABLE 3: Numerical values of new extension of Gauss hypergeometric function ${}^{MC}F_m(2,1;3;x)$.

x	$m=-2$	$m=-1$	$m=0$	$m=1$	$m=2$
0.0000	0.7248	0.8489	1.0000	1.1846	1.4107
0.1000	0.7734	0.9080	1.0721	1.2728	1.5188
0.2000	0.8308	0.9778	1.1572	1.3768	1.6463
0.3000	0.9000	1.0618	1.2594	1.5017	1.7993
0.4000	0.9855	1.1653	1.3853	1.6552	1.9870
0.5000	1.0947	1.2972	1.5452	1.8495	2.2240
0.6000	1.2410	1.4730	1.7572	2.1061	2.5356
0.7000	1.4512	1.7236	2.0570	2.4664	2.9701
0.8000	1.7914	2.1235	2.5295	3.0270	3.6382
0.9000	2.4976	2.9335	3.4632	4.1089	4.8982
1.0000	6.7829	7.5302	8.4142	9.4648	10.7195

TABLE 4: Numerical values of new extension of confluent hypergeometric function ${}^{MC}{}_1F_{1,m}(1;3;x)$.

x	$m=-2$	$m=-1$	$m=0$	$m=1$	$m=2$
0.0000	0.7248	0.8489	1.0000	1.1846	1.4107
0.1000	0.7478	0.8769	1.0342	1.2264	1.4620
0.2000	0.7721	0.9064	1.0701	1.2704	1.5159
0.3000	0.7976	0.9374	1.1080	1.3167	1.5726
0.4000	0.8245	0.9701	1.1478	1.3654	1.6324
0.5000	0.8528	1.0045	1.1898	1.4167	1.6953
0.6000	0.8826	1.0408	1.2340	1.4707	1.7616
0.7000	0.9141	1.0790	1.2806	1.5277	1.8314
0.8000	0.9473	1.1194	1.3298	1.5879	1.9051
0.9000	0.9824	1.1620	1.3817	1.6513	1.9828
1.0000	1.0195	1.2070	1.4366	1.7183	2.0649

of new extension of Gauss hypergeoemtric function and new extension of confluent hypergeoemtric function in Table 3 and Table 4 for $m = -2 : 1 : 2$. Further their graphs are plotted in Figure 3 and Figure 4, respectively. When $m = 0$ we have the values of Gauss hypergeoemtric function and confluent hypergeoemtric function.

6.4. Integral Representation of New Extension of Gauss Hypergeometric Function and New Extension of Confluent Hypergeometric Function. The new extension of Gauss hypergeometric function and new extension of confluent hypergeometric function can be provided with an integral representation by using the definition of the new extension of classical beta function (16); we have the following.

Theorem 16. *For the new extension of Gauss hypergeometric function ${}^{MC}F_m(\alpha,\beta;\gamma;z)$, we have the following integral representations:*

$$ {}^{MC}F_m(\alpha,\beta;\gamma;z) := \frac{1}{B(\beta,\gamma-\beta)}\int_0^1 t^{\beta-1}(1-t)^{\gamma-\beta-1} \cdot (1-zt)^{-\alpha} e^{mt(1-t)}dt, \tag{69}$$

$$ {}^{MC}F_m(\alpha,\beta;\gamma;z) := \frac{1}{B(\beta,\gamma-\beta)}\int_0^1 u^{\beta-1}(1+u)^{\alpha-\gamma-1} \cdot (1+(1-z)u)^{-\alpha} e^{mu/(1+u)^2}du, \tag{70}$$

$$ {}^{MC}F_m(\alpha,\beta;\gamma;z) := \frac{2}{B(\beta,\gamma-\beta)} \cdot \int_0^{\pi/2} \sin^{2\beta-1}\theta\cos^{2\gamma-2\beta-1}\theta\left(1-z\sin^2\theta\right)^{-\alpha} \cdot e^{m\sin^2\theta\cos^2\theta}d\theta. \tag{71}$$

Proof. Equation (67) can be written as

$$ {}^{MC}F_m(\alpha,\beta;\gamma;z) := \sum_{n=0}^{\infty}(\alpha)_n \frac{{}^{MC}B_m(\beta+n,\gamma-\beta)}{B(\beta,\gamma-\beta)}\frac{z^n}{n!} = \frac{1}{B(\beta,\gamma-\beta)}\sum_{n=0}^{\infty}(\alpha)_n \cdot \int_0^1 t^{\beta+n-1}(1-t)^{\gamma-\beta-1}e^{mt(1-t)}\frac{z^n}{n!}dt$$

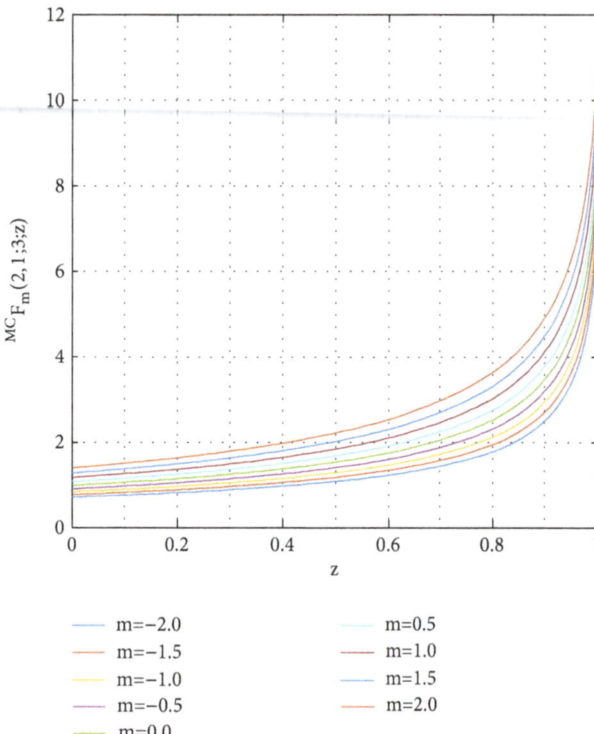

FIGURE 3: Graph of the new extension of Gauss hypergeometric function for $\alpha = 2, \beta = 1, \gamma = 3$, and $z = 0 : 0.1 : 1$.

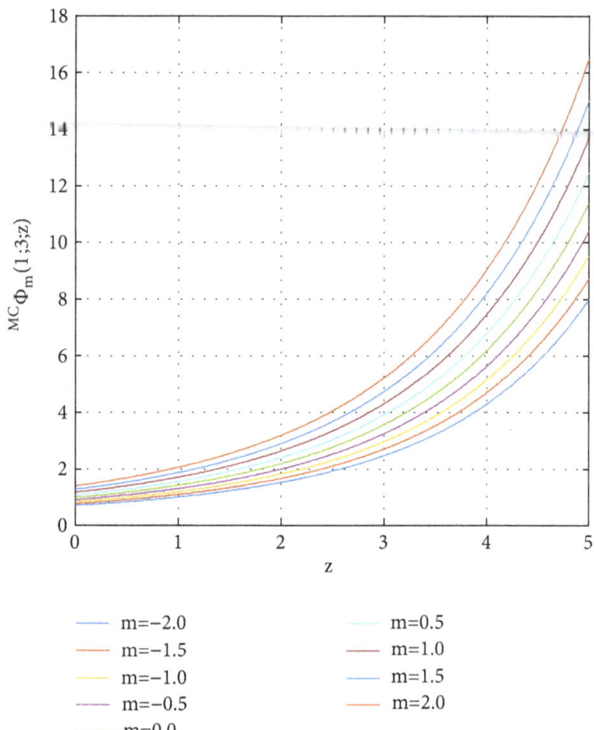

FIGURE 4: Graph of the new extension of confluent hypergeometric function for $\beta = 1, \gamma = 3$, and $z = 0 : 0.1 : 1$.

$$= \frac{1}{B(\beta, \gamma - \beta)} \cdot \int_0^1 t^{\beta-1} (1-t)^{\gamma-\beta-1} e^{mt(1-t)} \sum_{n=0}^{\infty} (\alpha)_n \frac{(tz)^n}{n!} dt = \frac{1}{B(\beta, \gamma - \beta)} \cdot \int_0^1 t^{\beta-1} (1-t)^{\gamma-\beta-1} (1-tz)^{-\alpha} e^{mt(1-t)}. \quad (72)$$

Setting $u = t/(1-t)$ in (69), we have the required result (70).

Again if we choose $t = \sin^2\theta$, we obtain the result (71).

Remark 17. Choosing $z = 1$ in (69), we have the following relation between new extensions of Gauss hypergeometric function:

$$ {}^{MC}F_m(\alpha, \beta; \gamma; 1) := \frac{1}{B(\beta, \gamma - \beta)} \int_0^1 t^{\beta-1} (1-t)^{\gamma-\alpha-\beta-1} e^{mt(1-t)} dt = \frac{{}^{MC}B_m(\beta, \gamma - \alpha - \beta)}{B(\beta, \gamma - \beta)}. \quad (73)$$

Theorem 18. *For the new extension of confluent hypergeometric function ${}^{MC}\Phi_m(\alpha; \beta; x)$, we have the following integral representations:*

$$ {}^{MC}\Phi_m(\beta; \gamma; z) := \frac{1}{B(\beta, \gamma - \beta)} \cdot \int_0^1 t^{\beta-1} (1-t)^{\gamma-\beta-1} e^{zt} e^{mt(1-t)} dt, \quad (74)$$

$$ {}^{MC}\Phi_m(\beta; \gamma; z) := \frac{1}{B(\beta, \gamma - \beta)} \cdot \int_0^1 u^{\beta-1} (1+u)^{-\gamma} e^{zu/(1+u)} e^{mu/(1+u)^2} du, \quad (75)$$

$$ {}^{MC}\Phi_m(\beta; \gamma; z) := \frac{2}{B(\beta, \gamma - \beta)} \cdot \int_0^{\pi/2} \sin^{2\beta-1}\theta \cos^{2\gamma-2\beta-1}\theta e^{z\sin^2\theta} e^{m\sin^2\theta\cos^2\theta} d\theta. \quad (76)$$

Proof. The proof of this theorem would run parallel to those of Theorem 16, so we skip the proof of this theorem.

6.5. Differentiation Formulas for the Representation of the New Extension of Gauss Hypergeometric Function and New Extension of Confluent Hypergeometric Function. In the present section, by using the formulas $B(\beta, \gamma-\beta) = (\gamma/\beta)B(\beta+1, \gamma-\beta)$ and $(\alpha)_{n+1} = \alpha(\alpha+1)_n$, we obtain new formulas including derivatives of the new extension of Gauss hypergeometric function and new extension of confluent hypergeometric function with respect to the variable z; we have the following.

Theorem 19. *If $\alpha, \beta, \gamma \in \mathbb{C}$; $\Re(\gamma) > \Re(\beta) > 0$ and $|m| < M$ (where M is positive real number), then we have the following result:*

$$\frac{d^n}{dz^n}\left\{{}^{MC}F_m(\alpha,\beta;\gamma;z)\right\} = \frac{(\beta)_n(\alpha)_n}{(\gamma)_n}{}^{MC}F_m(\alpha+n,\beta+n;\gamma+n;z). \tag{77}$$

Proof. Taking the derivative of ${}^{MC}F_m(\alpha,\beta;\gamma;z)$ with respect to z, we have

$$\frac{d}{dz}\left\{{}^{MC}F_m(\alpha,\beta;\gamma;z)\right\} = \sum_{n=1}^{\infty}(\alpha)_n\frac{{}^{MC}B_m(\beta+n,\gamma-\beta)}{B(\beta,\gamma-\beta)}\frac{z^{n-1}}{(n-1)!}, \tag{78}$$

and replacing $n \longrightarrow n+1$, (78) reduces to

$$\begin{aligned}\frac{d}{dz}\left\{{}^{MC}F_m(\alpha,\beta;\gamma;z)\right\} &= \frac{\beta\alpha}{\gamma}\sum_{n=0}^{\infty}(\alpha+1)_n\frac{{}^{MC}B_m(\beta+n+1,\gamma-\beta)}{B(\beta+1,\gamma-\beta)}\frac{z^{n-1}}{(n-1)!} \\ &= \frac{\beta\alpha}{\gamma}{}^{MC}F_m(\alpha+1,\beta+1;\gamma+1;z),\end{aligned} \tag{79}$$

and with recursive application of this procedure in (79), we have the desired result (77).

Theorem 20. *If $\beta, \gamma \in \mathbb{C}$; $\Re(\gamma) > \Re(\beta) > 0$ and $|m| < M$ (where M is positive real number), then we have the following result:*

$$\frac{d^n}{dz^n}\left\{{}^{MC}\Phi_m(\beta;\gamma;z)\right\} = \frac{(\beta)_n(\alpha)_n}{(\gamma)_n}{}^{MC}\Phi_m(\beta+n;\gamma+n;z). \tag{80}$$

Proof. The proof of Theorem 20 is as that of Theorem 19, so it can be omitted here.

6.5.1. Generating Relations Associated with Hypergeometric Functions

Theorem 21. *If $\alpha, \beta, \gamma \in \mathbb{C}; k \in \mathbb{R}$; $\Re(\gamma) > \Re(\beta) > 0$ and $|m| < M$ (where M is positive real number), then the following generating functions hold:*

$$\sum_{n=0}^{\infty}(\alpha)_n{}^{MC}F_m(\alpha+n,\beta;\gamma;z)\frac{t^n}{n!} = (1-t)^{-\alpha}{}^{MC}F_m\left(\alpha,\beta;\gamma;;\frac{z}{1}-t\right), \quad (|t|<1). \tag{81}$$

Proof. Let the left-hand side of (81) be denoted by $\mathcal{S}$; then using the definition of new extension of Gauss hypergeometric function, we have

$$\mathcal{S} = \sum_{n=0}^{\infty}(\alpha)_n\left[\sum_{k=0}^{\infty}(\alpha+n)_k\frac{{}^{MC}B_m(\beta+k,\gamma-\beta)}{B(\beta,\gamma-\beta)}\frac{z^k}{k!}\right]\frac{t^n}{n!}. \tag{82}$$

Upon reversal of the order of summation and then using the identity $(\alpha)_n(\alpha+n)_k = (\alpha)_k(\alpha+k)_n$, (82) reduces to

$$\mathcal{S} = \sum_{k=0}^{\infty}(\alpha)_k\frac{{}^{MC}B_m(\beta+k,\gamma-\beta)}{B(\beta,\gamma-\beta)}\left[\sum_{n=0}^{\infty}(a+k)_n\frac{t^n}{n!}\right]\frac{z^k}{k!}, \tag{83}$$

and further using the definition of binomial $(1-t)^{-\alpha-k} = \sum_{n=0}^{\infty}(\alpha+k)_n(t^n/n!)$, $(|t|<1)$ in (83), we have

$$\mathcal{S} = \sum_{k=0}^{\infty}(\alpha)_k\frac{{}^{MC}B_m(\beta+k,\gamma-\beta)}{B(\beta,\gamma-\beta)}(1-t)^{-\alpha-k}\frac{z^k}{k!}, \tag{84}$$

and interpreting the above equation with the view of (67), we have the desired result (81).

Theorem 22. *If $\alpha, \beta, \gamma \in \mathbb{C}; k \in \mathbb{R}$; $\Re(\gamma) > \Re(\beta) > 0$ and $|m| < M$ (where M is positive real number), then the following generating functions hold:*

$$\sum_{k=0}^{\infty}\binom{\alpha+k-1}{k}{}^{MC}F_{m,k}(\alpha+k,\beta;\gamma;z)t^k = (1-t)^{-\alpha}{}^{MC}F_{m,k}\left(\alpha,\beta;\gamma;;\frac{z}{1}-t\right), \quad (|t|<1). \tag{85}$$

Proof. For convenience, let the left-hand side of (85) be denoted by $\mathcal{J}$. Applying the series of (67) to $\mathcal{J}$, we get

$$\mathcal{J} = \sum_{k=0}^{\infty}\binom{\alpha+k-1}{k} \cdot \left\{\sum_{n=0}^{\infty}(\alpha+k)_n\frac{{}^{MC}B_m(\beta+n,\gamma-\beta)}{B(\beta,\gamma-\beta)}\frac{z^n}{n!}\right\}t^k. \tag{86}$$

By changing the order of summation in (86) and using the known identity ([13, p.5]), namely,

$$\binom{\alpha}{n} = \frac{\Gamma(\alpha+1)}{n!\Gamma(\alpha-n+1)}, \quad (n \in \mathbb{N}_o; \alpha \in \mathbb{C}), \tag{87}$$

then, after little simplification, we obtain

$$\mathcal{J} = \sum_{n=0}^{\infty}(\alpha)_n \cdot \frac{{}^{MC}B_m(\beta+n,\gamma-\beta)}{B(\beta,\gamma-\beta)}\left\{\sum_{k=0}^{\infty}\binom{\alpha+n+k-1}{k}t^k\right\} \cdot \frac{z^n}{n!}. \tag{88}$$

Further, upon using the generalized binomial expansion, we find that the inner sum in (88) yields

$$\sum_{k=0}^{\infty}\binom{\alpha+n+k-1}{k}t^{k}=(1-t)^{-(\alpha+n)},\quad (|t|<0). \tag{89}$$

Finally in view of (88) and (89), we get the desired assertion (85) of Theorem 1.

A further generalized Gauss hypergeometric function (67) is given in the following definition.

Definition 23. Let us introduce a sequence $\{\zeta_k^{(\alpha,\rho)}\}$ defined by

$$\begin{aligned}\zeta_k^{(\alpha,\rho)}(z)&=\zeta_k^{(\alpha,\rho)}(\alpha,\beta;\gamma;z)\\&={}^{MC}F_{m,k}(\Delta(\rho;\alpha+k),\beta;\gamma;z)\end{aligned} \tag{90}$$

where $\alpha,\beta,\gamma\in\mathbb{C};k\in\mathbb{R};\rho\in\mathbb{N};\Re(\gamma)>\Re(\beta)>0$ and $|m|<M$ (where M is positive real number); for convenience, $\Delta(\rho;\alpha)$ abbreviates the array of ρ parameters

$$\frac{\alpha}{\rho},\frac{\alpha+1}{\rho},\frac{\alpha+2}{\rho},\ldots,\frac{\alpha+\rho-1}{\rho},\quad(\rho\in\mathbb{N}). \tag{91}$$

Now, we prove the following result, which provides the generating functions for the Gauss hypergeometric function defined above.

Theorem 24. *For each $\rho\in\mathbb{N}$, the following generating functions hold true:*

$$\begin{aligned}&\sum_{k=0}^{\infty}\binom{\alpha+l+k-1}{k}\zeta_k^{(\alpha,\rho)}(z)t^k\\&=(1-t)^{-\alpha-l}\zeta_l^{(\alpha,\rho)}\left(\frac{z}{(1-t)^{\rho}}\right),\quad(|t|<1)\end{aligned} \tag{92}$$

where $\alpha,\beta,\gamma\in\mathbb{C};k\in\mathbb{R};\rho\in\mathbb{N};l\in\mathbb{N}_0$ $\Re(\gamma)>\Re(\beta)>0$ and $|m|<M$ (where M is positive real number).

Proof. Using the definition introduced in (90) and the new extended Gauss hypergeometric function introduced in (67); then changing the order of summations, the left hand side of (92) (say I) leads to

$$\begin{aligned}I=&\sum_{n=0}^{\infty}\left(\frac{\alpha+l}{\rho}\right)_n\cdots\left(\frac{\alpha+l+\rho-1}{\rho}\right)_n\\&\cdot\frac{{}^{MC}B_m(\beta+n,\gamma-\beta)}{B(\beta,\gamma-\beta)}\left\{\sum_{k=0}^{\infty}\binom{\alpha+l+n\rho+k-1}{k}\right.\\&\left.\cdot t^k\right\}\frac{z^n}{n!}.\end{aligned} \tag{93}$$

Now taking (90) into account, one can easily arrive at the desired result (92).

Remark 25. It may be noted that if we set $\rho=1$ and replace α by $\alpha-l$ in (92), we are easily led to the result (85).

6.6. Extension of Riemann-Liouville Fractional Derivative. In this section, we introduce new extension of Riemann-Liouville fractional derivative operator:

$$\begin{aligned}&{}^{MC}D_z^{\mu,m}\{f(z)\}\\&:=\frac{1}{\Gamma(-\mu)}\int_0^z f(t)(z-t)^{-\mu-1}e^{mt(z-t)/z^2}dt\end{aligned} \tag{94}$$

and for $n-1<\Re(\mu)<n$, $n=1,2,..$

$$\begin{aligned}{}^{MC}D_z^{\mu,m}\{f(z)\}&:=\frac{d^n}{dz^n}{}^{MC}D_z^{\mu-n}\{f(z)\}\\&=\frac{d^n}{dz^n}\left\{\frac{1}{\Gamma(-\mu+n)}\right.\\&\left.\cdot\int_0^z f(t)(z-t)^{-\mu+n-1}e^{mt(z-t)/z^2}dt\right\},\end{aligned} \tag{95}$$

where the path of integration is a line from 0 to z in complex t–plane. For the case $m=0$, we obtain the classical Riemann-Liouville fractional operator.

We start our investigation by calculating the extended fractional derivative of some elementary functions.

Lemma 26. *Let $\xi,\mu\in\mathbb{C};\Re(\xi)>-1,\Re(\mu)<0$ and $|m|<M$ (where M is positive real number); then we have*

$${}^{MC}D_z^{\mu,m}\{z^{\xi}\}=\frac{{}^{MC}B_m(\xi+1,-\mu)}{\Gamma(-\mu)}z^{\xi-\mu}. \tag{96}$$

Proof. Employing the definition given in (94) in the left-hand side of (96), we have

$${}^{MC}D_z^{\mu,m}\{z^{\xi}\}:=\frac{1}{\Gamma(-\mu)}\int_0^z t^{\xi}(z-t)^{-\mu-1}e^{mt(z-t)/z^2}dt. \tag{97}$$

Choosing $t=zu$, (97) reduces to

$$\begin{aligned}&{}^{MC}D_z^{\mu,m}\{z^{\xi}\}\\&=\frac{1}{\Gamma(-\mu)}\int_0^1(zu)^{\xi}z^{-\mu-1}(1-u)^{-\mu-1}e^{mu(1-u)}z\,du\\&=\frac{z^{\xi-\mu}}{\Gamma(-\mu)}\int_0^1u^{\xi}(1-u)^{-\mu-1}e^{mu(1-u)}du\\&=\frac{{}^{MC}B_m(\xi+1,-\mu)}{\Gamma(-\mu)}z^{\xi-\mu}.\end{aligned} \tag{98}$$

Lemma 27. *Let $\xi,\alpha,\mu\in\mathbb{C};\Re(\xi)>0,\Re(\alpha)>0,\Re(\mu)<0$, $|z|<1$ and $|m|<M$ (where M is positive real number), then we have*

$$\begin{aligned}&{}^{MC}D_z^{\xi-\mu,m}\{z^{\xi}-1(1-z)^{-\alpha}\}\\&:==\frac{\Gamma(\xi)}{\Gamma(\mu)}z^{\mu-1MC}F_m(\alpha,\xi;\mu;z).\end{aligned} \tag{99}$$

Proof. Employing the definition given in (94) in the left-hand side of (99), we have

$$^{MC}D_z^{\xi-\mu,m}\left\{z^{\xi}-1(1-z)^{-\alpha}\right\}:=\frac{1}{\Gamma(\mu-\xi)}\cdot\int_0^z t^{\xi-1}(1-t)^{-\alpha}e^{mt(z-t)/z^2}(z-t)^{\mu-\xi-1}dt. \tag{100}$$

Choosing $t=zu$, (100) reduces to

$$\begin{aligned}^{MC}D_z^{\xi-\mu,m}\left\{z^{\xi}-1(1-z)^{-\alpha}\right\}&:=\frac{z^{\mu-1}}{\Gamma(\mu-\xi)}\cdot\int_0^1 t^{\xi-1}(1-uz)^{-\alpha}(1-u)^{\mu-\xi-1}e^{mu(1-u)}du\\&=\frac{z^{\mu-1}}{\Gamma(\mu-\xi)}\cdot\int_0^1 t^{\xi-1}(1-uz)^{-\alpha}(1-u)^{\mu-\xi-1}e^{mu(1-u)}du,\end{aligned} \tag{101}$$

and further employing the result in (69), after simplification, we have the required result (101).

Theorem 28. *Let $f(z)$ be an analytic function in the disc $|z|<\rho$ and it has the power series $f(z)=\sum_{n=0}^{\infty}a_nz^n$; then we have the following result:*

$$\begin{aligned}^{MC}D_z^{\mu,m}\left\{z^{\xi-1}f(z)\right\}&={}^{MC}D_z^{\mu,m}\left\{z^{\xi-1}\sum_{n=0}^{\infty}a_nz^n\right\}\\&=\frac{z^{\xi-\mu-1}}{\Gamma(-\mu)}\sum_{n=0}^{\infty}a_n{}^{MC}B_m(\xi+n,-\mu)z^n,\end{aligned} \tag{102}$$

where $\xi,\mu\in\mathbb{C};\Re(\xi)>0,\Re(\mu)<0$ and $|m|<M$ (where M is positive real number).

Proof.

$$\begin{aligned}^{MC}D_z^{\mu,m}\left\{z^{\xi-1}f(z)\right\}&={}^{MC}D_z^{\mu,m}\left\{z^{\xi-1}\sum_{n=0}^{\infty}a_nz^n\right\}\\&=\frac{1}{\Gamma(-\mu)}\int_0^z t^{\xi-1}\sum_{n=0}^{\infty}a_nt^n(z-t)^{-\mu-1}e^{mt(z-t)/z^2}dt\\&=\frac{1}{\Gamma(-\mu)}\int_0^1(z\xi)^{\xi-1}z^{-\mu-1}(1-\xi)^{-\mu-1}\\&\quad\cdot e^{m\xi(1-\xi)}\sum_{n=0}^{\infty}a_n(z\xi)^nz\,d\xi.\end{aligned} \tag{103}$$

The series $\sum_{n=0}^{\infty}a_n(z\xi)^n$ is uniformly convergent in the disc $|z|<\rho$ for $0\le\xi\le1$ and the integral $\int_0^1|\xi^{\xi+n-1}(1-\xi)^{-\mu-1}e^{m\xi(1-\xi)}|d\xi$ is convergent provided that $\Re(\xi)>0,\Re(\mu)<0$ and $|m|<M$ (where M is positive real number), therefore we can interchange the order of integration and summation; after simplification, above equation (103) reduces to

$$\begin{aligned}&^{MC}D_z^{\mu,m}\left\{z^{\xi-1}f(z)\right\}\\&=\frac{z^{\xi-\mu-1}}{\Gamma(-\mu)}\sum_{n=0}^{\infty}a_nz^n\int_0^1\xi^{\xi+n-1}(1-\xi)^{-\mu-1}e^{m\xi(1-\xi)}d\xi,\end{aligned} \tag{104}$$

and interpreting (104) with the view of the definition of new extension of classical beta function (16), we have the required result.

Theorem 29 (linear generating function). *Let $\alpha,\beta,\xi,\in\mathbb{C};\Re(\xi)>0,\Re(\beta)>\Re(\alpha)>0$ and $|m|<M$ (where M is positive real number), such that $|x|<\min\{1,|1-t|\}$; then we have the following result:*

$$\begin{aligned}&\sum_{n=0}^{\infty}\frac{(\xi)_n}{n!}{}^{MC}F_m(\xi+n,\alpha;\beta;x)t^n\\&=(1-t)^{-\xi}\,{}^{MC}F_m\left(\xi,\alpha;\beta;\frac{x}{1-t}\right).\end{aligned} \tag{105}$$

Proof. Let us consider the elementary identity

$$[(1-x)-t]^{-\xi}=(1-t)^{-\xi}\left[1-\frac{x}{1-t}\right]^{-\xi}. \tag{106}$$

Expanding the left-hand side of (106) for $|t|<|1-x|$, we have

$$\begin{aligned}&\sum_{n=0}^{\infty}\frac{(\xi)_n}{n!}(1-x)^{-\xi}\left(\frac{t}{1-x}\right)^n\\&=(1-t)^{-\xi}\left[1-\frac{x}{1-t}\right]^{-\xi}.\end{aligned} \tag{107}$$

Further, multiplying both sides of (107) and then applying the new extension of fractional derivative operator $D_x^{\alpha-\beta,m}$ on both sides, we have

$$\begin{aligned}&^{MC}D_x^{\alpha-\beta,m}\left\{\sum_{n=0}^{\infty}\frac{(\xi)_n}{n!}(1-x)^{-\xi}\left(\frac{t}{1-x}\right)^nx^{\alpha-1}\right\}\\&=(1-t)^{-\xi}\,{}^{MC}D_x^{\alpha-\beta,m}\left\{x^{\alpha-1}\left[1-\frac{x}{1-t}\right]^{-\xi}\right\}.\end{aligned} \tag{108}$$

Interchanging the order, which is valid for $\Re(\alpha)>0$ and $|t|<|1-x|$, we have

$$\begin{aligned}&\sum_{n=0}^{\infty}\frac{(\xi)_n}{n!}{}^{MC}D_x^{\alpha-\beta,m}\left\{x^{\alpha-1}(1-x)^{-\xi-n}\right\}t^n\\&=(1-t)^{-\xi}\,{}^{MC}D_x^{\alpha-\beta,m}\left\{x^{\alpha-1}\left[1-\frac{x}{1-t}\right]^{-\xi}\right\}.\end{aligned} \tag{109}$$

Now applying the result established in (99) in (109), after simplification, we have the required result (105).

6.7. Transformation Formulas

Theorem 30. *Let $\alpha, \beta, \xi, \in \mathbb{C}; \Re(\xi) > 0, \Re(\beta) > \Re(\alpha) > 0$ and $|m| < M$ (where M is positive real number), then we have the following transformation formula:*

$$\begin{aligned} &{}^{MC}F_m(\alpha, \beta; \gamma; z) \\ &= (1-z)^{-\alpha\, MC}F_m\left(\alpha, \gamma-\beta; \beta; \frac{z}{(z-1)}\right), \\ &\qquad |\arg(1-z)| < \pi, \end{aligned} \tag{110}$$

$$\begin{aligned} &{}^{MC}F_m\left(\alpha, \beta; \gamma; 1-\frac{1}{z}\right) \\ &= z^{\alpha MC}F_m(\alpha, \gamma-\beta; \beta; 1-z), \quad |\arg(z)| < \pi, \end{aligned} \tag{111}$$

$$\begin{aligned} &{}^{MC}F_m\left(\alpha, \beta; \gamma; \frac{z}{1+z}\right) \\ &= (1+z)^{\alpha\, MC}F_m(\alpha, \gamma-\beta; \beta; z), \\ &\qquad |\arg(1+z)| < \pi. \end{aligned} \tag{112}$$

Proof. To prove the theorem, we consider the following identity:

$$[1-z(1-t)]^{-\alpha} = (1-z)^{-\alpha}\left(1+\frac{z}{1-z}t\right)^{-\alpha}. \tag{113}$$

Replacing t into $1-t$, then using the result (113) in (69), after simplification, we have

$$\begin{aligned} {}^{MC}F_m(\alpha, \beta; \gamma; z) &:= \frac{(1-z)^{-\alpha}}{B(\beta, \gamma-\beta)} \\ &\cdot \int_0^1 t^{\gamma-\beta-1}(1-t)^{\beta-1}\left(1+\frac{z}{1-z}t\right)^{-\alpha} e^{mt(1-t)}dt, \end{aligned} \tag{114}$$

further interpreted with the view of (69), we obtain the desired result (110).

7. Fractional Integration of New Extension of Hypergeometric Functions

The concept of the Hadamard products (see [14]) is very useful in our investigation.

Definition 31 (Hadamard products [14]). Let $f(z) := \sum_{n=0}^{\infty} a_n z^n$ and $g(z) := \sum_{n=0}^{\infty} b_n z^n$ be two power series whose radii of convergence are given by R_f and R_g, respectively. Then their Hadamard product is power series defined by

$$(f * g)(z) := \sum_{n=0}^{\infty} a_n b_n z^n, \tag{115}$$

whose radius of convergence R satisfies $R_f.R_g \le R$.

In particular, let us consider the function ${}_pF_{p+r}^{(\alpha,\beta;\rho,\xi)}[z;b]$. Its decomposition is illustrative. That is

$$\begin{aligned} &{}_pF_{p+r}^{(\alpha,\beta;\rho,\xi)}\left[\begin{matrix} x_1,\ldots,x_p \\ y_1,\ldots,y_{p+r}\end{matrix}; z; b\right] \\ &= {}_1F_r\left[\begin{matrix} 1 \\ y_1,\ldots,y_r\end{matrix}; z; b\right] \\ &* {}_pF_p^{(\alpha,\beta;\rho,\xi)}\left[\begin{matrix} x_1,\ldots,x_p \\ y_{1+r},\ldots,y_{p+r}\end{matrix}; z; b\right]. \end{aligned} \tag{116}$$

The above-mentioned detailed and systematic investigation by many authors (see, for example, [4, 15]) has largely motivated our present study. Therefore, the results established in this paper are of general character and hence encompass several cases of interest.

In this section, we will establish certain fractional integral formulas involving the new extension of Gauss hypergeometric function and new extension of confluent hypergeometric function. To do this, we need to recall the following pair of Saigo hypergeometric fractional integral operators.

For $x > 0, \xi, \zeta, \eta \in \mathbb{C}$ and $\Re(\xi) > 0$, we have

$$\begin{aligned} \left(I_{0,x}^{\xi,\zeta,\eta} f(t)\right)(x) &= \frac{x^{-\xi-\zeta}}{\Gamma(\xi)} \\ &\cdot \int_0^x (x-t)^{\xi-1}\, {}_2F_1\left(\xi+\zeta, -\eta; \xi; 1-\frac{t}{x}\right) f(t)\,dt \end{aligned} \tag{117}$$

and

$$\begin{aligned} \left(J_{x,\infty}^{\xi,\zeta,\eta} f(t)\right)(x) &= \frac{1}{\Gamma(\xi)}\int_x^{\infty}(t-x)^{\xi-1} \\ &\cdot t^{-\xi-\zeta}\, {}_2F_1\left(\xi+\zeta, -\eta; \xi; 1-\frac{x}{t}\right) f(t)\,dt, \end{aligned} \tag{118}$$

where the function ${}_2F_1(.)$ is a special case of the generalized hypergeometric function, the Gauss hypergeometric function.

The operator $I_{0,x}^{\xi,\zeta,\eta}(.)$ contains the Riemann-Liouville $R_{0,x}^{\xi}(.)$ fractional integral operators by means of the following relationships:

$$\begin{aligned} \left(R_{0,x}^{\xi} f(t)\right)(x) &= \left(I_{0,x}^{\xi,-\xi,\eta} f(t)\right)(x) \\ &= \frac{1}{\Gamma(\xi)}\int_0^x (x-t)^{\xi-1} f(t)\,dt \end{aligned} \tag{119}$$

and

$$\begin{aligned} \left(W_{x,\infty}^{\xi} f(t)\right)(x) &= \left(J_{x,\infty}^{\xi,-\xi,\eta} f(t)\right)(x) \\ &= \frac{1}{\Gamma(\xi)}\int_x^{\infty}(t-x)^{\xi-1} f(t)\,dt. \end{aligned} \tag{120}$$

It is noted that the operator (118) unifies the Erdélyi-Kober fractional integral operators as follows:

$$\left(E_{0,x}^{\xi,\eta} f(t)\right)(x) = \left(I_{0,x}^{\xi,0,\eta} f(t)\right)(x) = \frac{x^{-\xi-\eta}}{\Gamma(\xi)} \int_0^x (x-t)^{\xi-1} t^{\eta} f(t)\,dt \tag{121}$$

and

$$\left(K_{x,\infty}^{\xi,\eta} f(t)\right)(x) = \left(J_{x,\infty}^{\xi,0,\eta} f(t)\right)(x) = \frac{x^{\eta}}{\Gamma(\xi)} \int_x^{\infty} (t-x)^{\xi-1} t^{-\xi-\eta} f(t)\,dt. \tag{122}$$

The following lemmas proved in Kilbas and Sebastin [16] are useful to prove our main results.

Lemma 32 (Kilbas and Sebastian 2008). *Letting $\xi, \zeta, \eta \in \mathbb{C}$ be such that $\Re(\xi) > 0, \Re(\rho) > \max[0, \Re(\zeta - \eta)]$, then*

$$\left(I_{0,x}^{\xi,\zeta,\eta} t^{\rho-1}\right)(x) = \frac{\Gamma(\rho)\Gamma(\rho+\eta-\zeta)}{\Gamma(\rho-\zeta)\Gamma(\rho+\xi+\eta)} x^{\rho-\zeta-1}. \tag{123}$$

Lemma 33 (Kilbas and Sebastian 2008). *Letting $\xi, \zeta, \eta \in \mathbb{C}$ be such that $\Re(\xi) > 0, \Re(\rho) < 1 + \min[\Re(\zeta), \Re(\eta)]$, then*

$$\left(J_{x,\infty}^{\xi,\zeta,\eta} t^{\rho-1}\right)(x) = \frac{\Gamma(\zeta-\rho+1)\Gamma(\eta-\rho+1)}{\Gamma(1-\rho)\Gamma(\xi+\zeta+\eta-\rho+1)} x^{\rho-\zeta-1}. \tag{124}$$

The main results are given in the following theorem.

Theorem 34. *Let $\xi, \zeta, \eta, \rho, \alpha, \beta, \gamma, m \in \mathbb{C}; \Re(\xi) > 0; \Re(\gamma) > \Re(\beta) > 0$ and $|m| < M$ (where M is positive real number), such that $\Re(\rho) > \max[0, \Re(\zeta - \eta)]$; then*

$$\left(I_{0,x}^{\xi,\zeta,\eta} t^{\rho-1}\,{}^{MC}F_m(\alpha,\beta;\gamma;t)\right)(x) = x^{\rho-\zeta-1} \frac{\Gamma(\rho)\Gamma(\rho+\eta-\zeta)}{\Gamma(\rho-\zeta)\Gamma(\rho+\xi+\eta)} \times {}^{MC}F_m(\alpha,\beta;\gamma;x) * {}_2F_2(\rho,\rho+\eta-\zeta;\rho-\zeta,\rho+\xi+\eta;x). \tag{125}$$

Proof. For convenience, we denote the left-hand side of the result (125) by $\mathcal{J}$. Using (67), and then interchanging the order of integration and summation, which is valid under the conditions of Theorem 34, then

$$\mathcal{J} = \sum_{n=0}^{\infty} (\alpha)_n \frac{{}^{MC}B_m(\beta+n,\gamma-\beta)}{B(\beta,\gamma-\beta)} \frac{1}{n!} \left(I_{0,x}^{\xi,\zeta,\eta} t^{n+\rho-1}\right) \cdot (x). \tag{126}$$

Applying the result (123), (126) reduced to

$$\mathcal{J} = \sum_{n=0}^{\infty} (\alpha)_n \frac{{}^{MC}B_m(\beta+n,\gamma-\beta)}{B(\beta,\gamma-\beta)} \cdot \frac{\Gamma(\rho+n)\Gamma(\rho+\eta-\zeta+n)}{\Gamma(\rho-\zeta+n)\Gamma(\rho+\xi+\eta+n)} \frac{x^{\rho+n-\zeta-1}}{n!}. \tag{127}$$

After simplification, (127) reduces to

$$\mathcal{J} = x^{\rho-\zeta-1} \sum_{n=0}^{\infty} (a)_n \frac{{}^{MC}B_m(\beta+n,\gamma-\beta)}{B(\beta,\gamma-\beta)} \cdot \frac{\Gamma(\rho+n)\Gamma(\rho+\eta-\zeta+n)}{\Gamma(\rho-\zeta+n)\Gamma(\rho+\xi+\eta+n)} \frac{x^n}{n!}. \tag{128}$$

Further using $(a)_n = \Gamma(a+n)/\Gamma(a)$, (128) reduces to the following form:

$$\mathcal{J} = x^{\rho-\zeta-1} \frac{\Gamma(\rho)\Gamma(\rho+\eta-\zeta)}{\Gamma(\rho-\zeta)\Gamma(\rho+\xi+\eta)} \sum_{n=0}^{\infty} (a)_n \cdot \frac{{}^{MC}B_m(\beta+n,\gamma-\beta)}{B(\beta,\gamma-\beta)} \times \frac{(\rho)_n(\rho+\eta-\zeta)_n}{(\rho-\zeta)_n(\rho+\xi+\eta)_n} \cdot \frac{x^n}{n!}, \tag{129}$$

$$\mathcal{J} = x^{\rho-\zeta-1} \cdot \frac{\Gamma(\rho)\Gamma(\rho+\eta-\zeta)}{\Gamma(\rho-\zeta)\Gamma(\rho+\xi\xi+\eta)} {}^{MC}{}_2F_{m,2}(\alpha,\beta,\rho,\rho+\eta-\zeta;\gamma,\rho-\zeta,\rho+\xi+\eta;x), \tag{130}$$

and interpreting the above equation with the help of the concept of Hadamard given in (116), we have the required result.

Theorem 35. *Let $\xi, \zeta, \eta, \rho, \alpha, \beta, \gamma, m \in \mathbb{C}; \Re(\xi) > 0; \Re(\gamma) > \Re(\beta) > 0$ and $|m| < M$ (where M is positive real number), such that $\Re(\rho) < 1 + \min[\Re(\zeta), \Re(\eta)]$; then*

$$\left(J_{x,\infty}^{\xi,\zeta,\eta} t^{\rho-1}\,{}^{MC}F_m\left(\alpha,\beta;\gamma;\frac{1}{t}\right)\right)(x) = x^{\rho-\zeta-1} \cdot \frac{\Gamma(\zeta-\rho+1)\Gamma(\eta-\rho+1)}{\Gamma(1-\rho)\Gamma(\xi+\zeta+\eta-\rho+1)} \times {}^{MC}F_m(\alpha,\beta;\gamma;x) * {}_2F_2(\zeta-\rho+1,\eta-\rho+1;1-\rho,\xi+\zeta+\eta-\rho+1;x). \tag{131}$$

Theorem 36. *Let $\xi, \zeta, \eta, \rho, \beta, \gamma, m \in \mathbb{C}; \Re(\xi) > 0; \Re(\gamma) > \Re(\beta) > 0$ and $|m| < M$ (where M is positive real number), such that $\Re(\rho) > \max[0, \Re(\zeta - \eta)]$; then*

$$\left(I_{0,x}^{\xi,\zeta,\eta} t^{\rho-1}\,{}^{MC}\Phi_m(\beta;\gamma;t)\right)(x) = x^{\rho-\zeta-1} \frac{\Gamma(\rho)\Gamma(\rho+\eta-\zeta)}{\Gamma(\rho-\zeta)\Gamma(\rho+\xi+\eta)} \times {}^{MC}\Phi_m(\beta;\gamma;x) * {}_2F_2(\rho,\rho+\eta-\zeta;\rho-\zeta,\rho+\xi+\eta;x). \tag{132}$$

Theorem 37. *Let $\xi,\zeta,\eta,\rho,\beta,\gamma,m \in \mathbb{C};\Re(\xi) > 0;\Re(\gamma) > \Re(\beta) > 0$ and $|m| < M$ (where M is positive real number), such that $\Re(\rho) < 1 + \min[\Re(\zeta),\Re(\eta)]$; then*

$$\left(J_{x,\infty}^{\xi,\zeta,\eta} t^{\rho-1MC}\Phi_m\left(\beta;\gamma;\frac{1}{t}\right)\right)(x) = x^{\rho-\zeta-1} \cdot \frac{\Gamma(\zeta-\rho+1)\Gamma(\eta-\rho+1)}{\Gamma(1-\rho)\Gamma(\xi+\zeta+\eta-\rho+1)} \times {}^{MC}\Phi_m(\beta;\gamma;x) * {}_2F_2(\zeta-\rho+1,\eta-\rho+1;1-\rho,\xi+\zeta+\eta-\rho+1;x). \quad (133)$$

Proof. The proofs of the Theorems 35, 36, and 37 are the same as those of Theorem 34.

7.1. Some Special Cases of the above Fractional Integral Formulas. By assigning the suitable values to the parameters involved in the results established in Theorems 34–37, we have the following special cases.

By putting $\zeta = -\xi$, the Saigo hypergeometric fractional integrals operators reduces to Riemann-Liouville fractional integral operators; then the results in (125), (131), (132) and (133) reduce to the following form.

Corollary 38. *Let $\xi,\rho,\alpha,\beta,\gamma,m \in \mathbb{C};\Re(\xi) > 0;\Re(\gamma) > \Re(\beta) > 0$ and $|m| < M$ (where M is positive real number), such that $\Re(\rho) > \max[0,\Re(-\xi)]$; then*

$$\left(R_{0,x}^{\xi} t^{\rho-1MC}F_m(\alpha,\beta;\gamma;t)\right)(x) = x^{\rho+\xi-1}\frac{\Gamma(\rho)}{\Gamma(\rho+\xi)} {}^{MC}F_m(\alpha,\beta;\gamma;x) * {}_1F_1(\rho;\rho+\xi;x). \quad (134)$$

Corollary 39. *Let $\xi,\rho,\alpha,\beta,\gamma,m \in \mathbb{C};\Re(\xi) > 0;\Re(\gamma) > \Re(\beta) > 0$ and $|m| < M$ (where M is positive real number), such that $\Re(\rho) < 1 + \Re(\xi)$; then*

$$\left(W_{x,\infty}^{\xi} t^{\rho-1MC}F_m\left(\alpha,\beta;\gamma;\frac{1}{t}\right)\right)(x) = x^{\rho+\xi-1}\frac{\Gamma(1-\xi-\rho)}{\Gamma(1-\rho)} {}^{MC}F_m(\alpha,\beta;\gamma;x) * {}_1F_1(1-\xi-\rho;1-\rho;x). \quad (135)$$

Corollary 40. *Let $\xi,\rho,\beta,\gamma,m \in \mathbb{C};\Re(\xi) > 0;\Re(\gamma) > \Re(\beta) > 0$ and $|m| < M$ (where M is positive real number), such that $\Re(\rho) > \max[0,\Re(-\xi)]$; then*

$$\left(R_{0,x}^{\xi} t^{\rho-1MC}\Phi_m(\beta;\gamma;t)\right)(x) = x^{\rho+\xi-1}\frac{\Gamma(\rho)}{\Gamma(\rho+\xi)} {}^{MC}\Phi_m(\beta;\gamma;x) * {}_1F_1(\rho;\rho+\xi;x). \quad (136)$$

Corollary 41. *Let $\xi,\rho,\beta,\gamma,m \in \mathbb{C};\Re(\xi) > 0;\Re(\gamma) > \Re(\beta) > 0$ and $|m| < M$ (where M is positive real number), such that $\Re(\rho) < 1 + \Re(\xi)$; then*

$$\left(W_{x,\infty}^{\xi} t^{\rho-1MC}\Phi_m\left(\beta;\gamma;\frac{1}{t}\right)\right)(x) = x^{\rho+\xi-1}\frac{\Gamma(1-\xi-\rho)}{\Gamma(1-\rho)} {}^{MC}\Phi_m(\beta;\gamma;x) * {}_1F_1(1-\xi-\rho;1-\rho;x). \quad (137)$$

By putting $\zeta = 0$, the Saigo hypergeometric fractional integrals operators reduce to the Erdélyi-Kober fractional integral operators; then the results in (125), (131), (132) and (133) reduce to the following form.

Corollary 42. *Let $\xi,\eta,\rho,\alpha,\beta,\gamma,m \in \mathbb{C};\Re(\xi) > 0;\Re(\gamma) > \Re(\beta) > 0$ and $|m| < M$ (where M is positive real number), such that $\Re(\rho) > \max[0,\Re(-\eta)]$; then*

$$\left(E_{0,x}^{\xi,\eta} t^{\rho-1MC}F_m(\alpha,\beta;\gamma;t)\right)(x) = x^{\rho-1}\frac{\Gamma(\rho+\eta)}{\Gamma(\rho+\xi+\eta)} {}^{MC}F_m(\alpha,\beta;\gamma;x) * {}_1F_1(\rho+\eta;\rho+\xi+\eta;x). \quad (138)$$

Corollary 43. *Let $\xi,\eta,\rho,\alpha,\beta,\gamma,m \in \mathbb{C};\Re(\xi) > 0;\Re(\gamma) > \Re(\beta) > 0$ and $|m| < M$ (where M is positive real number), such that $\Re(\rho) < 1 + \min[\Re(\zeta),\Re(\eta)]$; then*

$$\left(K_{x,\infty}^{\xi,\eta} t^{\rho-1MC}F_m\left(\alpha,\beta;\gamma;\frac{1}{t}\right)\right)(x) = x^{\rho-1}\frac{\Gamma(\eta-\rho+1)}{\Gamma(\xi+\eta-\rho+1)} {}^{MC}F_m(\alpha,\beta;\gamma;x) * {}_1F_1(\eta-\rho+1;\xi+\eta-\rho+1;x). \quad (139)$$

Corollary 44. *Let $\xi,\eta,\rho,\beta,\gamma,m \in \mathbb{C};\Re(\xi) > 0;\Re(\gamma) > \Re(\beta) > 0$ and $|m| < M$ (where M is positive real number), such that $\Re(\rho) > \max[0,\Re(-\eta)]$; then*

$$\left(E_{0,x}^{\xi,\eta} t^{\rho-1MC}\Phi_m(\beta;\gamma;t)\right)(x) = x^{\rho-1}\frac{\Gamma(\rho+\eta)}{\Gamma(\rho+\xi+\eta)} {}^{MC}\Phi_m(\beta;\gamma;x) * {}_1F_1(\rho+\eta;\rho+\xi+\eta;x). \quad (140)$$

Corollary 45. *Let $\xi,\eta,\rho,\beta,\gamma,m \in \mathbb{C};\Re(\xi) > 0;\Re(\gamma) > \Re(\beta) > 0$ and $|m| < M$ (where M is positive real number), such that $\Re(\rho) < 1 + \min[\Re(\zeta),\Re(\eta)]$; then*

$$\left(K_{x,\infty}^{\xi,\eta} t^{\rho-1MC}\Phi_m\left(\beta;\gamma;\frac{1}{t}\right)\right)(x) = x^{\rho-1}\frac{\Gamma(\eta-\rho+1)}{\Gamma(\xi+\eta-\rho+1)} {}^{MC}\Phi_m(\beta;\gamma;x) * {}_1F_1(\eta-\rho+1;\xi+\eta-\rho+1;x). \quad (141)$$

If we choose $m = 0$, then new extension of Gauss hypergeometric function and new extension of confluent hypergeometric function reduce to Gauss hypergeometric function and confluent hypergeometric function; then from the formulae establisehd in (125), (131), (132) and (133), we have the following results.

Corollary 46. *Let $\xi,\zeta,\eta,\rho,\alpha,\beta,\gamma \in \mathbb{C}; \Re(\xi) > 0; \Re(\gamma) > \Re(\beta) > 0$, such that $\Re(\rho) > \max[0, \Re(\zeta-\eta)]$; then*

$$\left(I_{0,x}^{\xi,\zeta,\eta} t^{\rho-1} {}_2F_1(\alpha,\beta;\gamma;t)\right)(x) = x^{\rho-\zeta-1} \frac{\Gamma(\rho)\Gamma(\rho+\eta-\zeta)}{\Gamma(\rho-\zeta)\Gamma(\rho+\xi+\eta)} \times {}_2F_1(\alpha,\beta;\gamma;x) * {}_2F_2(\rho,\rho+\eta-\zeta;\rho-\zeta,\rho+\xi+\eta;x). \quad (142)$$

Corollary 47. *Let $\xi,\zeta,\eta,\rho,\alpha,\beta,\gamma \in \mathbb{C}; \Re(\xi) > 0; \Re(\gamma) > \Re(\beta) > 0$, such that $\Re(\rho) < 1 + \min[\Re(\zeta), \Re(\eta)]$; then*

$$\left(J_{x,\infty}^{\xi,\zeta,\eta} t^{\rho-1} {}_2F_1\left(\alpha,\beta;\gamma;\frac{1}{t}\right)\right)(x) = x^{\rho-\zeta-1} \cdot \frac{\Gamma(\zeta-\rho+1)\Gamma(\eta-\rho+1)}{\Gamma(1-\rho)\Gamma(\xi+\zeta+\eta-\rho+1)} \times {}_2F_1(\alpha,\beta;\gamma;x) * {}_2F_2(\zeta-\rho+1,\eta-\rho+1;1-\rho,\xi+\zeta+\eta-\rho+1;x). \quad (143)$$

Corollary 48. *Let $\xi,\zeta,\eta,\rho,\beta,\gamma \in \mathbb{C}; \Re(\xi) > 0; \Re(\gamma) > \Re(\beta) > 0$, such that $\Re(\rho) > \max[0, \Re(\zeta-\eta)]$; then*

$$\left(I_{0,x}^{\xi,\zeta,\eta} t^{\rho-1} \Phi(\beta;\gamma;t)\right)(x) = x^{\rho-\zeta-1} \frac{\Gamma(\rho)\Gamma(\rho+\eta-\zeta)}{\Gamma(\rho-\zeta)\Gamma(\rho+\xi+\eta)} \times \Phi(\beta;\gamma;x) * {}_2F_2(\rho,\rho+\eta-\zeta;\rho-\zeta,\rho+\xi+\eta;x). \quad (144)$$

Corollary 49. *Let $\xi,\zeta,\eta,\rho,\beta,\gamma \in \mathbb{C}; \Re(\xi) > 0; \Re(\gamma) > \Re(\beta) > 0$, such that $\Re(\rho) < 1 + \min[\Re(\zeta), \Re(\eta)]$. Then*

$$\left(J_{x,\infty}^{\xi,\zeta,\eta} t^{\rho-1} \Phi\left(\beta;\gamma;\frac{1}{t}\right)\right)(x) = x^{\rho-\zeta-1} \cdot \frac{\Gamma(\zeta-\rho+1)\Gamma(\eta-\rho+1)}{\Gamma(1-\rho)\Gamma(\xi+\zeta+\eta-\rho+1)} \times \Phi(\beta;\gamma;x) * {}_2F_2(\zeta-\rho+1,\eta-\rho+1;1-\rho,\xi+\zeta+\eta-\rho+1;x). \quad (145)$$

8. Beta Transform

The Beta transform of $f(z)$ is defined as follows [17]:

$$B\{f(z):a,b\} = \int_0^1 z^{a-1}(1-z)^{b-1} f(z)\,dz. \quad (146)$$

Theorem 50. *Let $\xi,\zeta,\eta,\rho,\beta,\gamma,m \in \mathbb{C}; \Re(\xi) > 0; \Re(\gamma) > \Re(\beta) > 0$ and $|m| < M$ (where M is positive real number), such that $\Re(\rho) > \max[0, \Re(\zeta-\eta)]$; then*

$$B\left\{\left(I_{0,x}^{\xi,\zeta,\eta} t^{\rho-1} {}^{MC}F_m(\alpha,\beta;\gamma;t)\right)(x) : l, m\right\} = \Gamma(m)\, x^{\rho-\zeta-1} \frac{\Gamma(\rho)\Gamma(\rho+\eta-\zeta)}{\Gamma(\rho-\zeta)\Gamma(\rho+\xi+\eta)} \times {}^{MC}F_m(\alpha,\beta;\gamma;x) * {}_3F_3(\rho,\rho+\eta-\zeta,l;\rho-\zeta,\rho+\xi+\eta,l+m;x). \quad (147)$$

Proof. For convenience, we denote the left-hand side of the result (147) by $\mathscr{B}$. Using the definition of beta transform, the LHS of (147) becomes

$$\mathscr{B} = \int_0^1 z^{l-1}(1-z)^{m-1}\left(I_{0,x}^{\xi,\zeta,\eta} t^{\rho-1} {}^{MC}F_m(\alpha,\beta;\gamma;tz)\right) \cdot (x)\,dz, \quad (148)$$

and further using (129) and then changing the order of integration and summation, which is valid under the conditions of Theorem 1, then

$$\mathscr{B} = x^{\rho-\zeta-1} \frac{\Gamma(\rho)\Gamma(\rho+\eta-\zeta)}{\Gamma(\rho-\zeta)\Gamma(\rho+\xi+\eta)} \sum_{n=0}^{\infty} (a)_n \cdot \frac{{}^{MC}B_m(\beta+n,\gamma-\beta)}{B(\beta,\gamma-\beta)} \times \frac{(\rho)_n(\rho+\eta-\zeta)_n}{(\rho-\zeta)_n(\rho+\xi+\eta)_n} \cdot \frac{x^n}{n!} \int_0^1 z^{l+n-1}(1-z)^{m-1}\,dz. \quad (149)$$

Applying the definition of beta transform, (149) reduced to

$$\mathscr{B} = x^{\rho-\zeta-1} \frac{\Gamma(\rho)\Gamma(\rho+\eta-\zeta)}{\Gamma(\rho-\zeta)\Gamma(\rho+\xi+\eta)} \sum_{n=0}^{\infty} (a)_n \cdot \frac{{}^{MC}B_m(\beta+n,\gamma-\beta)}{B(\beta,\gamma-\beta)} \frac{(\rho)_n(\rho+\eta-\zeta)_n}{(\rho-\zeta)_n(\rho+\xi+\eta)_n} \times \frac{\Gamma(l+n)\Gamma(m)}{\Gamma(l+m+n)} \frac{x^n}{n!}, \quad (150)$$

and interpreting the above equation with the help of (67), we have

$$\mathscr{B} = \Gamma(m)\, x^{\rho-\zeta-1} \cdot \frac{\Gamma(\rho)\Gamma(\rho+\eta-\zeta)}{\Gamma(\rho-\zeta)\Gamma(\rho+\xi+\eta)} {}^{MC}{}_5F_{m,3}(\alpha,\beta,\rho,\rho+\eta-\zeta,l;\gamma,\rho-\zeta,\rho+\xi+\eta,l+m;x). \quad (151)$$

Further, interpreting (151) with the view of the concept of Hadamard (116), we have the required the result.

Theorem 51. *Let $\xi,\zeta,\eta,\rho,\beta,\gamma,m \in \mathbb{C};\Re(\xi) > 0;\Re(\gamma) > \Re(\beta) > 0$ and $|m| < M$ (where M is positive real number), such that $\Re(\rho) < 1 + \min[\Re(\zeta),\Re(\eta)]$; then*

$$B\left\{\left(J_{x,\infty}^{\xi,\zeta,\eta}t^{\rho-1MC}F_m\left(\alpha,\beta;\gamma;\frac{z}{t}\right)\right)(x):l,m\right\} = \Gamma(m) \cdot x^{\rho-\zeta-1}\frac{\Gamma(\zeta-\rho+1)\Gamma(\eta-\rho+1)}{\Gamma(1-\rho)\Gamma(\xi+\zeta+\eta-\rho+1)} \cdot {}^{MC}F_m(\alpha,\beta;\gamma;x) * {}_3F_3(\zeta-\rho+1,\eta-\rho+1,l;1-\rho,\xi+\zeta+\eta-\rho+1,l+m;x). \tag{152}$$

Theorem 52. *Let $\xi,\zeta,\eta,\rho,\beta,\gamma,m \in \mathbb{C};\Re(\xi) > 0;\Re(\gamma) > \Re(\beta) > 0$ and $|m| < M$ (where M is positive real number), such that $\Re(\rho) > \max[0,\Re(\zeta-\eta)]$; then*

$$B\left\{\left(I_{0,x}^{\xi,\zeta,\eta}t^{\rho-1MC}\Phi_m(\beta;\gamma;t)\right)(x):l,m\right\} = \Gamma(m)x^{\rho-\zeta-1}\frac{\Gamma(\rho)\Gamma(\rho+\eta-\zeta)}{\Gamma(\rho-\zeta)\Gamma(\rho+\xi+\eta)} \times {}^{MC}\Phi_m(\beta;\gamma;x) * {}_3F_3(\rho,\rho+\eta-\zeta,l;\rho-\zeta,\rho+\xi+\eta,l+m;x). \tag{153}$$

Theorem 53. *Let $\xi,\zeta,\eta,\rho,\beta,\gamma,m \in \mathbb{C};\Re(\xi) > 0;\Re(\gamma) > \Re(\beta) > 0$ and $|m| < M$ (where M is positive real number), such that $\Re(\rho) < 1 + \min[\Re(\zeta),\Re(\eta)]$; then*

$$B\left\{\left(J_{x,\infty}^{\xi,\zeta,\eta}t^{\rho-1MC}\Phi_m\left(\beta;\gamma;\frac{z}{t}\right)\right)(x):l,m\right\} = \Gamma(m) \cdot x^{\rho-\zeta-1}\frac{\Gamma(\zeta-\rho+1)\Gamma(\eta-\rho+1)}{\Gamma(1-\rho)\Gamma(\xi+\zeta+\eta-\rho+1)} \cdot {}^{MC}\Phi_m(\beta;\gamma;x) * {}_3F_3(\zeta-\rho+1,\eta-\rho+1,l;1-\rho,\xi+\zeta+\eta-\rho+1,l+m;x). \tag{154}$$

Proof. The proofs of the Theorems 51, 52, and 53 are parallel to those of Theorem 50.

9. Laplace Transform

The Laplace transform of $f(z)$ is defined as follows [17]:

$$L\{f(z)\} = \int_0^\infty e^{-sz}f(z)\,dz. \tag{155}$$

Theorem 54. *Let $\xi,\zeta,\eta,\rho,\beta,\gamma,m \in \mathbb{C};\Re(\xi) > 0;\Re(\gamma) > \Re(\beta) > 0$ and $|m| < M$ (where M is positive real number), such that $\Re(\rho) > \max[0,\Re(\zeta-\eta)]$; then*

$$L\left\{z^{l-1}\left(I_{0,x}^{\xi,\zeta,\eta}t^{\rho-1MC}F_m(\alpha,\beta;\gamma;tz)\right)(x)\right\} = \frac{x^{\rho-\zeta-1}}{s^l}\frac{\Gamma(\rho)\Gamma(\rho+\eta-\zeta)}{\Gamma(\rho-\zeta)\Gamma(\rho+\xi+\eta)} \times {}^{MC}F_m(\alpha,\beta;\gamma;x) * {}_3F_2(\rho,\rho+\eta-\zeta,l;\rho-\zeta,\rho+\xi+\eta;x). \tag{156}$$

Proof. For convenience, we denote the left-hand side of the result (156) by $\mathscr{L}$. Then, applying the Laplace, we have

$$\mathscr{L} = \int_0^\infty e^{-sz}z^{l-1}\left(I_{0,x}^{\xi,\zeta,\eta}t^{\rho-1MC}F_m(\alpha,\beta;\gamma;tz)\right)(x)\,dz, \tag{157}$$

and further using (129) and then changing the order of integration and summation, which is valid under the conditions of Theorem 1, then

$$\mathscr{L} = x^{\rho-\zeta-1}\frac{\Gamma(\rho)\Gamma(\rho+\eta-\zeta)}{\Gamma(\rho-\zeta)\Gamma(\rho+\xi+\eta)}\sum_{n=0}^{\infty}(a)_n \cdot \frac{{}^{MC}B_m(\beta+n,\gamma-\beta)}{B(\beta,\gamma-\beta)} \times \frac{(\rho)_n(\rho+\eta-\zeta)_n}{(\rho-\zeta)_n(\rho+\xi+\eta)_n} \cdot \frac{x^n}{n!}\int_0^\infty e^{-sz}z^{n+l-1}dz. \tag{158}$$

After simplification, (158) reduces to

$$\mathscr{L} = x^{\rho-\zeta-1}\frac{\Gamma(\rho)\Gamma(\rho+\eta-\zeta)}{\Gamma(\rho-\zeta)\Gamma(\rho+\xi+\eta)}\sum_{n=0}^{\infty}(a)_n \cdot \frac{{}^{MC}B_m(\beta+n,\gamma-\beta)}{B(\beta,\gamma-\beta)}\frac{(\rho)_n(\rho+\eta-\zeta)_n}{(\rho-\zeta)_n(\rho+\xi+\eta)_n}\frac{x^n}{n!} \cdot \frac{\Gamma(n+l)}{s^{n+l}} = x^{\rho-\zeta-1} \cdot \frac{\Gamma(\rho)\Gamma(\rho+\eta-\zeta)\Gamma(l)}{\Gamma(\rho-\zeta)\Gamma(\rho+\xi+\eta)}{}^{MC}{}_3F_{m,2}(\alpha,\beta,\rho,\rho+\eta-\zeta,l;\rho-\zeta,\rho+\xi+\eta,\gamma;x), \tag{159}$$

and interpreting (159) with the view of the concept of Hadamard (116), we have the required the result.

Theorem 55. *Let $\xi,\zeta,\eta,\rho,\beta,\gamma,m \in \mathbb{C};\Re(\xi) > 0;\Re(\gamma) > \Re(\beta) > 0$ and $|m| < M$ (where M is positive real number), such that $\Re(\rho) < 1 + \min[\Re(\zeta),\Re(\eta)]$; then*

$$L\left\{z^{l-1}\left(J_{x,\infty}^{\xi,\zeta,\eta}t^{\rho-1MC}F_m\left(\alpha,\beta;\gamma;\frac{z}{t}\right)\right)(x)\right\} = \frac{x^{\rho-\zeta-1}}{s^l}\frac{\Gamma(\zeta-\rho+1)\Gamma(\eta-\rho+1)}{\Gamma(1-\rho)\Gamma(\xi+\zeta+\eta-\rho+1)} \cdot {}^{MC}F_m(\alpha,\beta;\gamma;x) * {}_3F_2(\zeta-\rho+1,\eta-\rho+1,l;1-\rho,\xi+\zeta+\eta-\rho+1;x). \tag{160}$$

Theorem 56. *Let $\xi,\zeta,\eta,\rho,\beta,\gamma,m \in \mathbb{C}; \Re(\xi) > 0; \Re(\gamma) > \Re(\beta) > 0$ and $|m| < M$ (where M is positive real number), such that $\Re(\rho) > \max[0, \Re(\zeta-\eta)]$; then*

$$L\left\{z^{l-1}\left(I_{0,x}^{\xi,\zeta,\eta}t^{\rho-1MC}\Phi_m(\beta;\gamma;tz)\right)(x)\right\} = \frac{x^{\rho-\zeta-1}}{s^l}\frac{\Gamma(\rho)\Gamma(\rho+\eta-\zeta)}{\Gamma(\rho-\zeta)\Gamma(\rho+\xi+\eta)} \times {}^{MC}\Phi_m(\beta;\gamma;x) * {}_3F_2(\rho,\rho+\eta-\zeta,l;\rho-\zeta,\rho+\xi+\eta;x). \quad (161)$$

Theorem 57. *Let $\xi,\zeta,\eta,\rho,\beta,\gamma,m \in \mathbb{C}; \Re(\xi) > 0; \Re(\gamma) > \Re(\beta) > 0$ and $|m| < M$ (where M is positive real number), such that $\Re(\rho) < 1 + \min[\Re(\zeta), \Re(\eta)]$; then*

$$L\left\{z^{l-1}\left(J_{x,\infty}^{\xi,\zeta,\eta}t^{\rho-1MC}\Phi_m\left(\beta;\gamma;\frac{z}{t}\right)\right)(x)\right\} = \frac{x^{\rho-\zeta-1}}{s^l} \cdot \frac{\Gamma(\zeta-\rho+1)\Gamma(\eta-\rho+1)}{\Gamma(1-\rho)\Gamma(\xi+\zeta+\eta-\rho+1)}{}^{MC}\Phi_m(\beta;\gamma;x) * {}_3F_2(\zeta-\rho+1,\eta-\rho+1,l;1-\rho,\xi+\zeta+\eta-\rho+1;x). \quad (162)$$

Proof. The proofs of Theorems 55, 56, and 57 would run parallel to those of Theorem 54, so the proofs of these theorems are omitted here.

10. Whittaker Transform

Theorem 58. *Let $\xi,\zeta,\eta,\rho,\beta,\gamma,m \in \mathbb{C}; \Re(\xi) > 0; \Re(\gamma) > \Re(\beta) > 0$ and $|m| < M$ (where M is positive real number) and $\Re(\sigma\pm\omega) > -1/2$, such that $\Re(\rho) > \max[0, \Re(\zeta-\eta)]$; then*

$$\int_0^\infty z^{\sigma-1}e^{-\delta z/2}W_{\tau,\omega}(\kappa z)W_1(z)\,dz = \frac{x^{\rho-\zeta-1}}{\kappa^{\sigma-1}} \cdot \frac{\Gamma(\rho)\Gamma(\rho+\eta-\zeta)}{\Gamma(\rho-\zeta)\Gamma(\rho+\xi+\eta)}{}^{MC}F_m\left(\alpha,\beta;\gamma;\frac{x}{\kappa}\right) * {}_4F_3\left(\rho,\rho+\eta-\zeta,\frac{1}{2}+\omega+\sigma,\frac{1}{2}-\omega+\sigma;\rho-\zeta,\rho+\xi+\eta,\frac{1}{2}-\tau+\sigma;\frac{x}{\kappa}\right), \quad (163)$$

where $W_1(z) = \{(I_{0,x}^{\xi,\zeta,\eta}t^{\rho-1}\,{}^{MC}F_m(\alpha,\beta;\gamma;tz))(x)\}$.

Proof. For convenience, we denote the left-hand side of the result (163) by $\mathscr{W}$. Then using the result from (128), after changing the order of integration and summation, we get

$$\mathscr{W} = x^{\rho-\zeta-1}\frac{\Gamma(\rho)\Gamma(\rho+\eta-\zeta)}{\Gamma(\rho-\zeta)\Gamma(\rho+\xi+\eta)}\sum_{n=0}^\infty (a)_n \cdot \frac{{}^{MC}B_m(\beta+n,\gamma-\beta)}{B(\beta,\gamma-\beta)} \times \frac{(\rho)_n(\rho+\eta-\zeta)_n}{(\rho-\zeta)_n(\rho+\xi+\eta)_n}\frac{x^n}{n!} \cdot \int_0^\infty z^{n+\sigma-1}e^{-\kappa z/2}W_{\tau,\omega}(\kappa z)\,dz. \quad (164)$$

By substituting $\kappa z = \varsigma$, (164) becomes

$$\mathscr{W} = x^{\rho-\zeta-1}\frac{\Gamma(\rho)\Gamma(\rho+\eta-\zeta)}{\Gamma(\rho-\zeta)\Gamma(\rho+\xi+\eta)}\sum_{n=0}^\infty (a)_n \cdot \frac{{}^{MC}B_m(\beta+n,\gamma-\beta)}{B(\beta,\gamma-\beta)} \times \frac{(\rho)_n(\rho+\eta-\zeta)_n}{(\rho-\zeta)_n(\rho+\xi+\eta)_n} \cdot \frac{x^n}{n!}\frac{1}{\kappa^{n+\sigma-1}}\int_0^\infty \varsigma^{n+\sigma-1}e^{-\varsigma/2}W_{\tau,\omega}(\varsigma)\,d\varsigma. \quad (165)$$

Now we use the following integral formula involving Whittaker function:

$$\int_0^\infty t^{\nu-1}e^{-t/2}W_{\tau,\omega}(t)\,dt = \frac{\Gamma(1/2+\omega+\nu)\Gamma(1/2-\omega+\nu)}{\Gamma(1/2-\tau+\nu)}, \quad \left(\Re(\nu\pm\omega) > \frac{-1}{2}\right). \quad (166)$$

Then we have

$$\mathscr{W} = \frac{x^{\rho-\zeta-1}}{\kappa^{\sigma-1}}\frac{\Gamma(\rho)\Gamma(\rho+\eta-\zeta)}{\Gamma(\rho-\zeta)\Gamma(\rho+\xi+\eta)}\sum_{n=0}^\infty (a)_n \cdot \frac{{}^{MC}B_m(\beta+n,\gamma-\beta)}{B(\beta,\gamma-\beta)}\frac{(\rho)_n(\rho+\eta-\zeta)_n}{(\rho-\zeta)_n(\rho+\xi+\eta)_n} \times \frac{\Gamma(1/2+\omega+\sigma+n)\Gamma(1/2-\omega+\sigma+n)}{\Gamma(1/2-\tau+\sigma+n)} \cdot \frac{1}{n!}\left(\frac{x}{\kappa}\right)^n, \quad (167)$$

and interpreting (167) with the help of (67), then, with the concept of Hadamard (116), we have the desired the result.

Theorem 59. *Let $\xi,\zeta,\eta,\rho,\beta,\gamma,m \in \mathbb{C}; \Re(\xi) > 0; \Re(\gamma) > \Re(\beta) > 0$ and $|m| < M$ (where M is positive real number) and $\Re(\sigma\pm\omega) > -1/2$, such that $\Re(\rho) < 1+\min[\Re(\zeta), \Re(\eta)]$; then*

$$\int_0^\infty z^{\sigma-1}e^{-\delta z/2}W_{\tau,\omega}(\kappa z)W_2(z)\,dz = \frac{x^{\rho-\zeta-1}}{\kappa^{\sigma-1}} \cdot \frac{\Gamma(\zeta-\rho+1)\Gamma(\eta-\rho+1)}{\Gamma(1-\rho)\Gamma(\xi+\zeta+\eta-\rho+1)}$$

$$\cdot {}^{MC}F_m\left(\alpha, \beta; \gamma; \frac{x}{\kappa}\right) * {}_4F_3\left(\zeta - \rho + 1, \eta - \rho + 1, \frac{1}{2} + \omega + \sigma, \frac{1}{2} - \omega + \sigma; 1 - \rho, \xi + \zeta + \eta - \rho + 1, \frac{1}{2} - \tau + \sigma; \frac{x}{\kappa}\right), \quad (168)$$

where $W_2(z) = \{(J_{x,\infty}^{\xi,\zeta,\eta} t^{\rho-1}\ {}^{MC}F_m(\alpha, \beta; \gamma; z/t))(x)\}$.

Theorem 60. *Let $\xi, \zeta, \eta, \rho, \beta, \gamma, m \in \mathbb{C}; \Re(\xi) > 0; \Re(\gamma) > \Re(\beta) > 0$ and $|m| < M$ (where M is positive real number) and $\Re(\sigma \pm \omega) > -1/2$, such that $\Re(\rho) > \max[0, \Re(\zeta - \eta)]$; then*

$$\int_0^\infty z^{\sigma-1} e^{-\delta z/2} W_{\tau,\omega}(\kappa z) W_3(z)\, dz = \frac{x^{\rho-\zeta-1}}{\kappa^{\sigma-1}} \cdot \frac{\Gamma(\rho)\Gamma(\rho+\eta-\zeta)}{\Gamma(\rho-\zeta)\Gamma(\rho+\xi+\eta)} {}^{MC}\Phi_m\left(\beta; \gamma; \frac{x}{\kappa}\right) * {}_4F_3\left(\rho, \rho+\eta-\zeta, \frac{1}{2} + \omega + \sigma, \frac{1}{2} - \omega + \sigma; \rho - \zeta, \rho + \xi + \eta, \frac{1}{2} - \tau + \sigma; \frac{x}{\kappa}\right), \quad (169)$$

where $W_3(z) = \{(I_{0,x}^{\xi,\zeta,\eta} t^{\rho-1}\ {}^{MC}\Phi_m(\beta; \gamma; tz))(x)\}$.

Theorem 61. *Let $\xi, \zeta, \eta, \rho, \beta, \gamma, m \in \mathbb{C}; \Re(\xi) > 0; \Re(\gamma) > \Re(\beta) > 0$ and $|m| < M$ (where M is positive real number) and $\Re(\sigma \pm \omega) > -1/2$, such that $\Re(\rho) < 1 + \min[\Re(\zeta), \Re(\eta)]$; then*

$$\int_0^\infty z^{\sigma-1} e^{-\delta z/2} W_{\tau,\omega}(\kappa z) W_4(z)\, dz = \frac{x^{\rho-\zeta-1}}{\kappa^{\sigma-1}} \cdot \frac{\Gamma(\zeta-\rho+1)\Gamma(\eta-\rho+1)}{\Gamma(1-\rho)\Gamma(\xi+\zeta+\eta-\rho+1)} {}^{MC}\Phi_m\left(\beta; \gamma; \frac{x}{\kappa}\right) * {}_4F_3\left(\zeta-\rho+1, \eta-\rho+1, \frac{1}{2} + \omega + \sigma, \frac{1}{2} - \omega + \sigma; 1 - \rho, \xi + \zeta + \eta - \rho + 1, \frac{1}{2} - \tau + \sigma; \frac{x}{\kappa}\right), \quad (170)$$

where $W_4(z) = \{(J_{x,\infty}^{\xi,\zeta,\eta} t^{\rho-1}\ {}^{MC}\Phi_m(\beta; \gamma; z/t))(x)\}$.

Proof. The proofs of Theorems 59, 60, and 61 would run parallel to those of Theorem 58, so the proofs of these theorems are omitted here.

11. Fractional Kinetic Equations

The importance of fractional differential equations in the field of applied science has gained more attention not only in mathematics but also in physics, dynamical systems, control systems, and engineering, to create the mathematical model of many physical phenomena. The kinetic equations especially describe the continuity of motion of substance. The extension and generalization of fractional kinetic equations involving many fractional operators were found in [18–31].

In view of the effectiveness and a great importance of the kinetic equation in certain astrophysical problems the authors develop a further generalized form of the fractional kinetic equation involving new extensions of Gauss hypergeometric function and confluent hypergeometric function.

The fractional differential equation between rate of change of the reaction, the destruction rate, and the production rate was established by Haubold and Mathai [23], given as follows:

$$\frac{dN}{dt} = -d(N_t) + p(N_t), \quad (171)$$

where $N = N(t)$ the rate of reaction, $d = d(N)$ the rate of destruction, $p = p(N)$ the rate of production, and N_t denotes the function defined by $N_t(t^*) = N(t - t^*), t^* > 0$.

In the special case of (171) for spatial fluctuations and inhomogeneities in $N(t)$ the quantities are neglected, that is, the equation

$$\frac{dN}{dt} = -c_i N_i(t), \quad (172)$$

with the initial condition that $N_i(t = 0) = N_0$ is the number density of the species i at time $t = 0$ and $c_i > 0$. If we remove the index i and integrate the standard kinetic equation (172), we have

$$N(t) - N_0 = -c\, {}_0D_t^{-1} N(t), \quad (173)$$

where ${}_0D_t^{-1}$ is the special case of the Riemann-Liouville integral operator ${}_0D_t^{-\nu}$ defined as

$${}_0D_t^{-\nu} f(t) = \frac{1}{\Gamma(\nu)} \int_0^t (t-s)^{\nu-1} f(s)\, ds, \quad (t > 0, R(\nu) > 0). \quad (174)$$

The fractional generalization of the standard kinetic equation (173) is given by Haubold and Mathai[23] as follows:

$$N(t) - N_0 = -c^\nu\, {}_0D_t^{-1} N(t) \quad (175)$$

and obtained the solution of (175) as follows:

$$N(t) = N_0 \sum_{k=0}^\infty \frac{(-1)^k}{\Gamma(\nu k + 1)} (ct)^{\nu k}. \quad (176)$$

Further, (Saxena and Kalla [27]) considered the following fractional kinetic equation:

$$N(t) - N_0 f(t) = -c^\nu\, {}_0D_t^{-\nu} N(t), \quad (\Re(\nu) > 0), \quad (177)$$

where $N(t)$ denotes the number density of a given species at time t, $N_0 = N(0)$ is the number density of that species at time $t = 0$, c is a constant, and $f \in \mathscr{L}(0, \infty)$.

By applying the Laplace transform to (177), we have

$$L\{N(t);p\} = N_0 \frac{F(p)}{1+c^\nu p^{-\nu}} = N_0\left(\sum_{n=0}^{\infty}(-c^\nu)^n p^{-\nu n}\right)F(p), \quad \left(n \in N_0, \left|\frac{c}{p}\right| < 1\right). \tag{178}$$

Where the Laplace transform [32] is given by

$$F(p) = L\{N(t);p\} = \int_0^\infty e^{-pt} f(t)\,dt, \quad (\mathscr{R}(p) > 0). \tag{179}$$

11.1. Solutions of Generalized Fractional Kinetic Equations. In this section, we investigated the solutions of the generalized fractional kinetic equations involving the new extension of Gauss hypergeometric function and confluent hypergeometric function.

Remark 62. The solutions of the fractional kinetic equations in this section are obtained in terms of the generalized Mittag-Leffler function $E_{\alpha,\beta}(x)$ (Mittag-Leffler[33]), which is defined as

$$E_{\alpha,\beta}(z) = \sum_{n=0}^{\infty} \frac{z^n}{\Gamma(\alpha n+\beta)}, \quad \Re(\alpha) > 0, \Re(\beta) > 0. \tag{180}$$

Theorem 63. *If $a > 0, d > 0, \nu > 0$; $\alpha, \beta, \gamma, m \in \mathbb{C}$; $\Re(\gamma) > \Re(\beta) > 0$ and $|m| < M$ (where M is positive real number), then the solution of the equation*

$$N(t) - N_0 {}^{MC}F_m(\alpha,\beta;\gamma;d^\nu t^\nu) = -a^\nu {}_0D_t^{-\nu} N(t) \tag{181}$$

is given by the following formula:

$$N(t) = N_0 \sum_{n=0}^{\infty} (\alpha)_n \frac{{}^{MC}B_m(\beta+n,\gamma-\beta)}{B(\beta,\gamma-\beta)} \frac{(d^\nu t^\nu)^n}{n!} \cdot \Gamma(\nu n+1) E_{\nu,\nu n+1}(-a^\nu t^\nu). \tag{182}$$

Proof. Laplace transform of Riemann-Liouville fractional integral operator is given by the following (Erdelyi et al. [34], Srivastava and Saxena[35]):

$$L\{{}_0D_t^{-\nu} f(t);p\} = p^{-\nu} F(p), \tag{183}$$

where $F(p)$ is defined in (179). Now, applying Laplace transform on (181) gives

$$L\{N(t);p\} = N_0 L\{{}^{MC}F_m(a,b;c;d^\nu t^\nu);p\} - a^\nu L\{{}_0D_t^{-\nu} N(t);p\} \tag{184}$$

$$\text{i.e.} \quad N(p) = N_0 \int_0^\infty e^{-pt} \sum_{n=0}^{\infty} (\alpha)_n \frac{{}^{MC}B_m(\beta+n,\gamma-\beta)}{B(\beta,\gamma-\beta)} \frac{(d^\nu t^\nu)^n}{n!} dt - a^\nu p^{-\nu} N(p) \tag{185}$$

and interchanging the order of integration and summation in (185), we have

$$N(p) + a^\nu p^{-\nu} N(p) = N_0 \sum_{n=0}^{\infty} (\alpha)_n \cdot \frac{{}^{MC}B_m(\beta+n,\gamma-\beta)}{B(\beta,\gamma-\beta)} \frac{(d^\nu)^n}{n!} \int_0^\infty e^{-pt} t^{\nu n} dt \tag{186}$$

$$= N_0 \sum_{n=0}^{\infty} (\alpha)_n \frac{{}^{MC}B_m(\beta+n,\gamma-\beta)}{B(\beta,\gamma-\beta)} \frac{(d^\nu)^n}{n!} \frac{\Gamma(\nu n+1)}{p^{\nu n+1}}. \tag{187}$$

This leads to

$$N(p) = N_0 \sum_{n=0}^{\infty} (\alpha)_n \frac{{}^{MC}B_m(\beta+n,\gamma-\beta)}{B(\beta,\gamma-\beta)} \frac{(d^\nu)^n}{n!} \cdot \Gamma(\nu n+1) \left\{p^{-(\nu n+1)} \sum_{l=0}^{\infty} \left[-\left(\frac{p}{a}\right)^{-\nu}\right]^l\right\}. \tag{188}$$

Taking Laplace inverse of (188), and by using

$$L^{-1}\{p^{-\nu};t\} = \frac{t^{\nu-1}}{\Gamma(\nu)}, \quad (R(\nu) > 0), \tag{189}$$

we have

$$L^{-1}\{N(p)\} = N_0 \sum_{n=0}^{\infty} (\alpha)_n \frac{{}^{MC}B_m(\beta+n,\gamma-\beta)}{B(\beta,\gamma-\beta)} \frac{(d^\nu)^n}{n!} \times \Gamma(\nu r+1) L^{-1}\left\{\sum_{l=0}^{\infty} (-1)^l a^{\nu l} p^{-[\nu(n+l)+1]}\right\} \tag{190}$$

$$\text{i.e.} \quad N(t) = N_0 \sum_{n=0}^{\infty} (\alpha)_n \frac{{}^{MC}B_m(\beta+n,\gamma-\beta)}{B(\beta,\gamma-\beta)} \frac{(d^\nu)^n}{n!} \times \Gamma(\nu n+1) \left\{\sum_{l=0}^{\infty} (-1)^l a^{\nu l} \frac{t^{\nu(n+l)}}{\Gamma(\nu(n+l)+1)}\right\} \tag{191}$$

$$= N_0 \sum_{n=0}^{\infty} (\alpha)_n \frac{{}^{MC}B_m(\beta+n,\gamma-\beta)}{B(\beta,\gamma-\beta)} \frac{(d^\nu t^\nu)^n}{n!} \times \Gamma(\nu n+1) \left\{\sum_{l=0}^{\infty} (-1)^l \frac{(a^\nu t^\nu)^l}{\Gamma(\nu(n+l)+1)}\right\}. \tag{192}$$

Equation (192) can be written as

$$N(t) = N_0 \sum_{n=0}^{\infty} (\alpha)_n \frac{{}^{MC}B_m(\beta+n,\gamma-\beta)}{B(\beta,\gamma-\beta)} \frac{(d^\nu t^\nu)^n}{n!} \cdot \Gamma(\nu n+1) E_{\nu,\nu n+1}(-a^\nu t^\nu). \quad (193)$$

Theorem 64. *If $a > 0, d > 0, \nu > 0$; $\alpha, \beta, \gamma, m \in \mathbb{C}$; $\Re(\gamma) > \Re(\beta) > 0$ and $|m| < M$ (where M is positive real number), then the solution of the equation*

$$N(t) - N_0 {}^{MC}\Phi_m(\beta;\gamma;d^\nu t^\nu) = -a^\nu {}_0D_t^{-\nu} N(t) \quad (194)$$

is given by the following formula:

$$N(t) = N_0 \sum_{n=0}^{\infty} \frac{{}^{MC}B_m(\beta+n,\gamma-\beta)}{B(\beta,\gamma-\beta)} \frac{(d^\nu t^\nu)^n}{n!} \Gamma(\nu n+1) \cdot E_{\nu,\nu n+1}(-a^\nu t^\nu). \quad (195)$$

Theorem 65. *If $d > 0, \nu > 0$; $\alpha, \beta, \gamma, m \in \mathbb{C}$; $\Re(\gamma) > \Re(\beta) > 0$ and $|m| < M$ (where M is positive real number), then the solution of the equation*

$$N(t) - N_0 {}^{MC}F_m(\alpha,\beta;\gamma;d^\nu t^\nu) = -d^\nu {}_0D_t^{-\nu} N(t) \quad (196)$$

is given by the following formula:

$$N(t) = N_0 \sum_{n=0}^{\infty} (\alpha)_n \frac{{}^{MC}B_m(\beta+n,\gamma-\beta)}{B(\beta,\gamma-\beta)} \frac{(d^\nu t^\nu)^n}{n!} \cdot \Gamma(\nu n+1) E_{\nu,\nu n+1}(-d^\nu t^\nu). \quad (197)$$

Theorem 66. *If $d > 0, \nu > 0$; $\alpha, \beta, \gamma, m \in \mathbb{C}$; $\Re(\gamma) > \Re(\beta) > 0$ and $|m| < M$ (where M is positive real number), then the solution of the equation*

$$N(t) - N_0 {}^{MC}\Phi_m(\alpha,\beta;\gamma;d^\nu t^\nu) = -d^\nu {}_0D_t^{-\nu} N(t) \quad (198)$$

is given by the following formula:

$$N(t) = N_0 \sum_{n=0}^{\infty} \frac{{}^{MC}B_m(\beta+n,\gamma-\beta)}{B(\beta,\gamma-\beta)} \frac{(d^\nu t^\nu)^n}{n!} \Gamma(\nu n+1) \cdot E_{\nu,\nu n+1}(-d^\nu t^\nu). \quad (199)$$

Theorem 67. *If $d > 0, \nu > 0$; $\alpha, \beta, \gamma, m \in \mathbb{C}$; $\Re(\gamma) > \Re(\beta) > 0$ and $|m| < M$ (where M is positive real number), then the solution of the equation*

$$N(t) - N_0 {}^{MC}F_m(\alpha,\beta;\gamma;t) = -d^\nu {}_0D_t^{-\nu} N(t) \quad (200)$$

is given by the following formula:

$$N(t) = N_0 \sum_{n=0}^{\infty} (\alpha)_n \frac{{}^{MC}B_m(\beta+n,\gamma-\beta)}{B(\beta,\gamma-\beta)} t^n E_{\nu,n+1}(-d^\nu t^\nu). \quad (201)$$

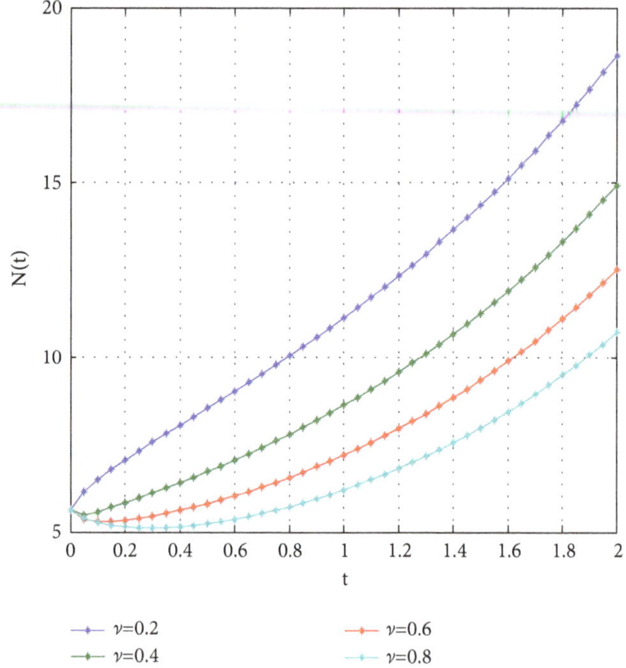

FIGURE 5: Solution of fractional kinetic equation given in (182) for $N_0 = 1, \alpha = 2, \beta = 1, \gamma = 3, m = 2, d = 1, a = 2, \nu = 0.2 : .2 : 0.8$.

Theorem 68. *If $d > 0, \nu > 0$; $\alpha, \beta, \gamma, m \in \mathbb{C}$; $\Re(\gamma) > \Re(\beta) > 0$ and $|m| < M$ (where M is positive real number), then the solution of the equation*

$$N(t) - N_0 {}^{MC}\Phi_m(\alpha,\beta;\gamma;t) = -d^\nu {}_0D_t^{-\nu} N(t) \quad (202)$$

is given by the following formula:

$$N(t) = N_0 \sum_{n=0}^{\infty} \frac{{}^{MC}B_m(\beta+n,\gamma-\beta)}{B(\beta,\gamma-\beta)} t^n E_{\nu,n+1}(-d^\nu t^\nu). \quad (203)$$

Proof. The proofs of Theorems 64, 65, 66, 67, and 68 are the same as those of Theorem 63, so we skip the proof of these theorems.

11.1.1. Numerical results and their graphs of solution of fractional kinetic equation. In this section, we present the numerical results in Table 5 for $N_0 = 1, \alpha = 2, \beta = 1, \gamma = 3, m = 2, d = 1, a = 2$ and $\nu = 0.2 : 1 : 4.2$ of equation (126) and the graph of solution of fractional kinetic equation given in (126) are presented in Figures 5, 6 and 7 for $N_0 = 1$, $\alpha = 2, \beta = 1, \gamma = 3, m = 2, d = 1, a = 2$; the values of $\nu$are chosen as $\nu = 0.2 : .2 : 0.8$, $\nu = 1.2 : .2 : 1.8$ and $\nu = 2.2 : .2 : 2.8$, respectively. Figure 8 presents the Mesh-plot of the same solution fractional kinetic equation. For different values of the parameters, it can be easily interpreted and can be observed that $N(t) > 0$ for different values of the parameters. As the solution of fractional kinetic equations are presented in the form of summation. For the numerical results and their graphs, sum first 500 terms have been taken. If we choose more than 500 terms the results are same.

Table 5: Numerical values of solution of fractional kinetic equation given in (182) for $N_0 = 1, \alpha = 2, \beta = 1, \gamma = 3, m = 2, d = 1, a = 2$.

t	$\nu = 0.2$	$\nu = 1.2$	$\nu = 2.2$	$\nu = 3.2$	$\nu = 4.2$
0.00	5.6427	5.6427	5.6427	5.6427	5.6427
0.10	6.5072	5.4044	5.6143	5.6408	5.6426
0.20	7.0748	5.1461	5.5127	5.6249	5.6410
0.30	7.5810	4.9199	5.3291	5.5773	5.6330
0.40	8.0669	4.7403	5.0624	5.4788	5.6101
0.50	8.5489	4.6124	4.7172	5.3088	5.5595
0.60	9.0358	4.5376	4.3026	5.0468	5.4638
0.70	9.5338	4.5149	3.8322	4.6727	5.3011
0.80	10.0472	4.5423	3.3237	4.1680	5.0448
0.90	10.5800	4.6174	2.7980	3.5166	4.6637
1.00	11.1353	4.7378	2.2788	2.7061	4.1222
1.10	11.7165	4.9012	1.7917	1.7296	3.3808
1.20	12.3265	5.1058	1.3630	0.5864	2.3968
1.30	12.9684	5.3500	1.0193	-0.7158	1.1249
1.40	13.6456	5.6329	0.7860	-2.1593	-0.4809
1.50	14.3612	5.9541	0.6861	-3.7152	-2.4656
1.60	15.1187	6.3138	0.7398	-5.3414	-4.8712
1.70	15.9220	6.7125	0.9628	-6.9812	-7.7330
1.80	16.7748	7.1514	1.3661	-8.5628	-11.0764
1.90	17.6815	7.6323	1.9551	-9.9980	-14.9121
2.00	18.6466	8.1573	2.7289	-11.1825	-19.2305

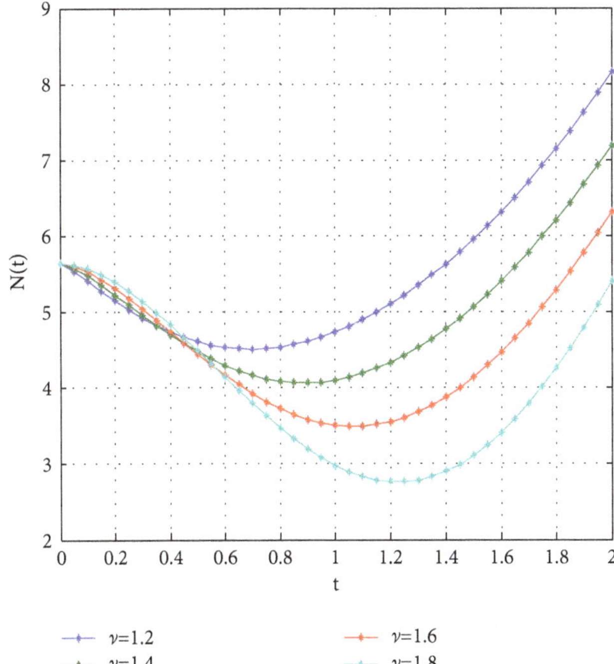

Figure 6: Solution of fractional kinetic equation given in (182) for $N_0 = 1, \alpha = 2, \beta = 1, \gamma = 3, m = 2, d = 1, a = 2, \nu = 1.2 : .2 : 1.8$.

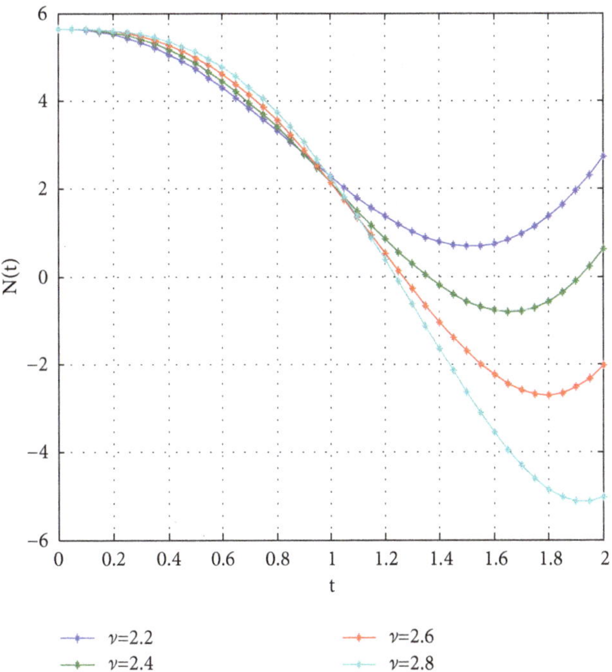

Figure 7: Solution of fractional kinetic equation given in (182) for $N_0 = 1, \alpha = 2, \beta = 1, \gamma = 3, m = 2, d = 1, a = 2, \nu = 2.2 : .2 : 2.8$.

11.2. Special Cases of Fractional Kinetic Equations. Choosing $m = 0$, Theorems 63, 64, 65, 66, 67, and 68 reduce to the following form.

Corollary 69. *If $a > 0, d > 0, \nu > 0$; $\alpha, \beta, \gamma \in \mathbb{C}$; $\Re(\gamma) > \Re(\beta) > 0$, then the solution of the equation*

$$N(t) - N_0\,{}_2F_1(\alpha, \beta; \gamma; d^\nu t^\nu) = -a^\nu\,{}_0D_t^{-\nu}N(t) \tag{204}$$

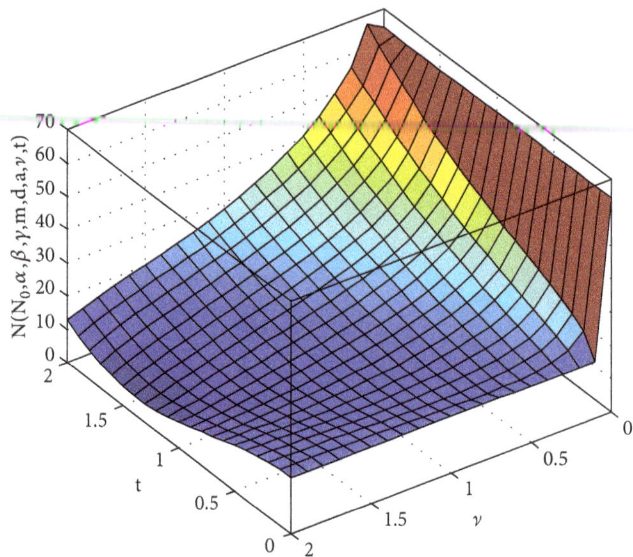

FIGURE 8: Mesh-Plot of solution of fractional kinetic equation given in (182) for $N_0 = 3, \alpha = 2, \beta = 1, \gamma = 3, m = 2, d = 1, a = 2$.

is given by the following formula:

$$N(t) = N_0 \sum_{n=0}^{\infty} (\alpha)_n \frac{B(\beta + n, \gamma - \beta)}{B(\beta, \gamma - \beta)} \frac{(d^\nu t^\nu)^n}{n!} \cdot \Gamma(\nu n + 1) E_{\nu,\nu n+1}(-a^\nu t^\nu). \tag{205}$$

Corollary 70. *If $a > 0, d > 0, \nu > 0; \alpha, \beta, \gamma \in \mathbb{C}; \Re(\gamma) > \Re(\beta) > 0$, then the solution of the equation*

$$N(t) - N_0 \Phi(\beta; \gamma; d^\nu t^\nu) = -a^\nu {}_0D_t^{-\nu} N(t) \tag{206}$$

is given by the following formula:

$$N(t) = N_0 \sum_{n=0}^{\infty} \frac{B(\beta + n, \gamma - \beta)}{B(\beta, \gamma - \beta)} \frac{(d^\nu t^\nu)^n}{n!} \Gamma(\nu n + 1) \cdot E_{\nu,\nu n+1}(-a^\nu t^\nu). \tag{207}$$

Corollary 71. *If $d > 0, \nu > 0; \alpha, \beta, \gamma \in \mathbb{C}; \Re(\gamma) > \Re(\beta) > 0$, then the solution of the equation*

$$N(t) - N_0 \, {}_2F_1(\alpha, \beta; \gamma; d^\nu t^\nu) = -d^\nu {}_0D_t^{-\nu} N(t) \tag{208}$$

is given by the following formula:

$$N(t) = N_0 \sum_{n=0}^{\infty} (\alpha)_n \frac{B(\beta + n, \gamma - \beta)}{B(\beta, \gamma - \beta)} \frac{(d^\nu t^\nu)^n}{n!} \cdot \Gamma(\nu n + 1) E_{\nu,\nu n+1}(-d^\nu t^\nu). \tag{209}$$

Corollary 72. *If $d > 0, \nu > 0; \alpha, \beta, \gamma \in \mathbb{C}; \Re(\gamma) > \Re(\beta) > 0$, then the solution of the equation*

$$N(t) - N_0 \Phi(\beta; \gamma; d^\nu t^\nu) = -d^\nu {}_0D_t^{-\nu} N(t) \tag{210}$$

is given by the following formula:

$$N(t) = N_0 \sum_{n=0}^{\infty} \frac{B(\beta + n, \gamma - \beta)}{B(\beta, \gamma - \beta)} \frac{(d^\nu t^\nu)^n}{n!} \Gamma(\nu n + 1) \cdot E_{\nu,\nu n+1}(-d^\nu t^\nu). \tag{211}$$

Corollary 73. *If $d > 0, \nu > 0; \alpha, \beta, \gamma \in \mathbb{C}; \Re(\gamma) > \Re(\beta) > 0$, then the solution of the equation*

$$N(t) - N_0 \, {}_2F_1(\alpha, \beta; \gamma; t) = -d^\nu {}_0D_t^{-\nu} N(t) \tag{212}$$

is given by the following formula:

$$N(t) = N_0 \sum_{n=0}^{\infty} (\alpha)_n \frac{B(\beta + n, \gamma - \beta)}{B(\beta, \gamma - \beta)} t^n E_{\nu,n+1}(-d^\nu t^\nu). \tag{213}$$

Corollary 74. *If $d > 0, \nu > 0; \alpha, \beta, \gamma \in \mathbb{C}; \Re(\gamma) > \Re(\beta) > 0$, then the solution of the equation*

$$N(t) - N_0 \Phi(\beta; \gamma; t) = -d^\nu {}_0D_t^{-\nu} N(t) \tag{214}$$

is given by the following formula:

$$N(t) = N_0 \sum_{n=0}^{\infty} \frac{B(\beta + n, \gamma - \beta)}{B(\beta, \gamma - \beta)} t^n E_{\nu,n+1}(-d^\nu t^\nu). \tag{215}$$

12. Concluding Remarks

In the present article authors introduced the new the generalization of the classical beta function. It has been further used to study the various properties of the new extended beta function. Furthermore, on application of this new extended beta function, extension of Gauss hypergeometric function and confluent hypergeometric function are introduced. Fractional integrals of extended hypergeometric functions and their image formulas (in the form of beta transform, Laplace transform, and Whittaker transform) are established. We introduce new fractional generalizations of the standard kinetic equation and we derived the solutions for the same. Their numerical results and graphs are established to study the nature of these fractional kinetic equations involving new extended Gauss hypergeometric function and confluent hypergeometric function. From the closed relationship of the hypergeometric functions with various special functions, we can easily construct various new known results.

Conflicts of Interest

The author declares that they have no conflicts of interest.

References

[1] M. A. Chaudhry, A. Qadir, M. Rafique, and S. M. Zubair, "Extension of Euler's beta function," *Journal of Computational and Applied Mathematics*, vol. 78, no. 1, pp. 19–32, 1997.

[2] H. M. Srivastava, A. Cetinkaya, and I. O. Kiymaz, "A certain generalized pochhammer symbol and its applications to hypergeometric functions," *Applied Mathematics and Computation*, vol. 226, pp. 484–491, 2014.

[3] R. K. Parmar, "A new generalization of gamma, beta, hypergeometric and confluent hypergeometric functions," *Le Matematiche*, vol. 69, no. 2, pp. 33–52, 2013.

[4] E. Ozergin, M. A. Ozarslan, and A. Altin, "Extension of gamma, beta and hypergeometric functions," *Journal of Computational and Applied Mathematics*, vol. 235, no. 16, pp. 4601–4610, 2011.

[5] Y. Simsek, "Beta-type polynomials and their generating functions," *Applied Mathematics and Computation*, vol. 254, pp. 172–182, 2015.

[6] Y. Simsek, "q–beta polynomials and their applications," *Applied Mathematics & Information Sciences*, vol. 7, no. 6, pp. 2539–2547, 2013.

[7] M. A. Chaudhry, A. Qadir, H. M. Srivastava, and R. B. Paris, "Extended hypergeometric and confluent hypergeometric functions," *Applied Mathematics and Computation*, vol. 159, no. 2, pp. 589-602, 2004.

[8] E. D. Rainville, "Special functions," Macmillan Company, New York, 1960; Reprinted by Chelsea Publishing Company, Bronx, New York, 1971.

[9] H. M. Srivastava and J. Cho, *Zeta and q-Zeta Functions and Associated Series and Integrals*, Elsevier Science Publishers, Amsterdam, London and New York, 2012.

[10] H.-M. Srivastava and P.-W. Karlsson, *Multiple Gaussian hypergeometric series. Halsted Press (Ellis Horwood Limited, Chichester)*, John Wiley and Sons, New York, Chichester, Brisbane and Toronto, 1985.

[11] I. S. Gradshteyn and I. M. Ryzhik, *Table of Integrals, Series, and Products*, Academic Press Publication, New York, NY, USA, 7th edition, 2007.

[12] V. K. Rohatgi, *An Introduction to Probability Theory and Mathematical Statistics*, John Wiley & Sons, New York, NY, USA, 1976.

[13] H. Srivastava and J. Choi, *Zeta and q-Zeta Functions and Associated Series and Integrals*, Elsevier Science Publishers, Amsterdam, London and New York, 2012.

[14] T. Pohlen, *The Hadamard Product and Universal Power Series*, Universität Trier Trier Germany, Germany, 2009.

[15] E. Ozergin, *Some Properties of Hypergeometric Functions*, Eastern Mediterranean University, North Cyprus, 2011.

[16] A. A. Kilbas and N. Sebastian, "Generalized fractional integration of Bessel function of the first kind," *Integral Transforms and Special Functions*, vol. 19, no. 12, pp. 869–883, 2008.

[17] I. N. Sneddon, *The Use of Integral Transforms*, Tata McGraw-Hill, 1979.

[18] V. B. L. Chaurasia and S. C. Pandey, "On the new computable solution of the generalized fractional kinetic equations involving the generalized function for the fractional calculus and related functions," *Astrophysics and Space Science*, vol. 317, no. 3-4, pp. 213–219, 2008.

[19] A. Chouhan and S. Sarswat, "On solution of generalized Kinetic equation of fractional order," *International Journal of Mathematical Sciences and Applications*, vol. 2, no. 2, pp. 813–818, 2012.

[20] A. Chouhan, S. D. Purohit, and S. Saraswat, "An alternative method for solving generalized differential equations of fractional order," *Kragujevac Journal of Mathematics*, vol. 37, no. 2, pp. 299–306, 2013.

[21] V. G. Gupta, B. Sharma, and F. B. M. Belgacem, "On the solutions of generalized fractional kinetic equations," *Applied Mathematical Sciences*, vol. 5, no. 19, pp. 899–910, 2011.

[22] A. Gupta and C. L. Parihar, "On solutions of generalized kinetic equations of fractional order," *Boletim da Sociedade Paranaense de Matemática*, vol. 32, no. 1, pp. 183–191, 2014.

[23] H. J. Haubold and A. M. Mathai, "The fractional kinetic equation and thermonuclear functions," *Astrophysics and Space Science*, vol. 273, no. 1–4, pp. 53–63, 2000.

[24] R. K. Saxena, A. M. Mathai, and H. J. Haubold, "On fractional kinetic equations," *Astrophysics and Space Science*, vol. 282, no. 1, pp. 281–287, 2002.

[25] R. K. Saxena, A. M. Mathai, and H. J. Haubold, "On generalized fractional kinetic equations," *Physica A: Statistical Mechanics and its Applications*, vol. 344, no. 3-4, pp. 657–664, 2004.

[26] R. K. Saxena, A. M. Mathai, and H. J. Haubold, "Solution of generalized fractional reaction-diffusion equations," *Astrophysics and Space Science*, vol. 305, no. 3, pp. 305–313, 2006.

[27] R. K. Saxena and S. L. Kalla, "On the solutions of certain fractional kinetic equations," *Applied Mathematics and Computation*, vol. 199, no. 2, pp. 504–511, 2008.

[28] A. I. Saichev and G. M. Zaslavsky, "Fractional kinetic equations: solutions and applications," *Chaos: An Interdisciplinary Journal of Nonlinear Science*, vol. 7, no. 4, pp. 753–764, 1997.

[29] G. M. Zaslavsky, "Fractional kinetic equation for Hamiltonian chaos," *Physica D: Nonlinear Phenomena*, vol. 76, no. 1–3, pp. 110–122, 1994.

[30] P. Agarwal, S. Ntouyas, S. Jain, M. Chand, and G. Singh, "Fractional kinetic equations involving generalized k-Bessel function via Sumudu transform," *Alexandria Engineering Journal*, vol. 57, no. 3, pp. 1937–1942, 2016.

[31] M. Chand, J. C. Prajapati, and E. Bonyah, "Fractional integrals and solution of fractional kinetic equations involving k-mittag-leffler function," *Transactions of A. Razmadze Mathematical Institute*, vol. 171, no. 2, pp. 144–166, 2017.

[32] M. R. Spiegel, *Laplace Transforms*, McGraw-Hill, New York, USA, 1965.

[33] G. Mittag-Leffler, "Sur la représentation analytique d'une branche uniforme d'une fonction monogène," *Acta Mathematica*, vol. 29, no. 1, pp. 101–181, 1905.

[34] H. Bateman and A. Erdelyi, *Tables of Integral Transforms*, vol. 1, Mcgraw-Hill Book Co, 1954.

[35] H. M. Srivastava and R. K. Saxena, "Operators of fractional integration and their applications," *Applied Mathematics and Computation*, vol. 118, no. 1, pp. 1–52, 2001.

20

(p, q)-Growth of an Entire GASHE Function and the Coefficient $\beta_n = |a_n| / \Gamma(n + \mu + 1)$

Mohammed Harfaoui, Abdellah Mourassil, and Loubna Lakhmaili

University Hassan II Mohammedia, Laboratory of Mathematics, Criptography and Mechanical F. S.T., BP 146, Mohammedia 20650, Morocco

Correspondence should be addressed to Mohammed Harfaoui; mharfaoui04@yahoo.fr

Academic Editor: Ram N. Mohapatra

The main purpose of this paper is to extend the work concerning the measures of growth of an entire function solution of the generalized axially symmetric Helmholtz equation $\partial^2 u/\partial x^2 + \partial^2 u/\partial y^2 + (2\mu/y)(\partial u/\partial y) = 0$, $\mu > 0$, by studying the general measures of growth ((p,q)-order, lower (p,q)-order, (p,q)-type, and lower (p,q)-type) in terms of coefficients $|a_n|/\Gamma(n+\mu+1)$ and the ratios of these successive coefficients.

1. Introduction

The partial differential equation

$$\frac{\partial^2 u}{\partial x^2} + \frac{\partial^2 u}{\partial y^2} + \frac{2\mu}{y}\frac{\partial u}{\partial y} + k^2 u = 0 \quad (1)$$

is called generalized axially symmetric Helmholtz equation (GASHE) and the solutions of (1) are called GASHE functions. The GASHE function u is regular about the origin and has the following Bessel-Gegenbauer series expansion:

$$u(r,\theta) = \Gamma(2\mu)(kr)^{-\mu} \sum_{n=0}^{\infty} \frac{a_n n!}{\Gamma(2\mu+n)} J_{\mu+n}(kr) C_n^{\mu}(\cos(\theta)), \quad (2)$$

where $x = r\cos(\theta)$ and $y = r\sin(\theta)$, $J_{\mu+n}$ are Bessel functions of first kind, C_n^{μ} are Gegenbauer polynomials, and $\Gamma(2\mu) = 2^{\mu-1}(\mu-1)!$.

When series (2) converges absolutely and uniformly on the compact subsets of the whole complex plane, then the GASHE function u is said to be entire. For u being entire, it is known [1, page 214] that

$$\limsup_{r\longrightarrow\infty} \left(\frac{|a_n|}{\Gamma(n+\mu+1)}\right)^{1/n} = 0. \quad (3)$$

The concept of order $\rho(f)$ and lower order $\lambda(f)$ of an entire function was introduced by R. P. Boas [2] as follows:

$$\begin{matrix}\rho(f)\\ \lambda(f)\end{matrix} = \lim_{r\longrightarrow+\infty} \begin{matrix}\sup\\ \inf\end{matrix} \frac{\log\log(M(r,f))}{\log(r)} \quad (4)$$

and the concept of type $T(f)$ and lower type $t(f)$ has been introduced to give more precise description of growth of entire functions when they have the same nonzero finite order. An entire function, of order ρ, $0 < \rho < +\infty$, is said to be of type $T(f)$ and lower type $t(f)$ if

$$\begin{matrix}T(f)\\ t(f)\end{matrix} = \lim_{r\longrightarrow+\infty} \begin{matrix}\sup\\ \inf\end{matrix} \frac{\log(M(r,f))}{r^{\rho(f)}}, \quad 0 < \rho(f) < \infty \quad (5)$$

where $M(r,f) = \max_{0\le\theta\le 2\pi} |f(r,\theta)|$.

Gilbert and Howard [3] have studied the order $\rho(u)$ of an entire GASHE function u in terms of the coefficients a_n occurring in the series expansion (2) of u. McCoy [4] studied the rapid growth of entire function solution of Helmholtz equation using the concept of index. Kumar [5, 6] extended and improved this result and studied the growth using the concept of index pair. Khan and Ali [7] studied the generalized order and type of entire GASHE function. Kumar and Singh [8] have studied the lower order and lower type of entire GASHE function in terms of the coefficients in its

Bessel-Gegenbauer series expansion (2) when the order of u is a finite nonzero number. But, for the class of order $\rho(u) = 0$ and $\rho(u) = \infty$, we cannot define a type of u. For this reason, numerous attempts have been made to refine the concept of order and type. Therefore, the (p,q)-order and (p,q)-type of an entire function have been defined [9, 10]. In this paper, we extend the work of Kumar and Singh [8] to this new classification of entire function.

For $(p,q) \neq (1,1)$ and $p \geq q \geq 1$, we define the (p,q)-order and lower (p,q)-order as

$$\begin{matrix} \rho(p,q,u) \\ \lambda(p,q,u) \end{matrix} = \lim_{r\longrightarrow+\infty} \begin{matrix} \sup \\ \inf \end{matrix} \frac{\log^{[p]}(M(r,u))}{\log^{[q]}(r)}, \quad (6)$$

where p and q are integers such that $b \leq \rho(p,q,u) \leq \infty$ where $b = 0$ if $p > q$ and $b = 1$ if $p = q$.

The (p,q)-type and lower (p,q)-type are defined as

$$\begin{matrix} T(p,q,u) \\ t(p,q,u) \end{matrix} = \lim_{r\longrightarrow+\infty} \begin{matrix} \sup \\ \inf \end{matrix} \frac{\log^{[p-1]}(M(r,u))}{\log^{[q-1]}(r)^{\rho(p,q,u)}}, \quad (7)$$

and $\log^{[0]}(x) = x$ and $\log^{[m]}(x) = \log^{[m-1]}\log(x)$ for $m \geq 1$ and we use the notations

$$P(\alpha) = P(\alpha,p,q) = \begin{cases} \alpha & \text{if } p > q \\ 1+\alpha & \text{if } p = q = 2 \\ \max(1,\alpha) & \text{if } 3 \leq p = q < \infty \\ \infty & \text{if } p = q = \infty, \end{cases} \quad (8)$$

and

$$M(\alpha) = M(\alpha,p,q) = \begin{cases} \dfrac{1}{e.\alpha} & \text{if } (p,q) = (2,1) \\ \dfrac{(\alpha-1)^{(\alpha-1)}}{\alpha^{\alpha}} & \text{if } (p,q) = (2,2) \\ 1 & \text{if } p \geq 3, \end{cases} \quad (9)$$

$$N(\alpha) = N(\alpha,p,q) = \begin{cases} \dfrac{1}{\alpha} & \text{if } p = 2 \\ 1 & \text{if } p \neq 2. \end{cases}$$

We note that the smallest integer p is 2 ($p \geq 2$) since, for example, the order is given by $\rho = \inf\{\mu > 0 : |f(z)| = O(e^{|z|^{\mu}}), |z| \longrightarrow +\infty\}$.

To prove that $P(\alpha) = P(\beta)$, $M(\alpha) = M(\beta)$, or $N(\alpha) = N(\beta)$ we can prove that $\alpha = \beta$ for the different values of p and q. From [9], we define the relation between (p,q)-order, lower (p,q)-order, the coefficients of u, and the ratios of these successive coefficients as follows.

Theorem 1. *Let $u(z) = \sum_{n=0}^{\infty} a_n z^n$ be an entire function of (p,q)-order $\rho(p,q,u)$, and then*

$$\rho(p,q,u) = P(L(p,q)) \quad (10)$$

where

$$L(p,q) = \limsup_{n\longrightarrow+\infty} \frac{\log^{[p-1]}(n)}{\log^{[q-1]}(-(1/n)\log(|a_n|))} \quad (11)$$

Theorem 2. *Let $u(z) = \sum_{n=0}^{\infty} a_n z^n$ be an entire function of (p,q)-order $\rho(p,q,u)$, and then*

$$\rho(p,q,u) = P(L^*(p,q)) \quad (12)$$

where

$$L^*(p,q) = \limsup_{n\longrightarrow\infty} \frac{\log^{[p-1]}(n)}{\log^{[q]}(|a_n/a_{n+1}|)} \quad (13)$$

Theorem 3. *Let $u(z) = \sum_{n=0}^{\infty} a_n z^n$ be an entire function of (p,q)-order $\rho(p,q,u)$ and $(|a_n/a_{n+1}|)_n$ a nondecreasing function of n for $n > n_0$ and then*

$$\lambda(p,q,u) = P(l(p,q)), \quad (14)$$

where

$$l(p,q) = \liminf_{n\longrightarrow+\infty} \frac{\log^{[p-1]}(n)}{\log^{[q-1]}(-(1/n)\log(|a_n|))}. \quad (15)$$

Theorem 4. *Let $u(z) = \sum_{n=0}^{\infty} a_n z^n$ be an entire function of (p,q)-order $\rho(p,q,u)$ and $(|a_n/a_{n+1}|)$ a nondecreasing function of n for $n > n_0$ and then*

$$\lambda(p,q,u) = P(l^*(p,q)), \quad (16)$$

where

$$l^*(p,q) = \liminf_{n\longrightarrow\infty} \frac{\log^{[p-1]}(n)}{\log^{[q]}(|a_n/a_{n+1}|)}. \quad (17)$$

From [10], we define the relation between (p,q)-type, lower (p,q)-type, the coefficients of u, and the ratios of these successive coefficients as follows.

Theorem 5. *Let $u(z) = \sum_{n=0}^{\infty} a_n z^n$ be an entire function. The function $f(z)$ is of (p,q)-order $\rho(p,q,u)$ and (p,q)-type $T(p,q,u)$ if and only if $T = MV$, where $b = 1$ if $p = q$ and $b = 0$ if $p > q$, and V is defined as*

$$V(p,q,u) = \limsup_{n\longrightarrow\infty} \frac{\log^{[p-2]}(n)}{\log^{[q-2]}(-(1/n)\log(|a_n|))^{\rho-C}} \quad (18)$$

with $C = 1$ if $(p,q) = (2,2)$ and $C = 0$ if $(p,q) \neq (2,2)$.

Theorem 6. *Let $u(z) = \sum_{n=0}^{\infty} a_n z^n$ be an entire function of (p,q)-order ρ and lower (p,q)-type $t(p,q,u)$ and $(|a_n/a_{n+1}|)_n$ a nondecreasing function of n for $n > n_0$ and then $t = Mv$, where*

$$v(p,q,u) = \liminf_{n\longrightarrow\infty} \frac{\log^{[p-2]}(n)}{\log^{[q-2]}(-(1/n)\log(|a_n|))^{\rho-C}} \quad (19)$$

with $C = 1$ if $(p,q) = (2,2)$ and $C = 0$ if $(p,q) \neq (2,2)$.

Theorem 7. *Let $u(z) = \sum_{n=0}^{\infty} a_n z^n$ be an entire function of (p,q)-order $\rho(p,q,u)$ and lower (p,q)-type $t(p,q,u)$, and $|a_n/a_{n+1}|$ forms a nondecreasing function of k for $k > k_0$; then*

$$N * R \leq t \leq T \leq N * Q, \quad (20)$$

where

$$Q(p,q) = \limsup_{n\to\infty} \frac{\log^{[p-2]}(n)}{\log^{[q-1]}(|a_n/a_{n+1}|)^{\rho-C}}, \quad (21)$$

and

$$R(p,q) = \liminf_{n\to\infty} \frac{\log^{[p-2]}(n)}{\log^{[q-1]}(|a_n/a_{n+1}|)^{\rho-C}}, \quad (22)$$

with $C = 1$ *if* $(p,q) = (2,2)$ *and* $C = 0$ *if* $(p,q) \neq (2,2)$.

2. Auxiliary Results

Let f and g be two functions defined as

$$f(z) = \sum_{n=0}^{+\infty} \frac{|a_n|}{n!}\left(\frac{k}{2}\right)^n z^n,$$
$$g(z) = \sum_{n=0}^{+\infty} \frac{|a_n|}{n!}\left(\frac{k}{2r_*}\right)^n z^n. \quad (23)$$

According to [3], we know that if u is an entire GASHE function, then f and g are also entire functions of the complex variable z, and

$$\frac{\Gamma(\mu+1/2)}{k_*\pi 2^{2-\mu}} m(r,g) \le M(r,u) \le kM(r,f) \quad (24)$$

where $m(r,g) = \max_{n\ge 0}[(|a_n|/n!)(k/2r_*)^n]$ and $M(r,f) = \max_{|z|\le r}|f(z)|$. In this section, we shall prove some auxiliary results which will be used in the sequel.

Lemma 8. *Let f and g be entire functions of particular form defined above. Then the (p,q)-orders and the (p,q)-types of f and g, respectively, are identical.*

Proof. Let $\Phi(z) = \sum_{n=0}^{\infty} a_n z^n$ be an entire function, and then, according to Theorem 1, the (p,q)-order of ϕ is given as

$$\rho(p,q,\phi) = \limsup_{n\to+\infty} \frac{\log^{[p-1]}(n)}{\log^{[q-1]}(-(1/n)\log(|a_n|))}, \quad (25)$$

and the (p,q)-type is defined in view of Theorem 5 as

$$T(p,q,\phi) = \limsup_{n\to+\infty} \frac{\log^{[p-2]}(n)}{\log^{[q-2]}((-1/n)\log(|a_n|))^{\rho(p,q,\phi)}}. \quad (26)$$

In the consequence of [3], we have

$$\rho(2,1,f) = \rho(2,1,g),$$
$$T(2,1,f) = T(2,1,g). \quad (27)$$

Here we consider the case when $q \ge 2$.
We have

$$\begin{aligned}
\frac{1}{\rho(2,1,f)} &= \liminf_{n\to+\infty} \frac{(-1/n)\log((k/2)^n(|a_n|/n!))}{\log(n)} \\
&= \liminf_{n\to+\infty} \frac{\log(|a_n|^{-1}) + \log(n!) - \log(k/2)^n}{n\log(n)} \\
&= \liminf_{n\to+\infty} \left(\frac{\log(|a_n|^{-1})}{n\log(n)} + \frac{\log(n!)}{n\log(n)} - \frac{\log(k/2)^n}{n\log(n)}\right) \\
&= \liminf_{n\to+\infty} \frac{\log(|a_n|^{-1})}{n\log(n)} + 1,
\end{aligned} \quad (28)$$

and then $1/\rho(2,1,f) \ge 1$ and $\rho(2,1,f) \le 1$.

This implies that we will necessarily have $\rho(2,1,f) = 0$ to define $\rho(p,q,f)$. And we have

$$\begin{gathered}
\rho(2,1,f) = 0 \Longrightarrow \\
\liminf_{n\to+\infty} \frac{1}{\rho(2,1,f)} = +\infty \Longrightarrow \\
\liminf_{n\to+\infty} \frac{\log(|a_n|^{-1})}{n\log(n)} = +\infty \Longrightarrow \\
\limsup_{n\to+\infty} \frac{n\log(n)}{\log(|a_n|^{-1})} = 0 \Longrightarrow \\
\lim_{n\to+\infty} \frac{\log(n!)}{\log(|a_n|^{-1})} = 0, \\
|a_n|^{1/n} \longrightarrow 0.
\end{gathered} \quad (29)$$

Hence, for the function f we have

$$\begin{aligned}
\frac{1}{\rho(p,q,f)} &= \liminf_{n\to+\infty} \frac{\log^{[q-1]}((-1/n)\log((k/2)^n(|a_n|/n!)))}{\log^{[p-1]}(n)} \\
&= \liminf_{n\to+\infty} \frac{\log^{[q-1]}(\log(|a_n|^{-1})/n + \log(n!)/n - (1/n)\log(k/2)^n)}{\log^{[p-1]}(n)} \\
&= \liminf_{n\to+\infty} \frac{1}{\log^{[p-1]}(n)} * \log^{[q-1]}\left(\frac{\log(|a_n|^{-1})}{n}\left(1 + \frac{\log(n!)}{\log(|a_n|^{-1})} - \frac{1}{\log(|a_n|^{-1})}\log\left(\frac{k}{2}\right)^n\right)\right) \\
&= \liminf_{n\to+\infty} \left(\frac{\log^{[q-1]}((-1/n)\log(|a_n|))}{\log^{[p-1]}(n)} + \frac{\log(1+o(1))}{\log^{[p-1]}(n)}\right) = \liminf_{n\to+\infty} \frac{\log^{[q-1]}((-1/n)\log(|a_n|))}{\log^{[p-1]}(n)}
\end{aligned} \quad (30)$$

and

$$\begin{aligned}\frac{1}{\rho(p,q,g)} &= \liminf_{n\to+\infty}\frac{\log^{[q-1]}\left((-1/n)\log\left((k/2r_*)^n(|a_n|/n!)\right)\right)}{\log^{[p-1]}(n)}\\ &= \liminf_{n\to+\infty}\frac{\log^{[q-1]}\left(\log\left(|a_n|^{-1}\right)/n+\log(n!)/n-(1/n)\log(k/2r_*)^n\right)}{\log^{[p-1]}(n)}\\ &= \liminf_{n\to+\infty}\left(\frac{\log^{[q-1]}\left((-1/n)\log(|a_n|)\right)}{\log^{[p-1]}(n)}+\frac{\log(1+o(1))}{\log^{[p-1]}(n)}\right) = \liminf_{n\to+\infty}\frac{\log^{[q-1]}\left((-1/n)\log(|a_n|)\right)}{\log^{[p-1]}(n)}\end{aligned} \tag{31}$$

Since f and g have the same (p,q)-order it follows that

$$\rho(p,q,f)=\rho(p,q,g)=\rho. \tag{32}$$

Now we will prove that f and g have the same (p,q)-type for $q=2$.

$$\begin{aligned}\frac{1}{T(p,2,f)} &= \liminf_{n\to+\infty}\frac{\left((-1/n)\log\left((k/2)^n(|a_n|/n!)\right)\right)^\rho}{\log^{[p-2]}(n)}\\ &= \liminf_{n\to+\infty}\frac{\left((-1/n)\log(|a_n|)+\log(n!)/n-(1/n)\log(k/2)^n\right)^\rho}{\log^{[p-2]}(n)}\\ &= \liminf_{n\to+\infty}\frac{\left((-1/n)\log(|a_n|)\right)^\rho}{\log^{[p-2]}(n)}\\ &\quad *\left(1+\frac{\log(n!)}{\log(|a_n|)^{-1}}-\frac{1}{\log(|a_n|)^{-1}}\log\left(\frac{k}{2}\right)^n\right)^\rho\\ &= \liminf_{n\to+\infty}\frac{\left((-1/n)\log(|a_n|)\right)^\rho}{\log^{[p-2]}(n)}(1+o(1))^\rho\\ &= \liminf_{n\to+\infty}\frac{\left((-1/n)\log(|a_n|)\right)^\rho}{\log^{[p-2]}(n)}.\end{aligned} \tag{33}$$

In the same way we prove that

$$\frac{1}{T(p,2,g)}=\liminf_{n\to+\infty}\frac{\left((-1/n)\log(|a_n|)\right)^\rho}{\log^{[p-2]}(n)}. \tag{34}$$

Now, for the case $q\ge 3$, we have

$$\begin{aligned}\frac{1}{T(p,q,f)} &= \liminf_{n\to+\infty}\frac{\log^{[q-2]}\left((-1/n)\log\left((k/2)^n(|a_n|/n!)\right)\right)^\rho}{\log^{[p-2]}(n)}\\ &= \liminf_{n\to+\infty}\frac{\log^{[q-2]}\left((-1/n)\log(|a_n|)+\log(n!)/n-(1/n)\log(k/2)^n\right)^\rho}{\log^{[p-2]}(n)}\\ &= \liminf_{n\to+\infty}\frac{1}{\log^{[p-2]}(n)}\cdot\log^{[q-2]}\left(\frac{-\log(|a_n|)}{n}\left(1+\frac{\log(n!)}{\log(|a_n|)^{-1}}-\frac{\log(k/2)^n}{\log(|a_n|)^{-1}}\right)\right)^\rho\\ &= \liminf_{n\to+\infty}\frac{\log^{[q-2]}\left((-1/n)\log(|a_n|)(1+o(1))\right)^\rho}{\log^{[p-2]}(n)} = \liminf_{n\to+\infty}\frac{\left(\log^{[q-2]}\left((-1/n)\log(|a_n|)\right)+\log(1+o(1))\right)^\rho}{\log^{[p-2]}(n)}\\ &= \liminf_{n\to+\infty}\frac{\log^{[q-2]}\left((-1/n)\log(|a_n|)\right)^\rho}{\log^{[p-2]}(n)}.\end{aligned} \tag{35}$$

The same is true for

$$\frac{1}{T(p,q,g)}=\liminf_{n\to+\infty}\frac{\log^{[q-2]}\left((-1/n)\log(|a_n|/n!)\right)^\rho}{\log^{[p-2]}(n)} \tag{36}$$

since f and g have identical (p,q)-order and (p,q)-type.

Lemma 9. *For an entire GASHE function u of (p,q)-order $\rho(p,q,u)$, lower (p,q)-order $\lambda(p,q,u)$, (p,q)-type $T(p,q,u)$, and lower (p,q)-type $t(p,q,u)$. If f and g are entire functions as defined above, then*

$$\rho(p,q,f)=\rho(p,q,u)=\rho(p,q,g), \tag{37}$$

$$\lambda(p,q,f)\le\lambda(p,q,u)\le\lambda(p,q,g), \tag{38}$$

$$T(p,q,f)=T(p,q,u)=T(p,q,g), \tag{39}$$

$$t(p,q,f)\le t(p,q,u)\le t(p,q,g). \tag{40}$$

Proof. Using (24), we get

$$\frac{\log^{[p]}m(r,g)}{\log^{[q]}(r)}\le\frac{\log^{[p]}M(r,u)}{\log^{[q]}(r)}\le\frac{\log^{[p]}M(r,f)}{\log^{[q]}(r)}. \tag{41}$$

From the above relation we obtain

$$\rho(p,q,f) \le \rho(p,q,u) \le \rho(p,q,g), \quad \lambda(p,q,f) \le \lambda(p,q,u) \le \lambda(p,q,g), \tag{42}$$

and since $\rho(p,q,f) = \rho(p,q,g)$ it proves (37) and (38).

Denoting by ρ the common value of (p,q)-order of f, g, and u, we have from (24)

$$\frac{\log^{[p-1]} m(r,g)}{\log^{[q-1]}(r)^{\rho}} \le \frac{\log^{[p-1]} M(r,u)}{\log^{[q-1]}(r)^{\rho}} \le \frac{\log^{[p-1]} M(r,f)}{\log^{[q-1]}(r)^{\rho}}. \tag{43}$$

This proves (39) and (40).

Before we start the next section, let us define $\beta_n = |a_n|/\Gamma(n+\mu+1)$, $\gamma_n = (|a_n|/n!)(k/2)^n$, and $\delta_n = (|a_n|/n!)(k/2r_*)^n$.

It is known, according to [3], that if (β_n/β_{n+1}) is a nondecreasing function of n then (Γ_n/Γ_{n+1}) and (δ_n/δ_{n+1}) also is a nondecreasing function of n.

3. Main Results

Theorem 10. *Let u be an entire GASHE function of (p,q)-order $\rho(p,q,u)$ and (p,q)-type $T(p,q,u)$. If (β_n/β_{n+1}) is a nondecreasing function of n for $n > n_0$, then*

$$\rho(p,q,u) = \limsup_{n\to\infty} \frac{\log^{[p-1]}(n)}{\log^{[q-1]}((-1/n)\log(\beta_n))}, \tag{44}$$

$$T(p,q,u) = \limsup_{n\to\infty} \frac{\log^{[p-2]}(n)}{\log^{[q-2]}((-1/n)\log(\beta_n))^{\rho}}. \tag{45}$$

Proof. For an entire function $\phi(z) = \sum_0^{\infty} a_n z^n$ and according to Theorem 1, we have

$$\rho(p,q,\phi) = \limsup_{n\to\infty} \frac{\log^{[p-1]}(n)}{\log^{[q-1]}((-1/n)\log(|a_n|))}. \tag{46}$$

We know that if (β_n/β_{n+1}) is a nondecreasing function of n for $n > n_0$, and then also (γ_n/γ_{n+1}) and (δ_n/δ_{n+1}).

Applying (46) to f, we get

$$\begin{aligned}
\rho(p,q,f) &= \limsup_{n\to\infty} \frac{\log^{[p-1]}(n)}{\log^{[q-1]}((-1/n)\log((|a_n|/n!)(k/2)^n)))} \\
&= \limsup_{n\to\infty} \log^{[p-1]}(n)\cdot\left(\log^{[q-1]}\left(\frac{-1}{n}\cdot\log\left(\frac{|a_n|}{n!}\frac{\Gamma(n+\mu+1)}{\Gamma(n+\mu+1)}\left(\frac{k}{2}\right)^n\right)\right)\right)^{-1} \\
&= \limsup_{n\to\infty} \log^{[p-1]}(n)\cdot\left(\log^{[q-1]}\left(\frac{-1}{n}\cdot\log\left(\beta_n(n+1)^{\mu}\left(\frac{k}{2}\right)^n\right)\right)\right)^{-1} \\
&= \limsup_{n\to\infty} \log^{[p-1]}(n)\cdot\left(\log^{[q-1]}\left(\frac{-1}{n}\cdot\log(\beta_n) - \frac{\log((n+1)^{\mu}(k/2)^n)}{n}\right)\right)^{-1} \\
&= \limsup_{n\to\infty} \frac{\log^{[p-1]}(n)}{\log^{[q-1]}((-1/n)\log(\beta_n)) + o(1)} \\
&= \limsup_{n\to\infty} \frac{\log^{[p-1]}(n)}{\log^{[q-1]}((-1/n)\log(\beta_n))}.
\end{aligned} \tag{47}$$

Similarly, applying (46) to g, we prove

$$\begin{aligned}
\rho(p,q,g) &= \limsup_{n\to\infty} \frac{\log^{[p-1]}(n)}{\log^{[q-1]}((-1/n)\log((|a_n|/n!)(k/2r_*)^n)))} \\
&= \limsup_{n\to\infty} \frac{\log^{[p-1]}(n)}{\log^{[q-1]}((-1/n)\log(\beta_n))}.
\end{aligned} \tag{48}$$

Then result (44) is found from the two relations $\rho(p,q,f)$ and $\rho(p,q,g)$ above and relation (37).

Let ρ be the common (p,q)-order of f and g.

The (p,q)-type of ϕ is defined according to Theorem 5 as

$$T(p,q,\phi) = \limsup_{n\to\infty} \frac{\log^{[p-2]}(n)}{\log^{[q-2]}((-1/n)\log(|a_n|))^{\rho(p,q,\phi)}}, \tag{49}$$

and we can easily prove that

$$T(p,q,f) = \limsup_{n\to\infty} \frac{\log^{[p-2]}(n)}{\log^{[q-2]}((-1/n)\log(\beta_n))^{\rho}}, \tag{50}$$

and

$$T(p,q,g) = \limsup_{n\to\infty} \frac{\log^{[p-2]}(n)}{\log^{[q-2]}((-1/n)\log(\beta_n))^{\rho}}. \tag{51}$$

Equation (45) now follows in view of (37) and (39).

Hence the proof is completed.

Theorem 11. *Let u be an entire GASHE function of (p,q)-order $\rho(p,q,u)$, and $|\beta_n/\beta_{n+1}|$ is a nondecreasing function of n for $n > n_0$. Then*

$$\rho(p,q,u) = \limsup_{n\to\infty} \frac{\log^{[p-1]}(n)}{\log^{[q]}(\beta_n/\beta_{n+1})}. \tag{52}$$

Proof. For an entire function $\phi(z) = \sum_{n=0}^{\infty} a_n z^n$, according to Theorem 2,

$$\rho(p,q,\phi) = \limsup_{n\to\infty} \frac{\log^{[p-1]}(n)}{\log^{[q]}(|a_n/a_{n+1}|)}, \tag{53}$$

provided $|a_n/a_{n+1}|$ is a nondecreasing function of n for $n > n_0$.

Applying this equation on f we get

$$\rho(p,q,f) = \limsup_{n\to\infty} \frac{\log^{[p-1]}(n)}{\log^{[q]}(\gamma_n/\gamma_{n+1})} = \limsup_{n\to\infty} \frac{\log^{[p-1]}(n)}{\log^{[q]}\left(\left(|a_n|/|a_{n+1}|\right)((n+1)!/n!)(2/k)\right)}$$
$$= \limsup_{n\to\infty} \log^{[p-1]}(n)\cdot\left(\log^{[q]}\left(\frac{|a_n|\cdot\Gamma(n+\mu+2)\cdot\Gamma(n+\mu+1)}{|a_{n+1}|\cdot\Gamma(n+\mu+1)\cdot\Gamma(n+\mu+2)}(n+1)\frac{2}{k}\right)\right)^{-1} \quad (54)$$
$$= \limsup_{n\to\infty} \frac{\log^{[p-1]}(n)}{\log^{[q]}\left((\beta_n/\beta_{n+1})(\Gamma(n+\mu+1)/\Gamma(n+\mu+2))(n+1)(2/k)\right)}.$$

Since $(\Gamma(n+\mu+1)/\Gamma(n+\mu+2))(n+1) \approx (n+1)/(n+\mu+1) \approx 1$ as $n \longrightarrow \infty$ then

$$\rho(p,q,f) = \limsup_{n\to\infty} \frac{\log^{[p-1]}(n)}{\log^{[q]}((\beta_n/\beta_{n+1})(2/k))} = \limsup_{n\to\infty} \frac{\log^{[p-1]}(n)}{\log^{[q]}(\beta_n/\beta_{n+1})}. \quad (55)$$

By the same way, we prove

$$\rho(p,q,g) = \limsup_{n\to\infty} \frac{\log^{[p-1]}(n)}{\log^{[q]}(\beta_n/\beta_{n+1})}. \quad (56)$$

Relation (52) now follows on using (37). Hence the proof is completed.

Theorem 12. *Let u be an entire GASHE function of (p,q)-order $\rho(p,q,u)$, lower (p,q)-order $\lambda(p,q,u)$, and lower (p,q)-type $t(p,q,u)$ and let (β_n/β_{n+1}) be a nondecreasing function of n for $n > n_0$. Then*

$$\lambda(p,q,u) = \liminf_{n\to\infty} \frac{\log^{[p-1]}(n)}{\log^{[q-1]}((-1/n)\log(\beta_n))}, \quad (57)$$

$$t(p,q,u) = \liminf_{n\to\infty} \frac{\log^{[p-2]}(n)}{\log^{[q-2]}((-1/n)\log(\beta_n))^{\rho}}. \quad (58)$$

Proof. For an entire function $\phi(z) = \sum_0^\infty a_n z^n$, and according to Theorem 3,

$$\lambda(p,q,\phi) = \liminf_{n\to\infty} \frac{\log^{[p-1]}(n)}{\log^{[q-1]}((-1/n)\log(|a_n|))}. \quad (59)$$

We know that if (β_n/β_{n+1}) form a nondecreasing function of n for $n > n_0$, then, also, (γ_n/γ_{n+1}) and (δ_n/δ_{n+1}).

Applying (59) to f, we get

$$\lambda(p,q,f) = \liminf_{n\to\infty} \frac{\log^{[p-1]}(n)}{\log^{[q-1]}\left((-1/n)\log\left(\left(|a_n|/n!\right)(k/2)^n\right)\right)}$$
$$= \liminf_{n\to\infty} \log^{[p-1]}(n) * \log^{[q-1]}\left(\frac{-1}{n} \cdot\log\left(\frac{|a_n|}{\Gamma(n+\mu+1)}\frac{\Gamma(n+\mu+1)}{n!}\left(\frac{k}{2}\right)^n\right)\right)^{-1} \quad (60)$$
$$= \liminf_{n\to\infty} \frac{\log^{[p-1]}(n)}{\log^{[q-1]}((-1/n)\log(\beta_n)) + o(1)}$$
$$= \liminf_{n\to\infty} \frac{\log^{[p-1]}(n)}{\log^{[q-1]}((-1/n)\log(\beta_n))}$$

Similarly, applying (59) to g, we prove that

$$\lambda(p,q,g) = \liminf_{n\to\infty} \frac{\log^{[p-1]}(n)}{\log^{[q-1]}\left((-1/n)\log\left(\left(|a_n|/n!\right)(k/2r_*)^n\right)\right)} \quad (61)$$
$$= \liminf_{n\to\infty} \frac{\log^{[p-1]}(n)}{\log^{[q-1]}((-1/n)\log(\beta_n))}.$$

Then result (57) is now followed by (38) and above by the two relations for $\lambda(p,q,f)$ and $\lambda(p,q,g)$.

If $\phi(z) = \sum_0^\infty a_n z^n$ is an entire function of (p,q)-order $\rho(p,q,\phi)$ and lower (p,q)-type $t(p,q,\phi)$, and if $|a_n/a_{n+1}|$ is a nondecreasing function of n for $n > n_0$, then, according to Theorem 6, we have

$$t(p,q,\phi) = \liminf_{n\to\infty} \frac{\log^{[p-2]}(n)}{\log^{[q-2]}((-1/n)\log(|a_n|))^{\rho(p,q,\phi)}}. \quad (62)$$

We can easily prove that

$$t(p,q,f) = \liminf_{n\to\infty} \frac{\log^{[p-2]}(n)}{\log^{[q-2]}((-1/n)\log(\beta_n))^{\rho(p,q,f)}}, \quad (63)$$

and

$$t(p,q,g) = \liminf_{n\to\infty} \frac{\log^{[p-2]}(n)}{\log^{[q-1]}\left((-1/n)\log\left(|\beta_n|\right)\right)^{\rho(p,q,g)}}. \quad (64)$$

Equation (58) now follows in view of (37) and (40). Hence the proof is completed.

Theorem 13. *Let u be an entire GASHE function of lower (p,q)-order $\lambda(p,q,u)$, and let (β_n/β_{n+1}) be a nondecreasing function of n for $n > n_0$. Then*

$$\lambda(p,q,u) = \liminf_{n\to\infty} \frac{\log^{[p-1]}(n)}{\log^{[q]}(\beta_n/\beta_{n+1})}. \quad (65)$$

Proof. Let $\phi(z) = \sum_0^\infty a_n z^n$ be an entire function and $|a_n/a_{n+1}|$ form a nondecreasing function of n for $n > n_0$, and, according to Theorem 4, we have

$$\lambda(p,q,\phi) = \liminf_{n\to\infty} \frac{\log^{[p-1]}(n)}{\log^{[q]}|a_n/a_{n+1}|}. \quad (66)$$

Applying this on f and g, we can easily prove that

$$\lambda(p,q,f) = \liminf_{n\to\infty} \frac{\log^{[p-1]}(n)}{\log^{[q]}(\beta_n/\beta_{n+1})}, \quad (67)$$

and

$$\lambda(p,q,g) = \liminf_{n\to\infty} \frac{\log^{[p-1]}(n)}{\log^{[q]}(\beta_n/\beta_{n+1})}. \quad (68)$$

Thus, we obtain relation (65) by using (38) and the two equalities above.

Theorem 14. *Let u be an entire GASHE function of (p,q)-order ρ, $(0 < \rho < \infty)$, (p,q)-type $T(p,q,u)$, and lower (p,q)-type $t(p,q,u)$. Then*

$$\begin{aligned} &\liminf_{n\to+\infty} N(\rho)\cdot\frac{\log^{[p-2]}(n)}{\log^{[q-1]}(\beta_n/\beta_{n+1})^{\rho}} \le t(u,p,q) \\ &\quad \le T(u,p,q) \le \limsup_{n\to+\infty} N(\rho)\cdot\frac{\log^{[p-2]}(n)}{\log^{[q-1]}(\beta_n/\beta_{n+1})^{\rho}}. \end{aligned} \quad (69)$$

Proof. If $\phi(z) = \sum_{n=0}^\infty a_n z^n$ is an entire function of (p,q)-type $T(p,q,g)$ and lower (p,q)-type $t(p,q,g)$, then in view of Theorem 7 we have

$$\liminf_{n\to\infty} N(\rho)\cdot\frac{\log^{[p-2]}(n)}{\log^{[q-1]}\left(|a_n/a_{n+1}|\right)^{\rho-C}} \le t \le T. \quad (70)$$

$$T \le \limsup_{n\to\infty} N(\rho) * \frac{\log^{[p-2]}(n)}{\log^{[q-1]}\left(|a_n/a_{n+1}|\right)^{\rho-C}}. \quad (71)$$

Applying inequality (71) to $f(z) = \sum_0^\infty \gamma_n z^n$, we get

$$\begin{aligned} T(f) &\le \limsup_{n\to\infty} N(\rho)\cdot\frac{\log^{[p-2]}(n)}{\log^{[q-1]}(\gamma_n/\gamma_{n+1})^{\rho-C}} \\ &\le \limsup_{n\to\infty} N(\rho) \\ &\cdot \frac{\log^{[p-2]}(n)}{\log^{[q-1]}\left(\left(|a_n|/|a_{n+1}|\right)\left((n+1)!/n!\right)(2/k)\right)^{\rho-C}} \\ &\le \limsup_{n\to\infty} N(\rho)\cdot\frac{\log^{[p-2]}(n)}{\log^{[q-1]},(\beta_n/\beta_{n+1})^{\rho-C}}, \end{aligned} \quad (72)$$

and then

$$\begin{aligned} T(u) &= T(f) \\ &\le \limsup_{n\to\infty} N(\rho)\cdot\frac{\log^{[p-2]}(n)}{\log^{[q-1]}(\beta_n/\beta_{n+1})^{\rho-C}} \end{aligned} \quad (73)$$

and applying (70) to the function $g(z) = \sum_0^\infty \delta_n z^n$ we get

$$\liminf_{n\to+\infty} N(\rho)\cdot\frac{\log^{[p-2]}(n)}{\log^{[q-1]}(\delta_n/\delta_{n+1})^{\rho}} \le t(g) \le t(u), \quad (74)$$

and we can easily prove

$$\begin{aligned} &\liminf_{n\to+\infty}\frac{\log^{[p-2]}(n)}{\log^{[q-1]}(\delta_n/\delta_{n+1})^{\rho}} \\ &\quad = \liminf_{n\to+\infty}\frac{\log^{[p-2]}(n)}{\log^{[q-1]}(\beta_n/\beta_{n+1})^{\rho}}, \end{aligned} \quad (75)$$

and, thus,

$$\liminf_{n\to+\infty} N(\rho)\cdot\frac{\log^{[p-2]}(n)}{\log^{[q-1]}(\beta_n/\beta_{n+1})^{\rho(p,q,f)}} \le t(u), \quad (76)$$

and thus the proof is completed.

Conflicts of Interest

The authors declare that there are no conflicts of interest regarding the publication of this paper.

References

[1] R. P. Gilbert, *Function Theoretic Methods in Partial Differential Equations*, Academic Press, NY, USA, 1996.

[2] R. P. Boas, *Entire Functions*, Academic Press. New York. NY, USA, 1954.

[3] R. P. Gilbert and H. C. Howard, "On solutions of the generalized axially symmetric wave equation represented by bergman operators," *Proceedings of the London Mathematical Society*, vol. 3-15, no. 1, pp. 346–360, 1965.

[4] P. A. McCoy, "Solutions of the helmholtz equation having rapid growth," *Complex Variables, Theory and Application: An International Journal*, vol. 18, no. 1-2, pp. 91–101, 2007.

[5] D. Kumar, "On the (p, q), growth of entire function solutions of Helmholtz equation," *J. Nonlinear Sci. Appl*, vol. 4, no. 1, 514 pages, 2011.

[6] D. Kumar, "Growth and Chebyshev approximation of entire function solutions of Helmholtz equation in R2," *European Journal of Pure and Applied Mathematics*, vol. 3, no. 6, pp. 1062–1069, 2010.

[7] H. H. Khan and R. Ali, "Slow Growth and Approximation of Entire Solution of Generalized Axially Symmetric Helmholtz Equation," *Asian Journal of Mathematics & Statistics*, vol. 5, no. 4, pp. 104–120, 2012.

[8] D. Kumar and R. Singh, "Measures of Growth of Entire Solutions of Generalized Axially Symmetric Helmholtz Equation," *Journal of Complex Analysis*, vol. 13.

[9] O. P. Juneja, G. P. Kapoor, and S. K. Bajpai, "On the(p, q)-order and lower(p, q)-order of an entire function," *J. Angew reine. Math*, vol. 282, pp. 53–67, 1976.

[10] O. P. Juneja, G. P. Kapoor, and S. K. Bajpai, "On the(p, q)-type and lower(p, q)-type of an entire function," *J. Angew reine. Math*, vol. 290, pp. 180–190, 1977.

Solving Volterra Integrodifferential Equations via Diagonally Implicit Multistep Block Method

Nur Auni Baharum,[1] Zanariah Abdul Majid,[1,2] and Norazak Senu[1,2]

[1] *Institute for Mathematical Research, Universiti Putra Malaysia, 43400 Serdang, Selangor, Malaysia*
[2] *Mathematics Department, Faculty of Science, Universiti Putra Malaysia, 43400 Serdang, Selangor, Malaysia*

Correspondence should be addressed to Zanariah Abdul Majid; zana_majid99@yahoo.com

Academic Editor: Theodore E. Simos

The performance of the numerical computation based on the diagonally implicit multistep block method for solving Volterra integrodifferential equations (VIDE) of the second kind has been analyzed. The numerical solutions of VIDE will be computed at two points concurrently using the proposed numerical method and executed in the predictor-corrector (PECE) mode. The strategy to obtain the numerical solution of an integral part is discussed and the stability analysis of the diagonally implicit multistep block method was investigated. Numerical results showed the competence of diagonally implicit multistep block method when solving Volterra integrodifferential equations compared to the existing methods.

1. Introduction

Consider the Volterra integrodifferential equation of the second kind

$$y'(x) = F(x, y(x), z(x)), \quad y(x_0) = y_0, \tag{1}$$

where

$$z(x) = \int_0^x K(x,s)\, y(s)\, ds, \quad 0 \le s \le x. \tag{2}$$

The numerical methods are generated to solve (1) which is a standard algorithm for ordinary differential equations and Newton-Cote integration formulae are required for solving the integral part since it cannot be solved explicitly. These equations usually appeared in physics, biology, and engineering applications such as biological models, neutron diffusion, wind ripple in the desert, heat transfer, and many more.

For many years, several methods had been applied to solve first-order problem of VIDE. Day [1] proposed Newton-Cotes integration formula of the trapezoidal rule for the solutions of outer and inner integral to obtain approximate solutions of integrodifferential equations. A comparison between the variational iteration method and trapezoidal rule revealing that the variational iteration method is more efficient and convenient to solve linear VIDE was discussed by Saadati et al. [2].

In [3], finite difference method is used for solving linear VIDE by Raftari. He transforms the Volterra integrodifferential equation in a matrix form and solved it by using finite difference method based on Simpson's rule and trapezoidal rule. A fourth-order robust numerical method was presented by Filiz [4] with a combination of the trapezoidal rule and Simpson's $1/3$ rule to evaluate the solution of VIDE for kernel equal to one. Then, he extended his work with a Runge-Kutta-Verner method in [5] and used higher rules of numerical integration method for solving the integral part.

The extended trapezoidal method [6] was proposed for the numerical solution of VIDE of the second kind and implemented the method in *PECE* scheme. Mohamed and Majid [7] had solved the second kind of VIDE using one-step block method and the Newton-Cotes quadrature formula was employed for finding the solution of the integral part. The multistep block method in [8] had implemented two approaches for solving VIDE for $K(x,s) = 1$ and $K(x,s) \neq 1$.

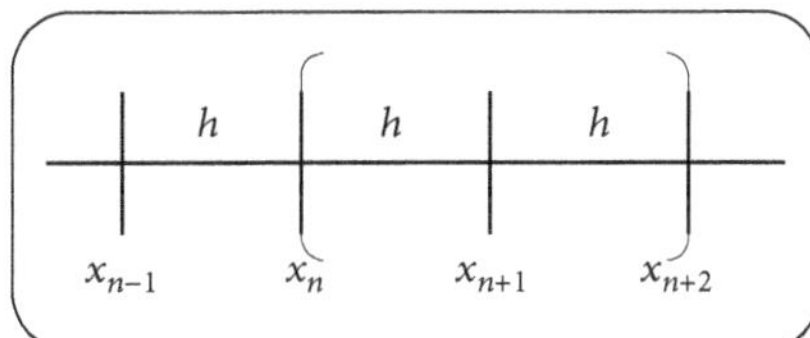

FIGURE 1: Two-point multistep block method.

2. Numerical Method

The proposed numerical method is in the form of block method and it will generate two or more solutions at the same time. The proposed method is a two-point block method; hence it will generate two solutions in one block.

In Figure 1, the two approximate values of y_{n+1} and y_{n+2} will be computed simultaneously in a block. The approximate values of y_{n+1} can be developed by integrating (1) over the interval $[x_n, x_{n+1}]$ while the interval for values y_{n+2} is $[x_n, x_{n+2}]$. Hence, the formulae of y_{n+1} and y_{n+2} can be obtained as

$$\int_{x_n}^{x_{n+r}} y'(x)\,dx = \int_{x_n}^{x_{n+r}} F(x,y,z)\,dx. \tag{3}$$

Therefore,

$$y_{n+r} - y_n = \int_{x_n}^{x_{n+r}} F(x,y,z)\,dx, \tag{4}$$

where $r = 1, 2$. Then, function $F(x,y,z)$ in (4) will be approximated using Lagrange interpolating polynomial, and the interpolation points involved in obtaining the corrector formula of y_{n+1} are $\{x_{n-1}, x_n, x_{n+1}\}$. Taking $x = x_{n+1} + sh$, $dx = hds$, and replacing into (4), the limit of integration in (4) will be -1 to 0.

The formulation of y_{n+2} can be obtained when three points are involved in the interpolation polynomial, that is, $\{x_{n-1}, x_{n+1}, x_{n+2}\}$. Considering $x = x_{n+2} + sh$, $dx = hds$ in (4) and the limit of integration will be changed from -2 to 0. The corrector formulae of y_{n+1} and y_{n+2} will be obtained using *MAPLE* software. The predictor formulae are one order less than the corrector formulae and the same process of derivation is applied.

Diagonally Implicit Multistep Block Method

Predictor

$$\begin{aligned} y_{n+1}^{p} &= y_n + h\left[\frac{3}{2}F_n - \frac{1}{2}F_{n-1}\right], \\ y_{n+2}^{p} &= y_n + h\left[4F_n - 2F_{n-1}\right]. \end{aligned} \tag{5}$$

Corrector

$$\begin{aligned} y_{n+1}^{c} &= y_n + h\left[\frac{5}{12}F_{n+1} + \frac{2}{3}F_n - \frac{1}{12}F_{n-1}\right], \\ y_{n+2}^{c} &= y_n + h\left[\frac{2}{9}F_{n+2} + \frac{5}{3}F_{n+1} + \frac{1}{9}F_{n-1}\right]. \end{aligned} \tag{6}$$

The matrix form of the corrector formulae is

$$\begin{aligned} \begin{bmatrix} 1 & 0 \\ 0 & 1 \end{bmatrix} \begin{bmatrix} y_{n+1} \\ y_{n+2} \end{bmatrix} &= \begin{bmatrix} 0 & -1 \\ 0 & -1 \end{bmatrix} \begin{bmatrix} y_{n-1} \\ y_n \end{bmatrix} \\ &+ h \begin{bmatrix} -\frac{1}{12} & \frac{2}{3} \\ \frac{1}{9} & \frac{5}{3} \end{bmatrix} \begin{bmatrix} F_{n-1} \\ F_n \end{bmatrix} \\ &+ h \begin{bmatrix} \frac{5}{12} & 0 \\ \frac{5}{3} & \frac{2}{9} \end{bmatrix} \begin{bmatrix} F_{n+1} \\ F_{n+2} \end{bmatrix} \end{aligned} \tag{7}$$

which is equivalent to the difference equations

$$A_0 Y_m = A_1 Y_{m-1} + h\left(B_0 F_{m-1} + B_1 F_m\right), \tag{8}$$

where A_0, A_1, B_0, and B_1 are the coefficients with m-vectors Y_m, Y_{m-1}, F_{m-1}, and F_m defined as

$$\begin{aligned} Y_m &= \begin{bmatrix} y_{n+1} \\ y_{n+2} \end{bmatrix}, \\ Y_{m-1} &= \begin{bmatrix} y_{n-1} \\ y_n \end{bmatrix}, \\ F_{m-1} &= \begin{bmatrix} F_{n-1} \\ F_n \end{bmatrix}, \\ F_m &= \begin{bmatrix} F_{n+1} \\ F_{n+2} \end{bmatrix}. \end{aligned} \tag{9}$$

3. Analysis of Diagonally Implicit Multistep Block Method

3.1. Order and Convergence of the Method. The order of the method can be obtained by referring [9]

$$\begin{aligned} &\sum_{j=0}^{k} \left[\alpha_j y(x+jh) - h\beta_j y'(x+jh)\right] \\ &\quad = C_p y^p + O\left(h^{p+1}\right), \end{aligned} \tag{10}$$

where p is the order of the linear multistep method, $O(h^{p+1})$ is the local truncation error, and C_p is defined as

$$C_p = \sum_{j=0}^{k} \frac{j^p \alpha_j}{p!} - \frac{j^{(p-1)}\beta_j}{(p-1)!}. \tag{11}$$

Definition 1. The numerical method is said to be in order p if the linear operator of numerical method is

$$C_0 = C_1 = C_2 = \cdots = C_p = 0, \quad C_{p+1} \neq 0, \tag{12}$$

where C_{p+1} is called as an error constant of the method.

The order of diagonally implicit multistep block method in (7) can be determined by applying the formula in (11); hence the values of α and β are obtained as follows:

$$\alpha_0 = \begin{bmatrix} 0 \\ 0 \end{bmatrix}, \quad \alpha_1 = \begin{bmatrix} -1 \\ -1 \end{bmatrix}, \quad \alpha_2 = \begin{bmatrix} 1 \\ 0 \end{bmatrix}, \quad \alpha_3 = \begin{bmatrix} 0 \\ 1 \end{bmatrix},$$
$$\beta_0 = \begin{bmatrix} -\frac{1}{12} \\ \frac{1}{9} \end{bmatrix}, \quad \beta_1 = \begin{bmatrix} \frac{2}{3} \\ 0 \end{bmatrix}, \quad \beta_2 = \begin{bmatrix} \frac{5}{12} \\ \frac{5}{3} \end{bmatrix}, \quad \beta_3 = \begin{bmatrix} 0 \\ \frac{2}{9} \end{bmatrix}. \tag{13}$$

Substitute the values of α and β into (11) and obtain

$$C_0 = \sum_{j=0}^{3} \frac{1}{0!} j^0 \alpha_j = \begin{bmatrix} 0 \\ 0 \end{bmatrix},$$
$$C_1 = \sum_{j=0}^{3} \frac{1}{1!} j^1 \alpha_j - \sum_{j=0}^{3} \frac{1}{0!} j^0 \beta_j = \begin{bmatrix} 0 \\ 0 \end{bmatrix},$$
$$C_2 = \sum_{j=0}^{3} \frac{1}{2!} j^2 \alpha_j - \sum_{j=0}^{3} \frac{1}{1!} j^1 \beta_j = \begin{bmatrix} 0 \\ 0 \end{bmatrix}, \tag{14}$$
$$C_3 = \sum_{j=0}^{3} \frac{1}{3!} j^3 \alpha_j - \sum_{j=0}^{3} \frac{1}{2!} j^2 \beta_j = \begin{bmatrix} 0 \\ 0 \end{bmatrix},$$
$$C_4 = \sum_{j=0}^{3} \frac{1}{4!} j^4 \alpha_j - \sum_{j=0}^{3} \frac{1}{3!} j^3 \beta_j = \begin{bmatrix} -\frac{1}{24} \\ \frac{1}{9} \end{bmatrix}.$$

Therefore, the diagonally implicit multistep block method is third-order where the coefficient of error constant is

$$C_{p+1} = C_4 = \begin{bmatrix} -\frac{1}{24} & \frac{1}{9} \end{bmatrix}^T \neq [0 \ \ 0]^T. \tag{15}$$

Definition 2. The local truncation error at x_{n+k} of the method is defined to be expression $L[y(x_n); h]$, when $y(x)$ is the theoretical solution of the initial value problem

$$L[y(x_n); h] = \sum_{j=0}^{k} [\alpha_j y(x + jh) - h\beta_j y'(x + jh)]. \tag{16}$$

For the formula y_{n+1}^c,

$$y_{n+1}^c = y_n + h\left[\frac{5}{12}F_{n+1} + \frac{2}{3}F_n - \frac{1}{12}F_{n-1}\right]. \tag{17}$$

Since $F_{n+1} = y'(x_{n+1})$, Taylor expansion will be applied to the derivatives $y'(x_{n+1})$ and $y'(x_{n-1})$ where $F_{n-1} = y'(x_{n-1})$,

$$\begin{aligned} y'(x_{n+1}) &= y'(x_n) + hy''(x_n) + \frac{h^2}{2!}y'''(x_n) + O(h^3), \\ y'(x_{n-1}) &= y'(x_n) - hy''(x_n) + \frac{h^2}{2!}y'''(x_n) + O(h^3). \end{aligned} \tag{18}$$

Then, since $F_n = y'(x_n)$ we have

$$y_{n+1} = y_n + hy_n' + \frac{h^2}{2}y_n'' + \frac{h^3}{6}y_n''' + O(h^4). \tag{19}$$

So, the local truncation error for y_{n+1} is $O(h^4)$. For the formula y_{n+2}^c,

$$y_{n+2}^c = y_n + h\left[\frac{2}{9}F_{n+2} + \frac{5}{3}F_{n+1} + \frac{1}{9}F_{n-1}\right]. \tag{20}$$

The Taylor expansion for $y'(x_{n+2}) = F_{n+2}$ is given as

$$y'(x_{n+2}) = y'(x_n) + 2hy''(x_n) + \frac{(2h)^2}{2!}y'''(x_n) + O(h^3). \tag{21}$$

Then, we will have

$$y_{n+2} = y_n + 2hy_n' + 2h^2 y_n'' + \frac{4}{3}h^3 y_n''' + O(h^4). \tag{22}$$

This shows that the local truncation error for y_{n+2} is $O(h^4)$.

Definition 3. The numerical method is said to be consistent if the order of method is $p \geq 1$ and the method is consistent if and only if

$$\sum_{j=0}^{k} \alpha_j = 0, \qquad \sum_{j=0}^{k} j\alpha_j = \sum_{j=0}^{k} \beta_j. \tag{23}$$

Proof. (i) $\sum_{j=0}^{k}\alpha_j = 0$,

$$\sum_{j=0}^{3}\alpha_j = \alpha_0 + \alpha_1 + \alpha_2 + \alpha_3 = \begin{bmatrix}0\\0\end{bmatrix} + \begin{bmatrix}-1\\-1\end{bmatrix} + \begin{bmatrix}1\\0\end{bmatrix} + \begin{bmatrix}0\\1\end{bmatrix} = \begin{bmatrix}0\\0\end{bmatrix}. \quad (24)$$

(ii) $\sum_{j=0}^{k} j\cdot\alpha_j = \sum_{j=0}^{k}\beta_j$,

$$\sum_{j=0}^{3} j\cdot\alpha_j = 0\cdot\alpha_0 + 1\cdot\alpha_1 + 2\cdot\alpha_2 + 3\cdot\alpha_3 = 0\begin{bmatrix}0\\0\end{bmatrix} + 1\begin{bmatrix}-1\\-1\end{bmatrix} + 2\begin{bmatrix}1\\0\end{bmatrix} + 3\begin{bmatrix}0\\1\end{bmatrix} = \begin{bmatrix}1\\2\end{bmatrix},$$

$$\sum_{j=0}^{3}\beta_j = \beta_0 + \beta_1 + \beta_2 + \beta_3 = \begin{bmatrix}-\frac{1}{12}\\ \frac{1}{9}\end{bmatrix} + \begin{bmatrix}\frac{2}{3}\\ 0\end{bmatrix} + \begin{bmatrix}\frac{5}{12}\\ \frac{5}{3}\end{bmatrix} + \begin{bmatrix}0\\ \frac{2}{9}\end{bmatrix} = \begin{bmatrix}1\\2\end{bmatrix}. \quad (25)$$

Therefore,

$$\sum_{j=0}^{3} j\cdot\alpha_j = \sum_{j=0}^{3}\beta_j = \begin{bmatrix}1\\2\end{bmatrix}. \quad (26)$$

By Definitions 1 and 3, the diagonally implicit multistep block method is consistent.

Definition 4. A block method is said to be zero-stable if and only if providing the roots of R_j, $j = 1(1)k$ of the first characteristic polynomial, $\rho(R)$, specified as

$$\rho(R) = \det\left[\sum_{j=0}^{k} A^{(i)}R^{(k-i)}\right] = 0 \quad (27)$$

satisfies $|R_j| \leq 1$ and those roots with $|R_j| = 1$.

Proof. The values of A can be obtained in (7):

$$\rho(r) = \det\left[RA_0 - A_1\right] = \det\left[R\begin{bmatrix}1 & 0\\0 & 1\end{bmatrix} - \begin{bmatrix}0 & -1\\0 & -1\end{bmatrix}\right] = \det\begin{bmatrix}R & 1\\0 & R+1\end{bmatrix} = R(R+1). \quad (28)$$

The diagonally implicit multistep block method is zero stable since $|R| \leq 1$.

Theorem 5. *The method is said to be convergent if and only if the method is consistent and zero-stable.*

Proof. By Definitions 1, 3, and 4, the diagonally implicit multistep block method is convergent.

3.2. Stability Region of the Method. In this section, the stability region of the diagonally implicit multistep block method of order three is discussed for the numerical solution of VIDE. The test equation for first-order VIDE of the second kind [10] is

$$y'(x) = \xi y(x) + \eta\int_0^x y(t)\,dt, \quad (29)$$

where ξ and η are real constants, $\xi = \lambda + \mu$ and $\eta = -\lambda\mu$. Therefore,

$$y'(x) = (\lambda + \mu)\,y(x) - \lambda\mu\int_0^x y(t)\,dt. \quad (30)$$

Definition 6. The method is said to be A-stable if and only if the region of absolute stability contains at the quarter plane $h\xi < 0$, $h^2\eta < 0$.

From the proposed method for the numerical solution, the characteristics polynomials $\rho(r)$, $\sigma(r)$, $\tilde{\rho}(r)$, and $\tilde{\sigma}(r)$ can be developed as follows.

Corrector formula for y_{n+1} is

$$\rho(r) = r^2 - r, \qquad \sigma(r) = \frac{5}{12}r^2 + \frac{2}{3}r - \frac{1}{12}. \quad (31)$$

Corrector formula for y_{n+2} is

$$\rho(r) = r^3 - r, \qquad \sigma(r) = \frac{2}{9}r^3 + \frac{5}{3}r^2 + \frac{1}{9}. \quad (32)$$

Simpson's rule is

$$\tilde{\rho}(r) = r^2 - 1, \qquad \tilde{\sigma}(r) = \frac{1}{3}r^2 + \frac{4}{3}r + \frac{1}{3}. \quad (33)$$

The stability polynomial of the diagonally implicit multistep block method can be determined by substituting (31), (32), and (33) into this particular formula,

$$\pi\left(r, h\xi, h^2\eta\right) = \tilde{\rho}(r)\left[\rho(r) - H_1\sigma(r)\right] - H_2\tilde{\sigma}(r)\,\sigma(r), \quad (34)$$

where $H_1 = h\xi$ and $H_2 = h^2\mu$. Thus, the stability polynomial of the proposed method is obtained:

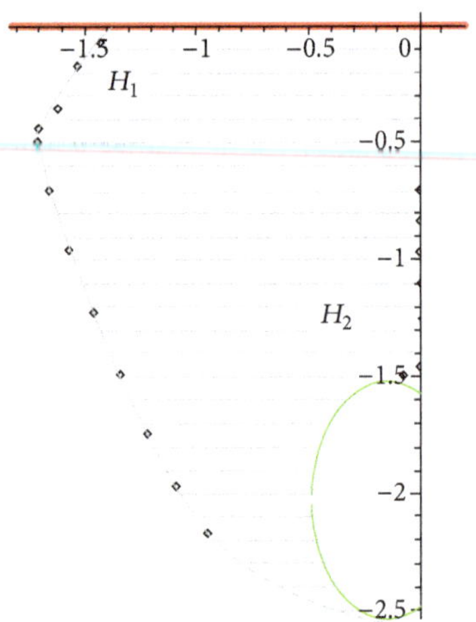

FIGURE 2: Stability region in H_1, H_2 plane.

$$\begin{aligned}
&-\frac{7}{36}H_1+\frac{7}{108}H_2-\frac{2}{27}H_1^2-\frac{2}{243}H_2^2+\frac{5}{54}H_1^2r^4\\
&-\frac{71}{54}H_1^2r^3+\frac{41}{18}H_1^2r^2-\frac{53}{54}H_1^2r+\frac{3}{2}H_1r^2\\
&-\frac{7}{9}H_1r-\frac{23}{36}H_1r^4+\frac{1}{9}H_1r^3+\frac{4}{81}H_1H_2\\
&+\frac{5}{486}H_2^2r^4-\frac{131}{486}H_2^2r^3+\frac{95}{54}H_2^2r^2-\frac{5}{486}H_2^2r \qquad (35)\\
&-\frac{23}{108}H_2r^4-\frac{205}{54}H_2r^3+\frac{116}{27}H_2r^2-\frac{19}{54}H_2r\\
&-\frac{61}{81}H_1H_2r^3+\frac{61}{81}H_1H_2r-\frac{1}{9}H_1H_2r^2\\
&+\frac{5}{81}H_1H_2r^4+r^4-3r^3+3r^2-r=0.
\end{aligned}$$

The region of stability polynomial can be illustrated in Figure 2.

Regarding Definition 6, the third-order of diagonally implicit multistep block method in Figure 2 is A-stable within the shaded region.

4. Implementation

The one-step methods are required for finding the first starting point at x_{n+1} since Volterra integrodifferential equations of the second kind have two types of kernels. For $K(x,s)=1$, Runge-Kutta method is involved in solving differential part of VIDE, while, midpoint method is needed to solve VIDE when $K(x,s)\neq 1$. Hence, the predictor and the corrector formula can be implemented until the end of the interval

$$\begin{aligned}
y_{n+1}^{p}&=y_n+h\left[\frac{3}{2}F_n-\frac{1}{2}F_{n-1}\right],\\
y_{n+2}^{p}&=y_n+h\left[4F_n-2F_{n-1}\right],\\
y_{n+1}^{c}&=y_n+h\left[\frac{5}{12}F_{n+1}+\frac{2}{3}F_n-\frac{1}{12}F_{n-1}\right],\\
y_{n+2}^{c}&=y_n+h\left[\frac{2}{9}F_{n+2}+\frac{5}{3}F_{n+1}+\frac{1}{9}F_{n-1}\right].
\end{aligned} \qquad (36)$$

Since $z(x)$ in $F(x,y(x),z(x))$ is the integral term in VIDE and cannot be solved explicitly, therefore, Simpson's rule is adapted for solving the integral part.

(i) For $K(x,s)=1$, Simpson's 1/3 rule is applied to solve the integral term in VIDE:

$$z_{n+2}=z_n+h\left[\frac{1}{3}y_n+\frac{4}{3}y_{n+1}+\frac{1}{3}y_{n+2}\right]. \qquad (37)$$

(ii) For $K(x,s)\neq 1$, composite Simpson's rule is employed for solving the integral part:

$$\begin{aligned}
z_{n+2}&=\frac{h}{3}\sum_{i=0}^{n+2}\omega_i^s K\left(x_{n+2},x_i,y_i\right),\\
z_{n+3}&=\frac{h}{3}\sum_{i=0}^{n+2}\omega_i^s K\left(x_{n+3},x_i,y_i\right)\\
&+\frac{h}{6}\left[K\left(x_{n+3},x_{n+2},y_{n+2}\right)\right.\\
&\left.+4K\left(x_{n+2},x_{n+5/2},y_{n+5/2}\right)+K\left(x_{n+3},x_{n+3},y_{n+3}\right)\right],
\end{aligned} \qquad (38)$$

where ω_i^s are Simpson's rule weights $1,4,2,4,\ldots,2,4,1$. The unknown value of $y_{n+5/2}$ in (38) can be estimated by using Lagrange interpolation at the point $\{x_n,x_{n+1},x_{n+2},x_{n+3}\}$,

$$y_{n+5/2}=\frac{1}{16}y_n-\frac{5}{16}y_{n+1}+\frac{15}{16}y_{n+2}+\frac{5}{16}y_{n+3}. \qquad (39)$$

4.1. Algorithm of the Method. The input of the programming is the endpoints of a and b and the integer, N. The developed algorithm for the method is given as follows.

Step 1. Set

$x_0=a;$

$y_0=\alpha;$

$z_0=0;$

$h=(b-a)/N;$

OUTPUT: (x_0,y_0,z_0).

Step 2. For $i=1$,

When $K(x,s)=1$, using RK3 to evaluate the value of y.

When $K(x,s)\neq 1$, applying Midpoint Method.

Step 3. For $i=2,\ldots,(N/2)$, do Steps 4–6.

Step 4. Set $x=a+ih$.

TABLE 1: Numerical results for Example 1.

h	0.025	0.0125	0.00625	0.003125
		MAXE		
RK3	8.3229(−08)	1.0910(−08)	1.3953(−09)	1.7637(−10)
ABM3	1.9468(−06)	2.4548(−07)	3.0813(−08)	3.8593(−09)
2PMBM	4.2079(−07)	4.9165(−08)	5.9272(−09)	7.2715(−10)
DIMBM	4.1354(−07)	4.7815(−08)	5.8390(−09)	7.2152(−10)
		TFC		
RK3	120	240	480	960
ABM3	82	162	322	642
2PMBM	43	83	163	323
DIMBM	42	82	162	322
		TS		
RK3	40	80	160	320
ABM3	40	80	160	320
2PMBM	21	41	81	161
DIMBM	21	41	81	161
		Time		
RK3	0.1035	0.1840	0.2970	0.4527
ABM3	0.0860	0.1410	0.2264	0.3454
2PMBM	0.0670	0.1200	0.2024	0.3204
DIMBM	0.0450	0.1120	0.1600	0.2660

Step 5. Calculate for y^p_{n+1} and z^p_{n+1}, y^p_{n+2} and z^p_{n+2}.

Step 6. Compute the solution for y^c_{n+1} and z^c_{n+1}, y^c_{n+2} and z^c_{n+2}.

Step 7. Calculate the error.

Step 8. OUTPUT: (x, y, z) and the absolute error.

Step 9. STOP.

5. Numerical Results

Four tested problems of first-order Volterra integrodifferential equations were considered in order to study the performance of the diagonally implicit multistep block method.

Example 1 ($(K(x,s) = 1)$ linear VIDE).

$$y'(x) = 1 - \int_0^x y(s)\,ds, \quad y(0) = 0, \ 0 \le x \le 1. \tag{40}$$

Exact solution is $y(x) = \sin(x)$.
Source: [4].

Example 2 ($(K(x,s) \ne 1)$ linear VIDE).

$$y'(x) = -\sin(x) - \cos(x) + \int_0^x 2\cos(x-s)\,y(s)\,ds, \quad y(0) = 1, \ 0 \le x \le 5. \tag{41}$$

Exact solution is $y(x) = \exp(-x)$.
Source: [11].

Example 3 ($(K(x,s) \ne 1)$ nonlinear VIDE).

$$y'(x) = x\exp(1 - y(x)) - \frac{1}{(1+x)^2} - x - \int_0^x \frac{x}{(1+s)^2}\exp(1 - y(s))\,ds \tag{42}$$

$$y(0) = 1, \ 0 \le x \le 4.$$

Exact solution is $y(x) = 1/(1+x)$.
Source: [12].

Example 4 ($(K(x,s) \ne 1)$ nonlinear VIDE).

$$y'(x) = 2x - \frac{1}{2}\sin\left(x^4\right) + \int_0^x x^2 s\cos\left(x^2 y(s)\right) ds, \quad y(0) = 0, \ 0 \le x \le 2. \tag{43}$$

Exact solution is $y(x) = x^2$.
Source: [13].

Notations used in Tables 1–4 are as follows:

h: step size

MAXE: maximum error

TABLE 2: Numerical results for Example 2.

h	0.25	0.125	0.0625	0.03125
		MAXE		
BVMs	2.1607(−01)	2.8411(−02)	3.6378(−03)	4.6011(−04)
ABM3	1.5401(−02)	3.1321(−03)	5.3778(−04)	7.9127(−05)
2PMBM	5.7923(−02)	1.2209(−03)	3.7500(−04)	8.4172(−05)
DIMBM	1.8211(−02)	3.4489(−03)	4.6803(−04)	6.0807(−05)
		TFC		
BVMs	-	-	-	-
ABM3	84	164	324	644
2PMBM	50	90	170	330
DIMBM	22	42	82	162
		TS		
BVMs	-	-	-	-
ABM3	20	40	80	160
2PMBM	11	21	41	81
DIMBM	11	21	41	81
		Time		
BVMs	-	-	-	-
ABM3	0.0715	0.1546	0.2433	0.3546
2PMBM	0.0460	0.0780	0.1710	0.2738
DIMBM	0.0140	0.0410	0.1170	0.1480

TABLE 3: Numerical results for Example 3.

h	0.025	0.0125	0.00625	0.003125
		MAXE		
ABM3	2.3797(−06)	3.2252(−07)	4.2061(−08)	5.3727(−09)
2PMBM	8.0020(−06)	9.7319(−07)	1.2000(−07)	1.4862(−08)
DIMBM	3.6545(−06)	4.6274(−07)	5.8216(−08)	7.2999(−09)
		TFC		
ABM3	644	1284	2564	5124
2PMBM	330	650	1290	2570
DIMBM	322	642	1282	2562
		Time		
ABM3	0.3543	0.5677	1.1650	1.9855
2PMBM	0.2810	0.4210	0.7496	1.3570
DIMBM	0.2610	0.3010	0.6740	1.1050

TS: total steps

TFC: total functions call

Time: the execution time taken

RK3: Runge-Kutta method of order 3 with Simpson's 1/3 rule by Filiz [4]

ABM3: Adam Bashforth Moulton Order 3 with Simpson's rule.

BVMs: Combination of BVMs and third-order Generalized Adams Method by Chen and Zhang [11]

2PMBM: Two points Multistep Block Method of Order 3 with Simpson's rule by Mohamed and Majid [8]

DIMBM: Diagonally implicit multistep block method with Simpson's rule proposed in this paper

Tables 1–4 display the numerical results for the four tested problems when solved using the proposed block method and the code was written in C language.

The numerical results for Examples 1–4 displayed in Tables 1–4 are solved numerically using the proposed numerical method with Simpson's rule. In Table 1, the numerical

TABLE 4: Numerical results for Example 4.

h	2/9	2/17	2/33	2/65
		MAXE		
ABM3	5.0218(−02)	1.8761(−03)	1.1046(−04)	5.8996(−06)
2PMBM	2.7425(−02)	6.5258(−04)	1.7781(−04)	6.6888(−06)
DIMBM	8.8008(−03)	8.6068(−04)	1.7703(−04)	6.6994(−06)
		TFC		
ABM3	40	72	136	246
2PMBM	22	38	70	134
DIMBM	10	18	34	66
		Time		
ABM3	0.0670	0.0723	0.1291	0.2030
2PMBM	0.0140	0.0352	0.0662	0.1336
DIMBM	0.0050	0.0280	0.0350	0.0420

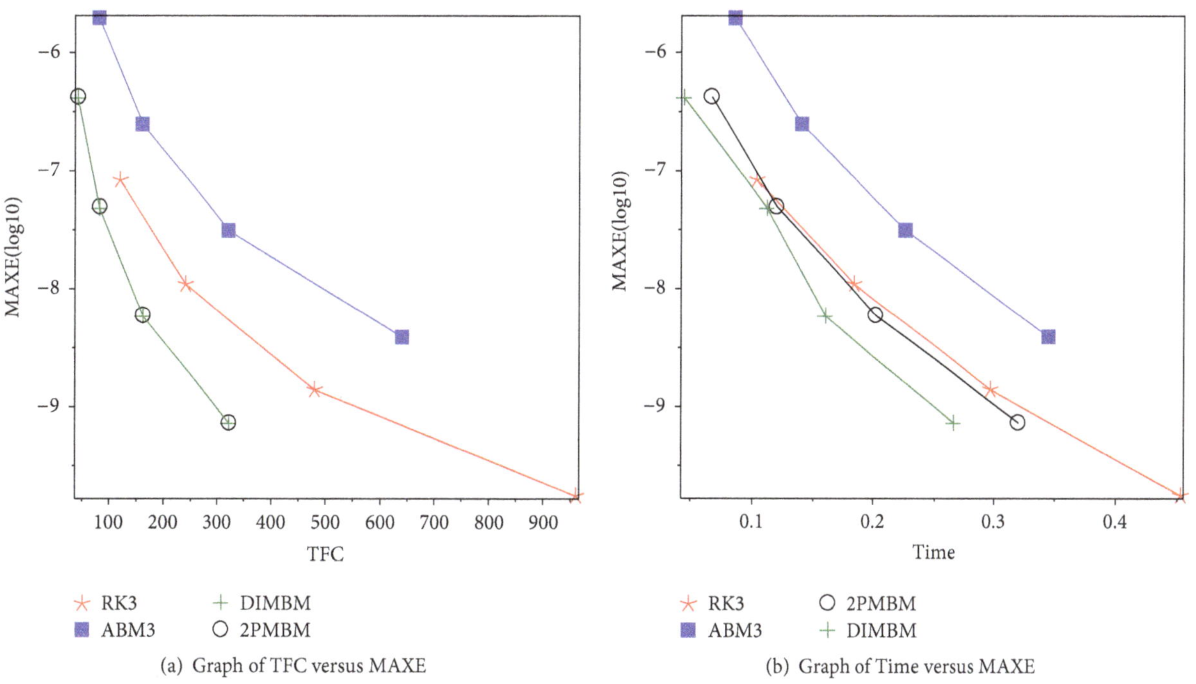

(a) Graph of TFC versus MAXE

(b) Graph of Time versus MAXE

FIGURE 3: Graph of the numerical results when solving Example 1.

results will be obtained when the step size $h = 0.025$, 0.0125, 0.00625, and 0.003125 for the case when $K(x,s) = 1$. The maximum error of DIMBM is comparable compared to 2PMBM at all tested h but the order of accuracy is the same or one order less compared to RK3 and ABM3. The performance of DIMBM is better in terms of total functions call and total number of steps compared to RK3 and ABM3.

In Tables 2, 3, and 4, the numerical results are solved using the proposed numerical method via composite Simpson's for the integral part when $K(x,s) \neq 1$. In Table 2, the maximum error of DIMBM is comparable compared to 2PMBM and ABM3. The DIMBM manage to obtain less total functions call compared to ABM3 and 2PMBM. For the nonlinear Examples 3 and 4, we could observe that ABM3 and 2PMBM are expensive in terms of total functions call, respectively. Figures 3–6 display the numerical results of maximum error versus total functions call when solving the tested problems. This has shown the advantage of DIMBM in the form of a standard multistep method because the cost per step is cheaper and the numerical results are more accurate when the step size is reduced. In terms of timing, DIMBM gave faster results compared to ABM3 and 2PMBM.

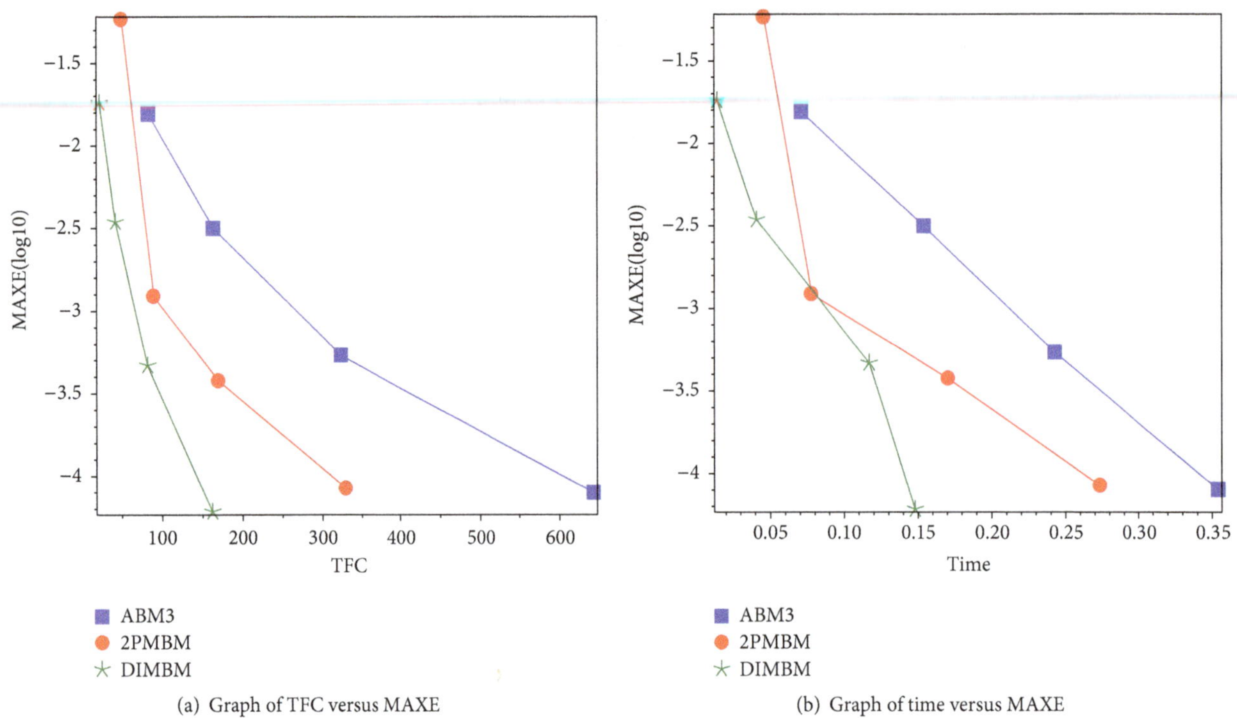

(a) Graph of TFC versus MAXE

(b) Graph of time versus MAXE

FIGURE 4: Graph of the numerical results when solving Example 2.

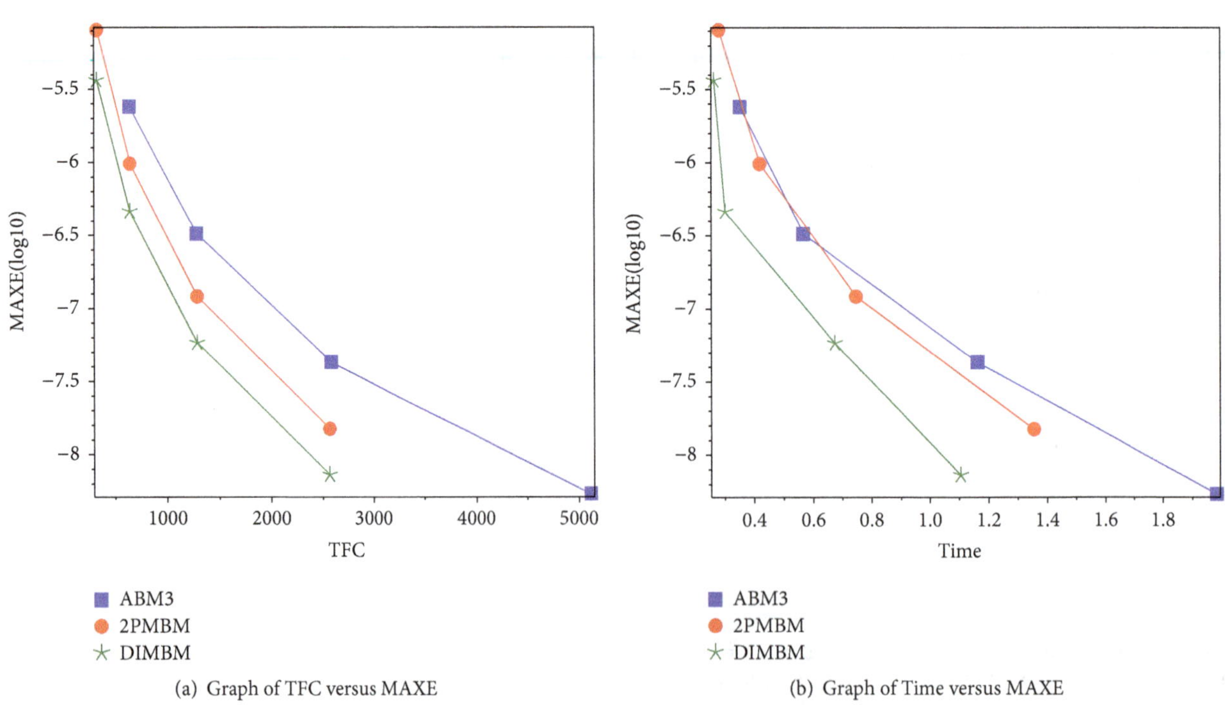

(a) Graph of TFC versus MAXE

(b) Graph of Time versus MAXE

FIGURE 5: Graph of the numerical results when solving Example 3.

6. Conclusion

In this research, we proposed the diagonally implicit multistep block method for solving linear and nonlinear Volterra integrodifferential equations and comparisons were made with the existing method. Comparisons with existing method reveal that the diagonally implicit multistep block method is more efficient and cheaper.

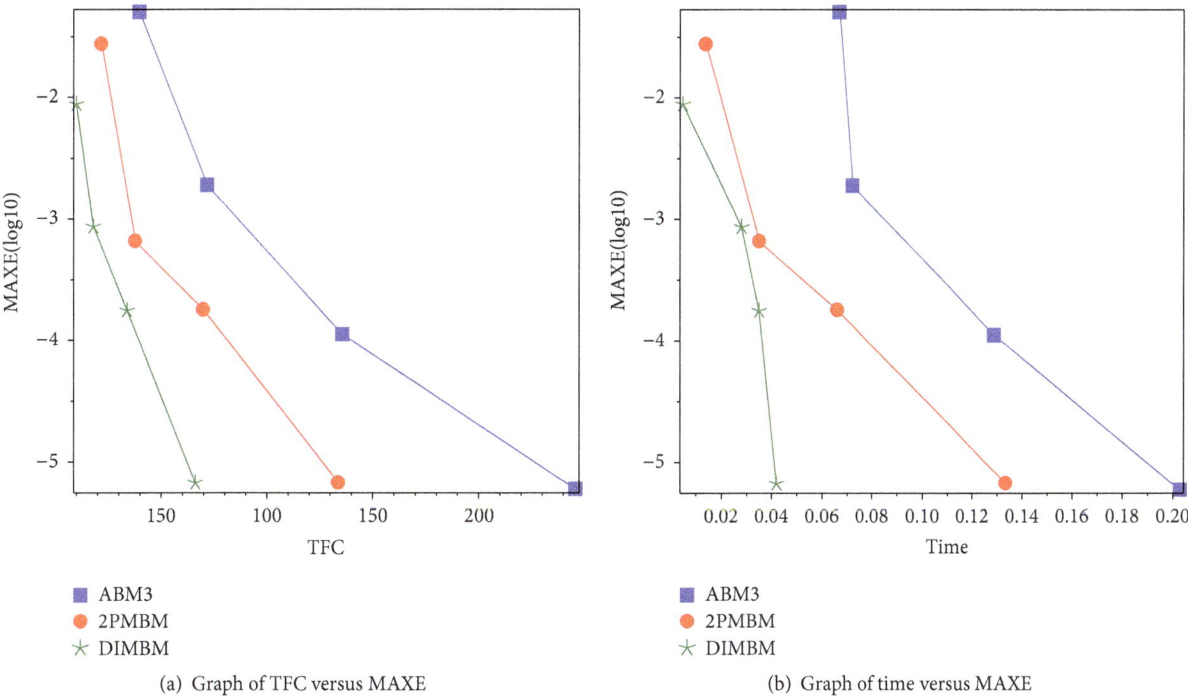

(a) Graph of TFC versus MAXE

(b) Graph of time versus MAXE

FIGURE 6: Graph of the numerical results when solving Example 4.

Conflicts of Interest

The authors declare that there are no conflicts of interest regarding the publication of this paper.

Acknowledgments

The authors gratefully acknowledged the financial support of Fundamental Research Grant Scheme (FRGS:5524973) and Graduate Research Fund (GRF) from Universiti Putra Malaysia.

References

[1] J. T. Day, "Note on the Numerical Solution of Integro-Differential Equations," *The Computer Journal*, vol. 9, no. 4, pp. 394-395, 1967.

[2] R. Saadati, B. Raftari, H. Adibi, S. M. Vaezpour, and S. Shakeri, "A comparison between the Variational Iteration method and Trapezoidal rule for solving linear integro-differential equations," *World Applied Sciences Journal*, vol. 4, no. 3, pp. 321–325, 2008.

[3] B. Raftari, "Numerical solutions of the linear volterra integro-differential equations: homotopy perturbation method and finite Difference method," *World Applied Sciences Journal*, vol. 9, pp. 7–12, 2010.

[4] A. Filiz, "Fourth-Order Robust Numerical Method for Integro-differential Equations," *Asian Journal of Fuzzy and Applied Mathematics*, vol. 1, no. 1, pp. 28–33, 2013.

[5] A. Filiz, "Numerical Solution of a Non-linear Volterra Integro-differential Equation via Runge-Kutta-Verner Method," *International Journal of Scientific and Research Publications*, vol. 3, article 9, 2013.

[6] F. Ishak and S. N. Ahmad, "Development of Extended Trapezoidal Method for Numerical Solution of Volterra Integro-Differential Equations," *International Journal of Mathematics, Computational, Physical, Electrical and Computer Engineering*, vol. 10, no. 11, article 52856, 2016.

[7] N. A. B. Mohamed and Z. A. Majid, "One-step block method for solving Volterra integro-differential equations," *AIP Conference Proceedings*, vol. 1682, no. 1, Article ID 020018, 2015.

[8] N. A. Mohamed and Z. A. Majid, "Multistep block method for solving volterra integro-differential equations," *Malaysian Journal of Mathematical Sciences*, vol. 10, pp. 33–48, 2016.

[9] J. D. Lambert, *Computational Methods in Ordinary Differential Equations*, John Wiley & Sons, New York, NY, USA, 1973.

[10] H. Brunner and J. D. Lambert, "Stability of numerical methods for volterra integro-differential equations," *Computing*, vol. 12, no. 1, pp. 75–89, 1974.

[11] H. Chen and C. Zhang, "Boundary value methods for Volterra integral and integro-differential equations," *Applied Mathematics and Computation*, vol. 218, no. 6, pp. 2619–2630, 2011.

[12] R. E. Shaw, "A parallel algorithm for nonlinear volterra integro-differential equations," in *Proceedings of the 2000 ACM Symposium on Applied Computing, SAC 2000*, pp. 86–88, Como, Italy, March 2000.

[13] M. Dehghan and R. Salehi, "The numerical solution of the non-linear integro-differential equations based on the meshless method," *Journal of Computational and Applied Mathematics*, vol. 236, no. 9, pp. 2367–2377, 2012.

22

New Modified Adomian Decomposition Recursion Schemes for Solving Certain Types of Nonlinear Fractional Two-Point Boundary Value Problems

Sekson Sirisubtawee and Supaporn Kaewta

Department of Mathematics, Faculty of Applied Science, King Mongkut's University of Technology North Bangkok, Bangkok 10800, Thailand

Correspondence should be addressed to Sekson Sirisubtawee; sekson.s@sci.kmutnb.ac.th

Academic Editor: A. Zayed

We apply new modified recursion schemes obtained by the Adomian decomposition method (ADM) to analytically solve specific types of two-point boundary value problems for nonlinear fractional order ordinary and partial differential equations. The new modified recursion schemes, which sometimes utilize the technique of Duan's convergence parameter, are derived using the Duan-Rach modified ADM. The Duan-Rach modified ADM employs all of the given boundary conditions to compute the remaining unknown constants of integration, which are then embedded in the integral solution form before constructing recursion schemes for the solution components. New modified recursion schemes obtained by the method are generated in order to analytically solve nonlinear fractional order boundary value problems with a variety of two-point boundary conditions such as Robin and separated boundary conditions. Some numerical examples of such problems are demonstrated graphically. In addition, the maximal errors (ME_n) or the error remainder functions ($\mathrm{ER}_n(x)$) of each problem are calculated.

1. Introduction

During the last three decades, fractional order differential equations (FDEs) have played an important role in modelling many phenomena in engineering [1, 2], applied sciences [3–6], and biological systems [7, 8]. This is because the behavior of most systems appear after effects or memory which can be explained better by using fractional order derivatives [5]. Several methods have been proposed to analytically and numerically solve nonlinear fractional order differential equations including initial value problems (IVPs) and boundary value problems (BVPs). These methods include the Adomian decomposition method (ADM) [9–14], the multi-step generalized differential transform method (MSGDTM) [15], the Adams-Bashforth-Moulton type predictor-corrector scheme [16], and the Haar wavelet method [17].

The ADM has been extensively utilized to solve IVPs and BVPs for nonlinear ordinary or partial differential equations, integral equations, or integrodifferential equations since it can provide approximate analytic solutions without linearization, perturbation, discretization, guessing the initial term, or using Green's functions which are quite difficult to determine in most cases. The ADM has been used as a tool to investigate analytical and numerical solutions of real-world problems by Hashim et al. [18] and Sweilam and Khader [19]. Modifications of the ADM have been developed for different purposes for IVPs for both integer order and fractional order differential equations.

Several researchers proposed expressing the initial solution component u_0 as a series of orthogonal polynomials, such as Chebyshev polynomials [20], or Legendre polynomials [21, 22]. In 2013 Duan et al. [11] combined the ADM with convergence acceleration techniques such as diagonal Padé approximants and iterated Shanks transforms to solve nonlinear fractional ordinary differential equations. It was found that the modified techniques can efficiently extend the convergence region of the decomposition series solution. In 2014 Ramana and Prasad [23] modified the ADM to solve parabolic equations and the results obtained by their modified method converged very quickly and were more accurate than the standard ADM results.

There are now several different resolution techniques using the ADM for solving BVPs of nonlinear integer order differential equations. These techniques were developed by many authors as follows. Tatari and Dehghan [24] gave the solution of the general form of multipoint BVPs using the ADM. Al-Hayani [25] used the ADM with Green's function to solve sixth-order BVPs. Duan and Rach [26] proposed the Duan-Rach modified decomposition method for solving BVPs for higher order nonlinear differential equations. Duan et al. [27] developed the multistage ADM for solving BVPs for second-order nonlinear ordinary differential equations with Robin boundary conditions. In particular, modified approaches for solving nonlinear fractional order BVPs can be found in [10, 14, 28]. Most of the resolution techniques are fundamentally based on two principles of using the ADM. The first approach is the method of undetermined coefficients (see, e.g., [10, 14, 29, 30]) in the ADM which approximates the constants of integration embedded in the recursion scheme of the ADM by solving numerically a sequence of nonlinear algebraic equations obtained by employing boundary conditions. The Duan-Rach modified ADM (see, e.g., [26, 27, 30, 31]), which is the second approach, determines the remaining unknown constants of integration of the solution by using the remaining boundary conditions before designing a suitable modified recursion scheme. In the Duan-Rach modified ADM, the constants of integration are calculated simultaneously along with the solution components. Details of the Duan-Rach modified ADM will be discussed in Section 2.3.

In our work, we study the use of the Duan-Rach modified ADM to solve nonlinear high fractional order boundary value problems with a variety of boundary conditions such as Robin and separated boundary conditions. We will study these fractional BVPs for the Caputo fractional derivative which allows some boundary conditions to be included into the formulas of the solutions. To the best of the authors' knowledge, our paper will be the first to develop formulas for the recursion schemes obtained by using the Duan-Rach modified ADM for the above types of problems.

This paper is organized as follows. In Section 2, we review necessary definitions and important properties of the fractional order integrals and derivatives that are needed in our work. The principal reviews of the ADM and the Duan-Rach modified ADM are briefly given in this section as well. In Section 3, we give the formulas for the recursion schemes obtained by using the Duan-Rach modified ADM for selected types of nonlinear fractional BVPs. In Section 4, we give numerical examples of solutions obtained using the proposed recursion schemes for some scientific fractional BVPs with mixed sets of Dirichlet, Neumann, Robin, and separated boundary conditions. These examples include a Bratu-type fractional BVP, an oscillating base temperature equation, and an elastic beam problem.

2. Mathematical Preliminaries

2.1. Review of Fractional Order Integrals and Derivatives. In this section, we present basic definitions and important theorems of the fractional calculus (see [5, 32–35]) required in this paper.

Definition 1 (see [5]). A function $f(x)$ $(x > 0)$ is said to be in the space C_α $(\alpha \in \mathbb{R})$ if it can be expressed as $f(x) = x^p f_1(x)$ for some $p > \alpha$, where $f_1(x)$ is continuous in $[0, \infty)$, and it is said to be in the space C_α^m if $f^{(m)} \in C_\alpha$, $m \in \mathbb{N}$. Clearly, $C_\alpha \subseteq C_\beta$ if $\beta, \alpha \in \mathbb{R}$ and $\beta \leq \alpha$.

Definition 2 (see [5]). The Riemann-Liouville fractional integral operator of order $\alpha > 0$ of a function $f \in C_\alpha$ with $a \geq 0$ is defined as

$$\begin{aligned} {}_{\mathrm{RL}}J_a^\alpha f(x) &= \frac{1}{\Gamma(\alpha)} \int_a^x (x-\tau)^{\alpha-1} f(\tau)\, d\tau, \quad x > a, \\ {}_{\mathrm{RL}}J_a^0 f(x) &= f(x), \end{aligned} \tag{1}$$

where $\Gamma(\cdot)$ is the gamma function.

Definition 3 (see [5]). The Caputo fractional derivative of $f(x)$ of order $\alpha > 0$ with $a \geq 0$ is defined as

$$ {}_{C}D_a^\alpha f(x) = \begin{cases} {}_{\mathrm{RL}}J_a^{m-\alpha} f^{(m)}(x) = \dfrac{1}{\Gamma(m-\alpha)} \displaystyle\int_a^x \frac{f^{(m)}(\tau)}{(x-\tau)^{\alpha-m+1}} d\tau, & m-1 < \alpha < m,\ m \in \mathbb{N}, \\ f^{(m)}(x), & \alpha = m, \end{cases} \tag{2}$$

for $x \geq a$ and $f \in C_{-1}^m$.

For $0 < \beta \leq \alpha$, $a \geq 0$, and $m \in \mathbb{N}$, we have the following important properties [5]:

$$ {}_{\mathrm{RL}}J_a^\alpha\, {}_{\mathrm{RL}}J_a^\beta f(x) = {}_{\mathrm{RL}}J_a^\beta\, {}_{\mathrm{RL}}J_a^\alpha f(x) = {}_{\mathrm{RL}}J_a^{\alpha+\beta} f(x), \quad \text{where } f \in C_\alpha, \tag{3}$$

$$ {}_{\mathrm{RL}}J_a^\alpha (x-a)^\mu = \frac{\Gamma(\mu+1)}{\Gamma(\mu+\alpha+1)} (x-a)^{\mu+\alpha}, \quad \text{where } \mu > -1,\ x > a, \tag{4}$$

$$ {}_{C}D_a^\alpha (x-a)^\mu = \frac{\Gamma(\mu+1)}{\Gamma(\mu-\alpha+1)} (x-a)^{\mu-\alpha}, \quad \text{where } 0 \leq m-1 < \alpha \leq m < \mu+1,\ \mu > 0,\ x > a, \tag{5}$$

$$_{\mathrm{RL}}J_a^{\gamma}\,{}_{\mathrm{C}}D_a^{\gamma} f(x) = f(x) - \sum_{k=0}^{m-1} f^{(k)}(a)\frac{(x-a)^k}{k!}, \tag{6}$$

where $m-1<\gamma\leq m,\ f\in C_{\gamma}^{m},\ \gamma\geq -1$.

2.2. Review of the Decomposition Methods. We first review the important concepts of the Adomian decomposition method (ADM) [9] introduced by George Adomian who is an American physicist. The ADM combined with the use of symbolic algebra packages such as MATHEMATICA or MAPLE is a powerful method for solving nonlinear operator equations including ordinary or partial differential equations [14, 36]. Here we describe the ADM to solve integer order IVPs and BVPs as follows.

Consider the ordinary differential equation in the following operator form:

$$Lu + Ru + Nu = g, \tag{7}$$

where L is the highest order derivative; that is, $L = (d^p/dx^p)(\cdot)$, where p is the integer order of the derivative, which is assumed to be easily invertible, R is a linear differential operator of order less than L, N denotes a nonlinear operator assumed to be analytic, g is a source term, and u is the solution of the equation. The ADM decomposes the solution $u(x)$ and the analytic nonlinear term Nu of the nonlinear operator equation (7) into a rapidly convergent series of solution components and a series of the Adomian polynomials.

Applying the inverse linear operator L^{-1}, which is a p-fold definite integration, to both sides of (7) and using the given conditions, that is, the initial conditions or boundary conditions, and the fact that $L^{-1}Lu = u - \Phi$, we obtain

$$u = \Phi + L^{-1}g - L^{-1}(Ru) - L^{-1}(Nu), \tag{8}$$

where Φ denotes the terms arising from using the given conditions. The ADM decomposes the solution $u(x)$ into an infinite series

$$u = \sum_{n=0}^{\infty} u_n, \tag{9}$$

and then it decomposes the nonlinear term $Nu(x)$ into a series

$$Nu = \sum_{n=0}^{\infty} A_n, \tag{10}$$

where $A_n = A_n(u_0(x), u_1(x), \ldots, u_n(x))$ are the Adomian polynomials that are obtained by the following formula (see the derivation of the formula in [27, 37] and the references therein):

$$A_n = \frac{1}{n!}\frac{d^n}{d\lambda^n} N\left(\sum_{k=0}^{\infty} u_k \lambda^k\right)\Bigg|_{\lambda=0}, \quad n\geq 0, \tag{11}$$

where λ is a grouping parameter. The first six Adomian polynomials obtained by using (11) for the general analytic nonlinear term $Nu(x) = f(u(x))$ are as follows:

$$\begin{aligned}
A_0 &= f(u_0),\\
A_1 &= f'(u_0)u_1,\\
A_2 &= f'(u_0)u_2 + f''(u_0)\frac{u_1^2}{2!},\\
A_3 &= f'(u_0)u_3 + f''(u_0)u_1u_2 + f'''(u_0)\frac{u_1^3}{3!},\\
A_4 &= f'(u_0)u_4 + f''(u_0)\left(\frac{u_2^2}{2!} + u_1u_3\right)\\
&\quad + f'''(u_0)\frac{u_1^2u_2}{2!} + f^{(4)}(u_0)\frac{u_1^4}{4!},\\
A_5 &= f'(u_0)u_5 + f''(u_0)(u_2u_3 + u_1u_4)\\
&\quad + f^{(3)}(u_0)\left(\frac{u_3u_1^2}{2!} + \frac{u_2^2u_1}{2!}\right)\\
&\quad + f^{(4)}(u_0)\frac{u_2u_1^3}{3!} + f^{(5)}(u_0)\frac{u_1^5}{5!}.
\end{aligned} \tag{12}$$

We observe that the Adomian polynomials are of the following forms:

$$\begin{aligned}
A_0 &= f(u_0),\\
A_n &= \sum_{k=1}^{n} f^{(k)}(u_0)C_n^k, \quad n\geq 1,
\end{aligned} \tag{13}$$

where C_n^k are the sums of all possible products of k components from $u_1, u_2, \ldots, u_{n-k+1}$, whose subscripts sum to n, divided by the factorial of the number of repeated subscripts.

From (8), (9), and (10), we have

$$\sum_{n=0}^{\infty} u_n = \Phi + L^{-1}g - L^{-1}(Ru) - L^{-1}(Nu), \tag{14}$$

and the classic (or standard) Adomian recursion scheme [36, 38] is as follows:

$$\begin{aligned}
u_0(x) &= \Phi + L^{-1}g,\\
u_{n+1}(x) &= -L^{-1}(Ru_n) - L^{-1}(A_n), \quad n\geq 0.
\end{aligned} \tag{15}$$

Then the n-term approximation of the solution is

$$\varphi_n(x) = \sum_{k=0}^{n-1} u_k(x), \tag{16}$$

which in the $\lim_{n\to\infty}$ yields the exact solution to (7) as

$$u(x) = \lim_{n\to\infty}\varphi_n(x). \tag{17}$$

The choice of different initial solutions $u_0(x)$ can generate different recursion schemes which can remedy problems in the classic scheme caused by the difficulty of integration for $u_0(x)$ or the slowness of convergence to the series solution (see [39] for details). One criterion that can be used for choosing the initial solution is to achieve a simple integration for the initial solution component of the series solution. Then, to obtain a fast rate of convergence and an extended region of convergence, we can apply Duan's convergence parameter technique [40–42] to the recursion scheme for that initial solution component. Duan and coresearchers [26, 27] have shown that the parametrized recursion scheme usually gives a sequence of decreasing maximal errors which approach zero when the value of n in the approximation $\varphi_n(x)$ increases.

2.3. The Duan-Rach Modified Decomposition Method. In this section, we will provide the idea of the modified ADM called the Duan-Rach modified decomposition method to solve integer order BVPs. The method generates a recursion scheme for computing successive solution components without any undetermined coefficients. Unlike the method of undetermined coefficients, this new Duan-Rach modified decomposition method [26, 27, 30] does not require the solution of a sequence of the nonlinear algebraic equations obtained from the approximation $\varphi_n(x)$ for the constants of integration. In the Duan-Rach modified decomposition method, we first incorporate as many of the given boundary conditions as possible into the solution $u(x)$ in (8) of the BVP and then we determine the remaining unknown constants of integration before constructing the modified recursion scheme.

We will now give an example of the use of the Duan-Rach modified decomposition method to solve the following two-point BVP:

$$\begin{aligned} Lu + Ru + Nu &= g, \\ u(a) &= \beta, \\ u(b) &= \gamma, \end{aligned} \tag{18}$$

where the operator L in (18) is $L = (d^2/dx^2)(\cdot)$. Using the first condition of the boundary conditions in (18), we obtain the solution form as follows:

$$\begin{aligned} u(x) = \beta + u'(a)(x-a) + L^{-1}g - L^{-1}(Ru) \\ - L^{-1}(Nu), \end{aligned} \tag{19}$$

where $L^{-1} = \int_a^x \int_a^\tau (\cdot) ds\, d\tau$. We then apply the second condition of the boundary conditions in (18) to solve for $u'(a)$ which can be expressed as follows:

$$\begin{aligned} u'(a) = \frac{1}{b-a}\left[\gamma - \beta - \left[L^{-1}g\right]_{x=b} + \left[L^{-1}(Ru)\right]_{x=b} \right. \\ \left. + \left[L^{-1}(Nu)\right]_{x=b}\right]. \end{aligned} \tag{20}$$

Substituting (20) into (19) yields

$$\begin{aligned} u(x) = {} & \beta + \frac{\gamma-\beta}{b-a}(x-a) - \frac{x-a}{b-a}\left[L^{-1}g\right]_{x=b} \\ & + \frac{x-a}{b-a}\left[L^{-1}(Ru)\right]_{x=b} \\ & + \frac{x-a}{b-a}\left[L^{-1}(Nu)\right]_{x=b} + L^{-1}g - L^{-1}(Ru) \\ & - L^{-1}(Nu). \end{aligned} \tag{21}$$

Applying the Adomian polynomials for the nonlinear terms in (21), we then obtain the following modified recursion scheme:

$$\begin{aligned} u_0(x) = {} & \beta + \frac{\gamma-\beta}{b-a}(x-a) - \frac{x-a}{b-a}\left[L^{-1}g\right]_{x=b} \\ & + L^{-1}g, \\ u_{n+1}(x) = {} & \frac{x-a}{b-a}\left[L^{-1}(Ru)\right]_{x=b} \\ & + \frac{x-a}{b-a}\left[L^{-1}(A_n)\right]_{x=b} - L^{-1}(Ru) \\ & - L^{-1}(A_n), \quad n \geq 0. \end{aligned} \tag{22}$$

3. Description of the Proposed Recursion Schemes for Solving Certain Types of Fractional Order BVPs

The main advantage of using the Duan-Rach modified ADM for solving nonlinear integer order BVPs which we can see from Section 2.3 is that evaluating the inverse operator directly at the boundary points allows us to obtain the solution components without using numerical methods to calculate the values of unknown constants of integration as in the method of undetermined coefficients. In this section, we construct the recursion schemes developed by using the Duan-Rach modified ADM for solving fractional higher order two-point BVPs with their boundary conditions. We give examples for a set of Robin conditions and separated boundary conditions. We consider the following nonlinear fractional order differential equation:

$${}_C D_a^\alpha u(x) + Nu = g(x), \quad a \leq x \leq b, \tag{23}$$

where α is a fractional order of the differential equation with $m-1 \leq \alpha \leq m$, $m = 2, 4$, N is a nonlinear operator, $g(x)$ is a source term, and $u(x)$ is the solution of the equation. Comparing with (7), we see that the operator $L = {}_C D_a^\alpha$ is the Caputo fractional derivative operator of order α. Applying $L^{-1} = {}_{RL}J_a^\alpha$, which is the Riemann-Liouville fractional integral operator of order α, to both sides of (23) and using the property in (6) yield

$$\begin{aligned}
{}_{\mathrm{RL}}J_a^{\alpha}\,{}_{C}D_a^{\alpha}u(x) &= {}_{\mathrm{RL}}J_a^{\alpha}(g(x) - Nu),\\
u(x) - \sum_{k=0}^{m-1} u^{(k)}(a)\frac{(x-a)^k}{k!} &= {}_{\mathrm{RL}}J_a^{\alpha}(g(x)) - {}_{\mathrm{RL}}J_a^{\alpha}(Nu),\\
u(x) &= \sum_{k=0}^{m-1} u^{(k)}(a)\frac{(x-a)^k}{k!} + {}_{\mathrm{RL}}J_a^{\alpha}(g(x)) - {}_{\mathrm{RL}}J_a^{\alpha}(Nu).
\end{aligned} \tag{24}$$

The specific types of boundary conditions imposed on the fractional differential equation in (23) or the fractional integral equation in (24) depend upon the value of α.

3.1. The Fractional Order Differential Equation (23) with a Set of Robin Boundary Conditions. We consider a nonlinear fractional order differential equation of the form

$${}_{C}D_a^{\alpha}u + Nu = g(x), \quad a \le x \le b,\ 1 < \alpha \le 2, \tag{25}$$

subject to a set of Robin boundary conditions

$$pu(a) + qu'(a) = \beta, \tag{26}$$

$$ru(b) + su'(b) = \gamma, \tag{27}$$

where $N(u(x))$ is an analytic nonlinear term and p, q, r, and s satisfy the following condition:

$$ps - qr + pr(b-a) \neq 0. \tag{28}$$

In order to have the two boundary conditions required for the problem and to make the condition (28) hold, it is necessary to have p, q not both zero, r, s not both zero, and p, r not both zero.

From (24), we have that the solution of this BVP can be written in the form

$$u(x) = u(a) + u'(a)(x-a) + {}_{\mathrm{RL}}J_a^{\alpha}(g(x)) - {}_{\mathrm{RL}}J_a^{\alpha}(Nu). \tag{29}$$

We now apply the Duan-Rach modified ADM to the problem in (25). Using (29), we evaluate $u(x)$ at $x = b$ to obtain

$$u(b) = u(a) + u'(a)(b-a) + \left[{}_{\mathrm{RL}}J_a^{\alpha}(g(x))\right]_{x=b} - \left[{}_{\mathrm{RL}}J_a^{\alpha}(Nu)\right]_{x=b}, \tag{30}$$

where $[{}_{\mathrm{RL}}J_a^{\alpha}(\cdot)]_{x=b}$ is the Riemann-Liouville fractional integral operator of order α evaluated at $x = b$.

Differentiating (29) and then using the property that $(d/dx)\,{}_{\mathrm{RL}}J_a^{\alpha}(\cdot) = {}_{\mathrm{RL}}J_a^{\alpha-1}(\cdot)$ and evaluating $u'(x)$ at $x = b$, we obtain

$$u'(b) = u'(a) + \left[{}_{\mathrm{RL}}J_a^{\alpha-1}(g(x))\right]_{x=b} - \left[{}_{\mathrm{RL}}J_a^{\alpha-1}(Nu)\right]_{x=b}, \tag{31}$$

where $[{}_{\mathrm{RL}}J_a^{\alpha-1}(\cdot)]_{x=b}$ is the Riemann-Liouville fractional integral operator of order $\alpha - 1$ evaluated at $x = b$.

Substituting (30) and (31) into (27), we get

$$\begin{aligned}
& r\Big[u(a) + u'(a)(b-a) + \left[{}_{\mathrm{RL}}J_a^{\alpha}(g(x))\right]_{x=b} - \left[{}_{\mathrm{RL}}J_a^{\alpha}(Nu)\right]_{x=b}\Big] + s\Big[u'(a)\\
&+ \left[{}_{\mathrm{RL}}J_a^{\alpha-1}(g(x))\right]_{x=b} - \left[{}_{\mathrm{RL}}J_a^{\alpha-1}(Nu)\right]_{x=b}\Big] = \gamma.
\end{aligned} \tag{32}$$

After manipulating the terms in the above equation, we obtain

$$\begin{aligned}
& ru(a) + [r(b-a) + s]u'(a)\\
&= \gamma + r\left(\left[{}_{\mathrm{RL}}J_a^{\alpha}(Nu)\right]_{x=b} - \left[{}_{\mathrm{RL}}J_a^{\alpha}(g(x))\right]_{x=b}\right)\\
&\quad + s\left(\left[{}_{\mathrm{RL}}J_a^{\alpha-1}(Nu)\right]_{x=b} - \left[{}_{\mathrm{RL}}J_a^{\alpha-1}(g(x))\right]_{x=b}\right).
\end{aligned} \tag{33}$$

It is possible to solve the system of two linearly independent equations (26) and (33) for the two remaining undetermined coefficients $u(a)$ and $u'(a)$ if the determinant of the coefficient matrix denoted by Δ is not zero, that is, if

$$\Delta = \begin{vmatrix} p & q \\ r & s + r(b-a) \end{vmatrix} = ps - qr + pr(b-a) \neq 0. \tag{34}$$

For $\Delta \neq 0$, the values of $u(a)$ and $u'(a)$ can be expressed in terms of the specified values of the system parameters a, b, β, γ, p, q, r, and s as follows:

$$\begin{aligned}
u(a) &= \frac{1}{\Delta}\Big[r\beta(b-a) + s\beta - qr\\
&\quad + qs\left[{}_{\mathrm{RL}}J_a^{\alpha-1}(g(x))\right]_{x=b} + qr\left[{}_{\mathrm{RL}}J_a^{\alpha}(g(x))\right]_{x=b}\\
&\quad - qs\left[{}_{\mathrm{RL}}J_a^{\alpha-1}(Nu)\right]_{x=b} - qr\left[{}_{\mathrm{RL}}J_a^{\alpha}(Nu)\right]_{x=b}\Big],\\
u'(a) &= \frac{1}{\Delta}\Big[p\gamma - r\beta - ps\left[{}_{\mathrm{RL}}J_a^{\alpha-1}(g(x))\right]_{x=b}\\
&\quad - pr\left[{}_{\mathrm{RL}}J_a^{\alpha}(g(x))\right]_{x=b} + ps\left[{}_{\mathrm{RL}}J_a^{\alpha-1}(Nu)\right]_{x=b}\\
&\quad + pr\left[{}_{\mathrm{RL}}J_a^{\alpha}(Nu)\right]_{x=b}\Big].
\end{aligned} \tag{35}$$

Substituting (35) into (29), we obtain the following equivalent nonlinear Fredholm-Volterra integral equation:

$$\begin{aligned} u(x) &= \frac{1}{\Delta}\Big[s\beta - q\gamma + r\beta(b-a) + (p\gamma - r\beta)(x-a) \\ &+ (qs - ps(x-a))\left[{}_{\mathrm{RL}}J_a^{\alpha-1}(g(x))\right]_{x=b} \\ &+ (qr - pr(x-a))\left[{}_{\mathrm{RL}}J_a^{\alpha}(g(x))\right]_{x=b} \\ &- (qs - ps(x-a))\left[{}_{\mathrm{RL}}J_a^{\alpha-1}(Nu)\right]_{x=b} \\ &- (qr - pr(x-a))\left[{}_{\mathrm{RL}}J_a^{\alpha}(Nu)\right]_{x=b}\Big] \\ &+ {}_{\mathrm{RL}}J_a^{\alpha}(g(x)) - {}_{\mathrm{RL}}J_a^{\alpha}(Nu), \end{aligned} \tag{36}$$

which is free of any undetermined coefficients. Therefore (36) represents the solution of the fractional order nonlinear differential equation (25) subject to the Robin boundary conditions (26) and (27). Next we apply the decomposition to the solution $u(x)$ and the nonlinear term $Nu(x)$; that is,

$$\begin{aligned} u(x) &= \sum_{n=0}^{\infty} u_n(x), \\ Nu(x) &= \sum_{n=0}^{\infty} A_n(x), \end{aligned} \tag{37}$$

respectively, where $A_n(x)$ are the Adomian polynomials defined in (11).

Inserting the equations in (37) into (36), the solution components are determined by the following modified recursion scheme:

$$\begin{aligned} u_0(x) &= \frac{1}{\Delta}\Big[s\beta - q\gamma + r\beta(b-a) + (p\gamma - r\beta)(x-a) \\ &+ (qs - ps(x-a))\left[{}_{\mathrm{RL}}J_a^{\alpha-1}(g(x))\right]_{x=b} \\ &+ (qr - pr(x-a))\left[{}_{\mathrm{RL}}J_a^{\alpha}(g(x))\right]_{x=b}\Big] \\ &+ {}_{\mathrm{RL}}J_a^{\alpha}(g(x)), \\ u_{n+1}(x) &= \frac{1}{\Delta}\Big[(-qs + ps(x-a))\left[{}_{\mathrm{RL}}J_a^{\alpha-1}(A_n)\right]_{x=b} \\ &+ (-qr + pr(x-a))\left[{}_{\mathrm{RL}}J_a^{\alpha}(A_n)\right]_{x=b}\Big] \\ &- {}_{\mathrm{RL}}J_a^{\alpha}(A_n), \quad n \geq 0, \end{aligned} \tag{38}$$

where the resulting integrals are assumed to exist.

The n-term approximation of the solution to the BVP obtained by the ADM is the following truncated decomposition series:

$$\varphi_n(x) = \sum_{k=0}^{n-1} u_k(x). \tag{39}$$

With the above decomposition obtained by the Duan-Rach modified ADM, each approximation $\varphi_n(x)$, $n \geq 1$, must exactly satisfy the boundary conditions (26) and (27). In addition, other techniques such as partitioning initial terms into two appropriate terms [39, 43, 44] or using the Duan's convergence parameter [40–42] can be incorporated, if necessary, into the recursion scheme (38) for solving the BVP described in (25)–(27).

Using the general formulas given above, we can derive the equivalent nonlinear Fredholm-Volterra integral equations and their associated recursion schemes for (25) for the special cases of the boundary conditions in (26) and (27). The results are as follows.

Case 1. The nonlinear fractional BVP consists of (25) and the following Dirichlet boundary conditions:

$$\begin{aligned} u(a) &= \beta, \\ u(b) &= \gamma. \end{aligned} \tag{40}$$

The boundary conditions (40) correspond to the case of $p = r = 1$ and $q = s = 0$ in (26) and (27). Thus we have $\Delta = b - a \neq 0$ and then (36) is reduced to

$$\begin{aligned} u(x) &= \frac{1}{b-a}\Big[\beta(b-a) + (\gamma - \beta)(x-a) \\ &- (x-a)\left[{}_{\mathrm{RL}}J_a^{\alpha}(g(x))\right]_{x=b} \\ &+ (x-a)\left[{}_{\mathrm{RL}}J_a^{\alpha}(Nu)\right]_{x=b}\Big] + {}_{\mathrm{RL}}J_a^{\alpha}(g(x)) \\ &- {}_{\mathrm{RL}}J_a^{\alpha}(Nu), \end{aligned} \tag{41}$$

where ${}_{\mathrm{RL}}J_a^{\alpha}(\cdot)$ and $[{}_{\mathrm{RL}}J_a^{\alpha}(\cdot)]_{x=b}$ are the Riemann-Liouville fractional integral operator of order α and the Riemann-Liouville fractional integral operator of order α evaluated at $x = b$, respectively.

Substituting equations in (37) into (41), we can determine the solution components from the following modified recursion scheme:

$$\begin{aligned} u_0(x) &= \frac{1}{b-a}\Big[\beta(b-a) + (\gamma - \beta)(x-a) \\ &- (x-a)\left[{}_{\mathrm{RL}}J_a^{\alpha}(g(x))\right]_{x=b}\Big] + {}_{\mathrm{RL}}J_a^{\alpha}(g(x)), \\ u_{n+1}(x) &= \frac{x-a}{b-a}\left[{}_{\mathrm{RL}}J_a^{\alpha}(A_n)\right]_{x=b} - {}_{\mathrm{RL}}J_a^{\alpha}(A_n), \\ &\qquad n \geq 0, \end{aligned} \tag{42}$$

provided that the resulting integrals exist.

Case 2. The nonlinear fractional BVP consists of (25) and the following mixed set of Neumann and Dirichlet boundary conditions:

$$\begin{aligned} u'(a) &= \beta, \\ u(b) &= \gamma. \end{aligned} \tag{43}$$

The boundary conditions (43) correspond to the case of $p = s = 0$ and $q = r = 1$ in (26) and (27). Thus we have $\Delta = -1 \neq 0$ and then (36) becomes

$$u(x) = \gamma - \beta(b-a) + \beta(x-a) - \left[{}_{\mathrm{RL}}J_a^{\alpha}(g(x))\right]_{x=b} + \left[{}_{\mathrm{RL}}J_a^{\alpha}(Nu)\right]_{x=b} + {}_{\mathrm{RL}}J_a^{\alpha}(g(x)) - {}_{\mathrm{RL}}J_a^{\alpha}(Nu), \quad (44)$$

where ${}_{\mathrm{RL}}J_a^{\alpha}(\cdot)$ and $[{}_{\mathrm{RL}}J_a^{\alpha}(\cdot)]_{x=b}$ are the Riemann-Liouville fractional integral operator of order α and the Riemann-Liouville fractional integral operator of order α evaluated at $x = b$, respectively.

Substituting the equations in (37) into (44), we obtain the following modified recursion scheme:

$$\begin{aligned} u_0(x) &= \gamma - \beta(b-a) + \beta(x-a) \\ &\quad - \left[{}_{\mathrm{RL}}J_a^{\alpha}(g(x))\right]_{x=b} + {}_{\mathrm{RL}}J_a^{\alpha}(g(x)), \\ u_{n+1}(x) &= \left[{}_{\mathrm{RL}}J_a^{\alpha}(A_n)\right]_{x=b} - {}_{\mathrm{RL}}J_a^{\alpha}(A_n), \quad n \geq 0, \end{aligned} \quad (45)$$

where we assume the resulting integrals exist.

Case 3. The nonlinear fractional BVP consists of (25) and the following mixed set of Robin and Neumann boundary conditions:

$$\begin{aligned} pu(a) + qu'(a) &= \beta, \\ su'(b) &= \gamma. \end{aligned} \quad (46)$$

The boundary conditions (46) correspond to the case of $r = 0$ and $s = 1$ in (27). Thus we have $\Delta = p \neq 0$ and then E (36) is reduced to

$$\begin{aligned} u(x) &= \frac{1}{p}\left[\beta - q\gamma + p\gamma(x-a) \right.\\ &\quad + (q - p(x-a))\left[{}_{\mathrm{RL}}J_a^{\alpha-1}(g(x))\right]_{x=b} \\ &\quad \left. + (-q + p(x-a))\left[{}_{\mathrm{RL}}J_a^{\alpha-1}(Nu)\right]_{x=b}\right] \\ &\quad + {}_{\mathrm{RL}}J_a^{\alpha}(g(x)) - {}_{\mathrm{RL}}J_a^{\alpha}(Nu), \end{aligned} \quad (47)$$

where ${}_{\mathrm{RL}}J_a^{\alpha}(\cdot)$ and $[{}_{\mathrm{RL}}J_a^{\alpha-1}(\cdot)]_{x=b}$ are the Riemann-Liouville fractional integral operator of order α and the Riemann-Liouville fractional integral operator of order $\alpha - 1$ evaluated at $x = b$, respectively.

Insertion of the equations in (37) into (47) gives the following modified recursion scheme:

$$\begin{aligned} u_0(x) &= \frac{1}{p}\left[\beta - q\gamma + p\gamma(x-a) \right.\\ &\quad \left. + (q - p(x-a))\left[{}_{\mathrm{RL}}J_a^{\alpha-1}(g(x))\right]_{x=b}\right] \\ &\quad + {}_{\mathrm{RL}}J_a^{\alpha}(g(x)), \\ u_{n+1}(x) &= \frac{1}{p}\left[(-q + p(x-a))\left[{}_{\mathrm{RL}}J_a^{\alpha-1}(A_n)\right]_{x=b}\right] \\ &\quad - {}_{\mathrm{RL}}J_a^{\alpha}(A_n), \quad n \geq 0, \end{aligned} \quad (48)$$

where we assume the resulting integrals exist.

3.2. The Fractional Order Differential Equation (23) with Separated Boundary Conditions. We consider a nonlinear fractional order two-point BVP consisting of the fractional order differential equation

$${}_{C}D_a^{\alpha}u + Nu = g(x), \quad a \leq x \leq b, \ 3 < \alpha \leq 4, \quad (49)$$

and the following separated boundary conditions:

$$\begin{aligned} u(a) &= \rho, \\ pu''(a) - qu'''(a) &= \beta, \end{aligned} \quad (50)$$

$$\begin{aligned} u(b) &= \sigma, \\ ru''(b) + su'''(b) &= \gamma, \end{aligned} \quad (51)$$

where $N(u(x))$ is an analytic nonlinear term, $\rho, \sigma, \beta, \gamma \in \mathbb{R}$, and $p, q, r, s \geq 0$ satisfy the following condition

$$ps + qr + pr(b-a) \neq 0. \quad (52)$$

In order to satisfy four necessary boundary conditions required for the problem and to have condition (52), it is necessary to have p, q not both zero, r, s not both zero, and p, r not both zero.

From (24) and the first condition in (50), we have that the solution of this BVP can be written in the form

$$\begin{aligned} u(x) &= u(a) + u'(a)(x-a) + \frac{u''(a)}{2}(x-a)^2 \\ &\quad + \frac{u'''(a)}{6}(x-a)^3 + {}_{\mathrm{RL}}J_a^{\alpha}(g(x)) \\ &\quad - {}_{\mathrm{RL}}J_a^{\alpha}(Nu), \\ u(x) &= \rho + u'(a)(x-a) + \frac{u''(a)}{2}(x-a)^2 \\ &\quad + \frac{u'''(a)}{6}(x-a)^3 + {}_{\mathrm{RL}}J_a^{\alpha}(g(x)) \\ &\quad - {}_{\mathrm{RL}}J_a^{\alpha}(Nu). \end{aligned} \quad (53)$$

We now apply the Duan-Rach modified ADM to the problem in (49). Using (53) and the first condition in (51), we evaluate $u(x)$ at $x = b$ to obtain

$$\begin{aligned} \sigma &= u(b) \\ &= \rho + u'(a)(b-a) + \frac{u''(a)}{2}(b-a)^2 \\ &\quad + \frac{u'''(a)}{6}(b-a)^3 + \left[{}_{\mathrm{RL}}J_a^{\alpha}(g(x))\right]_{x=b} \\ &\quad - \left[{}_{\mathrm{RL}}J_a^{\alpha}(Nu)\right]_{x=b}. \end{aligned} \quad (54)$$

Differentiating (53) three times and then using the properties that $(d^n/dx^n)\,{}_{\mathrm{RL}}J_a^{\alpha}(\cdot) = {}_{\mathrm{RL}}J_a^{\alpha-n}(\cdot)$, $n = 1, 2, 3$, to the resulting equations, we obtain

$$\begin{aligned} u'(x) &= u'(a) + u''(a)(x-a) + \frac{u'''(a)}{2}(x-a)^2 \\ &\quad + {}_{\mathrm{RL}}J_a^{\alpha-1}(g(x)) - {}_{\mathrm{RL}}J_a^{\alpha-1}(Nu), \\ u''(x) &= u''(a) + u'''(a)(x-a) + {}_{\mathrm{RL}}J_a^{\alpha-2}(g(x)) \\ &\quad - {}_{\mathrm{RL}}J_a^{\alpha-2}(Nu), \\ u'''(x) &= u'''(a) + {}_{\mathrm{RL}}J_a^{\alpha-3}(g(x)) - {}_{\mathrm{RL}}J_a^{\alpha-3}(Nu). \end{aligned} \tag{55}$$

Evaluating (55) at $x = b$, we have

$$\begin{aligned} u''(b) &= u''(a) + u'''(a)(b-a) \\ &\quad + \left[{}_{\mathrm{RL}}J_a^{\alpha-2}(g(x))\right]_{x=b} \\ &\quad - \left[{}_{\mathrm{RL}}J_a^{\alpha-2}(Nu)\right]_{x=b}, \\ u'''(b) &= u'''(a) + \left[{}_{\mathrm{RL}}J_a^{\alpha-3}(g(x))\right]_{x=b} \\ &\quad - \left[{}_{\mathrm{RL}}J_a^{\alpha-3}(Nu)\right]_{x=b}. \end{aligned} \tag{56}$$

Insertion of (56) into (51) gives the following relation:

$$\begin{aligned} &ru''(a) + (r(b-a)+s)u'''(a) \\ &= \gamma + r\left(\left[{}_{\mathrm{RL}}J_a^{\alpha-2}(Nu)\right]_{x=b} - \left[{}_{\mathrm{RL}}J_a^{\alpha-2}(g(x))\right]_{x=b}\right) \\ &\quad + s\left(\left[{}_{\mathrm{RL}}J_a^{\alpha-3}(Nu)\right]_{x=b} - \left[{}_{\mathrm{RL}}J_a^{\alpha-3}(g(x))\right]_{x=b}\right). \end{aligned} \tag{57}$$

It is possible to solve the system consisting of the second condition of (50) and (57) for the two remaining undetermined coefficients $u''(a)$ and $u'''(a)$ if the determinant of the coefficient matrix denoted by Δ is not zero, that is, if

$$\Delta = \begin{vmatrix} p & -q \\ r & r(b-a)+s \end{vmatrix} = ps + qr + pr(b-a) \neq 0. \tag{58}$$

For $\Delta \neq 0$, the values of $u''(a)$ and $u'''(a)$ can be expressed in terms of the specified values of the system parameters $a, b, \beta, \gamma, p, q, r$, and s as follows:

$$\begin{aligned} u''(a) &= \frac{1}{\Delta}\Big[r\beta(b-a) + s\beta + q\gamma \\ &\quad + qr\left(\left[{}_{\mathrm{RL}}J_a^{\alpha-2}(Nu)\right]_{x=b} - \left[{}_{\mathrm{RL}}J_a^{\alpha-2}(g(x))\right]_{x=b}\right) \\ &\quad + qs\left(\left[{}_{\mathrm{RL}}J_a^{\alpha-3}(Nu)\right]_{x=b} - \left[{}_{\mathrm{RL}}J_a^{\alpha-3}(g(x))\right]_{x=b}\right)\Big], \\ u'''(a) &= \frac{1}{\Delta}\Big[p\gamma - r\beta \\ &\quad + pr\left(\left[{}_{\mathrm{RL}}J_a^{\alpha-2}(Nu)\right]_{x=b} - \left[{}_{\mathrm{RL}}J_a^{\alpha-2}(g(x))\right]_{x=b}\right) \\ &\quad + ps\left(\left[{}_{\mathrm{RL}}J_a^{\alpha-3}(Nu)\right]_{x=b} - \left[{}_{\mathrm{RL}}J_a^{\alpha-3}(g(x))\right]_{x=b}\right)\Big]. \end{aligned} \tag{59}$$

Substituting (59) for $u''(a)$ and $u'''(a)$ into (54) and then solving the resulting equation for $u'(a)$, we obtain the value of $u'(a)$ as follows:

$$\begin{aligned} u'(a) &= \frac{1}{b-a}\Big[\sigma - \rho - \left[{}_{\mathrm{RL}}J_a^{\alpha}(g(x))\right]_{x=b} \\ &\quad + \left[{}_{\mathrm{RL}}J_a^{\alpha}(Nu)\right]_{x=b} - \frac{(b-a)^2}{2\Delta}\Big(r\beta(b-a) + s\beta \\ &\quad + q\gamma + qr\Big(\left[{}_{\mathrm{RL}}J_a^{\alpha-2}(Nu)\right]_{x=b} \\ &\quad - \left[{}_{\mathrm{RL}}J_a^{\alpha-2}(g(x))\right]_{x=b}\Big) + qs\Big(\left[{}_{\mathrm{RL}}J_a^{\alpha-3}(Nu)\right]_{x=b} \\ &\quad - \left[{}_{\mathrm{RL}}J_a^{\alpha-3}(g(x))\right]_{x=b}\Big)\Big) + \frac{(b-a)^3}{6\Delta}\Big(r\beta - p\gamma \\ &\quad + pr\left(\left[{}_{\mathrm{RL}}J_a^{\alpha-2}(g(x))\right]_{x=b} - \left[{}_{\mathrm{RL}}J_a^{\alpha-2}(Nu)\right]_{x=b}\right) \\ &\quad + ps\Big(\left[{}_{\mathrm{RL}}J_a^{\alpha-3}(g(x))\right]_{x=b} \\ &\quad - \left[{}_{\mathrm{RL}}J_a^{\alpha-3}(Nu)\right]_{x=b}\Big)\Big)\Big]. \end{aligned} \tag{60}$$

For the computational convenience, we set

$$\begin{aligned} \Delta &= ps + qr + pr(b-a), \\ \kappa &= 2r\beta + p\gamma, \\ \lambda &= s\beta + q\gamma, \\ \mu &= 2br\beta + bp\gamma, \\ \nu &= p\gamma - r\beta, \\ \omega &= 3q + bp - 2ap. \end{aligned} \tag{61}$$

Then we substitute (60) and (59) into (53) to obtain the following equivalent nonlinear Fredholm-Volterra integral equation:

$$\begin{aligned} u(x) &= \rho - \frac{(x-a)}{6(b-a)\Delta}\Big(\kappa\left(b^3 - a^3\right) + 3\lambda b^2 \\ &\quad + 3(\lambda+\mu)a^2 - 3(2\lambda+\mu)ab + 6(\rho-\sigma)\Delta\Big) \\ &\quad + \frac{(x-a)^2}{2\Delta}\left(r\beta(b-a) + \lambda\right) + \frac{(x-a)^3}{6\Delta}\nu \\ &\quad - \frac{(x-a)(x-b)(px+\omega)}{6\Delta}\Big(s\left[{}_{\mathrm{RL}}J_a^{\alpha-3}(g(x))\right]_{x=b} \\ &\quad + r\left[{}_{\mathrm{RL}}J_a^{\alpha-2}(g(x))\right]_{x=b}\Big) \\ &\quad - \frac{(x-a)}{(b-a)}\left[{}_{\mathrm{RL}}J_a^{\alpha}(g(x))\right]_{x=b} + {}_{\mathrm{RL}}J_a^{\alpha}(g(x)) \\ &\quad + \frac{(x-a)(x-b)(px+\omega)}{6\Delta}\Big(s\left[{}_{\mathrm{RL}}J_a^{\alpha-3}(Nu)\right]_{x=b} \end{aligned}$$

$$+ r\left[{}_{\mathrm{RL}}J_a^{\alpha-2}(Nu)\right]_{x=b}\Big) + \frac{(x-a)}{(b-a)}\left[{}_{\mathrm{RL}}J_a^{\alpha}(Nu)\right]_{x=b} - {}_{\mathrm{RL}}J_a^{\alpha}(Nu), \tag{62}$$

where ${}_{\mathrm{RL}}J_a^{\alpha}(\cdot)$ is the Riemann-Liouville fractional integral operator of order α and where $[{}_{\mathrm{RL}}J_a^{\alpha}(\cdot)]_{x=b}$, $[{}_{\mathrm{RL}}J_a^{\alpha-2}(\cdot)]_{x=b}$, and $[{}_{\mathrm{RL}}J_a^{\alpha-3}(\cdot)]_{x=b}$ are the operators of orders α, $\alpha-2$, and $\alpha-3$ evaluated at $x=b$.

Substituting the equations in (37) into (62), we can determine the solution components from the following modified recursion scheme:

$$\begin{aligned}
u_0(x) &= \rho - \frac{(x-a)}{6(b-a)\Delta}\Big(\kappa\left(b^3-a^3\right) + 3\lambda b^2 \\
&\quad + 3(\lambda+\mu)a^2 - 3(2\lambda+\mu)ab + 6(\rho-\sigma)\Delta\Big) \\
&\quad + \frac{(x-a)^2}{2\Delta}\left(r\beta(b-a)+\lambda\right) + \frac{(x-a)^3}{6\Delta}\nu \\
&\quad - \frac{(x-a)(x-b)(px+\omega)}{6\Delta}\Big(s\left[{}_{\mathrm{RL}}J_a^{\alpha-3}(g(x))\right]_{x=b} \\
&\quad + r\left[{}_{\mathrm{RL}}J_a^{\alpha-2}(g(x))\right]_{x=b}\Big) \\
&\quad - \frac{(x-a)}{(b-a)}\left[{}_{\mathrm{RL}}J_a^{\alpha}(g(x))\right]_{x=b} + {}_{\mathrm{RL}}J_a^{\alpha}(g(x)), \\
u_{n+1}(x) &= \frac{(x-a)(x-b)(px+\omega)}{6\Delta}\Big(s\left[{}_{\mathrm{RL}}J_a^{\alpha-3}(A_n)\right]_{x=b} \\
&\quad + r\left[{}_{\mathrm{RL}}J_a^{\alpha-2}(A_n)\right]_{x=b}\Big) + \frac{(x-a)}{(b-a)}\left[{}_{\mathrm{RL}}J_a^{\alpha}(A_n)\right]_{x=b} \\
&\quad - \left[{}_{\mathrm{RL}}J_a^{\alpha}(A_n)\right], \quad n \geq 0,
\end{aligned} \tag{63}$$

where the resulting integrals are assumed to exist. The n-term approximation of the solution to the BVP can be obtained using (16). In addition, other techniques such as partitioning initial terms into two appropriate terms [39, 43, 44] or using Duan's convergence parameter [40–42] can be incorporated, if necessary, into the recursion scheme (63) for solving the BVP described in (49)–(51).

Using the general formulas derived in (62) and (63), we can derive the equivalent nonlinear Fredholm-Volterra integral equations and their associated recursion schemes for (49) for the special cases of the boundary conditions in (50) and (51). The results are as follows.

Case 1. The nonlinear fractional BVP consists of (49) and the following two-point boundary conditions:

$$\begin{aligned}
u(a) &= \rho, \\
u''(a) &= \beta, \\
u(b) &= \sigma, \\
u''(b) &= \gamma.
\end{aligned} \tag{64}$$

The derivative boundary conditions (64) correspond to the case of $p=r=1$ and $q=s=0$ in (50) and (51). Thus (62) is then reduced to

$$\begin{aligned}
u(x) &= \rho - \frac{(x-a)}{6(b-a)\Delta}\Big(\kappa\left(b^3-a^3\right) + 3\mu a^2 - 3\mu ab \\
&\quad + 6(\rho-\sigma)\Delta\Big) + \frac{\beta}{2}(x-a)^2 + \frac{(x-a)^3}{6\Delta}\nu \\
&\quad - \frac{(x-a)(x-b)(x+\omega)}{6\Delta}\left[{}_{\mathrm{RL}}J_a^{\alpha-2}(g(x))\right]_{x=b} \\
&\quad - \frac{(x-a)}{(b-a)}\left[{}_{\mathrm{RL}}J_a^{\alpha}(g(x))\right]_{x=b} + {}_{\mathrm{RL}}J_a^{\alpha}(g(x)) \\
&\quad + \frac{(x-a)(x-b)(x+\omega)}{6\Delta}\left[{}_{\mathrm{RL}}J_a^{\alpha-2}(Nu)\right]_{x=b} \\
&\quad + \frac{(x-a)}{(b-a)}\left[{}_{\mathrm{RL}}J_a^{\alpha}(Nu)\right]_{x=b} - {}_{\mathrm{RL}}J_a^{\alpha}(Nu),
\end{aligned} \tag{65}$$

where $\Delta = b-a \neq 0$, $\kappa = 2\beta+\gamma$, $\lambda = 0$, $\mu = 2b\beta+b\gamma$, $\nu = \gamma-\beta$, and $\omega = b-2a$ and where ${}_{\mathrm{RL}}J_a^{\alpha}(\cdot)$ is the Riemann-Liouville fractional integral operator of order α and $[{}_{\mathrm{RL}}J_a^{\alpha}(\cdot)]_{x=b}$ and $[{}_{\mathrm{RL}}J_a^{\alpha-2}(\cdot)]_{x=b}$ are the operators of orders α and $\alpha-2$ evaluated at $x=b$.

Insertion of (37) into (65) gives the following modified recursion scheme for solution components:

$$\begin{aligned}
u_0(x) &= \rho - \frac{(x-a)}{6(b-a)\Delta}\Big(\kappa\left(b^3-a^3\right) + 3\mu a^2 - 3\mu ab \\
&\quad + 6(\rho-\sigma)\Delta\Big) + \frac{\beta}{2}(x-a)^2 + \frac{(x-a)^3}{6\Delta}\nu \\
&\quad - \frac{(x-a)(x-b)(x+\omega)}{6\Delta}\left[{}_{\mathrm{RL}}J_a^{\alpha-2}(g(x))\right]_{x=b} \\
&\quad - \frac{(x-a)}{(b-a)}\left[{}_{\mathrm{RL}}J_a^{\alpha}(g(x))\right]_{x=b} + {}_{\mathrm{RL}}J_a^{\alpha}(g(x)), \\
u_{n+1}(x) &= \frac{(x-a)(x-b)(x+\omega)}{6\Delta}\left[{}_{\mathrm{RL}}J_a^{\alpha-2}(A_n)\right]_{x=b} \\
&\quad + \frac{(x-a)}{(b-a)}\left[{}_{\mathrm{RL}}J_a^{\alpha}(A_n)\right]_{x=b} - {}_{\mathrm{RL}}J_a^{\alpha}(A_n), \quad n \geq 0,
\end{aligned} \tag{66}$$

where we assume the resulting integrals exist.

Case 2. The nonlinear fractional BVP consists of (49) and the following two-point boundary conditions:

$$\begin{aligned}
u(a) &= \rho, \\
-u'''(a) &= \beta, \\
u(b) &= \sigma, \\
u''(b) &= \gamma.
\end{aligned} \tag{67}$$

The derivative boundary conditions (67) correspond to the case of $p = s = 0$ and $q = r = 1$ in (50) and (51). Thus we have $\Delta = 1 \neq 0$ and $\omega = 3$ and then (62) is reduced to

$$\begin{aligned} u(x) = {} & \rho - \frac{(x-a)}{6(b-a)}\left(\kappa\left(b^3 - a^3\right) + 3\lambda b^2 \right. \\ & \left. + 3(\lambda+\mu)a^2 - 3(2\lambda+\mu)ab + 6(\rho-\sigma)\right) \\ & + \frac{(x-a)^2}{2}(\beta(b-a)+\lambda) + \frac{(x-a)^3}{6}\nu \\ & - \frac{(x-a)(x-b)}{2}\left[{}_{\mathrm{RL}}J_a^{\alpha-2}(g(x))\right]_{x=b} \\ & - \frac{(x-a)}{(b-a)}\left[{}_{\mathrm{RL}}J_a^{\alpha}(g(x))\right]_{x=b} + {}_{\mathrm{RL}}J_a^{\alpha}(g(x)) \\ & + \frac{(x-a)(x-b)}{2}\left[{}_{\mathrm{RL}}J_a^{\alpha-2}(Nu)\right]_{x=b} \\ & + \frac{(x-a)}{(b-a)}\left[{}_{\mathrm{RL}}J_a^{\alpha}(Nu)\right]_{x=b} - {}_{\mathrm{RL}}J_a^{\alpha}(Nu), \end{aligned} \tag{68}$$

where $\kappa = 2\beta$, $\lambda = \gamma$, $\mu = 2b\beta$, and $\nu = -\beta$ and where ${}_{\mathrm{RL}}J_a^{\alpha}(\cdot)$ is the Riemann-Liouville fractional integral operator of order α and where $[{}_{\mathrm{RL}}J_a^{\alpha}(\cdot)]_{x=b}$ and $[{}_{\mathrm{RL}}J_a^{\alpha-2}(\cdot)]_{x=b}$ are the operators of orders α and $\alpha-2$ evaluated at $x = b$.

Insertion of (37) into (68) gives the following modified recursion scheme for solution components:

$$\begin{aligned} u_0(x) = {} & \rho - \frac{(x-a)}{6(b-a)}\left(\kappa\left(b^3 - a^3\right) + 3\lambda b^2 \right. \\ & \left. + 3(\lambda+\mu)a^2 - 3(2\lambda+\mu)ab + 6(\rho-\sigma)\right) \\ & + \frac{(x-a)^2}{2}(\beta(b-a)+\lambda) + \frac{(x-a)^3}{6}\nu \\ & - \frac{(x-a)(x-b)}{2}\left[{}_{\mathrm{RL}}J_a^{\alpha-2}(g(x))\right]_{x=b} \\ & - \frac{(x-a)}{(b-a)}\left[{}_{\mathrm{RL}}J_a^{\alpha}(g(x))\right]_{x=b} + {}_{\mathrm{RL}}J_a^{\alpha}(g(x)), \\ u_{n+1}(x) = {} & \frac{(x-a)(x-b)}{2}\left[{}_{\mathrm{RL}}J_a^{\alpha-2}(A_n)\right]_{x=b} \\ & + \frac{(x-a)}{(b-a)}\left[{}_{\mathrm{RL}}J_a^{\alpha}(A_n)\right]_{x=b} - {}_{\mathrm{RL}}J_a^{\alpha}(A_n), \quad n \geq 0, \end{aligned} \tag{69}$$

where we assume the resulting integrals exist.

4. Numerical Examples

In this section, we demonstrate a use of the proposed recursion schemes in Section 3 derived from the Duan-Rach modified decomposition method to analytically and numerically solve nonlinear fractional BVPs. Several nonlinear fractional BVPs presented in this section correspond to the problems and their formulas described in Section 3 and some of these problems include physical and engineering problems such as problems of the Bratu type, a problem of the periodic base temperature in convective longitudinal fins, and an elastic beam problem. Numerical results obtained by the method are demonstrated graphically. Moreover, if the presented nonlinear fractional BVPs have their exact solutions then we will compute their corresponding maximal errors. Otherwise, we will investigate the error remainder function for the remaining problems.

In general, we consider a nonlinear fractional BVP: ${}_{C}D_a^{\alpha}u(x) + Nu = g(x)$, $a \leq x \leq b$, where α is a fractional order of the equation imposed by some boundary conditions. If the exact solution $u^*(x)$ of the problem is known then we can examine the convergence of the n-term approximation $\varphi_n(x) = \sum_{k=0}^{n-1} u_k(x)$ from the error functions expressed as

$$E_n(x) = \varphi_n(x) - u^*(x) \tag{70}$$

and the maximal errors defined as

$$\mathrm{ME}_n = \max_{a \leq x \leq b} \left|E_n(x)\right|. \tag{71}$$

For each value of n, we can calculate ME_n using the MATHEMATICA command "NMaximize" over an interval of interest. Then the logarithmic plot of these values of ME_n can be made using the MATHEMATICA command "ListLogPlot"; however, if the exact solution $u^*(x)$ of such a problem is unknown, we compute the error remainder functions defined as

$$\mathrm{ER}_n(x) = {}_{C}D_a^{\alpha}\varphi_n(x) + N\varphi_n(x) - g(x), \quad a \leq x \leq b. \tag{72}$$

We observe that ER_n is the indicator for measuring how well the approximation $\varphi_n(x)$ satisfies the original nonlinear fractional differential equation.

Example 1. Consider the Bratu-type fractional BVP which is modified from the Bratu-type second-order equation in [12, 26, 45, 46] as follows:

$$\begin{gathered} {}_{C}D_0^{\alpha}u + e^u = 0, \quad 0 \leq x \leq 1, \ 1 < \alpha \leq 2, \\ u(0) = u(1) = 0. \end{gathered} \tag{73}$$

For $\alpha = 2$, it can be easily verified that the exact solution of the BVP is

$$\begin{gathered} u^*(x) = -2\ln\frac{\cosh((x-1/2)(C/2))}{\cosh(C/4)}, \\ \text{where } C \text{ satisfies } C = \sqrt{2}\cosh\left(\frac{C}{4}\right). \end{gathered} \tag{74}$$

Fundamentally, we decompose the solution u and the nonlinearity $Nu = e^u$ as $u(x) = \sum_{n=0}^{\infty} u_n(x)$ and $Nu = \sum_{n=0}^{\infty} A_n(x)$, where u_n are the solution components and A_n are the Adomian polynomials. We will show computational details for the proposed method, that is, the Duan-Rach modified

ADM. We here list the first four Adomian polynomials for the nonlinear term Nu as follows:

$$\begin{aligned} A_0 &= e^{u_0}, \\ A_1 &= e^{u_0} u_1, \\ A_2 &= \frac{1}{2} e^{u_0} \left(u_1^2 + 2u_2\right), \\ A_3 &= \frac{1}{6} e^{u_0} \left(u_1^3 + 6u_1 u_2 + 6u_3\right). \end{aligned} \tag{75}$$

The Duan-Rach Modified ADM. We can apply the formulas in (41) and (42) in Section 3.1 to this BVP. We have $a = 0$, $b = 1$, $p = 1$, $q = 0$, $r = 1$, $s = 0$, $\beta = 0$, $\gamma = 0$, and $\Delta = 1$. Equation (41) turns out to be

$$u(x) = x\left[{}_{\mathrm{RL}}J_0^\alpha (Nu)\right]_{x=1} - {}_{\mathrm{RL}}J_0^\alpha (Nu), \tag{76}$$

where $Nu = e^u$. Using (42) and Duan's convergence parameter technique, we obtain the following parametrized recursion scheme:

$$\begin{aligned} u_0(x) &= c, \\ u_1(x) &= -c + x\left[{}_{\mathrm{RL}}J_0^\alpha (A_0)\right]_{x=1} - {}_{\mathrm{RL}}J_0^\alpha (A_0), \\ u_{n+1}(x) &= x\left[{}_{\mathrm{RL}}J_0^\alpha (A_n)\right]_{x=1} - {}_{\mathrm{RL}}J_0^\alpha (A_n), \quad n \geq 1. \end{aligned} \tag{77}$$

Since the expressions of the solution components $u_4(x)$, $u_5(x), u_6(x), \ldots$ are very long, we show only the solution components $u_1(x)$, $u_2(x)$, and $u_3(x)$ using the Adomian polynomials A_0, A_1, and A_2 in (75) as follows:

$$\begin{aligned} u_1(x) &= -c + \frac{e^c}{\Gamma(\alpha+1)} x - \frac{e^c}{\Gamma(\alpha+1)} x^\alpha, \\ u_2(x) &= \frac{4^{-\alpha} e^c \left(4^\alpha \Gamma(\alpha+1/2)\left(e^c - c\Gamma(\alpha+2)\right) - \sqrt{\pi} e^c \Gamma(\alpha+2)\right)}{\alpha\Gamma(\alpha)\Gamma(\alpha+1/2)\Gamma(\alpha+2)} \\ &\quad \cdot x + \frac{ce^c}{\Gamma(\alpha+1)} x^\alpha - \frac{e^{2c}}{\Gamma(\alpha+1)\Gamma(\alpha+2)} x^{\alpha+1} + \frac{e^{2c}}{\Gamma(2\alpha+1)} \\ &\quad \cdot x^{2\alpha}, \\ u_3(x) &= \frac{e^c}{2}\left(2e^c\left(\frac{2c}{\Gamma(2\alpha+1)}\right.\right. \\ &\quad + \frac{(\alpha+1)\left((2\alpha+3)e^c - 2c\Gamma(\alpha+3)\right)}{\Gamma^2(\alpha+2)\Gamma(\alpha+3)} \\ &\quad \left.\left. - \frac{(\alpha(\alpha+5)+3)e^c}{\Gamma(\alpha+2)\Gamma(2\alpha+2)} + \frac{e^c}{\Gamma(3\alpha+1)}\right) + \frac{c^2}{\Gamma(\alpha+1)}\right. \\ &\quad \left. + \frac{4^\alpha e^{2c}\Gamma(\alpha+1/2)}{\sqrt{\pi}\Gamma(3\alpha+1)\Gamma(\alpha+1)}\right) x - \frac{c^2 e^c}{2\Gamma(\alpha+1)} x^\alpha \\ &\quad + \frac{e^{2c}}{\Gamma^3(\alpha+2)}\left(-(\alpha+1)e^c + \frac{e^c\Gamma^2(\alpha+2)}{\Gamma(2\alpha+1)} + 2(\alpha+1)c\Gamma(\alpha\right. \\ &\quad \left. + 2)\right) x^{\alpha+1} - \frac{e^{3c}}{\Gamma^2(\alpha+1)\Gamma(\alpha+3)} x^{\alpha+2} - \frac{2ce^{2c}}{\Gamma(2\alpha+1)} x^{2\alpha} \\ &\quad + \frac{(\alpha+2)e^{3c}}{\Gamma(\alpha+1)\Gamma(2\alpha+2)} x^{2\alpha+1} \\ &\quad - \frac{1}{2\Gamma(\alpha+1)\Gamma(3\alpha+1)}\left(2\alpha e^{3c}\Gamma(\alpha) + \frac{4^\alpha e^{3c}\Gamma(\alpha+1/2)}{\sqrt{\pi}}\right) \\ &\quad \cdot x^{3\alpha}. \end{aligned} \tag{78}$$

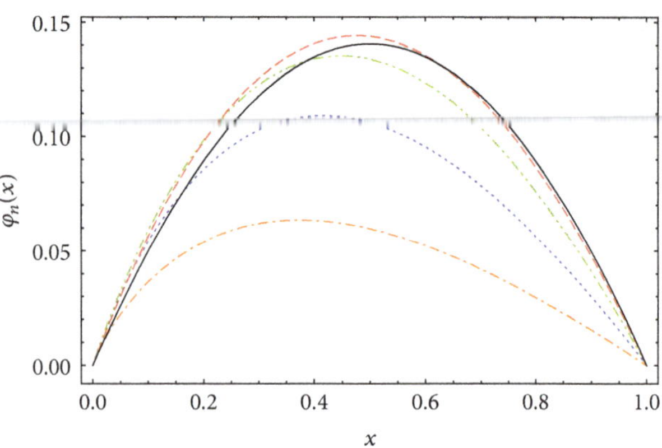

FIGURE 1: $\varphi_n(x)$ when $n = 10$ obtained by the Duan-Rach modified ADM for $\alpha = 1.2$ (dot-dash), $\alpha = 1.4$ (dot line), $\alpha = 1.6$ (dot-dot-dash), $\alpha = 1.8$ (dash line), and $\alpha = 2$ (solid line).

Here we select $c = 0.1$ for the above recursion scheme for the following values of $\alpha = 1.2, 1.4, 1.6, 1.8, 2$. We can compute the approximate solution $\varphi_n(x) = \sum_{k=0}^{n-1} u_k(x)$ from the solution components $u_0(x), u_1(x), u_2(x), \ldots, u_{n-1}(x)$. In particular, we compute $\varphi_{10}(x)$ for each selected value of α. Figure 1 shows curves of the approximation solutions $\varphi_{10}(x)$ for $\alpha = 1.2, 1.4, 1.6, 1.8, 2$.

For each specified value of $\alpha \neq 2$, since the exact solution to this BVP is not known, we calculate the corresponding error remainder function $\mathrm{ER}_n(x)$ for $n = 10$. Figure 2 displays the graphs of $\mathrm{ER}_{10}(x)$ for the specified values of α. We can deduce from the graphs in Figure 2 that the approximations $\varphi_{10}(x)$ obtained by this method give remarkable accuracy, as expected, since when n is sufficiently large the magnitude of each function $\mathrm{ER}_n(x)$ approaches zero.

For $\alpha = 2$, it is possible to compute both the error function $E_n(x)$ and the maximal errors ME_n; however, we only compute the values of ME_n for $n = 2, 3, \ldots, 9$ listed in Table 1. We show the logarithmic plots of ME_n versus n for $n = 2, 3, \ldots, 9$ in Figure 3. We can observe in Figure 3 that all of the data points after $n = 2$ lie almost on a straight line which demonstrates that the maximal errors ME_n are reduced approximately at an exponential rate.

Example 2. Consider the following nonlinear fractional order BVP which is developed from the second-order

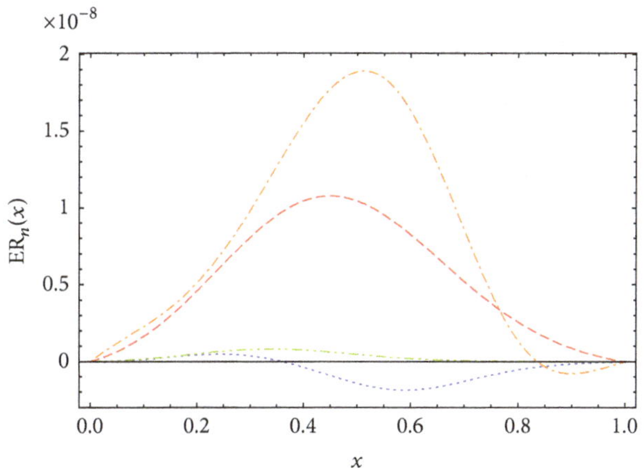

FIGURE 2: $ER_n(x)$ when $n = 10$ obtained by the Duan-Rach modified ADM for $\alpha = 1.2$ (dot-dash), $\alpha = 1.4$ (dot line), $\alpha = 1.6$ (dot-dot-dash), and $\alpha = 1.8$ (dash line).

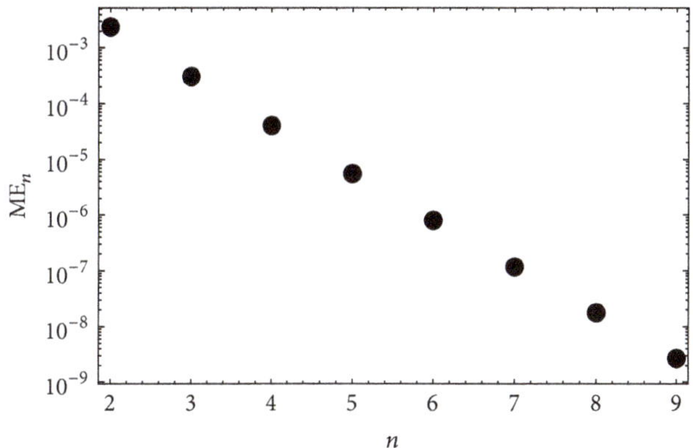

FIGURE 3: Logarithmic plots of the maximal errors ME_n versus n for $n = 2$ through 9 obtained by the Duan-Rach modified ADM.

homogeneous partial differential equation of the engineering model in [47] as follows:

$$\frac{\partial^\alpha u(x,t)}{\partial x^\alpha} + \frac{\varepsilon}{1+\varepsilon u(x,t)}\left(\frac{\partial u(x,t)}{\partial x}\right)^2 - K^2\frac{u(x,t)}{1+\varepsilon u(x,t)} - \left(\frac{1}{1+\varepsilon u(x,t)}\right)\frac{\partial u(x,t)}{\partial t} = 0, \quad (79)$$

where the notation $\partial^\alpha/\partial x^\alpha$ in (79) represents a fractional partial derivative with respect to the space x in the Caputo sense with the fractional order $\alpha \in (1,2]$ and where $u(x,t)$ has the domain of definition $x \in [0,1]$ and $t \in [0,\infty)$. The physical variables u, x, t are the dimensionless temperature, distance, and time, respectively. For (79), the following mixed set of homogeneous Neumann and inhomogeneous Dirichlet boundary conditions is given as

$$u_x(0,t) = 0, \qquad u(1,t) = 1 + S\cos(Bt). \quad (80)$$

TABLE 1: For $c = 0.1$ and $\alpha = 2$, maximal errors ME_n for $n = 2, 3, \ldots, 9$ obtained by the Duan-Rach modified ADM.

n	ME_n
2	0.00239285
3	0.000303804
4	0.0000401907
5	5.53739×10^{-6}
6	7.91736×10^{-7}
7	1.16592×10^{-7}
8	1.75801×10^{-8}
9	2.70207×10^{-9}

The above conditions consist of a sinusoidally varying boundary value. The parameters ε, K, S, and B in (79) and (80) represent thermal conductivity parameter, fin parameter, amplitude of oscillation, and frequency of oscillation, respectively. The BVP in (79) and (80) describes physically the periodic base temperature in convective longitudinal fins.

The BVP in (79) and (80) has the same form as the BVP defined in (25) and (43) in Section 3.1. Thus, we have the nonlinearity Nu as follows:

$$Nu = \frac{\varepsilon}{1+\varepsilon u(x,t)}(u_x(x,t))^2 - K^2\frac{u(x,t)}{1+\varepsilon u(x,t)} - \frac{u_t(x,t)}{1+\varepsilon u(x,t)} \quad (81)$$

and the source term $g(x,t) = 0$. We decompose the solution u and the nonlinearity Nu as $u(x,t) = \sum_{n=0}^{\infty} u_n(x,t)$ and $Nu(x,t) = \sum_{n=0}^{\infty} A_n(x,t)$, where u_n are the solution components and A_n are the Adomian polynomials. Using (81), we list the first two Adomian polynomials for this nonlinear term as follows:

$$\begin{aligned} A_0 &= -\frac{K^2 u_0}{\varepsilon u_0 + 1} + \frac{\varepsilon\left(u_0^{(1,0)}\right)^2}{\varepsilon u_0 + 1} - \frac{u_0^{(0,1)}}{\varepsilon u_0 + 1}, \\ A_1 &= \frac{\varepsilon K^2 u_0 u_1}{(\varepsilon u_0 + 1)^2} - \frac{K^2 u_1}{\varepsilon u_0 + 1} - \frac{\varepsilon^2 u_1\left(u_0^{(1,0)}\right)^2}{(\varepsilon u_0 + 1)^2} + \frac{\varepsilon u_1 u_0^{(0,1)}}{(\varepsilon u_0 + 1)^2} + \frac{2\varepsilon u_0^{(1,0)} u_1^{(1,0)}}{\varepsilon u_0 + 1} - \frac{u_1^{(0,1)}}{\varepsilon u_0 + 1}, \end{aligned} \quad (82)$$

where the superscript $(1,0)$ denotes the differentiation with respect to x, while the superscript $(0,1)$ denotes the differentiation with respect to t and so forth.

We employ the Duan-Rach modified ADM for the above BVP to which (44)-(45) are applied. Comparing with (43), we have $a = 0$, $b = 1$, $\beta = 0$, $\gamma = 1 + S\cos(Bt)$, $p = s = 0$, $q = r = 1$, and $\Delta = -1$. Then (44) becomes

$$u(x,t) = 1 + S\cos(Bt) + \left[{}_{RL}J_0^\alpha(Nu)\right]_{x=1} - {}_{RL}J_0^\alpha(Nu). \quad (83)$$

Using (45), we obtain the following recursion scheme:

$$u_0(x,t) = 1 + S\cos(Bt),$$
$$u_{n+1}(x,t) = \left[{}_{\mathrm{RL}}J_0^\alpha(A_n)\right]_{x=1} - {}_{\mathrm{RL}}J_0^\alpha(A_n). \quad (84)$$

Since the expressions of the solution components $u_3(x)$, $u_4(x), u_5(x), \ldots$ are complicated, we show only the solution components $u_0(x)$, $u_1(x)$, and $u_2(x)$ using the Adomian polynomials A_0 and A_1 in (82) as follows:

$$u_0(x,t) = 1 + S\cos(Bt),$$
$$u_1(x,t) = \frac{(x^\alpha - 1)\left(K^2 S\cos(Bt) - BS\sin(Bt) + K^2\right)}{\Gamma(\alpha+1)(\varepsilon S\cos(Bt) + \varepsilon + 1)},$$
$$u_2(x,t) = \frac{(x^\alpha - 1)\left(\sqrt{\pi}\Gamma(\alpha+1)(x^\alpha+1) - 4^\alpha\Gamma(\alpha+1/2)\right)}{2\sqrt{\pi}\Gamma(\alpha+1)\Gamma(2\alpha+1)(\varepsilon S\cos(Bt)+\varepsilon+1)^3}\left(-2S\left(B^2(\varepsilon+1) - K^4\right)\cos(Bt) - 3B^2\varepsilon S^2 + 2K^4\right.$$
$$\left. + BS\left(2K^2\sin(Bt)(\varepsilon S\cos(Bt) + \varepsilon - 2) + B\varepsilon S\cos(2Bt)\right)\right). \quad (85)$$

We can compute the approximate solution $\varphi_n(x,t) = \sum_{k=0}^{n-1} u_k(x,t)$ from the solution components $u_0(x,t), u_1(x,t)$, $u_2(x,t), \ldots, u_{n-1}(x,t)$. Since the time consumed for computing $\varphi_n(x,t)$ is long, we choose to calculate the approximate solution $\varphi_5(x,t)$. Here we take $\varepsilon = 0.2$, $K = 0.5$, $S = 0.1$, and $B = 1$ which are used to plot the approximate solutions and the error surfaces for $0 \le x \le 1$ and $0 \le t \le 4\pi$. The three-dimensional approximate solutions $\varphi_5(x,t)$ for the values of $\alpha = 1.2, 1.4, 1.6, 1.8, 2$ are plotted in Figures 4(a) and 4(b) with various orientations. Figures 4(c)–4(e) show the cross sections of the approximate solutions $\varphi_5(x,t)$ at $x = 0.2$, $x = 0.5$, and $x = 0.8$, respectively, for $\alpha = 1.2, 1.4, 1.6, 1.8, 2$.

Since we do not yet know what the exact solution is for this BVP for each specified value of α, the corresponding error remainder function

$$\mathrm{ER}_n(x,t) = \frac{\partial^\alpha \varphi_n(x,t)}{\partial x^\alpha} + \frac{\varepsilon}{1+\varepsilon\varphi_n(x,t)}\left(\frac{\partial\varphi_n(x,t)}{\partial x}\right)^2 - K^2\frac{\varphi_n(x,t)}{1+\varepsilon\varphi_n(x,t)} - \left(\frac{1}{1+\varepsilon\varphi_n(x,t)}\right)\frac{\partial\varphi_n(x,t)}{\partial t} \quad (86)$$

is computed. Here we choose to compute $\mathrm{ER}_n(x,t)$ for only $\alpha = 1.8$ by varying $n = 2,3,4,5$. Figures 5(a)–5(d) display the surfaces of the absolute error remainder functions $|\mathrm{ER}_n(x,t)|$ for $n = 2,3,4,5$. We can deduce from Figures 5(a)–5(d) that for $\alpha = 1.8$ the approximations $\varphi_n(x,t)$ obtained by this method provide the greater accuracy for larger n; that is, the maximum values of $|\mathrm{ER}_n(x,t)|$ decrease significantly as n increases.

Example 3. The following nonlinear fractional BVP is modified from the nonlinear second-order BVP in [27] with the sum of an exponential nonlinearity in the solution and a quadratic nonlinearity in the derivative of the solution and subject to a set of Robin boundary conditions. The modified BVP can be expressed as follows:

$${}_C D_0^\alpha u(x) = -\frac{1}{8}\left(e^{-2u} + 4\left(u'\right)^2\right), \quad 0 \le x \le 1,\ 1 < \alpha \le 2, \quad (87)$$

subject to the following set of Robin boundary conditions:

$$u(0) - 2u'(0) = -1,$$
$$u(1) + 2u'(1) = \frac{2}{3} + \ln\left(\frac{3}{2}\right). \quad (88)$$

Here we can rewrite (87) as

$${}_C D_0^\alpha u(x) + \frac{1}{8}\left(e^{-2u} + 4\left(u'\right)^2\right) = 0, \quad (89)$$

with the nonlinearity and the source term as follows:

$$Nu = \frac{1}{8}\left(e^{-2u} + 4\left(u'\right)^2\right),$$
$$g(x) = 0. \quad (90)$$

For $\alpha = 2$, it can be easily verified that the exact solution of the BVP is $u^*(x) = \ln((2+x)/2)$; however, the exact solution to this BVP is not known for $1 < \alpha < 2$. Next we decompose the solution u and the nonlinearity Nu as $u(x) = \sum_{n=0}^\infty u_n(x)$ and $Nu = \sum_{n=0}^\infty A_n(x)$, where u_n are the solution components and A_n are the Adomian polynomials. We here list the first four Adomian polynomials for the nonlinear term as follows:

$$A_0 = \frac{1}{8}\left(e^{-2u_0} + 4\left(u_0'\right)^2\right),$$
$$A_1 = -\frac{1}{4}e^{-2u_0}u_1 + u_0'u_1',$$
$$A_2 = \frac{1}{4}e^{-2u_0}\left(u_1^2 - u_2 + 2e^{2u_0}\left(\left(u_1'\right)^2 + 2u_0'u_2'\right)\right), \quad (91)$$
$$A_3 = \frac{1}{12}e^{-2u_0}\left(-2u_1^3 + 6u_1u_2 - 3\left(u_3 - 4e^{2u_0}\left(u_1'u_2' + u_0'u_3'\right)\right)\right).$$

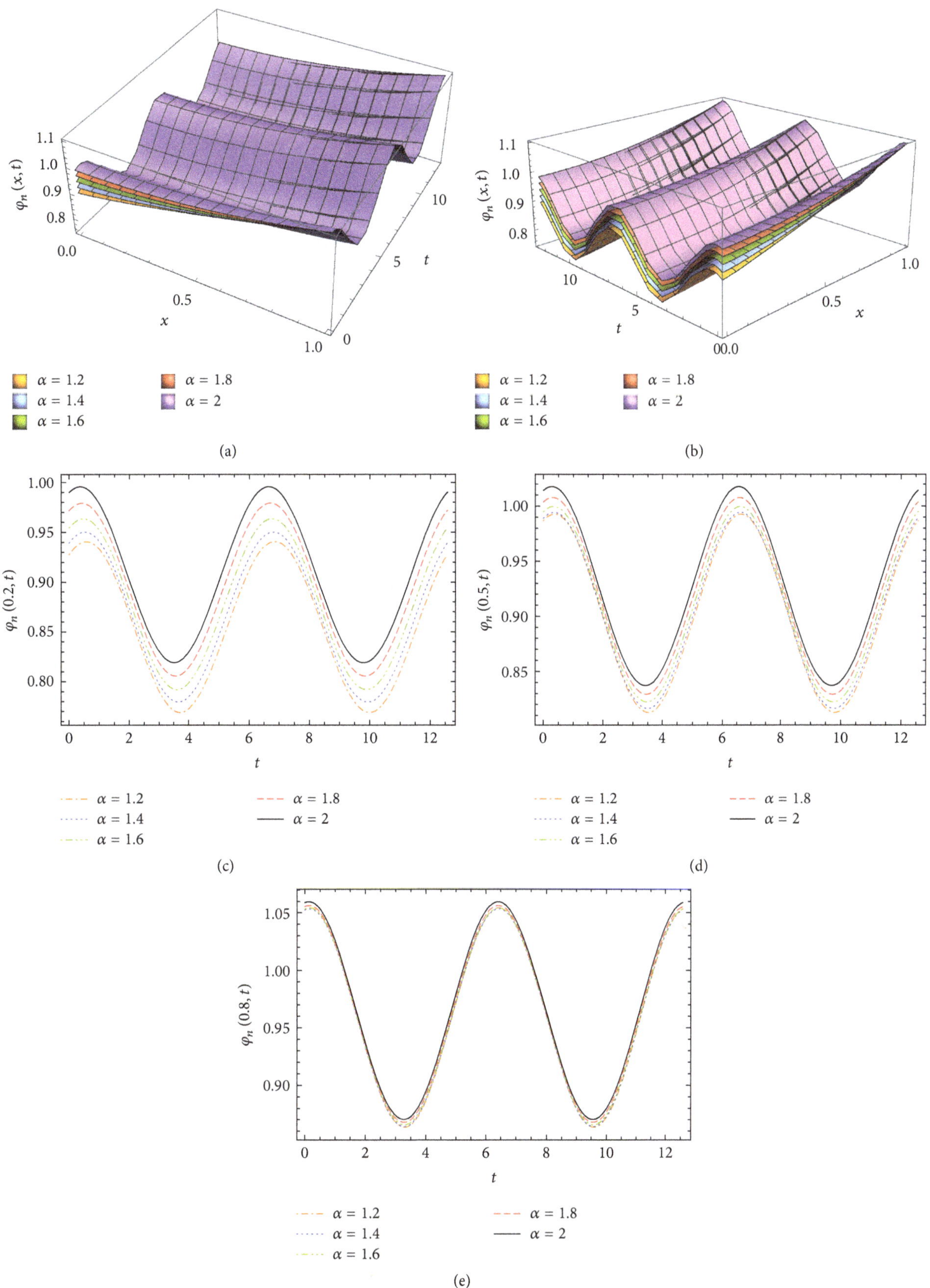

FIGURE 4: ((a)-(b)) Surfaces of approximate solutions $\varphi_5(x,t)$ for the values of $\alpha = 1.2, 1.4, 1.6, 1.8, 2$ with different angles of view. ((c)–(e)) Cross sections of the surfaces of $\varphi_5(x,t)$ at $x = 0.2$, $x = 0.5$, and $x = 0.8$, respectively, for $\alpha = 1.2$ (dot-dash), $\alpha = 1.4$ (dot line), $\alpha = 1.6$ (dot-dot-dash), $\alpha = 1.8$ (dash line), and $\alpha = 2$ (solid line).

The Duan-Rach Modified ADM. We can now apply the formulas in (36) and (38) in Section 3.1 to this BVP. For the parameters described in such a section, we have $a = 0, b = 1, p = 1, q = -2, r = 1, s = 2, \beta = -1, \gamma = 2/3 + \ln(3/2)$, and $\Delta = 5$. Equation (36) turns out to be

$$u(x) = \frac{1}{5}\left[\left(-\frac{5}{3} + 2\ln\left(\frac{3}{2}\right)\right) + \left(\frac{5}{3} + \ln\left(\frac{3}{2}\right)\right)x + (2x+4)\left[{}_{\mathrm{RL}}J_0^{\alpha-1}(Nu)\right]_{x=1} + (x+2)\left[{}_{\mathrm{RL}}J_0^{\alpha}(Nu)\right]_{x=1}\right] - {}_{\mathrm{RL}}J_0^{\alpha}(Nu). \quad (92)$$

Using (38) and Duan's convergence parameter technique, we obtain the following parametrized recursion scheme:

$$\begin{aligned} u_0(x) &= c, \\ u_1(x) &= -c + \frac{1}{5}\left(-\frac{5}{3} + 2\ln\left(\frac{3}{2}\right)\right) + \frac{1}{5}\left(\frac{5}{3} + \ln\left(\frac{3}{2}\right)\right)x + \frac{(2x+4)}{5}\left[{}_{\mathrm{RL}}J_0^{\alpha-1}(A_0)\right]_{x=1} + \frac{(x+2)}{5}\left[{}_{\mathrm{RL}}J_0^{\alpha}(A_0)\right]_{x=1} - {}_{\mathrm{RL}}J_0^{\alpha}(A_0), \\ u_{n+1}(x) &= \frac{(2x+4)}{5}\left[{}_{\mathrm{RL}}J_0^{\alpha-1}(A_n)\right]_{x=1} + \frac{(x+2)}{5}\left[{}_{\mathrm{RL}}J_0^{\alpha}(A_n)\right]_{x=1} - {}_{\mathrm{RL}}J_0^{\alpha}(A_n), \quad n \geq 1. \end{aligned} \quad (93)$$

Since the expressions of the solution components $u_n(x)$, $n \geq 2$, are quite long, we show only the first two solution components $u_1(x)$ and $u_2(x)$ computed using the Adomian polynomials A_0 and A_1 in (91) as follows:

$$\begin{aligned} u_1(x) &= -c - \frac{1}{3} + \frac{2}{5}\ln\left(\frac{3}{2}\right) + \frac{(2\alpha+1)e^{-2c}}{20\Gamma(\alpha+1)} + \left(\frac{1}{3} + \frac{1}{5}\ln\left(\frac{3}{2}\right) + \frac{(2\alpha+1)e^{-2c}}{40\Gamma(\alpha+1)}\right)x - \frac{e^{-2c}}{8\Gamma(\alpha+1)}x^{\alpha}, \\ u_2(x) &= \frac{e^{-4c}}{1200}\left(\frac{60\alpha+15}{\Gamma(2\alpha+1)} - \frac{3(\alpha+1)(2\alpha+1)(4\alpha(\alpha+2)+5)}{\Gamma^2(\alpha+2)} + \frac{8e^{2c}\left(\alpha\left(2\alpha(5-6\ln(3)+\ln(64))+5-48\coth^{-1}(5)\right)+15(\alpha+1)(2\alpha+1)c-10\left(1+3\coth^{-1}(5)\right)\right)}{\Gamma(\alpha+2)}\right) \\ &\quad + \frac{e^{-4c}}{2400}\left(\frac{60\alpha+15}{\Gamma(2\alpha+1)} - \frac{3(\alpha+1)(2\alpha+1)(4\alpha(\alpha+2)+5)}{\Gamma^2(\alpha+2)} + \frac{8e^{2c}\left(\alpha\left(2\alpha(5-6\ln(3)+\ln(64))+5-48\coth^{-1}(5)\right)+15(\alpha+1)(2\alpha+1)c-10\left(1+3\coth^{-1}(5)\right)\right)}{\Gamma(\alpha+2)}\right)x \\ &\quad + \frac{e^{-4c}\left(6\alpha^2+9\alpha-4e^{2c}\Gamma(\alpha+2)(15c+5-6\ln(3)+\ln(64))+3\right)}{240\alpha\Gamma(\alpha)\Gamma(\alpha+2)}x^{\alpha} \\ &\quad + \frac{e^{-4c}\left(3(\alpha+1)(2\alpha+1)+8e^{2c}(5+\ln(27/8))\Gamma(\alpha+2)\right)}{480\Gamma^2(\alpha+2)}x^{\alpha+1} - \frac{e^{-4c}}{32\Gamma(2\alpha+1)}x^{2\alpha}. \end{aligned} \quad (94)$$

Throughout this method, we use $c = 0.5$ and the following values of $\alpha = 1.2, 1.4, 1.6, 1.8, 2$ for the recursion scheme in (93). We can compute the approximate solution $\varphi_n(x) = \sum_{k=0}^{n-1} u_k(x)$ from the solution components $u_0(x), u_1(x), u_2(x), \ldots, u_{n-1}(x)$. In particular, we calculate $\varphi_{15}(x)$ for each selected value of α using the Duan-Rach modified ADM. Figure 6 shows curves of the approximate solutions $\varphi_{15}(x)$ for $\alpha = 1.2, 1.4, 1.6, 1.8, 2$.

For each specified value of $\alpha \neq 2$, since the exact solution to this BVP is not known, we calculate the corresponding error remainder functions $\mathrm{ER}_n(x)$ for $n = 15$. Figure 7 displays the graphs of $\mathrm{ER}_{15}(x)$ for the specified values of α. We can deduce from the graphs in Figure 7 that the approximations $\varphi_{15}(x)$ obtained by this method give the remarkable accuracy as expected that when n is sufficiently large then the magnitude of each function $\mathrm{ER}_n(x)$ approaches zero.

For $\alpha = 2$, it is possible to compute both of the error functions $E_n(x)$ and the maximal errors ME_n; however, we only compute the values of ME_n for $n = 2, 3, \ldots, 15$ listed in Table 2. We show the logarithmic plots of ME_n versus n for $n = 2, 3, \ldots, 15$ in Figure 8. We can observe in Figure 8

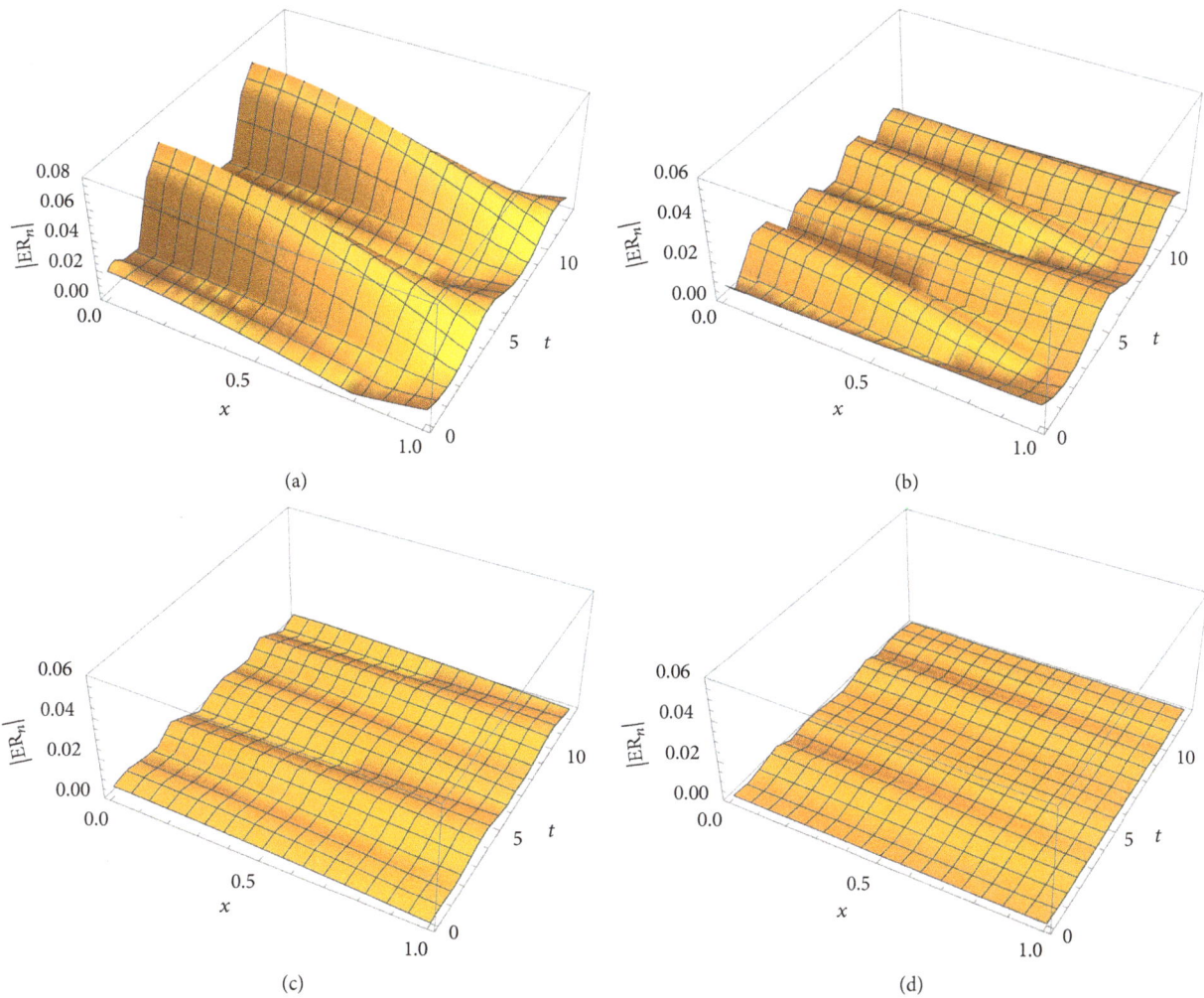

FIGURE 5: Surfaces of the absolute error remainder functions for $\alpha = 1.8$, (a) $|\mathrm{ER}_2(x,t)|$, (b) $|\mathrm{ER}_3(x,t)|$, (c) $|\mathrm{ER}_4(x,t)|$, and (d) $|\mathrm{ER}_5(x,t)|$.

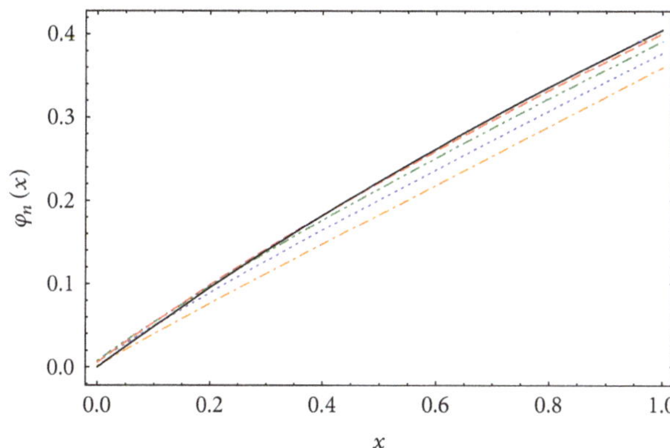

FIGURE 6: $\varphi_n(x)$ when $n = 15$ obtained by the Duan-Rach modified ADM for $\alpha = 1.2$ (dot-dash), $\alpha = 1.4$ (dot line), $\alpha = 1.6$ (dot-dot-dash), $\alpha = 1.8$ (dash line), and $\alpha = 2$ (solid line).

that all of the data points after $n = 7$ lie almost on a straight line which demonstrates the maximal errors ME_n are reduced approximately at an exponential rate.

Example 4. Consider the beam-type fractional boundary value problem

$$\begin{gathered} {}_CD_0^{\alpha}u - 2u'' + u = g(x), \quad 0 \le x \le 1, \ 3 < \alpha \le 4, \\ u(0) = 0, \\ u''(0) = 0, \\ u(1) = 0, \\ u''(1) = -12, \end{gathered} \tag{95}$$

where $g(x) = -24 + x + 24x^2 - x^4$. For $\alpha = 4$, it can be easily verified that the exact solution of the BVP is $u^*(x) = x(1 - x^3)$. Fundamentally, we decompose the solution u and the nonlinearity $Nu = u - 2u''$ as $u(x) = \sum_{n=0}^{\infty} u_n(x)$ and $Nu = \sum_{n=0}^{\infty} A_n(x)$, where u_n are the solution components and A_n are the Adomian polynomials. We here list the first four Adomian polynomials for this nonlinear term as follows:

$$A_0 = u_0 - 2u_0'',$$

$$A_1 = u_1 - 2u_1'',$$

$$A_2 = u_2 - 2u_2'',$$
$$A_3 = u_3 - 2u_3''. \tag{96}$$

The Duan-Rach Modified ADM. We can apply the formulas in (65) and (66) in Section 3.2 to this BVP. We have $a = 0, b = 1$, $p = 1, q = 0, r = 1, s = 0, \rho = 0, \sigma = 0, \beta = 0, \gamma = -12$, and $\Delta = 1$. Then, we have $\kappa = -12, \lambda = 0, \mu = -12, \nu = -12$, and $\omega = 1$. Thus, (65) turns out to be

$$\begin{aligned} u(x) = {} & 2x - 2x^3 - \frac{1}{6}\left(x^3 - x\right)\left[{}_{RL}J_0^{\alpha-2}(g(x))\right]_{x=1} \\ & - x\left[{}_{RL}J_0^{\alpha}(g(x))\right]_{x=1} + {}_{RL}J_0^{\alpha}(g(x)) \\ & + \frac{1}{6}\left(x^3 - x\right)\left[{}_{RL}J_0^{\alpha-2}(Nu)\right]_{x=1} \\ & + x\left[{}_{RL}J_0^{\alpha}(Nu)\right]_{x=1} - {}_{RL}J_0^{\alpha}(Nu), \end{aligned} \tag{97}$$

where Nu is given as above. Using (66), we obtain the following recursion scheme:

$$\begin{aligned} u_0(x) = {} & 2x - 2x^3 \\ & - \frac{1}{6}\left(x^3 - x\right)\left[{}_{RL}J_0^{\alpha-2}(g(x))\right]_{x=1} \\ & - x\left[{}_{RL}J_0^{\alpha}(g(x))\right]_{x=1} + {}_{RL}J_0^{\alpha}(g(x)), \\ u_{n+1}(x) = {} & \frac{1}{6}\left(x^3 - x\right)\left[{}_{RL}J_0^{\alpha-2}(A_n)\right]_{x=1} \\ & + x\left[{}_{RL}J_0^{\alpha}(A_n)\right]_{x=1} - {}_{RL}J_0^{\alpha}(A_n), \\ & n \geq 0. \end{aligned} \tag{98}$$

Since the expressions of the solution components $u_2(x)$, $u_3(x), u_4(x), \ldots$ are very long, we show only the first two solution components $u_0(x)$ and $u_1(x)$ as follows:

$$\begin{aligned} u_0(x) = {} & \left(2 + \frac{-24\alpha^6 - 215\alpha^5 - 398\alpha^4 + 1589\alpha^3 + 7028\alpha^2 + 7860\alpha + 864}{6\Gamma(\alpha+5)}\right)x + \left(-2\right. \\ & \left. + \frac{24\alpha^4 + 47\alpha^3 - 75\alpha^2 - 194\alpha - 72}{6\Gamma(\alpha+3)}\right)x^3 - \frac{24}{\Gamma(\alpha+1)}x^{\alpha} + \frac{x^{\alpha+1}}{\Gamma(\alpha+2)} + \frac{48}{\Gamma(\alpha+3)}x^{\alpha+2} - \frac{24}{\Gamma(\alpha+5)}x^{\alpha+4}, \\ u_1(x) = {} & \frac{1}{36}\left(\frac{432/\Gamma(\alpha+3) + 72}{\Gamma(\alpha+2)} - \frac{156\alpha^4 + 936\alpha^3 + 780\alpha^2 - 3744\alpha - 5184}{\Gamma(\alpha+4)}\right. \\ & + \frac{312\alpha^9 + 3731\alpha^8 + 13895\alpha^7 + 755\alpha^6 - 111451\alpha^5 - 216262\alpha^4 + 64052\alpha^3 + 576816\alpha^2 + 544464\alpha + 124416}{\Gamma(\alpha+3)\Gamma(\alpha+5)} \\ & \left. - \frac{73728\alpha^8 + 145920\alpha^7 - 420096\alpha^6 - 1221312\alpha^5 - 655872\alpha^4 + 576432\alpha^3 + 897120\alpha^2 + 474480\alpha + 74304}{\Gamma(2\alpha+5)}\right)x \\ & + \left(\frac{4\alpha^4 + 8\alpha^3 - 12\alpha^2 - 32\alpha - 2\Gamma(\alpha+3) - 12}{\Gamma(\alpha+2)\Gamma(\alpha+3)} + \frac{1536\alpha^6 - 2336\alpha^5 - 2880\alpha^4 + 1964\alpha^3 + 822\alpha^2 - 54\alpha + 84}{3\Gamma(2\alpha+3)}\right. \\ & \left. + \frac{13}{36\Gamma^2(\alpha)} + \frac{-26\alpha^6 - 234\alpha^5 - 586\alpha^4 + 249\alpha^3 + 2857\alpha^2 + 3448\alpha + 13\Gamma(\alpha+5) + 924}{3\Gamma(\alpha)\Gamma(\alpha+5)}\right)x^3 \\ & + \frac{\left(312\alpha^6 + 2795\alpha^5 + 6902\alpha^4 - 3449\alpha^3 - 34988\alpha^2 - 41844\alpha - 156\Gamma(\alpha+5) - 11232\right)}{6\Gamma(\alpha+2)\Gamma(\alpha+5)}x^{\alpha+1} \\ & + \frac{\left(-24\alpha^4 - 47\alpha^3 + 75\alpha^2 + 194\alpha + 12\Gamma(\alpha+3) + 72\right)}{\Gamma(\alpha+3)\Gamma(\alpha+4)}x^{\alpha+3} - \frac{48}{\Gamma(2\alpha-1)}x^{2\alpha-2} + \frac{2}{\Gamma(2\alpha)}x^{2\alpha-1} + \frac{120}{\Gamma(2\alpha+1)}x^{2\alpha} \\ & - \frac{x^{2\alpha+1}}{\Gamma(2\alpha+2)} - \frac{96}{\Gamma(2\alpha+3)}x^{2\alpha+2} + \frac{24}{\Gamma(2\alpha+5)}x^{2\alpha+4}. \end{aligned} \tag{99}$$

In particular, we calculate $\varphi_{10}(x)$ for $\alpha = 3.2, 3.4, 3.6, 3.8$, 4. Figure 9 shows curves of the approximate solutions $\varphi_{10}(x)$ for the selected values of α.

For $\alpha = 3.2, 3.4, 3.6, 3.8$, we do not know what the exacts solutions are, and thus we calculate the error remainder functions $\mathrm{ER}_n(x) = {}_{C}D_0^{\alpha}\varphi_n(x) - 2\varphi_n''(x) + \varphi_n(x) - g(x)$, $0 \leq x \leq 1$, for each value of α. We display the functions $\mathrm{ER}_n(x)$ for $n = 10$ in Figure 10. The approximations of this order give significant accuracy, as expected, since the limit of the functions $\mathrm{ER}_n(x)$ approaches zero.

For $\alpha = 4$, it is possible to compute the error functions E_n in (70) as $E_n = \varphi_n(x) - u^*(x)$ and the maximal errors ME_n in

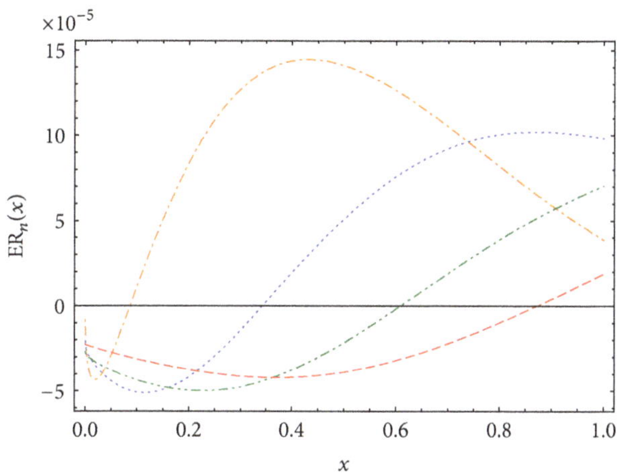

FIGURE 7: $\mathrm{ER}_n(x)$ when $n = 15$ obtained by the Duan-Rach modified ADM for $\alpha = 1.2$ (dot-dash), $\alpha = 1.4$ (dot line), $\alpha = 1.6$ (dot-dot-dash), $\alpha = 1.8$ (dash line), and $\alpha = 2$ (solid line).

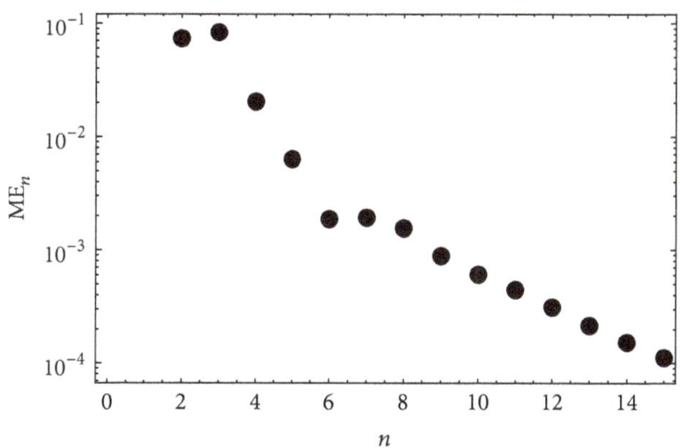

FIGURE 8: Logarithmic plots of the maximal errors ME_n versus n for $n = 2$ through 15 obtained by the Duan-Rach modified ADM.

(71) as $\mathrm{ME}_n = \max_{0\le x\le 1}|E_n(x)|$. However, we only compute the values of ME_n for $n = 2, 3, 4, \ldots, 9$ listed in Table 3. In Figure 11 we show the logarithmic plots of ME_n versus n for $n = 2, 3, 4, \ldots, 9$ obtained by the method. All of the data points lie almost on a straight line so the maximal errors are decreasing approximately at an exponential rate.

5. Conclusions

We have established new recursion schemes using the Duan-Rach modified decomposition method to solve a variety of nonlinear fractional BVPs. The obtained recursion schemes have been derived for solving the nonlinear fractional BVPs with a set of Robin boundary conditions (order $1 < \alpha \le 2$) and with separated boundary conditions (order $3 < \alpha \le 4$). We have applied the new recursion schemes to four numerical expository examples. Example 1 is a nonlinear Dirichlet fractional BVP of order $1 < \alpha \le 2$ with the exponential nonlinearity. Example 2 is a nonlinear Neumann and Dirichlet fractional BVP of order $1 < \alpha \le 2$ with the first derivative nonlinearity. Example 3 is a fractional BVP with the sum of an exponential nonlinearity in the solution and a quadratic nonlinearity in the derivative of the solution

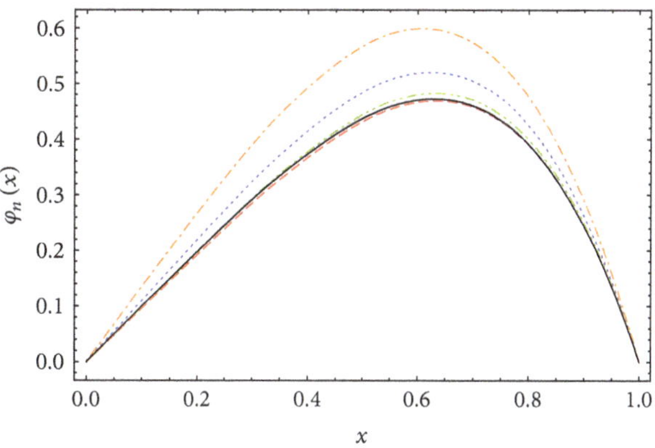

FIGURE 9: $\varphi_n(x)$ when $n = 10$ obtained by the Duan-Rach modified ADM for $\alpha = 3.2$ (dot-dash), $\alpha = 3.4$ (dot line), $\alpha = 3.6$ (dot-dot-dash), $\alpha = 3.8$ (dash line), and $\alpha = 4$ (solid line).

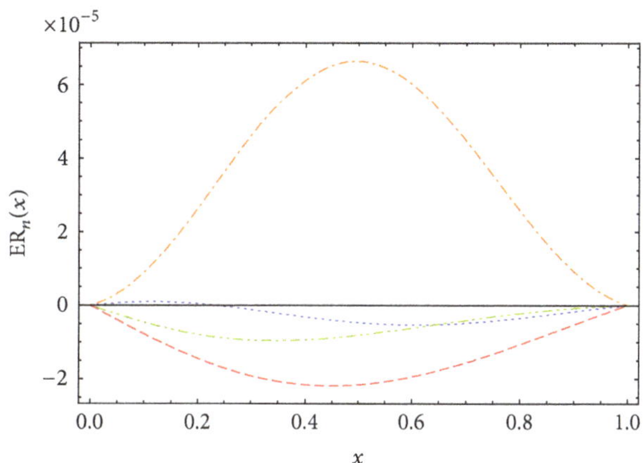

FIGURE 10: $\mathrm{ER}_n(x)$ when $n = 10$ obtained by the Duan-Rach modified ADM for $\alpha = 3.2$ (dot-dash), $\alpha = 3.4$ (dot line), $\alpha = 3.6$ (dot-dot-dash), and $\alpha = 3.8$ (dash line).

TABLE 2: For $c = 0.5$ and $\alpha = 2$, maximal errors ME_n for $n = 2, 3, \ldots, 15$ obtained by the Duan-Rach modified ADM.

n	ME_n
2	0.0734942
3	0.0830039
4	0.0203679
5	0.00633962
6	0.00186625
7	0.00192242
8	0.00155408
9	0.000886808
10	0.000608945
11	0.000445665
12	0.000313388
13	0.00021568
14	0.00015314
15	0.000111835

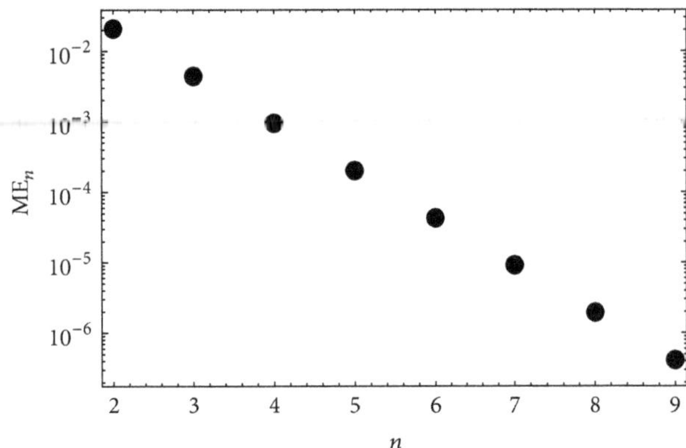

FIGURE 11: Logarithmic plots of the maximal errors ME_n versus n for $n = 2$ through 9 obtained by the Duan-Rach modified ADM.

TABLE 3: For $\alpha = 4$, maximal errors ME_n for $n = 2, 3, \ldots, 9$ obtained by the Duan-Rach modified ADM.

n	ME_n
2	0.0208592
3	0.00444248
4	0.000945886
5	0.000201388
6	0.0000428772
7	9.12892×10^{-6}
8	1.94362×10^{-6}
9	4.13814×10^{-7}

which is imposed with a set of Robin boundary conditions. The beam-type fractional BVP of order $3 < \alpha \leq 4$ with separated boundary conditions and the product nonlinearity is provided in Example 4. Besides the obtained approximate solutions, we also provided the maximal errors (ME_n) and the error remainder functions ($ER_n(x)$) for each problem if possible. The results for all examples shown confirm that increasing the number of solution components (i.e., the value of n) reduces the errors in the numerical solutions. Furthermore, unlike the method of undetermined coefficients in the ADM, the Duan-Rach modified decomposition method does not require solving a system of nonlinear algebraic equations obtained from using the n-term approximation $\varphi_n(x)$ for the remaining unknown constants of integration, which are sometimes multiple roots or nonphysical roots. Hence, the method is very efficient and has provided very accurate approximate solutions when compared with the exact solutions (if any).

Conflicts of Interest

The authors declare that there are no conflicts of interest regarding the publication of this paper.

References

[1] S. Kumar, "A new fractional modeling arising in engineering sciences and its analytical approximate solution," *Alexandria Engineering Journal*, vol. 52, no. 4, pp. 813–819, 2013.

[2] C. Tricaud and Y. Q. Chen, "An approximate method for numerically solving fractional order optimal control problems of general form," *Computers and Mathematics with Applications*, vol. 59, no. 5, pp. 1644–1655, 2010.

[3] R. Magin, X. Feng, and D. Baleanu, "Solving the fractional order Bloch equation," *Concepts in Magnetic Resonance Part A: Bridging Education and Research*, vol. 34, no. 1, pp. 16–23, 2009.

[4] J. He, "Nonlinear oscillation with fractional derivative and its applications," in *Proceedings of the International conference on vibrating engineering*, pp. 288–291, 1998.

[5] I. Podlubny, *Fractional Differential Equations*, vol. 198 of *Mathematics in Science and Engineering*, Academic Press, San Diego, Calif, USA, 1999.

[6] C. Lederman, J.-M. Roquejoffre, and N. Wolanski, "Mathematical justification of a nonlinear integrodifferential equation for the propagation of spherical flames," *Annali di Matematica Pura ed Applicata. Series IV*, vol. 183, no. 2, pp. 173–239, 2004.

[7] A. Carvalho and C. M. Pinto, "A delay fractional order model for the co-infection of malaria and HIV/AIDS," *International Journal of Dynamics and Control*, vol. 5, no. 1, pp. 168–186, 2017.

[8] N. Ozalp and E. Demirci, "A fractional order SEIR model with vertical transmission," *Mathematical and Computer Modelling*, vol. 54, no. 1-2, pp. 1–6, 2011.

[9] G. Adomian, "Solution of physical problems by decomposition," *Computers and Mathematics with Applications*, vol. 27, no. 9-10, pp. 145–154, 1994.

[10] H. Jafari and V. Daftardar-Gejji, "Positive solutions of nonlinear fractional boundary value problems using Adomian decomposition method," *Applied Mathematics and Computation*, vol. 180, no. 2, pp. 700–706, 2006.

[11] J.-S. Duan, T. Chaolu, R. Rach, and L. Lu, "The Adomian decomposition method with convergence acceleration techniques for nonlinear fractional differential equations," *Computers and Mathematics with Applications*, vol. 66, no. 5, pp. 728–736, 2013.

[12] A.-M. Wazwaz, "Adomian decomposition method for a reliable treatment of the Bratu-type equations," *Applied Mathematics and Computation*, vol. 166, no. 3, pp. 652–663, 2005.

[13] S. Momani and N. Shawagfeh, "Decomposition method for solving fractional Riccati differential equations," *Applied Mathematics and Computation*, vol. 182, no. 2, pp. 1083–1092, 2006.

[14] J. S. Duan, R. Rach, D. Baleanu, and A. M. Wazwaz, "A review of the Adomian decomposition method and its applications to fractional differential equations," *Communications in Fractional Calculus*, vol. 3, no. 2, 2012.

[15] V. S. Ertürk, G. Zaman, and S. Momani, "A numeric-analytic method for approximating a giving up smoking model containing fractional derivatives," *Computers and Mathematics with Applications*, vol. 64, no. 10, pp. 3065–3074, 2012.

[16] I. Petras, "Modeling and numerical analysis of fractional-order bloch equations," *Computers Mathematics with Applications*, vol. 61, no. 2, pp. 341–356, 2011.

[17] U. Saeed and M. ur Rehman, "Haar wavelet operational matrix method for fractional oscillation equations," *International Journal of Mathematics and Mathematical Sciences*, Article ID 174819, Art. ID 174819, 8 pages, 2014.

[18] I. Hashim, M. S. M. Noorani, R. Ahmad, S. A. Bakar, E. S. Ismail, and A. M. Zakaria, "Accuracy of the adomian decomposition method applied to the Lorenz system," *Chaos, Solitons & Fractals*, vol. 28, no. 5, pp. 1149–1158, 2006.

[19] N. H. Sweilam and M. M. Khader, "Approximate solutions to the nonlinear vibrations of multiwalled carbon nanotubes using

Adomian decomposition method," *Applied Mathematics and Computation*, vol. 217, no. 2, pp. 495–505, 2010.

[20] M. M. Hosseini, "Adomian decomposition method with Chebyshev polynomials," *Applied Mathematics and Computation*, vol. 175, no. 2, pp. 1685–1693, 2006.

[21] Y. Liu, "Adomian decomposition method with orthogonal polynomials: Legendre polynomials," *Mathematical and Computer Modelling*, vol. 49, no. 5-6, pp. 1268–1273, 2009.

[22] W.-C. Tien and C.-K. Chen, "Adomian decomposition method by Legendre polynomials," *Chaos, Solitons and Fractals*, vol. 39, no. 5, pp. 2093–2101, 2009.

[23] P. V. Ramana and B. K. R. Prasad, "Modified Adomian Decomposition Method for Van der Pol equations," *International Journal of Non-Linear Mechanics*, vol. 65, pp. 121–132, 2014.

[24] M. Tatari and M. Dehghan, "The use of the Adomian decomposition method for solving multipoint boundary value problems," *Physica Scripta*, vol. 73, no. 6, article no. 023, pp. 672–676, 2006.

[25] W. Al-Hayani, "Adomian decomposition method with Green's function for sixth-order boundary value problems," *Computers & Mathematics with Applications*, vol. 61, no. 6, pp. 1567–1575, 2011.

[26] J.-S. Duan and R. Rach, "A new modification of the Adomian decomposition method for solving boundary value problems for higher order nonlinear differential equations," *Applied Mathematics and Computation*, vol. 218, no. 8, pp. 4090–4118, 2011.

[27] J.-S. Duan, R. Rach, A.-M. Wazwaz, T. Chaolu, and Z. Wang, "A new modified Adomian decomposition method and its multistage form for solving nonlinear boundary value problems with Robin boundary conditions," *Applied Mathematical Modelling. Simulation and Computation for Engineering and Environmental Systems*, vol. 37, no. 20-21, pp. 8687–8708, 2013.

[28] Y. Lin, Y. Liu, and Z. Li, "Symbolic computation of analytic approximate solutions for nonlinear fractional differential equations," *Computer Physics Communications*, vol. 184, no. 1, pp. 130–141, 2013.

[29] O. Kelesoglu, "The solution of fourth order boundary value problem arising out of the beam-column theory using Adomian decomposition method," *Mathematical Problems in Engineering*, Article ID 649471, 6 pages, 2014.

[30] J.-S. Duan, R. Rach, and A.-M. Wazwaz, "Solution of the model of beam-type micro- and nano-scale electrostatic actuators by a new modified Adomian decomposition method for nonlinear boundary value problems," *International Journal of Non-Linear Mechanics*, vol. 49, pp. 159–169, 2013.

[31] B. Jang, "Two-point boundary value problems by the extended Adomian decomposition method," *Journal of Computational and Applied Mathematics*, vol. 219, no. 1, pp. 253–262, 2008.

[32] N. T. Shawagfeh, "Analytical approximate solutions for nonlinear fractional differential equations," *Applied Mathematics and Computation*, vol. 131, no. 2-3, pp. 517–529, 2002.

[33] S. G. Samko, A. A. Kilbas, and O. I. Marichev, *Fractional Integrals and Derivatives, Theory and Applications*, Gordon and Breach, Yverdon, Switzerland, 1993.

[34] A. A. Kilbas, H. M. Srivastava, and J. J. Trujillo, "Preface," *North-Holland Mathematics Studies*, vol. 204, no. C, pp. vii–x, 2006.

[35] S. Das, *Functional Fractional Calculus*, Springer, New Delhi, India, 2011.

[36] G. Adomian, "A review of the decomposition method and some recent results for nonlinear equations," *Computers & Mathematics with Applications*, vol. 21, no. 5, pp. 101–127, 1991.

[37] G. Adomian and R. Rach, "Inversion of nonlinear stochastic operators," *Journal of Mathematical Analysis and Applications*, vol. 91, no. 1, pp. 39–46, 1983.

[38] G. Adomian, "A review of the decomposition method and some recent results for nonlinear equations," *Mathematical and Computer Modelling*, vol. 13, no. 7, pp. 17–43, 1990.

[39] A.-M. Wazwaz, "A reliable modification of Adomian decomposition method," *Applied Mathematics and Computation*, vol. 102, no. 1, pp. 77–86, 1999.

[40] J.-S. Duan, "Recurrence triangle for Adomian polynomials," *Applied Mathematics and Computation*, vol. 216, no. 4, pp. 1235–1241, 2010.

[41] J.-S. Duan and R. Rach, "On the domain of convergence of the Adomian series solution," *Applied Mathematics and Computation*, vol. 66, no. 5, pp. 728–736, 2013.

[42] J. S. Duan, R. Rach, and Z. Wang, "On the effective region of convergence of the decomposition series solution," *Journal of Algorithms Computational Technology*, vol. 7, no. 2, pp. 287-247, 2012.

[43] A. Wazwaz, "Approximate solutions to boundary value problems of higher order by the modified decomposition method," *Computers & Mathematics with Applications*, vol. 40, no. 6-7, pp. 679–691, 2000.

[44] A. Wazwaz, *Partial Differential Equations and Solitary Waves Theory*, Higher Education Press, Beijing, China; Springer, Berlin, Germany, 2009.

[45] U. M. Ascher, R. M. Mattheij, and R. D. Russell, *Numerical Solution of Boundary Value Problems for Ordinary Differential Equations*, SIAM, Philadelphia, Pa, USA, 1995.

[46] J. P. Boyd, "Chebyshev polynomial expansions for simultaneous approximation of two branches of a function with application to the one-dimensional Bratu equation," *Applied Mathematics and Computation*, vol. 143, no. 2-3, pp. 189–200, 2003.

[47] Y.-T. Yang, S.-K. Chien, and C.-K. Chen, "A double decomposition method for solving the periodic base temperature in convective longitudinal fins," *Energy Conversion and Management*, vol. 49, no. 10, pp. 2910–2916, 2008.

A 4-Point Block Method for Solving Higher Order Ordinary Differential Equations Directly

Nazreen Waeleh[1] and Zanariah Abdul Majid[2,3]

[1] *Faculty of Electronic & Computer Engineering, Universiti Teknikal Malaysia Melaka (UTeM), 76100 Melaka, Malaysia*
[2] *Department of Mathematics, Faculty of Science, Universiti Putra Malaysia, 43400 Serdang, Selangor, Malaysia*
[3] *Institute for Mathematical Research, Universiti Putra Malaysia, 43400 Serdang, Selangor, Malaysia*

Correspondence should be addressed to Nazreen Waeleh; nazreen@utem.edu.my

Academic Editor: Harvinder S. Sidhu

An alternative block method for solving fifth-order initial value problems (IVPs) is proposed with an adaptive strategy of implementing variable step size. The derived method is designed to compute four solutions simultaneously without reducing the problem to a system of first-order IVPs. To validate the proposed method, the consistency and zero stability are also discussed. The improved performance of the developed method is demonstrated by comparing it with the existing methods and the results showed that the 4-point block method is suitable for solving fifth-order IVPs.

1. Introduction

Many natural processes or real-world problems can be translated into the language of mathematics [1–4]. The mathematical formulation of physical phenomena in science and engineering often leads to a differential equation, which can be categorized as an ordinary differential equation (ODE) and a partial differential equation (PDE). This formulation will explain the behavior of the phenomenon in detail. The search for solutions of real-world problems requires solving ODEs and thus has been an important aspect of mathematical study. For many interesting applications, an exact solution may be unattainable, or it may not give the answer in a convenient form. The reliability of numerical approximation techniques in solving such problems has been proven by many researchers as the role of numerical methods in engineering problems solving has increased dramatically in recent years. Thus a numerical approach has been chosen as an alternative tool for approximating the solutions consistent with the advancement in technology.

Commonly, the formulation of real-world problems will take the form of a higher order differential equation associated with its initial or boundary conditions [4]. In the literature, a mathematical model in the form of a fifth-order differential equation, known as Korteweg-de Vries (KdV) equation, has been used to describe several wave phenomena depending on the values of its parameters [2, 3, 5, 6]. The KdV equation is a PDE and researchers have tackled the problem analytically and numerically. It is also noted that in certain cases by using different approaches the KdV might be transformed into a higher order ODE [7]. To date, there are a number of studies that have proposed solving fifth-order ODE directly [8, 9]. Hence, the purpose of the present paper is to solve directly the fifth-order IVPs with the implementation of a variable step size strategy. The fifth-order IVP with its initial conditions is defined as

$$y^{\text{v}} = f\left(x, y, y', y'', y''', y^{\text{iv}}\right), \quad y(a) = y_0, \; y'(a) = y_1, \; y''(a) = y_2, \; y'''(a) = y_3, \; y^{\text{iv}}(a) = y_4, \; x \in [a, b]. \tag{1}$$

Conventionally, (1) will be converted to a system of first-order ODEs by a simple change of variables. However, it will increase the computational cost in terms of function evaluation and thus will affect the computational time. This drawback is obviously seen when dealing with a higher order problem. Furthermore, [10] also has remarked that the block method is far more cost-effective when it is implemented in direct integration. Hence, several researchers [11–16] have shown an interest in the development of direct integration methods. A direct integration method of variable order and step size for solving systems of nonstiff higher order ODEs has been discussed in [11] whereby [12] has proposed an algorithm based on collocation of the differential system at selected grid points for direct solution of general second-order ODEs. In addition, [13] has used the Gaussian method in order to solve fourth-order differential equations directly. However, it requires a tedious computation as well, since it consists of higher order partial derivatives of Taylor series algorithm which supplies the starting values. Jator and Li [15] have proposed the linear multistep method (LMM) for solving general second-order IVPs directly. The method is self-starting, so it involves less computational time by avoiding incorporating subroutines to supply the starting values.

Thus far, a number of researchers have concerned themselves with developing a numerical method based on block features, and the characteristic feature of the block method is that in each application it generates a set of solutions concurrently [10]. Rosser [10] also has remarked that the implementation of block method in numerical computation will reduce the computational cost by reducing the number of function evaluations. Shampine and Watts [17] have constructed an A-stable implicit one-step block method and Cash [18] has studied block methods based upon the Runge-Kutta method for the numerical solution of nonstiff IVPs. Furthermore [19] has used the self-starting LMM to solve second-order ODEs in a block-by-block fashion and recently [20] has constructed a predictor-corrector scheme 3-point block method with the implementation of variable step size. This research is an extension of the work in [20] in which the solution is computed at three points concurrently and it shows the satisfactory numerical results obtained when solving general higher order ODEs.

An increasing amount of literature is devoted to variable step size implementations of numerical methods [11, 21, 22]. The practicality of varying the step size for block method has been justified by [10]. This strategy is an attempt to reduce the computational cost as well as maintaining the accuracy. The Falkner method with variable step size implementation for the numerical solution of second-order IVPs has been employed in [21]. Although the implementation of the method involves varying the step size and solving directly, the computation is still tedious since the coefficients of the formulae must be calculated every time the step size is changed. On the contrary, the present work will store all the integration coefficients in the code in order to avoid the tedious calculations of the divided differences.

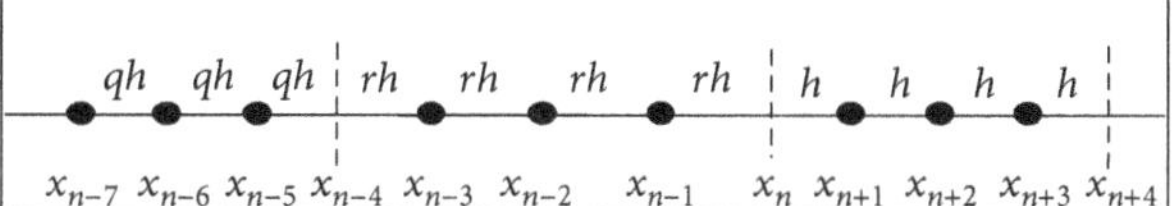

FIGURE 1: 4-point block method.

2. Methodology

2.1. Derivation of 4-Point Block Method. The basic approach of numerical methods for integration is performed by subdividing the interval of integration into certain subintervals. The proposed method was based on concurrent computation; hence the closed finite interval was subdivided into a series of blocks and each block contains four equal subintervals as illustrated in Figure 1.

Initially, (1) was integrated five times over the corresponding interval: $[x_n, x_{n+1}]$, $[x_n, x_{n+2}]$, $[x_n, x_{n+3}]$, $[x_n, x_{n+4}]$ for first, second, third, and fourth point, respectively. The integration was started by replacing the function $f(x, y, y', y'', y''', y^{\text{iv}})$ with the interpolating function which was generated from Lagrange polynomials. A set of points $\{(x_{n-7}, f_{n-7}), \ldots, (x_n, f_n)\}$, $\{(x_{n-4}, f_{n-4}), \ldots, (x_{n+4}, f_{n+4})\}$ was interpolated for deriving predictor and corrector formulae, respectively. Let the Lagrange polynomial, $P_w(x)$, be written as

$$P_w(x) = L_{w,0}(x) f(x_{n+4}) + L_{w,1} f(x_{n+3}) + \cdots + L_{w,w} f(x_{n+4-w}) = \sum_{j=0}^{w} L_{w,j}(x) f(x_{n+4-j}), \tag{2}$$

where

$$L_{w,j}(x) = \prod_{\substack{i=0 \\ i \neq j}}^{w} \frac{(x - x_{n+4-i})}{(x_{n+4-j} - x_{n+4-i})} \quad \text{for each } j = 0, 1, \ldots, w. \tag{3}$$

Nine points were interpolated in (2) with w set to be eight for deriving the corrector and thus one point less for the predictor formula. Then, the integration process was proceeded by substituting $z = (x - x_{n+4})/h$ and $dx = h dz$ in (2). Consistent with the number of interpolation points involved in deriving the formulae, predictor and corrector formulae were obtained in terms of variables r and q. The variables r and q refer to the distance ratio between current and previous point as a result of implementation variable step size strategy in the proposed method.

In this work, the selection of the next step size could be increased by a factor of $(r = 0.5, q = 0.5)$ or maintained by $((r = 1, q = 1), (r = 1, q = 2), (r = 1, q = 0.5))$ and $(r = 2, q = 2)$ for halving the current step size. This step size changing technique was limited in order to minimize the number of coefficients to be stored as well as reducing the computational storage. The increment of step size was also limited to doubling in order to control the accuracy of the computation [10]. The compact form of the 4-point block method is presented in

$$y_{n+d}^{(v-g)} = \sum_{k=0}^{g-1} \frac{(dh)^k}{k!} y_n^{(v-g+k)} + \frac{h^g}{(g-1)!} \sum_{j=s}^{t} \gamma_{d,j}^{g} f_{n+j}, \quad (4)$$

where $\gamma_{d,j}^{g}$ are the coefficients of the formulae to be calculated, d is the number of points ($d = 1, 2, 3, 4$), g is the number of times (1) will be integrated, and k is the number of terms when the equation is integrated. The values of $s = -7$, $t = 0$ and $s = -4$, $t = 4$ were considered for deriving the predictor and corrector formulae, respectively. After further simplification, the associated corrector formulae of the 4-point block method when $r = 1$ are represented below.

Integrate once:

$$\begin{bmatrix} 0 & 0 & 0 & 1 \\ 0 & 0 & 1 & 0 \\ 0 & 1 & 0 & 0 \\ 1 & 0 & 0 & 0 \end{bmatrix} \begin{bmatrix} y_{n+4}^{\text{iv}} \\ y_{n+3}^{\text{iv}} \\ y_{n+2}^{\text{iv}} \\ y_{n+1}^{\text{iv}} \end{bmatrix} = \begin{bmatrix} 1 & 0 & 0 & 0 \\ 1 & 0 & 0 & 0 \\ 1 & 0 & 0 & 0 \\ 1 & 0 & 0 & 0 \end{bmatrix} \begin{bmatrix} y_{n}^{\text{iv}} \\ y_{n-1}^{\text{iv}} \\ y_{n-2}^{\text{iv}} \\ y_{n-3}^{\text{iv}} \end{bmatrix} + \frac{h}{3628800} \left(\begin{bmatrix} -3233 & 36394 & -216014 & 1909858 \\ 4064 & -63232 & 1422272 & 4541696 \\ -29889 & 1312362 & 4667058 & 2789154 \\ 1040128 & 5779456 & 62464 & 8384512 \end{bmatrix} \begin{bmatrix} f_{n+4} \\ f_{n+3} \\ f_{n+2} \\ f_{n+1} \end{bmatrix} \right.$$

$$\left. + \begin{bmatrix} 2224480 & -425762 & 126286 & -25706 \\ 1391360 & -27904 & -15808 & 5888 \\ 2708640 & -782946 & 278478 & -63018 \\ -2324480 & 2363392 & -1012736 & 249856 \end{bmatrix} \begin{bmatrix} f_{n} \\ f_{n-1} \\ f_{n-2} \\ f_{n-3} \end{bmatrix} + \begin{bmatrix} 2497 & 0 & 0 & 0 \\ -736 & 0 & 0 & 0 \\ 6561 & 0 & 0 & 0 \\ -27392 & 0 & 0 & 0 \end{bmatrix} \begin{bmatrix} f_{n-4} \\ f_{n-5} \\ f_{n-6} \\ f_{n-7} \end{bmatrix} \right). \quad (5)$$

Integrate twice:

$$\begin{bmatrix} 0 & 0 & 0 & 1 \\ 0 & 0 & 1 & 0 \\ 0 & 1 & 0 & 0 \\ 1 & 0 & 0 & 0 \end{bmatrix} \begin{bmatrix} y_{n+4}''' \\ y_{n+3}''' \\ y_{n+2}''' \\ y_{n+1}''' \end{bmatrix} = \begin{bmatrix} 1 & 0 & 0 & 0 \\ 1 & 0 & 0 & 0 \\ 1 & 0 & 0 & 0 \\ 1 & 0 & 0 & 0 \end{bmatrix} \begin{bmatrix} y_{n}''' \\ y_{n-1}''' \\ y_{n-2}''' \\ y_{n-3}''' \end{bmatrix} + h \begin{bmatrix} 1 & 0 & 0 & 0 \\ 2 & 0 & 0 & 0 \\ 3 & 0 & 0 & 0 \\ 4 & 0 & 0 & 0 \end{bmatrix} \begin{bmatrix} y_{n}^{\text{iv}} \\ y_{n-1}^{\text{iv}} \\ y_{n-2}^{\text{iv}} \\ y_{n-3}^{\text{iv}} \end{bmatrix}$$

$$+ \frac{h^2}{7257600} \left(\begin{bmatrix} -3057 & 34208 & -197216 & 1258488 \\ -3008 & 20480 & 370944 & 8341504 \\ -22599 & 659016 & 7027560 & 15769728 \\ 433152 & 8093696 & 12378112 & 26050560 \end{bmatrix} \begin{bmatrix} f_{n+4} \\ f_{n+3} \\ f_{n+2} \\ f_{n+1} \end{bmatrix} \right. \quad (6)$$

$$\left. + \begin{bmatrix} 2875850 & -444560 & 128472 & -25864 \\ 6492800 & -909312 & 244480 & -47104 \\ 10485450 & -1650456 & 478224 & -97200 \\ 11724800 & -622592 & -24576 & 32768 \end{bmatrix} \begin{bmatrix} f_{n} \\ f_{n-1} \\ f_{n-2} \\ f_{n-3} \end{bmatrix} + \begin{bmatrix} 2497 & 0 & 0 & 0 \\ 4416 & 0 & 0 & 0 \\ 9477 & 0 & 0 & 0 \\ -5120 & 0 & 0 & 0 \end{bmatrix} \begin{bmatrix} f_{n-4} \\ f_{n-5} \\ f_{n-6} \\ f_{n-7} \end{bmatrix} \right).$$

Integrate thrice:

$$\begin{bmatrix} 0 & 0 & 0 & 1 \\ 0 & 0 & 1 & 0 \\ 0 & 1 & 0 & 0 \\ 1 & 0 & 0 & 0 \end{bmatrix}\begin{bmatrix} y''_{n+4} \\ y''_{n+3} \\ y''_{n+2} \\ y''_{n+1} \end{bmatrix} = \begin{bmatrix} 1 & 0 & 0 & 0 \\ 1 & 0 & 0 & 0 \\ 1 & 0 & 0 & 0 \\ 1 & 0 & 0 & 0 \end{bmatrix}\begin{bmatrix} y''_{n} \\ y''_{n-1} \\ y''_{n-2} \\ y''_{n-3} \end{bmatrix} + h\begin{bmatrix} 1 & 0 & 0 & 0 \\ 2 & 0 & 0 & 0 \\ 3 & 0 & 0 & 0 \\ 4 & 0 & 0 & 0 \end{bmatrix}\begin{bmatrix} y'''_{n} \\ y'''_{n-1} \\ y'''_{n-2} \\ y'''_{n-3} \end{bmatrix} + \frac{h^2}{2}\begin{bmatrix} 1 & 0 & 0 & 0 \\ 4 & 0 & 0 & 0 \\ 9 & 0 & 0 & 0 \\ 16 & 0 & 0 & 0 \end{bmatrix}\begin{bmatrix} y^{\text{iv}}_{n} \\ y^{\text{iv}}_{n-1} \\ y^{\text{iv}}_{n-2} \\ y^{\text{iv}}_{n-3} \end{bmatrix}$$

$$+\frac{h^3}{19958400}\left(\begin{bmatrix} -4872 & 53782 & -304397 & 1693482 \\ -29632 & 313088 & -1292928 & 25524992 \\ -63423 & 1004562 & 15809823 & 93765438 \\ 218112 & 20512768 & 74186752 & 202702848 \end{bmatrix}\begin{bmatrix} f_{n+4} \\ f_{n+3} \\ f_{n+2} \\ f_{n+1} \end{bmatrix}\right. \tag{7}$$

$$\left.+\begin{bmatrix} 5791735 & -751598 & 213153 & -42578 \\ 32478080 & -4915968 & 1396352 & -278272 \\ 77715045 & -11123082 & 3059613 & -599238 \\ 144240640 & -20774912 & 5701632 & -1114112 \end{bmatrix}\begin{bmatrix} f_{n} \\ f_{n-1} \\ f_{n-2} \\ f_{n-3} \end{bmatrix} + \begin{bmatrix} 4093 & 0 & 0 & 0 \\ 26688 & 0 & 0 & 0 \\ 56862 & 0 & 0 & 0 \\ 105472 & 0 & 0 & 0 \end{bmatrix}\begin{bmatrix} f_{n-4} \\ f_{n-5} \\ f_{n-6} \\ f_{n-7} \end{bmatrix}\right).$$

Integrate four times:

$$\begin{bmatrix} 0 & 0 & 0 & 1 \\ 0 & 0 & 1 & 0 \\ 0 & 1 & 0 & 0 \\ 1 & 0 & 0 & 0 \end{bmatrix}\begin{bmatrix} y'_{n+4} \\ y'_{n+3} \\ y'_{n+2} \\ y'_{n+1} \end{bmatrix} = \begin{bmatrix} 1 & 0 & 0 & 0 \\ 1 & 0 & 0 & 0 \\ 1 & 0 & 0 & 0 \\ 1 & 0 & 0 & 0 \end{bmatrix}\begin{bmatrix} y'_{n} \\ y'_{n-1} \\ y'_{n-2} \\ y'_{n-3} \end{bmatrix} + h\begin{bmatrix} 1 & 0 & 0 & 0 \\ 2 & 0 & 0 & 0 \\ 3 & 0 & 0 & 0 \\ 4 & 0 & 0 & 0 \end{bmatrix}\begin{bmatrix} y''_{n} \\ y''_{n-1} \\ y''_{n-2} \\ y''_{n-3} \end{bmatrix} + \frac{h^2}{2}\begin{bmatrix} 1 & 0 & 0 & 0 \\ 4 & 0 & 0 & 0 \\ 9 & 0 & 0 & 0 \\ 16 & 0 & 0 & 0 \end{bmatrix}\begin{bmatrix} y'''_{n} \\ y'''_{n-1} \\ y'''_{n-2} \\ y'''_{n-3} \end{bmatrix} + \frac{h^3}{6}$$

$$\cdot\begin{bmatrix} 1 & 0 & 0 & 0 \\ 8 & 0 & 0 & 0 \\ 27 & 0 & 0 & 0 \\ 64 & 0 & 0 & 0 \end{bmatrix}\begin{bmatrix} y^{\text{iv}}_{n} \\ y^{\text{iv}}_{n-1} \\ y^{\text{iv}}_{n-2} \\ y^{\text{iv}}_{n-3} \end{bmatrix} + \frac{h^4}{159667200}\left(\begin{bmatrix} -25143 & 276056 & -1542812 & 7955976 \\ -437504 & 4788224 & -25525248 & 255250432 \\ -1358127 & 14924088 & 74244276 & 1604715624 \\ -3342336 & 190840832 & 1093402624 & 5052039168 \end{bmatrix}\begin{bmatrix} f_{n+4} \\ f_{n+3} \\ f_{n+2} \\ f_{n+1} \end{bmatrix}\right. \tag{8}$$

$$\left.+\begin{bmatrix} 679888370 & -66798970 & 18463200 & -3649810 \\ 455836160 & -15732736 & 19158016 & -3837952 \\ 1734575310 & -251470008 & 70150212 & -13839336 \\ 4381736960 & -643825664 & 180092928 & -35651584 \end{bmatrix}\begin{bmatrix} f_{n} \\ f_{n-1} \\ f_{n-2} \\ f_{n-3} \end{bmatrix} + \begin{bmatrix} 21649 & 0 & 0 & 0 \\ 369408 & 0 & 0 & 0 \\ 1318761 & 0 & 0 & 0 \\ 3407872 & 0 & 0 & 0 \end{bmatrix}\begin{bmatrix} f_{n-4} \\ f_{n-5} \\ f_{n-6} \\ f_{n-7} \end{bmatrix}\right).$$

Integrate five times:

$$
\begin{aligned}
&\begin{bmatrix} 0&0&0&1\\ 0&0&1&0\\ 0&1&0&0\\ 1&0&0&0 \end{bmatrix}
\begin{bmatrix} y_{n+4}\\ y_{n+3}\\ y_{n+2}\\ y_{n+1} \end{bmatrix}
=
\begin{bmatrix} 1&0&0&0\\ 1&0&0&0\\ 1&0&0&0\\ 1&0&0&0 \end{bmatrix}
\begin{bmatrix} y_{n}\\ y_{n-1}\\ y_{n-2}\\ y_{n-3} \end{bmatrix}
+h
\begin{bmatrix} 1&0&0&0\\ 2&0&0&0\\ 3&0&0&0\\ 4&0&0&0 \end{bmatrix}
\begin{bmatrix} y'_{n}\\ y'_{n-1}\\ y'_{n-2}\\ y'_{n-3} \end{bmatrix}
+\frac{h^2}{2}
\begin{bmatrix} 1&0&0&0\\ 4&0&0&0\\ 9&0&0&0\\ 16&0&0&0 \end{bmatrix}
\begin{bmatrix} y''_{n}\\ y''_{n-1}\\ y''_{n-2}\\ y''_{n-3} \end{bmatrix}
+\frac{h^3}{6}\\
&\cdot\begin{bmatrix} 1&0&0&0\\ 8&0&0&0\\ 27&0&0&0\\ 64&0&0&0 \end{bmatrix}
\begin{bmatrix} y'''_{n}\\ y'''_{n-1}\\ y'''_{n-2}\\ y'''_{n-3} \end{bmatrix}
+\frac{h^4}{24}
\begin{bmatrix} 1&0&0&0\\ 16&0&0&0\\ 81&0&0&0\\ 256&0&0&0 \end{bmatrix}
\begin{bmatrix} y^{\text{iv}}_{n}\\ y^{\text{iv}}_{n-1}\\ y^{\text{iv}}_{n-2}\\ y^{\text{iv}}_{n-3} \end{bmatrix}\\
&+\frac{h^5}{3632428800}\left(
\begin{bmatrix} -397695 & 4349090 & -24084760 & 118367466\\ -16628480 & 183347200 & -1022977536 & 7852902400\\ -93592665 & 995920434 & -1796664240 & 79986003930\\ -329908224 & 6980894720 & 40512389120 & 363276533800 \end{bmatrix}
\begin{bmatrix} f_{n+4}\\ f_{n+3}\\ f_{n+2}\\ f_{n+1} \end{bmatrix}\right.\\
&+\left.\begin{bmatrix} 679888370 & -66798970 & 18463200 & -3649810\\ 18183101440 & -2522204160 & 720401920 & -144287744\\ 109670461100 & -15912327690 & 4493865096 & -893148930\\ 375644487700 & -54760308740 & 1534525440 & -3038248960 \end{bmatrix}
\begin{bmatrix} f_{n}\\ f_{n-1}\\ f_{n-2}\\ f_{n-3} \end{bmatrix}
+\begin{bmatrix} 348869&0&0&0\\ 13889280&0&0&0\\ 85522635&0&0&0\\ 1095172096&0&0&0 \end{bmatrix}
\begin{bmatrix} f_{n-4}\\ f_{n-5}\\ f_{n-6}\\ f_{n-7} \end{bmatrix}\right).
\end{aligned}
\tag{9}
$$

2.2. Order and Convergence of the Method. The matrix differential equation of the derived method is given as

$$\alpha Y_m = h\beta Y'_m + h^2\lambda Y''_m + h^3\mu Y'''_m + h^4\sigma Y^{\text{iv}}_m + h^5\theta F_m, \tag{10}$$

where α, β, λ, μ, σ, and θ are the coefficients of the developed method. Consequently, the order of 4-point block method can be determined using the following formulae:

$$
\begin{aligned}
C_0 &= \sum_{j=0}^{19}\alpha_j,\\
C_1 &= \sum_{j=0}^{19}\left(j\alpha_j - \beta_j\right),\\
C_2 &= \sum_{j=0}^{19}\left(\frac{j^2}{2!}\alpha_j - j\beta_j - \lambda_j\right),\\
&\vdots\\
C_q &= \sum_{j=0}^{19}\left(\frac{j^q}{q!}\alpha_j - \frac{j^{q-1}}{(q-1)!}\beta_j - \frac{j^{q-2}}{(q-2)!}\lambda_j\right.\\
&\quad\left. - \frac{j^{q-3}}{(q-3)!}\mu_j - \frac{j^{q-4}}{(q-4)!}\sigma_j - \frac{j^{q-5}}{(q-5)!}\theta_j\right).
\end{aligned}
\tag{11}
$$

As a result, a 4-point block method of order nine is developed with $C_{p+5} \neq 0$ and the error constant obtained is given as

$$
\begin{aligned}
C_{14} = \Bigg[&\frac{2497}{7257600}, \frac{40321}{239500800}, \frac{47357}{479001600},\\
&\frac{2818273}{43589145600}, \frac{164971}{3632428800}, -\frac{23}{113400}, \frac{481}{1871100},\\
&\frac{1187}{1871100}, \frac{95143}{85135050}, \frac{8753}{4729725}, \frac{113}{89600}, \frac{689}{985600},\\
&\frac{2559}{1971200}, \frac{683073}{179379200}, \frac{497799}{44844800}, -\frac{94}{14175},\\
&-\frac{568}{467775}, \frac{1088}{467775}, \frac{429568}{42567525}, \frac{532736}{14189175}\Bigg]^T.
\end{aligned}
\tag{12}
$$

Hence, the consistency of 4-point block method is proven according to the definition in [23]. The analysis of zero stability for the developed method is tested using a similar approach as presented in [24] and the first characteristic polynomial obtained is $\rho(R) = R^3(R-1) = 0$. It is clearly seen that the roots are 0 and 1. Thus from Theorem 2.1 in [23], the convergence of the proposed method is asserted.

3. Implementation

Throughout this section, the implementation of 4-point block method for solving fifth-order IVPs will be explained in detail. The code starts by finding the values of y in the initial

Table 1: Numerical results for solving Problem 1.

TOL	MTD	TS	MAXE	AVERR	FC
	ode45	11	1.743 (−6)	2.244 (−6)	67
10^{-2}	4PIFI	23	1.640 (−4)	5.346 (−7)	210
	4PHODE	20	9.031 (−9)	1.611 (−10)	186
	ode45	11	1.641 (−6)	2.186 (−6)	67
10^{-4}	4PIFI	37	9.727 (−7)	6.988 (−9)	339
	4PHODE	27	7.059 (−10)	1.092 (−11)	243
	ode45	18	5.234 (−7)	8.596 (−7)	109
10^{-6}	4PIFI	50	5.582 (−9)	5.727 (−11)	447
	4PHODE	35	1.009 (−9)	3.401 (−11)	307
	ode45	44	5.485 (−9)	9.472 (−9)	265
10^{-8}	4PIFI	96	7.597 (−11)	2.881 (−13)	799
	4PHODE	42	1.838 (−10)	2.619 (−12)	363
	ode45	108	5.552 (−11)	9.551 (−11)	649
10^{-10}	4PIFI	788	1.834 (−14)	2.554 (−15)	6331
	4PHODE	50	1.023 (−10)	3.443 (−12)	439
KAYODE (a)		Not stated	1.638 (−6)	Not stated	Not stated
KAYODE (b)		100	5.082 (−7)	Not stated	Not stated

block using Euler method. However, it should be noted that Euler method will act only as a fundamental building block. Then the 4-point block method will be applied until the end of the interval. As stated earlier, the proposed method is implemented in the mode of predicting and correcting. In order to preserve the accuracy, the step will succeed if the local truncation error (LTE) is less than the specified error tolerance (TOL) such that

$$\text{LTE} = \left| y_{n+4}^{t} - y_{n+4}^{t-1} \right| < \text{TOL}, \quad (13)$$

where y_{n+4}^{t} and y_{n+4}^{t-1} are the corrector value of y at the last point for each block with t iterations. If (13) is satisfied, the new step size will be calculated via the step size increment formula. Otherwise, the current step size will be reduced by half. The step size increment formula is defined as

$$h_{\text{new}} = \delta \times h_{\text{old}} \times \left(\frac{\text{TOL}}{2 \times \text{LTE}} \right)^{1/m}, \quad (14)$$

where h_{new} and h_{old} denote the current and previous step size, respectively, with value of 0.5 for safety factor (δ) and m is the order of corrector formulae. To show the accuracy and efficiency of the proposed method, the computational errors will be reported, equal to

$$E_i = \left| \frac{y_i - y(x_i)}{A + B(y(x_i))} \right|. \quad (15)$$

Different values of A and B represent the type of error test which will be considered; namely, $A = 1$, $B = 0$ are for the absolute error test; $A = 1$, $B = 1$ represent the mixed error test; and $A = 0$, $B = 1$ correspond to the relative error test. Here we will use the mixed error test and the maximum error is calculated by

$$\text{MAXE} = \max_{1 \le i \le 4} (E_i). \quad (16)$$

4. Numerical Results

To illustrate the technique proposed in the preceding sections, two test problems are solved and the results obtained compared with the method proposed in [8, 9, 25] and the ODE solver in MATLAB ode45. The work done by [25] involved the solving of the first-order ODEs using 4-point block method using variable step size. This means that (1) needs to be reduced into a system of first-order IVPs whereby the methods proposed by [8, 9] are for solving (1) directly with the implementation of constant step size. The notations used in Tables 1 and 2 are listed below:

TOL: error tolerance limit;

MTD: method used;

TS: total steps taken;

MAXE: maximum error of the computed solution;

AVERR: average error of the computed solution;

FC: total function calls;

ode45: Runge-Kutta-Dormand-Prince ODE solver;

4PHODE: implementation of the 4-point block method in this research;

4PIFI: numerical results in [25];

KAYODE (a): numerical results in [8];

KAYODE (b): numerical results in [9].

TABLE 2: Numerical results for solving Problem 2.

TOL	MTD	TS	MAXE	AVERR	FC
	ode45	11	5.027 (−3)	3.806 (−3)	67
10^{-2}	4PIFI	58	2.528 (−7)	3.002 (−8)	490
	4PHODE	28	5.537 (−6)	2.245 (−7)	250
	ode45	13	1.967 (−4)	1.253 (−4)	79
10^{-4}	4PIFI	68	9.819 (−9)	1.117 (−9)	571
	4PHODE	36	4.193 (−7)	1.432 (−8)	315
	ode45	25	4.592 (−7)	1.870 (−7)	151
10^{-6}	4PIFI	86	1.219 (−10)	9.992 (−12)	715
	4PHODE	46	1.003 (−7)	4.290 (−9)	395
	ode45	58	1.861 (−9)	1.006 (−9)	349
10^{-8}	4PIFI	135	1.363 (−12)	1.597 (−13)	1107
	4PHODE	56	2.413 (−8)	1.012 (−9)	475
	ode45	142	4.555 (−11)	2.026 (−11)	853
10^{-10}	4PIFI	807	5.630 (−14)	1.250 (−14)	6483
	4PHODE	71	1.949 (−10)	7.564 (−12)	595
	KAYODE (a)	Not stated	2.225 (−7)	Not stated	Not stated
	KAYODE (b)	10	3.156 (−4)	Not stated	Not stated

Problem 1. Consider the following:

$$y^{(v)} = 2y'y'' - yy^{(iv)} - y'y''' - 8x + \left(x^2 - 2x - 3\right)e^x,$$
$$x \in [0, 2], \; y(0) = 1, \; y'(0) = 1, \; y''(0) = 3, \; y'''(0) = 1, \; y^{(iv)}(0) = 1. \quad (17)$$

Solution is as follows: $y(x) = e^x + x^2$.

Source is [8].

Problem 2. Consider the following:

$$y^{(v)} = 6\left(2\left(y'\right)^3 + 6yy'y'' + y^2y'''\right), \quad x \in [1, 3], \; y(1) = 1, \; y'(1) = -1, \; y''(1) = 2, \; y'''(1) = -6, \; y^{(iv)}(1) = 24. \quad (18)$$

Solution is as follows: $y(x) = 1/x$.
Source is [8].

5. Discussion

In this section the performances of 4PHODE, ode45, 4PIFI, KAYODE (a), and KAYODE (b) are discussed in terms of three parameters, namely, total steps taken, accuracy, and total function evaluations. It is apparent from these tables, mostly at tolerances 10^{-2}, 10^{-4}, and 10^{-6}, that 4PHODE gives better accuracy compared to ode45, whereas at tolerances 10^{-8} and 10^{-10}, ode45 is one decimal more accurate compared to 4PHODE for both problems. As both tables show, the total steps taken by the 4PHODE reduce by nearly half to ode45 at tolerance 10^{-10}. Although, at other tolerance, ode45 requires lesser steps to compute the solution, this result may be explained by the fact that the initial step size generated by 4PHODE is extremely small in order for the method control of the accuracy. From the data in Tables 1 and 2, it is apparent that the number of function calls is likely to be related to the number of steps taken.

In the comparison of block-by-block method, the numerical results obtained in Tables 1 and 2 demonstrate that the proposed method 4PHODE always requires fewer steps in converging to the exact solution compared to 4PIFI. This gain becomes more obvious as the tolerance decreases. However, the 4PIFI achieves better accuracy than 4PHODE. Even so, the maximum error for 4PHODE is still within the specified tolerance. Another issue that had to be addressed was the total number of function evaluations used by each method. The 4PHODE solved all the problems at much lower cost than 4PIFI and this superiority is most apparent at finer tolerance. This result is in agreement with [10] which states the drawback of using the reduction approach in the implementation of simultaneous computations.

Comparing the results with the method proposed by KAYODE (a) and KAYODE (b), for Problem 1, 4PHODE outperformed both KAYODE (a) and KAYODE (b) in terms of accuracy and total steps taken. While, for Problem 2, KAYODE (b) has lesser total steps taken, however, 4PHODE still has superiority in terms of accuracy.

6. Conclusion

The overall performance revealed that the 4-point block method is best to be implemented in a direct integration approach as it required much less storage than the reduction method while still maintaining an acceptable accuracy. Besides that, the results of this study also indicate that the developed method has better accuracy compared to the existing methods. Hence it can be said that 4PHODE is one of the alternative methods that can be used for solving fifth-order ODEs.

Competing Interests

The authors declare that there are no competing interests regarding the publication of this paper.

Acknowledgments

Grateful acknowledgement is made to the Ministry of Higher Education, Malaysia, for financial support with Grant no. RAGS/2013/FKEKK/SG04/01/B00037.

References

[1] M. T. Darvishi, S. Kheybari, and F. Khani, "A numerical solution of the Korteweg-de Vries equation by pseudospectral method using Darvishi's preconditionings," *Applied Mathematics and Computation*, vol. 182, no. 1, pp. 98–105, 2006.

[2] L. Jin, "Application of variational iteration method to the fifth-order KdV equation," *International Journal of Contemporary Mathematical Sciences*, vol. 3, no. 5, pp. 213–221, 2008.

[3] L. Kaur, "Generalized (G'/G)—expansion method for generalized fifth order KdV equation with time-dependent coefficients," *Mathematical Sciences Letters*, vol. 3, no. 3, pp. 255–261, 2014.

[4] M. Suleiman, Z. B. Ibrahim, and A. F. N. Bin Rasedee, "Solution of higher-order ODEs using backward difference method," *Mathematical Problems in Engineering*, vol. 2011, Article ID 810324, 18 pages, 2011.

[5] U. Goktas and W. Hereman, "Symbolic computation of conserved densities for systems of nonlinear evolution equations," *Journal of Symbolic Computation*, vol. 24, no. 5, pp. 591–621, 1997.

[6] N. Khanal, R. Sharma, J. Wu, and J.-M. Yuan, "A dual-Petrov-Galerkin method for extended fifth-order Korteweg-de Vries type equations," *Discrete and Continuous Dynamical Systems*, pp. 442–450, 2009.

[7] J. Li and Z. Qiao, "Explicit soliton solutions of the Kaup-Kupershmidt equation through the dynamical system approach," *Journal of Applied Analysis and Computation*, vol. 1, no. 2, pp. 243–250, 2011.

[8] S. J. Kayode and D. O. Awoyemi, "A multiderivative Collocation method for 5th order ordinary differential equations," *Journal of Mathematics and Statistics*, vol. 6, no. 1, pp. 60–63, 2010.

[9] S. J. Kayode, "An order seven continuous explicit method for direct solution of general fifth order ordinary differential equations," *International Journal of Differential Equations and Applications*, vol. 13, no. 2, pp. 71–80, 2014.

[10] J. B. Rosser, "A Runge-Kutta for all seasons," *SIAM Review*, vol. 9, no. 3, pp. 417–452, 1967.

[11] M. B. Suleiman, "Solving nonstiff higher order ODEs directly by the direct integration method," *Applied Mathematics and Computation*, vol. 33, no. 3, pp. 197–219, 1989.

[12] D. O. Awoyemi, "A new sixth-order algorithm for general second order ordinary differential equations," *International Journal of Computer Mathematics*, vol. 77, no. 1, pp. 117–124, 2001.

[13] S. J. Kayode, "An efficient zero-stable numerical method for fourth-order differential equations," *International Journal of Mathematics and Mathematical Sciences*, vol. 2008, Article ID 364021, 10 pages, 2008.

[14] B. T. Olabode and Y. Yusuph, "A new block method for special third order ordinary differential equations," *Journal of Mathematics and Statistics*, vol. 5, no. 3, pp. 167–170, 2009.

[15] S. N. Jator and J. Li, "A self-starting linear multistep method for a direct solution of the general second-order initial value problem," *International Journal of Computer Mathematics*, vol. 86, no. 5, pp. 827–836, 2009.

[16] N. Waeleh, Z. A. Majid, F. Ismail, and M. Suleiman, "Numerical solution of higher order ordinary differential equations by direct block code," *Journal of Mathematics and Statistics*, vol. 8, no. 1, pp. 77–81, 2011.

[17] L. F. Shampine and H. A. Watts, "Block implicit one-step methods," *Mathematics of Computation*, vol. 23, pp. 731–740, 1969.

[18] J. R. Cash, "Block Runge-Kutta methods for the numerical integration of initial value problems in ordinary differential equations. Part I. The nonstiff case," *Mathematics of Computation*, vol. 40, no. 161, pp. 175–191, 1983.

[19] S. N. Jator, "Solving second order initial value problems by a hybrid multistep method without predictors," *Applied Mathematics and Computation*, vol. 217, no. 8, pp. 4036–4046, 2010.

[20] N. Waeleh, Z. A. Majid, and F. Ismail, "A new algorithm for solving higher order IVPs of ODEs," *Applied Mathematical Sciences*, vol. 5, no. 53–56, pp. 2795–2805, 2011.

[21] J. Vigo-Aguiar and H. Ramos, "Variable stepsize implementation of multistep methods for $y''=f(x, y, y')$," *Journal of Computational and Applied Mathematics*, vol. 192, no. 1, pp. 114–131, 2006.

[22] Z. A. Majid, N. A. Azmi, M. Suleiman, and Z. B. Ibrahaim, "Solving directly general third order ordinary differential equations using two-point four step block method," *Sains Malaysiana*, vol. 41, no. 5, pp. 623–632, 2012.

[23] J. D. Lambert, *Computational Methods in Ordinary Differential Equations*, John Wiley & Sons, London, UK, 1973.

[24] S. Ola Fatunla, "Block methods for second order ODEs," *International Journal of Computer Mathematics*, vol. 41, no. 1-2, pp. 55–63, 1991.

[25] Z. A. Majid and M. B. Suleiman, "Implementation of four-point fully implicit block method for solving ordinary differential equations," *Applied Mathematics and Computation*, vol. 184, no. 2, pp. 514–522, 2007.

Permissions

All chapters in this book were first published in IJMMS, by Hindawi Publishing Corporation; hereby published with permission under the Creative Commons Attribution License or equivalent. Every chapter published in this book has been scrutinized by our experts. Their significance has been extensively debated. The topics covered herein carry significant findings which will fuel the growth of the discipline. They may even be implemented as practical applications or may be referred to as a beginning point for another development.

The contributors of this book come from diverse backgrounds, making this book a truly international effort. This book will bring forth new frontiers with its revolutionizing research information and detailed analysis of the nascent developments around the world.

We would like to thank all the contributing authors for lending their expertise to make the book truly unique. They have played a crucial role in the development of this book. Without their invaluable contributions this book wouldn't have been possible. They have made vital efforts to compile up to date information on the varied aspects of this subject to make this book a valuable addition to the collection of many professionals and students.

This book was conceptualized with the vision of imparting up-to-date information and advanced data in this field. To ensure the same, a matchless editorial board was set up. Every individual on the board went through rigorous rounds of assessment to prove their worth. After which they invested a large part of their time researching and compiling the most relevant data for our readers.

The editorial board has been involved in producing this book since its inception. They have spent rigorous hours researching and exploring the diverse topics which have resulted in the successful publishing of this book. They have passed on their knowledge of decades through this book. To expedite this challenging task, the publisher supported the team at every step. A small team of assistant editors was also appointed to further simplify the editing procedure and attain best results for the readers.

Apart from the editorial board, the designing team has also invested a significant amount of their time in understanding the subject and creating the most relevant covers. They scrutinized every image to scout for the most suitable representation of the subject and create an appropriate cover for the book.

The publishing team has been an ardent support to the editorial, designing and production team. Their endless efforts to recruit the best for this project, has resulted in the accomplishment of this book. They are a veteran in the field of academics and their pool of knowledge is as vast as their experience in printing. Their expertise and guidance has proved useful at every step. Their uncompromising quality standards have made this book an exceptional effort. Their encouragement from time to time has been an inspiration for everyone.

The publisher and the editorial board hope that this book will prove to be a valuable piece of knowledge for researchers, students, practitioners and scholars across the globe.

List of Contributors

Dieudonne Agbor
Department of Mathematics, Faculty of Science, University of Buea, Buea, Cameroon

Armand Armand
Lycee Mixte Antsiranana, Madagascar

André Totohasina
Department of Mathematics and Informatics, University of Antsiranana, Madagascar

Daniel Rajaonasy Feno
Department of Mathematics, Informatics and Applications, University of Toamasina, Madagascar

Hiroshi Inoue
Center for Advancing Pharmaceutical Education, Daiichi University of Pharmacy, 22-1 Tamagawa-cho, Minami-ku, Fukuoka 815-8511, Japan

Om P. Ahuja and Jay M. Jahangiri
Mathematical Sciences, Kent State University, Burton, OH 44021-9500, USA

Marat V. Markin
Department of Mathematics, California State University, Fresno, 5245 N. Backer Avenue, M/S PB 108, Fresno, CA 93740-8001, USA

Karim Chaira and Samih Lazaiz
Laboratory of Algebra, Analysis and Applications, Faculty of Sciences Ben M'sik, University of Hassan II Casablanca, Casablanca, Morocco

Abderrahim Eladraoui and Mustapha Kabil
Laboratory of Mathematics and Applications, Faculty of Sciences and Technologies, Mohammedia, University of Hassan II Casablanca, Casablanca, Morocco

Artur R. Valiullin, Albert R. Valiullin and Vladimir V. Galatenko
Faculty of Mechanics and Mathematics, Lomonosov Moscow State University, Russia

Chaiporn Thangthong and Anchalee Khemphet
Center of Excellence in Mathematics and Applied Mathematics, Department of Mathematics, Faculty of Science, Chiang Mai University, Chiang Mai 50200, Thailand

Ratchata Theinchai, Siriwan Chankan and Weera Yukunthorn
Faculty of Science and Technology, Kanchanaburi Rajabhat University, Kanchanaburi 71000, Thailand

S. B. Damelin
Mathematical Reviews, The American Mathematical Society, 416 Fourth Street, Ann Arbor, MI 48104, USA

N. S. Hoang
Department of Mathematics, University of Oklahoma, Norman, OK 73019-3103, USA

M. P. Markakis and P. S. Douris
Department of Electrical & Computer Engineering, University of Patras, 26504 Patras, Greece

Safia Meftah
Operators Theory and DPE Foundations and Applications Laboratory, Science Exact Faculty, Echahid Hamma Lakhdar University, El Oued 39000, Algeria

Aaisha Farzana Habibullah and Adolf Stephen Bhaskaran
Department of Mathematics, Madras Christian College, Tambaram, Chennai, Tamil Nadu 600 059, India

Jeyaraman Muthusamy Palani
Department of Mathematics, L. N. Government College, Ponneri, Chennai, Tamil Nadu 601 204, India

Taksaporn Sirirut and Pattanapong Tianchai
Faculty of Science, Maejo University, Chiangmai 50290, Thailand

Radhanath Rath
VSSUT, Burla, Sambalpur, 768018 Odisha, India
Former Principal Khallikote Autonomous College, Berhampur, 760001 Odisha, India

Chittaranjan Behera
Department of Mathematics, Silicon Institute of Technology, Bhubaneswar, Odisha, India

Y. Yousfi and A. Benbrik
Department of Mathematics, Faculty of Science, University Mohammed 1*st*, Oujda, Morocco

I. Hadi
Moroccan Ministry of National Education, Higher Education and Research, Morocco

Gerald Wanjala
Department of Mathematics and Statistics, Sultan Qaboos University, Al-Khod, 123 Muscat, Oman

Chalongchai Klanarong
Department of Mathematics, Faculty of Science, Mahasarakham University, Mahasarakham 44150,Thailand

Tanadon Chaobankoh
Center of Excellence in Mathematics and Applied Mathematics, Department of Mathematics, Faculty of Science, Chiang Mai University, Chiang Mai 50200,Thailand

Mehar Chand
Department of Mathematics, Baba Farid College, Bathinda 151001, India

Hanaa Hachimi
BOSS Team, GS laboratory, ENSA, Ibn Tofail University, Kenitra 14000, Morocco

Rekha Rani
Department of Applied Sciences, Gurukashi University, Bathinda 151302, India

Mohammed Harfaoui, Abdellah Mourassil and Loubna Lakhmaili
University Hassan II Mohammedia, Laboratory of Mathematics, Criptography and Mechanical F. S.T., BP 146, Mohammedia 20650, Morocco

Nur Auni Baharum
Institute for Mathematical Research, Universiti Putra Malaysia, 43400 Serdang, Selangor, Malaysia

Zanariah Abdul Majid and Norazak Senu
Institute for Mathematical Research, Universiti Putra Malaysia, 43400 Serdang, Selangor, Malaysia
Mathematics Department, Faculty of Science, Universiti Putra Malaysia, 43400 Serdang, Selangor, Malaysia

Sekson Sirisubtawee and Supaporn Kaewta
Department of Mathematics, Faculty of Applied Science, King Mongkut's University of Technology North Bangkok, Bangkok 10800, Thailand

Nazreen Waeleh
Faculty of Electronic & Computer Engineering, Universiti Teknikal Malaysia Melaka (UTeM), 76100 Melaka, Malaysia

Zanariah Abdul Majid
Department of Mathematics, Faculty of Science, Universiti Putra Malaysia, 43400 Serdang, Selangor, Malaysia
Institute for Mathematical Research, Universiti Putra Malaysia, 43400 Serdang, Selangor, Malaysia

Index

A

Abstract Evolution Equation, 32, 45
Adomian Decomposition Method, 65, 190, 208-209
Adomian Polynomials, 193, 199-200, 205, 209
Alfuraidan And Khamsi Theorem, 46
Analytic Convex Functions, 28

B

Banach Space, 32, 45
Bergman Spaces, 11
Biharmonic Functions, 70, 73-75
Biorthogonal Pair, 27
Boundary Value Problems, 190, 208-209

C

Caristi's Fixed Point Theorem, 46, 51
Coincidence Point, 64, 146
Confluent Hypergeometric Function, 153, 156, 163, 166, 170
Constrained Convex Minimization Problem, 106, 114-115
Cyclic Mapping, 141, 145-146

D

Deflection Curve, 65, 67, 69
Differential Equation, 65, 91, 99, 128, 166, 193, 199, 210
Differential Equations, 46, 51, 69, 91, 98, 105, 123, 166, 171, 180, 189-190, 208-211, 217

E

Euler-bernoulli Beam Theory, 65

F

Finite Difference Approximation, 72-73
Finite-dimensional Spaces, 52-53
Fixed Point Theory, 46, 51, 64, 115, 146
Fractional Calculus, 69, 171, 208-209
Fractional Differential Equation, 166
Fractional Integral Operators, 160-161
Fractional Integrals, 171, 209
Fractional Order Differential Equation, 193
Fractional Order Model, 208

G

Gauss Hypergeometric Function, 147, 156, 160, 170
Generalized Axially Symmetric Helmholtz Equation, 179
Generalized Riesz Systems, 21, 27
Generalized Struve Function, 99
Gradient Projection Method, 106, 115

H

Helmholtz Equation, 179
Hilbert Space, 32, 45, 52-54, 57, 106, 134, 139
Hilbert Spaces, 11, 27, 52-53, 57, 114-115, 134
Homeomorphism, 19
Hopf Bifurcation, 85

I

Implicit Multistep Block Method, 180, 185, 188
Initial Value Problem, 65, 217
Interpolation, 73-74, 77, 139, 184, 211
Interpolation Problems, 139

K

Kirk-saliga Fixed Point Theorem, 46, 51
Krein Space, 134

L

Laplace Transform, 65, 164
Lewy's Theorem, 28
Lindstedt Method, 91
Linear Differential Equation, 65
Local Truncation Error, 215

M

Multistep Block Method, 180, 185-186, 188-189

N

Neumann Boundary Condition, 124
Nonlinear Differential Equation, 91
Nonlinear Fractional Order, 190

O

Ordinary Differential Equation, 210
Ordinary Differential Equations, 51, 180, 190, 210, 217
Orthogonal Sum Decomposition, 134

P

Partial Differential Equations, 190, 209
Perturbation Method, 91, 124, 189
Poisson's Equation, 72
Polynomial, 67, 184, 209, 211, 214
Polynomials, 171, 190, 193, 199-200, 205, 209, 211
Positive Integers, 46
Positive Real Number, 106, 153, 159-161, 163-164, 166

Proximity Point Theorems, 140, 145-146

R

Reflexive Digraph, 46

Regular Biorthogonal Sequences, 21

Runge-kutta Method, 186, 211

S

Scalar Type Spectral Operator, 32, 45

Semisharp Proximal Pair, 140-141, 145

Split Feasibility Problem, 106, 114-115

T

Toeplitz Operators, 11

Topology, 45, 124, 134, 146

Tridiagonal Matrix, 72

V

Variable Lebesgue Spaces, 11

Volterra Integrodifferential Equations, 180, 184-185

www.ingramcontent.com/pod-product-compliance
Lightning Source LLC
Chambersburg PA
CBHW080624200326
41458CB00013B/4493
9781632387301